65-

PLASMA DIAGNOSTICS

Editors

A.A. Ovsyannikov

Institute of Petrochemical Synthesis, Russian Academy of Sciences, Moscow

M.F. Zhukov

Institute of Thermophysics, Suberian Division of Russian Academy of Sciences, Novosibirsk

CAMBRIDGE INTERNATIONAL SCIENCE PUBLISHING

Published by
Cambridge International Science Publishing
7 Meadow Walk, Great Abington, Cambridge CB1 6AZ, UK
http://www.demon.co.uk/cambsci/homepage.htm

First published March 2000

©   Cambridge International Science Publishing

British Library Cataloguing in Publication Data
A catalogue record for this book is available from the British Library

ISBN   1 898326231

Translated by *Victor Riecansky*
Production *Irina Stupak*
Printed by *Pear Tree Press Ltd, Stevenage, England*

Preface

It is believed that the term 'diagnostics' was borrowed by the investigators of electric discharge plasma in the Forties and Fifties of the previous the century from medicine, obviously owing to the fact that the investigation methods used at that time made it possible, as is usually the case in medicine, to obtain only highly approximate information on the state of the examined object. In the last 50 years, there have been not only significant improvements in the traditional methods but also new, interesting and greatly different methods have been developed which can now be used to measure the plasma parameters with the accuracy sufficient for solving many fundamental and applied problems. Some of these methods are discussed in this monograph, and in many cases the application of these methods makes it possible to solve the main task of diagnostics, i.e. find the concentration of components and the form of energy distribution in the plasma.

There are a relatively large number of reviews and monographs on plasma diagnostics (for example, the well-known book by H. Grim [1], the collective monographs edited by R. Huddlestone and S. Leonard [2], W. Lochte-Holtgreven [3] and S. Frish [4], and the books published by A.A. Dushin and O.S. Pavlichenko [5], N.G. Preobrazhenskii [6], L.N. Pyatnitskii [7], P. Chang, L. Tolbot and K. Turyan [8], Yu.A. Ivanov, Yu.A. Lebedev and L.S. Polak [9], V.K. Zhivotov, V.D. Rusanov and A.A. Fridman [10], collection 'Plasma Diagnostics' and others). Many of these books discuss mostly the experience obtained by foreign investigators. The task of this book is to present to the reader information on the achievements of Russian diagnostics. Each section of the monograph was prepared by experts concerned directly with the given range of problems. The nature of the sections differs: in some of them, the authors present more or less complete reviews of the problems in the given area, whereas in others they describe mainly the experimental procedure and results of their own investigations.

In the first part, we present articles relating to spectral and optical methods of diagnostics. In addition to the conventional methods, a number of relatively new laser methods are also discussed. The second part of the monograph is concerned with the probe methods; the methods of electrostatic probing, suitable for examination of plasma at both reduced and atmospheric pressure, in particular in the conditions of occurrence of chemical reactions in plasma, calorimetric probes for measuring the enthalpy of plasma flows, etc., are discussed. In part 3, the authors describe the methods of macroscopic diagnostics for determining the spatial–time characteristics of plasma and the distribution of electromagnetic fields in plasma. Finally, part 4, the authors discuss some important problems of processing the experimental data and automation of experiments.

The editors are grateful to all authors who have taken part in this monograph:

A. Abdrazakov (chapter 9,15,18)
R.Kh. Amirov (17),
M.K. Asanaliev (7,9,15,18)
E.I. Asinovskii (17)
M.S. Benilov (11)
P.A. Vlasov (12)
Zh.Zh. Zheenbaev (7,9,15,18)
Yu.A. Ivanov (13)
V.S. Klubnikin (14)
R.I. Konavko (9,15,18)
L.T. Lar'kin (1,9,19)
Yu.A. Lebedev (10)
V.V. Markovets (17)
G.V. Ostrovskaya (4)
D.K. Otorbaev (3,20)
V.N. Ochkin (2,5)
A.S. Panfilov (17)
I.S. Samoilov (17)
M.A. Samsonov (7,9)
A.I. Sedel'nikov (20)
O.P. Solonenko (8,21)
V.S. Engel'sht (1,7,9,15,18,19)
N.A. Yatsenko (16).

A.A. Ovsyannikov

Part I
<u>SPECTRAL AND OPTICAL METHODS</u>

i

Part III
<u>MACROSCOPIC DIAGNOSTICS</u>

Part IV
METHODS OF PROCESSING RESULTS AND AUTOMATION OF
DIAGNOSTICS

Chapter 1

SPECTRAL DIAGNOSTICS METHODS

The spectral diagnostics methods are based on measuring the intensity of the spectral lines of emission and absorption and the continuous spec trum, halfwidths and shifts of spectral lines.

The state of high-pressure plasma can be usually described using models of local thermodynamic (LTE) or partial local thermodynamic equilibrium (PLTE). LTE plasma has the following characteristics. All plasma particles, including free electrons, are distributed with respect to the velocity of random motion in accordance with the Maxwell law:

$$dN(v) = N4\pi \left(\frac{m}{2\pi kT} \right)^{3/2} \exp\left(-\frac{mv^2}{2kT} \right) v^2 dv. \tag{1.1}$$

Here N is the concentration of the particles of the given type (the number of particles in unit volume), $N(v)$ is the concentration of particles with velocity v; m is the particle mass; T is plasma temperature; k is the Boltzmann constant.

The population of the energy levels is determined by the Boltzmann law

$$N_n = N \frac{g_n}{S} \exp\left(-\frac{E_n}{kT} \right), \tag{1.2}$$

where N_n is the concentration of particles of the energy level n with energy E_n and the statistical weight g_n; S is the statistical sum with respect to the states

1

$$S = \sum_{n=1}^{n_{max}} g_n \exp\left(-\frac{E_n}{kT}\right), \tag{1.3}$$

where n_{max} is the maximum main quantum number of energy levels realised in plasma. The latter is determined from the condition $E_{n\,max} \leq U - \Delta U$, where U is the ionisation energy, ΔU is the reduction of the ionisation energy of atoms (ions) in plasma,

$$\Delta U_z = \frac{(1+z)e^2}{\rho_D}, \quad \rho_D = \left(\frac{kT}{8\pi e^2 N_e}\right)^{1/2}, \tag{1.4}$$

ρ_D is the radius of the Debye sphere; e is the electron charge; z is the ion charge (for a neutral atom $z = 0$, for a singly-ionised atom $z = 1$, etc.).

The relationship between the concentration of particles taking part in the chemical reaction $\alpha\beta \Leftrightarrow \alpha + \beta$, is determined in accordance with the law of acting masses:

$$\frac{N_\alpha N_\beta}{N_{\alpha\beta}} = \frac{S_\alpha S_\beta}{S_{\alpha\beta}} \left(\frac{2\pi m_\alpha m_\beta kT}{h^2 m_{\alpha\beta}}\right)^{3/2} \exp\left(-\frac{D_{\alpha\beta}}{kT}\right), \tag{1.5}$$

where N_α, N_β, $N_{\alpha\beta}$ is the concentration of particles of components α, β, $\alpha\beta$ of the chemical reaction, m_α, m_β, $m_{\alpha\beta}$ is the mass of the corresponding particles, $D_{\alpha\beta}$ is the energetic threshold of the reaction, h is the Planck constant. In particular, for the ionisation reaction $a \leftrightarrow i + e$ we can write the Saha equation

$$\frac{N_e N_i}{N_a} = \frac{2S_i}{S_a} \left(\frac{2\pi m_e kT}{h^2}\right)^{3/2} \exp\left(-\frac{U - \Delta U}{kT}\right), \tag{1.6}$$

where m_e is the electron mass.

If it is required to calculate the composition of plasma, the equations (1.5) and (1.6) supplement the following relationships:

the equation of quasineutrality of plasma

$$N_e = \sum_{\xi=1}^{p} \sum_{z=1}^{t} z N_{i,\xi}^{z}, \tag{1.7}$$

where $N_{i,\xi}^{z}$ is the concentration of ions of component ξ with the charge z;

the equation of state

$$P = kT \sum_{\xi=1}^{p} N_{\xi} + kTN_e, \tag{1.8}$$

where summation is carried out over all neutral or ionised plasma components;

The equation of conservation of the initial composition (the equation of elementary plasma balance)

$$\sum_{z=0}^{t} N_{\xi=1}^{z} : \sum_{z=0}^{t} N_{\xi=2}^{z} \dots = c_1 : c_2 \dots . \tag{1.9}$$

For example, for the plasma formed from steam, the initial composition $N_H : H_O = 2:1$; with an allowance made for all possible plasma components H_2, H, H^+, OH, O_2, O, O^+, equation (1.9) assumes the form

$$(2N_{H_2} + N_H + N_{H^+} + N_{OH}):(2N_{O_2} + N_O + N_{O^+} + N_{OH}) = 2:1 .$$

PLTE plasma differs from LTE plasma by the fact the kinetic energy of heavy particles (molecules, atoms, ions) is not equal to the electron temperature since the excitation (1.2) and ionisation (1.6) temperatures are equal to the temperature of free electrons.

The intensity of molecular bands, spectral lines and the continuous spectrum is calculated in the LTE and PLTE plasma on the basis of the equations of transfer of radiation using the probabilities of transitions and cross sections of the photoprocesses.

The diagnostics methods based on the emission spectra can be applied directly to optically thin plasma objects where the absorption of radiation is negligible. Verification of reabsorption can be carried out by measuring the optical thickness $\tau_v = \kappa_v l$, where κ_v is the absorption coefficient at frequency v; l is the effective length of the absorbing layer. For this purpose, the examined object is irradiated with an auxiliary source or inherent radiation.

The optical thickness can also be evaluated by comparing the intensity of examined radiation I_v with the radiation of an absolute black body B_v at the same temperature. For a homogeneous emitter

$$\tau_v = \kappa_v l = -\ln\left(1 - \frac{I_v}{B_v}\right).$$

In order to ignore absorption, the following condition must be fulfilled $\kappa_v l \ll 1$ (for example, ~0.01).

The emission spectra can be used for diagnostics and also in the case of marked self-absorption (optically dense plasma). However, to determine the local emission coefficients, it is necessary to take absorption into account.

A number of methods is based on measuring absorption, specially the methods of determining the temperature using the Kirchhoff law and the concentration of the atoms on the basis of the spectral absorption lines.

The local values of the emission and absorption coefficients in axisymmetric objects are found from the solution of corresponding integral inclusions. In operation with the contours of spectral lines, a similar procedure is used for selecting the frequencies inside the contour. This is followed by constructing the contour for a number of values of the radial coordinate and by determining of the half with (shift) of the spectral line as well as the function of the radial coordinate.

1.1 Integral relationships. Optical measurement circuits

We shall now present equations for a partial case of propagation of monochromatic radiation along the chord of the cross section of a cylindrical (axisymmetric) object. All local optical parameters are assumed to be functions of only the radial coordinate. The following concepts are used:

– the emission coefficient – the amount of energy emitted by the unit volume per unit time into the unit solid angle in the unit frequency range;

– the radiation intensity – the amount of energy in the unit frequency range passing through the unit area normal to the direction of propagation of radiation, per unit time in the unit solid angle;

– the absorption coefficient – its numerical value is equal to the value reciprocal to the distance along which the light flux is weakened $e \cong 2.72$ times;

– the transmittance coefficient – the fraction of radiant energy passing through the object.

An optically thin source. We shall examine a cylindrical source (Fig.1.1) whose emission coefficient $\varepsilon(r)$ depends only on the radial coordinate (self absorption can be ignored). The intensity of radiation leaving the source in plane XY along the chord $[-y_0, y_0]$ at distance x from the axis of symmetry is

$$I(x) = \int_{-y_0}^{y_0} \varepsilon(r)\, dy.$$

We set $\varepsilon\,(r \geq R) = 0$ and, consequently, $I(|x| \geq R) = 0$. In the coordinates (x, y), $y = \sqrt{r^2 - x^2}$

$$I(x) - 2\int_{|x|}^{R} \frac{\varepsilon(r)r\,dr}{\sqrt{r^2 - x^2}}. \tag{1.10}$$

The equation (1.10) is the integral Volterra equation of the first kind. It is also referred to as the Abel integral transform. The following procedure can be used for inversion of the Abel transform. We shall transform (1.10) using the operator:[1]

$$\frac{d}{dz^2} \int_{z^2}^{R^2} [\ldots] \frac{dx^2}{\sqrt{x^2 - z^2}}. \tag{1.11}$$

The order of integration on the right hand side will be changed

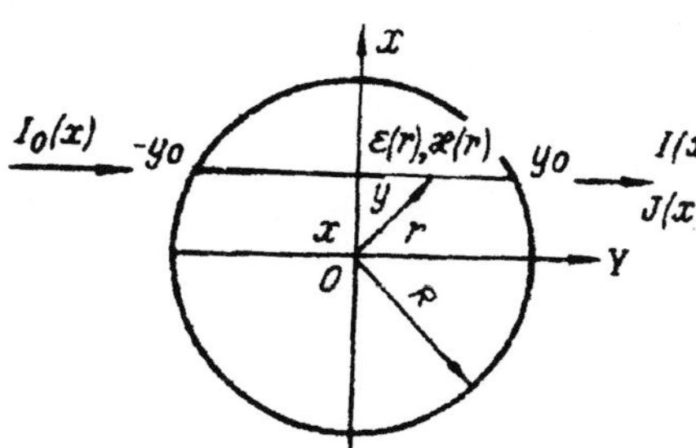

Fig.1.1 Cross-section of an axisymmetric object.

$$\frac{\mathrm{d}}{\mathrm{d}z^2}\int\limits_{z^2}^{R^2}\frac{I(x)\mathrm{d}x^2}{\sqrt{x^2-z^2}}=\frac{\mathrm{d}}{\mathrm{d}z^2}\int\limits_{z^2}^{R^2}\varepsilon(r)\left(\int\limits_{z^2}^{r^2}\frac{\mathrm{d}x^2}{\sqrt{r^2-x^2}\sqrt{x^2-z^2}}\right)\mathrm{d}r^2.$$

The internal integral on the right hand side is reduced as a result of substitution

$$x^2=z^2\cos^2\varphi+r^2\sin^2\varphi$$

to a tabulated integral and is equal to π. We shall write the result for $z=r$

$$\varepsilon(r)=-\frac{1}{\pi}\frac{\mathrm{d}}{r\mathrm{d}r}\int\limits_{r}^{R}\frac{I(x)x\mathrm{d}x}{\sqrt{x^2-r^2}}. \tag{1.12a}$$

Per parts integration of (1.12a) and subsequent differentiation with respect to the parameter give

$$\varepsilon(r)=-\frac{1}{\pi}\int\limits_{r}^{R}\frac{\mathrm{d}I(x)/\mathrm{d}x}{\sqrt{x^2-r^2}}\mathrm{d}x. \tag{1.12b}$$

The equation (1.12) are the inversions of the Abel transform (1.10).

An optically dense source. At an arbitrary value of coefficient $\kappa(r)$, the intensity of radiation can be determined from the transport equation

$$\frac{\mathrm{d}I}{\mathrm{d}y}=\varepsilon-\kappa I. \tag{1.13}$$

This is a linear differential equation of the first order whose solution has the form

$$I(y)=\exp\left(\int\kappa(r)\mathrm{d}y\right)\left[\int\varepsilon(r)\exp\left(\int\kappa(r)\mathrm{d}y\right)\mathrm{d}y+C\right]$$

Transferring to integration in the ranges $[-y_0, y_0]$ (Fig.1.1) and setting $\kappa\ (r \geq R) = 0$, $\varepsilon\ (r \geq R) = 0$, $I\ (-y_0) = I_0\ (x) = C$, $I\ (y_0) = J\ (x)$, we obtain

$$J(x) = I_0(x)\exp\left(-\int_{-y_0}^{y_0}\kappa(r)\,dy\right) + \int_{-y_0}^{y_0}\varepsilon(r)\exp\left(-\int_{y}^{y_0}\kappa(r)\,dy\right)\!,dy$$

or

$$J(x) = I_0(x)\omega(x) + I(x). \tag{1.14}$$

Here $I\ (x)$ is the intensity of inherent radiation of the source; $\omega\ (x)$ is the transmittance coefficient, $I_0\ (x)$ and $I_0\ (x)\omega\ (x)$ is the intensity of external emission prior to and after passage of the object. The equation for the transmittance coefficient

$$\omega(x) = \exp\left(-\int_{-y_0}^{y_0}\kappa(r)\,dy\right) \tag{1.15}$$

can be written in the form of Abel's integral equation

$$-\ln\ \omega(x) \equiv \tau(x) = 2\int_{|x|}^{R}\frac{\kappa(r)r\,dr}{\sqrt{r^2 - x^2}}, \tag{1.16}$$

where $\tau\ (x)$ is the optical thickness of the object. Equation (1.16) has the inversion of the type (1.12).

The expression for the intensity of inherent radiation of the source

$$I(x) = \int_{-y_0}^{y_0}\varepsilon(r)\exp\left(-\int_{y}^{y_0}\kappa(r)\,dy'\right)dy \tag{1.17}$$

is transformed as follows

$$I(x) = \exp\left(-\int_0^{y_0} \kappa(r)dy\right)\left[\int_{-y_0}^{0} \varepsilon(r)\exp\left(\int_0^{y} \kappa(r)\right)dy' + \int_0^{y_0} \varepsilon(r)\exp\left(\int_0^{y} \kappa(r)dy'\right)dy\right] =$$

$$\exp\left(-\int_0^{y_0} \kappa(r)dy\right)\int_0^{y_0} \varepsilon(r)\left[\exp\left(-\int_0^{y} \kappa(r)dy'\right)\right]dy,$$

which takes into account the equality

$$\int_{-y_0}^{0} \varepsilon(r)\exp\left(\int_0^{y} \kappa(r)dy'\right) = \int_0^{y_0} \varepsilon(r)\exp\left(-\int_0^{y} \kappa(r)dy'\right)dy.$$

Using the notations of the hyperbolic cosine and the transmittance coefficient (1.15), we can write

$$I(x) = 2\sqrt{\omega(x)}\int_0^{y_0} \omega(r)\,\mathrm{ch}\left(\int_0^{y} \kappa(r)dy'\right)dy$$

or in coordinates (x, r)

$$\frac{I(x)}{\sqrt{\omega(x)}} = 2\int_{|x|}^{R} \varepsilon(r)\,\mathrm{ch}\left(\int_{|x|}^{r} \frac{\kappa(r')r'dr'}{\sqrt{r'^2 - x^2}}\right)\frac{rdr}{\sqrt{r^2 - x^2}}$$

which represents the integral Volterra equation of the first kind.

Measurement of radiation and transmittance. Figure 1.2 shows some of the existing methods of measurement of radiation and transmittance.[1] The image of the examined object O is projected by the lens L_1 on the aperture of a spectral apparatus. When measuring the intensity $I(x)$, the translucent radiation is cut off by an obturator. In the circuits 1.2 a-c, the axis of the object should be normal to the direction of the aperture. If necessary, the image of the object is rotated through 90° using a prism of total internal reflection, a system of three or two mirrors placed between the lens L_1 and the object O. In photographic registration we measure the distribution of blackening across the spectrogram and

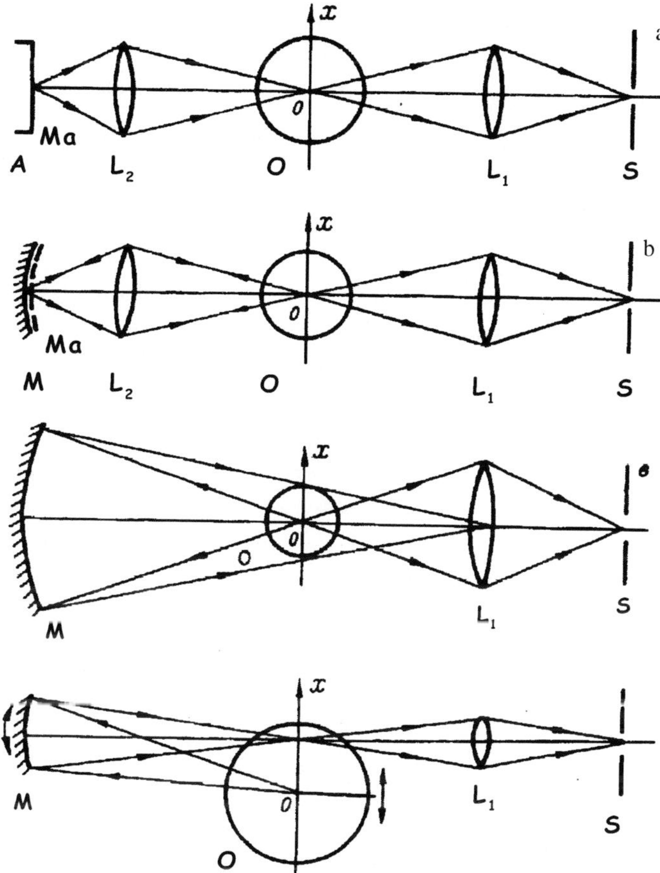

Fig.1.2 Optical circuit of measurement of radiation and transmittance. A) auxilliary source; L_1, L_2 (lenses); M) spherical mirror; Ma) mask; O) examined object; S) input aperture of the spectral apparatus.

then transferred to the intensity distribution $I(x)$ on the basis of the characteristic curve of the photolayer. In photoelectric recording, the intensity distribution $I(x)$ is measured directly. Scanning in the direction of the x axis is carried out by moving the diaphragm along the input or output aperture of the device. In the diagram shown in Fig.1.2d the object is scanned in the direction normal to the optical axis.

To measure the transmittance $\omega(x)$, the examined object is illuminated by the radiation of auxiliary source A (Fig.1.2a) or by natural radiation reflected from the mirror M (Fig.1.2b-d). The total intensity in illumination is

$$J(x) = I_0(x)\omega(x) + I(x),$$

where $I(x)$ is the intensity of radiation, $I(x)$ is the natural radiation of the source.

Therefore, knowing $J(x)$, $I_0(x)$, $I(x)$ we obtain

$$\omega(x) = \frac{J(x) - I(x)}{I_0(x)}.$$

The diagram in Fig. 1.2 b gives the total combination of the image with the object. The intensity of translucent radiation $I_0(x) = \gamma I(x)$, where γ are radiation losses on the path from the source to the mirror and back. The value of γ is determined from the section of the spectrum where absorption can be ignored, i.e. $\omega(x) \approx 1$ and $\gamma = (J(x) - I(x))/I(x)$. If the mask Ma with non-transparent strips, parallel to the axis of the source, is placed in front of the mirror M, in the focal plane of the device we obtain alternating values of $J(x)$ and $I(x)$. The main disadvantage of the setup shown in Fig. 1.2b is that the transmittance factor cannot be measured outside the zone of radiation of the source. This disadvantage is not found in the setup in Figs 1.2 a,c,d. In the circuit in Fig. 1.2 c, the image is projected by the mirror M in the plane of the lens L_1. The examined object is illuminated with a wide, relatively uniform beam. The circuit showed in Fig.1.2 d is even more efficient in this respect. The spherical mirror is rotated when scanning the object O in such a manner that the object is illuminated for any values of x by its brightest central part. In Fig.1.2 a–c, the value $I_0 = \text{const}$ and is determined from measurements outside the limits of the absorption zone of the object O or in the setup in Fig.1.2 a in the absence of the examined object. In photographic recording, the translucent radiation is often modulated by setting the receiving apparatus to the modulation frequency. This greatly increases the accuracy of measuring transmittance and also enables auxiliary sources (Fig.1.2 a) of relatively small brightness to be used.

The circuit shown in Fig.1.2 d is designed for examining stationary sources. Nonstationary processes can be examined using the setup in Fig.1.2 a-c with photographic registration. The time resolution is obtained by placing an instantaneous gate (Kerr cell, a system of rotating mirrors) in front of the aperture or using other measures.

Spatial resolution power. The integral equations presented above were obtained in the one-beam approximation. However, actual devices have a finite angular aperture. Figure 1.3 shows the geometry of beams in reflecting the object O on the aperture of the device by the lens L. The normal incidence beam on the objective has the intensity $I(x)$. The oblique beam, propagating under angle φ to the optical axis corresponds

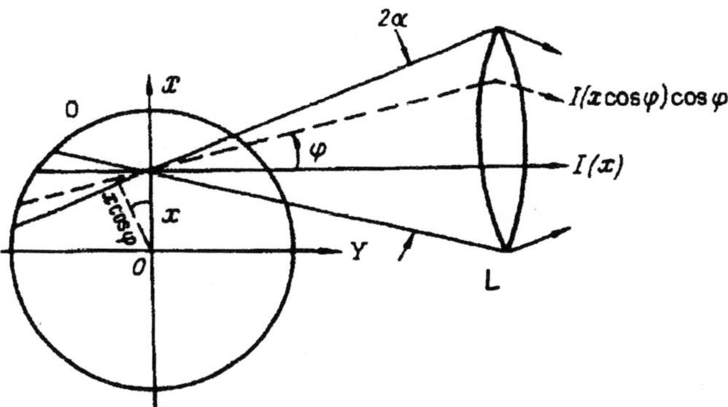

Fig.1.3 Diagram of recording radiation with a finite appature of the device.

to the chord whose distance from the centre of the object is $x \cos \varphi$ and intensity $I(x \cos \varphi)\cos \varphi$. The recorded intensity of radiation is averaged out with respect to the angle 2α and is equal to

$$I(x) = \frac{1}{\alpha}\int_0^\alpha I(x \cos \varphi)\cos \varphi \, d\varphi. \tag{1.21}$$

Replacing $I(x)$ over the averaging region by a segment of the straight line, for small angles we obtain α

$$\bar{I}(x) \approx \left(1 - \frac{\alpha^2}{6}\right)I\left(x\left(1 - \frac{\alpha^2}{6}\right)\right). \tag{1.22}$$

The difference between the measured intensity $\bar{I}(x)$ and true intensity is determined to a first approximation by the scale factor

$$\beta = 1 - \frac{\alpha^2}{6}.$$

The value of β differs only slightly from unity, even for relatively high angles α. Thus, $\beta = 0.998$ at $\alpha = 0.1$. This shows that the reduction of the spatial resolution power as a result of the finite angular aperture of the device is insignificant in the majority of cases.

Taking transverse oscillations into account. Electric arcs, plasma and flame jets are often show transverse oscillations over a wide frequency range. The direct path of diagnostics of these objects is to obtain

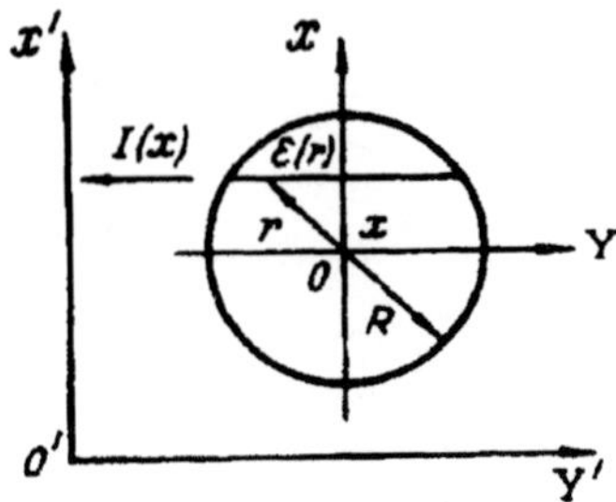

Fig.1.4 Coordinate system for examining an oscillating source.

instantaneous lateral intensity distributions. This requires using multichannel photoelectric registration or electron–optical converters with a good time resolution. Experiment formulation is greatly simplified if we use information on the transverse oscillations of the examined object.[2,3]

Let us assume that the properties of the cross section of the examined object are constant and that the cross section wanders in the $X'O'Y'$ plane around some most probable position, Fig.1.4. The input aperture of the spectral device is combined with the X' axis. It will be assumed that small displacements of the source in the direction of the Y' axis do not change the amount of light passing into the device. This is justified for a small relative orifice of the projecting lens, as is usually the case. Consequently, it is sufficient to take into account the displacement of the source in the direction of the axis X'. $\varphi(x')$ will denote the density of the distribution of probability of deviation of the axis of the source from the mean position $x' = 0$, coinciding with the centre of the inlet aperture of the device. Correspondingly, the probability of some chord x of the source being situated opposite the section $x' \div x'+\mathrm{d}x'$ of the inlet aperture is

$$\varphi(x' - x)\mathrm{d}x = \varphi(x' - x)(\mathrm{d}I / \mathrm{d}x)^{-1}\mathrm{d}I = \frac{\varphi(x' - x)}{I'(x)}\mathrm{d}I. \tag{1.23}$$

The total probability $\alpha(I, x')$ $\mathrm{d}I$ of passage through the section of the aperture $x' \div x'+\mathrm{d}x'$ with the intensity $I \div I + \mathrm{d}I$ is obtained by summing up equation (1.23) with respect to all chords x_j of the source with the same intensity I

$$\alpha(I, x')\mathrm{d}I = \sum_i \frac{\varphi(x' - x_i)}{I'(x_i)}\mathrm{d}I. \tag{1.24}$$

If the functions $\varphi(x')$ and $\varphi(I, x')$ are known, the determination of the distribution of intensity $I(x)$ is reduced to solving the problem of non-linear algebraic equations, following from (1.24)

$$\sum_{i=1}^{n} \frac{\varphi(x'_j - x_i)}{I'(x_i)} = \alpha(I, x'_j), \quad j = \overline{1, m}, \quad m = 2n. \tag{1.25}$$

At fixed values of I we exclude $I'(x_i)$ from the system (1.25) and determine x_i, which gives a discrete function $I(x)$. In a partial case of a symmetric and monotonically decreasing function $I(x)$ at $|x| \geq 0$, each intensity I corresponds to two symmetric chords x and $-x$. Consequently $I(-x) = I(x)$, $I'(-x) = I'(x)$ and the system (1.25) has the form

$$\frac{\varphi(x'_1 - x) + \varphi(x'_1 + x)}{I'(x)} = \alpha(I, x'_1),$$

$$\frac{\varphi(x'_2 - x) + \varphi(x'_2 - x)}{I'(x)} = \alpha(I, x'_2). \tag{1.25a}$$

Dividing the first equation (1.25a) by terms by the second equation, we obtain

$$\frac{\varphi(x'_1 - x) + \varphi(x'_1 + x)}{\varphi(x'_2 - x) + \varphi(x'_2 + x)} = \frac{\alpha(I, x'_1)}{\alpha(I, x'_2)}, \tag{1.25b}$$

We introduce the notations

$$\varphi(x'_1, x'_2, x) = \frac{\varphi(x'_1 - x) + \varphi(x'_1 + x)}{\varphi(x'_2 - x) + \varphi(x'_2 + x)},$$

$$\overline{\alpha}(I, x'_1, x'_2) = \frac{\alpha(I, x'_1)}{\alpha(I, x'_2)}.$$

We specify x and calculate $\overline{\varphi}(\overline{x}_1, \overline{x}_2, x)$. For the value of α equal to $\overline{\varphi}$, from the known dependence of $\overline{\alpha}$ on I, we determine the corresponding $I(x)$. This is the method of processing the results of measurements of the radiation intensity of a vibrating object proposed in Ref. 2.

The function $\varphi(x')$ is determined by experiments, and the examined object is displayed on the aperture of the photographic recording device in such a manner that the axis of the source is perpendicular to the direction of the aperture. On the photographic recording, with sufficient time resolution, we calculate the mean position of the axis of the source and the probability of its deviation to different distances from the mean distance. This gives the function $\varphi(x')$ after normalisation with respect to unity.

To determine $\alpha(I, x')$ by the photoelectric method with a good resolution, it is necessary to recorded the time dependence of the intensity of the analytical spectral line or the section of continuum consecutively for two or more co-ordinates x'. The function $\alpha(I, x')$ is calculated from the frequency with which the intensity fits into the fixed ranges $I \div I + dI$.

The authors of Ref. 3 proposed a different method of the spectroscopy of vibrating objects. Like in Ref. 2, using the method proposed in Ref. 3, we measure the function $\varphi(x')$ of vibration of the source but the spectrum is recorded without time resolution. The ratio between the distribution of intensity $q(x')$, averaged with respect to time, and the true lateral distribution of intensity $I(x)$ of the source is determined as follows. As mentioned previously, the probability of passage of intensity $I(x)$, corresponding to the chord x, through the section of the aperture $x \div x' + dx'$ is $\varphi(x' - x)\, dx$. The radiation flux from the chord x through the given section of the aperture, averaged of a relatively long period of time, is

$$dq(x') = I(x)\varphi(x' - x)dx. \tag{1.26}$$

The total flux $q(x')$ is obtained by integrating (1.26) throughout the entire range of variation $x \in [-R, R]$, in which $I(x)$ differs from 0

$$q(x') = \int_{-R}^{R} I(x)\varphi(x' - x)dx. \tag{1.27}$$

The required distribution of intensity $I(x)$ is the solution of the integral equation (1.27). The relationship (1.27) between the true $I(x)$ and the measured $q(x')$ intensity distributions is also observed in recording a stable source in a real device with the finite resolving power and the apparatus function $\alpha(x')$. The distortion of the intensity distribution in lateral examination as a result of the finite resolution of the recording device may be considerable for sources with high intensity gradients. If both reasons are valid, then $\varphi(x')$ in equation (1.27) is the convolution of the apparatus function of the device and the conditional apparatus function, chatacterising the vibrations of the source. To determine the apparatus function of the device, a uniformly illuminated narrow aperture is made at the optical axis instead of the examined object. The distribution of intensity in the focal plane of the device, measured for this auxiliary source, is also its apparatus function. Numerical realisation and examples of practical application of the method were published in, for example, Ref. 10–13.

Brightness reference sources

The measurement of the absolute intensity of the spectral lines and continuum and of the relative intensity of radiation in a wide spectral range is carried out using brightness reference sources (standards of radiation intensity) [4,5].

The primary intensity standard is the radiation of an absolutely black body. This radiation is determined only by its temperature and does not depend on the properties of the material from which it is made. The brightness and spectral composition of its radiation are calculated using Planck's equation. Artificially, the properties of the absolutely black body are reproduced by means of a small hole in the wall of a large closed cavity. The material of the walls is tungsten or carbon. In most cases, the device has the form of a pipe heated with electric current. A large number of diaphragms is placed inside the pipe. These diaphragms prevent the direct exit of radiation from the cavity through the examination hole. The pipe contains a small crucible with gold or platinum, and the intensity of radiation of the cavity is measured at the melting point of these materials. The melting point of gold is 1337.58 K, that of platinum is 2044.9 K.

Secondary intensity standards are calibrated at specific wavelengths by comparing the intensity of their radiation with the intensity of radiation of the absolutely black body. The calibration error with respect to intensity is $\pm(5$–$10)\%$.

Tungsten reference sources are represented tungsten lamps with a strip-shaped heating filament. The temperature of the filament is maintained at approximately 2700 K. The degree of blackness of tungsten has been measured over a wide range of wavelength and filament temperature, and its value is close to 0.45.

Another suitable secondary intensity standard is the crater of the anode of a carbon arc (Fig. 1.5). In the carbon arc where the electrodes are positioned under the almost right angle in relation to each other, at the current intensity in the arc slightly lower than the value corresponding to the start of hissing the temperature of the end surface of the anode facing the examiner is equal to the sublimation temperature of graphite (3995 ± 20 K), at a degree of blackness of 0.7–0.75, depending on

Fig.1.5 Distribution of electrodes in recording the crater of the carbon anode.

the wavelength. The brightness of radiation of the crater of the carbon anode is considerably higher in comparison with the tungsten strip heating lamp. Consequently, it can be used as a brightness reference in the wavelength range exceeding 2500 Å. The arc current is selected on the basis of the dimensions of the anode crater. For example, for a carbon anode with a right-angled cross-section of 5.5 × 3 mm, the arc current is 9–9.5 A, which is slightly lower than the current of the hissing arc. The carbon arc is not recommended for use in the range 3700–4200 Å which contains the emission bands of cyanogen CN, and in the range 4500–4740 Å containing the bands of molecular carbon C_2.

In these regions of the spectrum heating is more efficient when carried out using a high-pressure xenon lamp whose radiation consists of the high-intensity continuous spectrum and contains only several lines at the wavelengths below 6700 Å. The colour temperature in the range 3700–5300 Å is 5572±33K.

The electric arc, stabilised by the walls, is characterised by an even higher brightnes in the group of the standards. The source can be used for calibration in a wide range of the spectrum from the infrared to vacuum ultraviolet range. The diagram of stabilisation of the arc is shown in Fig. 1.6. At a relatively long arc column, the radiation in the centre of the spectral lines corresponds to Planck's radiation. This radiation is taken through end windows and is used as reference radiation. The temperature at the axis of the arc in relation to current and the internal diameter of the pipe may vary from 10 000 to 30 000 K. In order to eliminate the absorption of radiation in the cold regions, the end zones of the pipe are blown with a different gas (argon, helium).

1.2 Spectral emission line

The emission factor of the spectral line for plasma in a local thermal

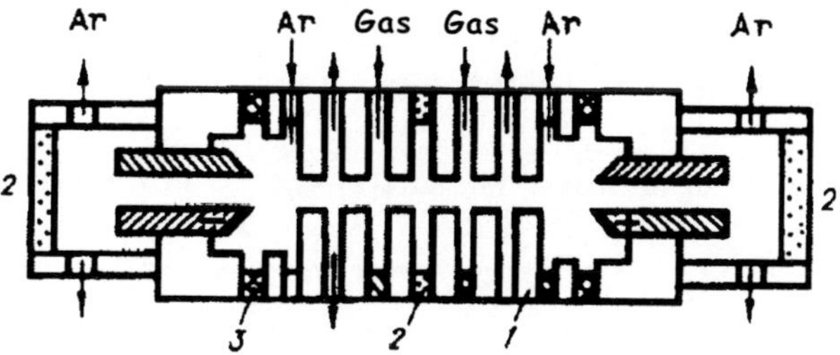

Fig.1.6 Diagram of the arc with stabilising walls. 1) water-cooled copper rings, hollow inside; 2) window; 3) insulating interlayers.

equilibrium is:

$$\varepsilon_{nk} = \frac{h\nu_{nk}}{4\pi} \frac{A_{nk} g_n}{S} N \exp(-E/kT).)$$
(1.28)

Here N is the concentration of the atoms (ions) of the emitting element; T is plasma temperature; ν_{nk}, A_{nk} is the emission frequency for the $n \to k$ transition and the transition probability, respectively.

The functional dependence (1.28) is used as a basis of the method of determining plasma temperature, particle concentration (atoms, ions) and transition probabilities of spectral lines.

Absolute intensity of spectral lines. At the known transition probabilities, initial composition and the pressure of LTE plasma, the temperature dependence of the particle concentration can be calculated in advance and the emission factor (1.28) is written in the form

$$\varepsilon = \varepsilon(T).$$
(1.29)

Consequently, by measuring the absolute intensity it is possible to determine the plasma temperature. Figure 1.7 shows the dependence of the intensity of the spectral line ArII 4806 Å on temperature for argon plasma at atmospheric pressure.

Varying (1.29) and taking into account (1.28), we obtain an estimate of the error of the temperature measuring method

$$\frac{\delta T}{T} = \frac{\varepsilon}{T} \frac{\partial T}{\partial \varepsilon} \left[\left(\frac{\delta \varepsilon}{\varepsilon} \right)_{ran} + \left(\frac{\delta \varepsilon}{\varepsilon} \right)_{sys} + \frac{\delta A}{A} \right].$$
(1.30)

The first term in the square brackets is the random error of measurement reproducibility, the second term is the systematic error of calibration of the reference brightness source, the third term is the error of the transition probability. The factor $\dfrac{\varepsilon}{T}\dfrac{\partial T}{\partial \varepsilon}$ is the coefficient of error transfer (Fig.1.7).

In most cases $(\delta\varepsilon/\varepsilon)_{\text{rand}} \approx 5\%$, $(\delta\varepsilon/\varepsilon)_{\text{syst}} \approx 10\%$, $\delta A/A \approx 10\%$, and for an example shown in Fig.1.7, where $(\varepsilon/T)(\partial T/\partial\varepsilon) < 0.1$, $\delta T/T \approx 3\%$.

If we know the transition probability and transfer temperature, equation (1.28) can be used to determine the atom (ion) concentration from the measured intensity. The corresponding estimate of the error follows from the variation of (1.28)

$$\frac{\delta N}{N} = \frac{\delta\varepsilon}{\varepsilon} + \frac{\delta A}{A} + \frac{E}{kT}\frac{\delta T}{T}, \tag{1.31}$$

which for the example shown in Fig.1.7 gives $\dfrac{\delta N}{N} \approx 15 + 10 + 10\cdot 3 \approx 50\%$.

The ratio of the intensities of spectral lines. For the ratio of the intensities of the spectral lines of some element of the same degree of ionisation, equation (1.28) gives

$$\frac{\varepsilon_2}{\varepsilon_1} = \frac{(Ag\nu)_2}{(Ag\nu)_1}\exp\left(-\frac{E_2 - E_1}{kT}\right), \tag{1.32}$$

and consequently

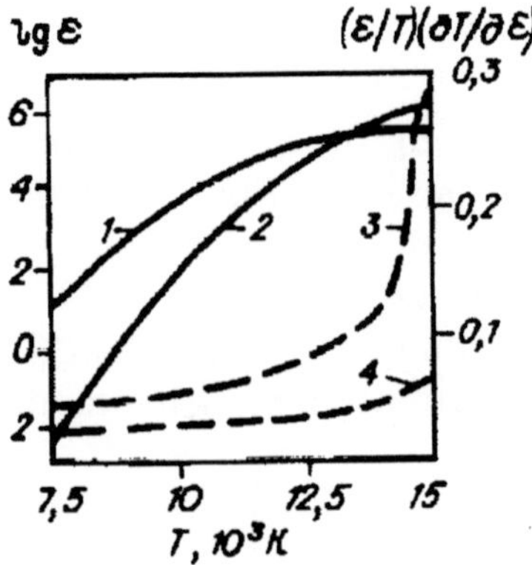

Fig.1.7 Dependence of the intensity of continuum $\lambda = 4810$ Å (I) (erg/(cm³ sterad Å)) and spectral line Ar(II) 4806 Å (2) (erg/(cm³ sterad) on temperature at $P = 1$ atm (3, 4 are the corresponding coefficients of transfer of the error).[6]

$$T = \frac{(E_2 - E_1)/k}{\ln\dfrac{(Agv)_2}{(Agv)_1} - \ln\dfrac{\varepsilon_2}{\varepsilon_1}}.$$

(1.33)

Measuring the ratio of intensities, from equation (1.33) we determine plasma temperature at a known ratio of the transition probabilities. The variation of equation (1.33) gives the following estimate of the temperature determination error:

$$\frac{\delta T}{T} = \frac{kT}{|E_2 - E_1|}\left[\frac{\delta(A_2/A_1)}{A_2/A_1} + \frac{\delta(\varepsilon_2/\varepsilon_1)}{\varepsilon_2/\varepsilon_1}\right].$$

(1.34)

In particular, equation (1.34) shows that to increase the accuracy it is necessary to use spectral lines with a higher excitation energy difference. If, for example,

$$\frac{kT}{|E_2 - E_1|} \approx \frac{1}{3}, \quad \frac{\delta(A_2/A_1)}{A_2/A_1} \approx 10\%, \quad \frac{\delta(\varepsilon_2/\varepsilon_1)}{\varepsilon_2/\varepsilon_1} \approx 5\%,$$

then $\dfrac{\delta T}{T} \approx 5\%..$

The Ornstein method of measuring temperature on the basis of the ratio of the intensities of two spectral lines can be extended to a case of using a group of spectral lines belonging to the same type of atoms. From equation (1.28), we obtain

$$\ln\left(\frac{\varepsilon}{Agv}\right)_i = y_i = a - \frac{E_i}{kT},$$

(1.35)

where a is the constant at fixed temperature T. Equation (1.35) gives a straight line whose slope is tan $\varphi = 1/kT$ (Fig.1.8). To determine temperature, we draw, through the experimental points y_i, $i = 1, n$, a section with a straight line by the methods of least squares.[7] We obtain

$$kT = \frac{n\sum\limits_{i=1}^{n} E_i^2 - \left(\sum\limits_{i=1}^{n} E_i\right)^2}{n\sum\limits_{i=1}^{n} y_i E_i - \sum\limits_{i=1}^{n} y_i \sum\limits_{i=1}^{n} E_i}.$$

(1.36)

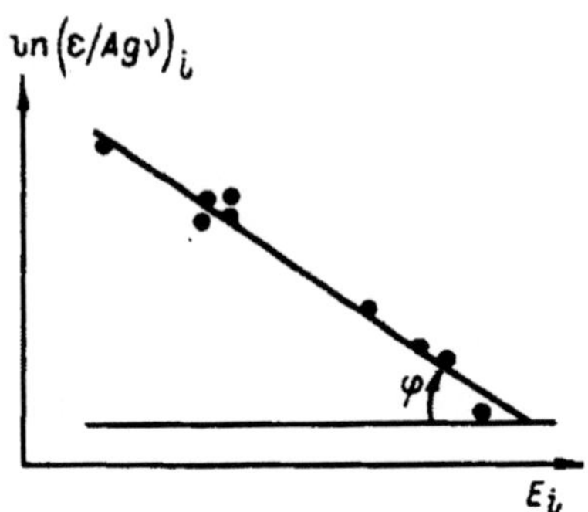

Fig.1.8 Graph $y_i = y(E_i)$ in the presence of Boltzmann distribution.

and the error estimate

$$\frac{\delta T}{T} = \frac{kT}{\left[\sum\limits_{i=1}^{n} E_i^2 - \frac{1}{n}\left(\sum\limits_{i=1}^{n} E_i\right)^2\right]^{1/2}}\left(\frac{\delta\varepsilon}{\varepsilon}+\frac{\delta A}{A}\right),$$

(1.37)

where $\delta\varepsilon/\varepsilon$ is the error of measurement of intensity measurements, $\delta A/A$ is the inaccuracy in the transition probabilities. When using two spectral lines, the equations (1.36) and (1.37) are usually replaced by (1.33) and (13.4), respectively. The increase of the number of lines increases the accuracy of temperature measurements and also gives an answer to the question as to whether there is a Boltzmann's distribution of atoms over the excited states. In this case, all points fit a single straight line, within the measurement error range. If there is a large deviation from Boltzmann's distribution, the experimental points are situated around some curve and not in the vicinity of a straight line.

For the ratio of the intensities of the atomic ε_a and ionic ε_i lines of the same element, equation (1.28) gives

$$\frac{\varepsilon_a}{\varepsilon_i} = \frac{(Ag\nu)_a}{(Ag\nu)_i}\frac{N_a}{N_i}\exp\left(-\frac{E_a-E_i}{kT}\right).$$

(1.38)

Consequently

$$\frac{N_i}{N_a} = \frac{x}{1-x} = \frac{(Ag\nu)_a}{(Ag\nu)_i}\frac{\varepsilon_i}{\varepsilon_a}\exp\left(-\frac{E_a-E_i}{kT}\right).$$

(1.39)

Measuring the ratio of the intensities of the atomic and ionic spectral lines, from (1.39) we determine the degree of ionisation of the atom at the known plasma temperature and the ratio of the transition probabilities. Using (1.39) together with the Saha equation (1.6), we determine the concentration of electrons in plasma

$$N_e = \frac{\varepsilon_a}{\varepsilon_i} \frac{(Ag\nu)_i}{(Ag\nu)_a} \, 2\left(\frac{2\pi mkT}{h^2}\right)^{3/2} \exp\left[-\frac{(U-\Delta U)-(E_a-E_i)}{kT}\right]. \qquad (1.40)$$

The error of determining N_e is obtain by varying (1.40)

$$\frac{\delta N_e}{N_e} = \frac{\delta(\varepsilon_a/\varepsilon_i)}{\varepsilon_a/\varepsilon_i} + \frac{\delta(A_a/A_i)}{A_a/A_i} + \left[\frac{3}{2} + \frac{(U-\Delta U)+|E_a-E_i|}{kT}\right]\frac{\delta T}{T}. \qquad (1.41)$$

Assuming that $E_a \approx E_i$, $U/kT \approx 10$, $\Delta U \ll U$, $\delta\,(\varepsilon_a/\varepsilon_i)/(\varepsilon_a/\varepsilon_i) \approx 5\%$, $\delta\,(A_a/A_i)/(A_a/A_i) \approx 10\%$, $\delta T/T \approx 3\%$, we have $\delta N_e/N_e \approx 50\%$.

Optimum temperature. With increasing temperature, the intensity of spectral lines initially increases as a result of the processes of excitation and ionisation of the atoms (ions), reaches a maximum value and then decreases. Figure 1.9 shows the temperature dependence of the intensity of a number of spectral lines. The temperature corresponding to the intensity maximum is referred to as optimum temperature. In stationary plasma with a temperature gradient, the maximum intensity can be observed in a separate region thus recording the temperature of this region. The spatial distribution of temperature is determined by comparing the measured intensity distribution with the calculated temperature dependence of intensity.

Optimum temperature depends of the ionisation energy of the atom (ion), the excitation energy of the spectral line and the electron concentration in plasma. The method of determining the temperature from the position of the maximum intensity of the line (Lorentz method) can be applied if the maximum temperature in a heterogeneous examined object exceeds the optimum temperature for the given line. The advantages of the Lorentz method is that it does not require data on the transition probabilities and it is not necessary to measure the absolute intensity. In addition, radiation readsorption has no significant effect in measuring the temperature by this method on the basis of resonance lines.[8]

The maximum intensity of the spectral line occupies a relatively wide

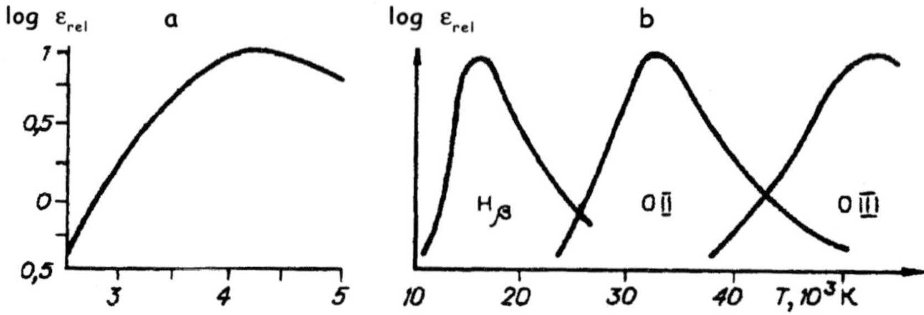

Fig.1.9 Intensities of the line NaI 5890 Å in air plasma at the atmospheric pressure and at a sodium concentration of $N = N_{a0} + N_i = 1 \times 10^{15}$ cm^{-3},[8] (a) and spectral lines H$_\beta$ 4861 Å, OII 4651 Å and OIII 3447 Å (b) in air plasma at the atmospheric pressure.[4]

temperature range (Fig.1.9) and this also determines the error of the method equalling 7–10%.

1.3 Width and shift of spectral lines

Doppler half width. As a result of random movement of the radiating atoms (ions), the spectral line broadens as a result of Doppler's effect. The intensity distribution in the Doppler contour has the form

$$P_D(\omega) = \frac{1}{\sqrt{\pi}\Delta\omega_D} \exp\left[-\frac{\Delta\omega^2}{\Delta\omega_D^2}\right]. \tag{1.42}$$

where $\Delta\omega = \omega - \omega_0$; $\Delta\omega_D = \dfrac{\omega_0 \bar{v}}{c}$; ω_0 is the centre of the line;

$\bar{v} = (2RT/\mu)^{1/2}$ is the most probable velocity of random movement of the atom; R is the gas constant; c is the speed of light; μ is the atomic weight of radiating particles. $\Delta\omega_D$ is the Doppler e-half width, i.e. at the distance $\Delta\omega = \Delta\omega_D$ from the centre of the line ω_0, the intensity is e times less than the maximum intensity. Doppler half width, i.e. the width of the line on the level of half of its intensity, expressed in wave length, is

$$\Delta\lambda_D = 2\sqrt{\ln 2}\,\frac{\bar{v}\lambda_0}{c} = 7.16 \cdot 10^{-7}\lambda_0\sqrt{\frac{T}{\mu}}. \tag{1.43}$$

Measuring the half width of the Doppler line, equation (1.43) can be used to determine the kinetic temperature of the radiating atoms (ions). The accuracy of temperature determination

$$\frac{\delta T}{T} = 2\frac{\delta(\Delta\lambda_D)}{\Delta\lambda_D} \tag{1.44}$$

is ~10%.

The Doppler contour (1.42) has the form of Gauss' distribution in contrast to the dispersion profiles determined by broadening at collisions. Therefore, it is quite easy to determine whether the contour is of the Doppler type. The Doppler half width can be separated if the collision broadening is of the same order as the Doppler broadening. Doppler broadening prevails in low-density plasma (small number of collision) and at high temperature. The Doppler half width is greater for the atoms with a low mass ($T = 10000$ K):

Element	H	O	Fe	Kr
$\Delta\lambda_D$, Å	0.36	0.09	0.05	0.04

For elements with unfilled internal shells, spectral lines with the purely Doppler contour are observed even at high pressure. These lines are determined by transitions between the levels not associated with the outer electrons, and do not broaden under the effect of collisions. The use of iron lines for determining the atom temperature from Doppler half width in an electric arc at the atmospheric pressure is a well known example.[4]

In experiments it is important to use apparatus with a high resolution.

Stark half width. Broadening of the spectral lines under the effect of charged particles (Stark broadening) is the main type of broadening in high-pressure plasma. The theory of Stark contours has been developed most extensively for hydrogen atoms. The error of theory is estimated for half width at 7–10%. The corresponding contours can be satisfactorily described by a dispersion (Lorentz) distribution

$$P_L(\omega) = \frac{\gamma/2\pi}{(\Delta\omega)^2 + \left(\dfrac{\gamma}{2}\right)^2}. \tag{1.45}$$

Here γ is the Stark half width. When expressed in the wavelength, it is equal to

$$\Delta\lambda_S = C^{-3/2}(N_e, T)N_e^{2/3}. \tag{1.46}$$

Figure 1.10 shows the dependences of the Stark half width of spectral lines H_α and H_β on the electron concentration with broadening parameters $C\,(N_e, T)$ from Ref.5. These graphs are used to determine the electron concentration from the measured half widths of the spectral hydrogen lines. The broadening parameter $C\,(N_e, T)$ for line H_β is almost independent of temperature. Since the contour H_β is not symmetric, it is recommended to use its short wave half (Fig.1.10). The method of determining the electron concentration from the Stark broadening of the spectral lines does not require the presence of Maxwell's velocity distribution of the electrons. The state of plasma can be evaluated by measuring the electron concentration by this method and comparing it with the results of measurements by another method assuming that plasma is equilibrium.

The error of measuring the electron concentration is determined by varying the equation (1.46):

$$\frac{\delta N_e}{N_e} = \frac{3}{2}\left[\left(\frac{\delta\Delta\lambda}{\Delta\lambda}\right)_{ran} + \left(\frac{\delta\Delta\lambda}{\Delta\lambda}\right)_{sys}\right]. \tag{1.47}$$

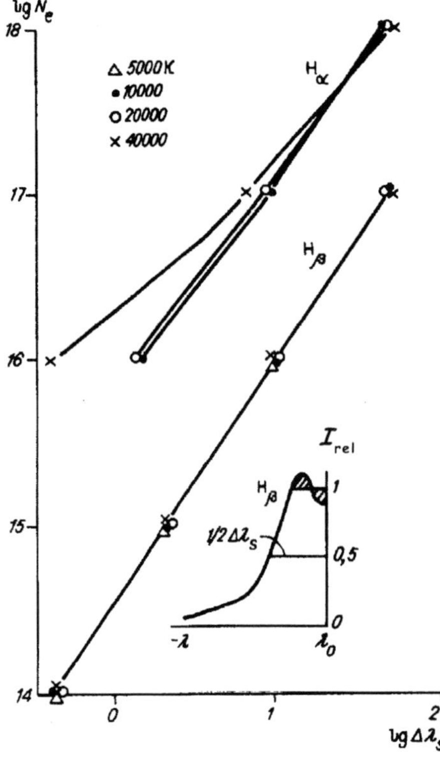

Fig.1.10 Dependence of the half width of spectral lines H_α and H_β (Å) on the electron concentration (cm^{-3}) at different plasma temperatures.[5]

Setting $(\delta\Delta\lambda/\Delta\lambda)_{sys} \approx 7 \div 10\%$, $(\delta\Delta\lambda/\Delta\lambda)_{ran} \approx 5\%$, we have $(\delta N_e/N_e) \approx 15 \div 20\%$. Determination of the electron concentration from the Stark half width of hydrogen lines is one of the most convenient and accurate methods. It should be noted that hydrogen radiation is almost always present in plasma due to the presence of small amounts of hydrogen or water impurities in gases and other substances. The hydrogen lines are wide so that they are easy to record and process.

The Stark broadening of the spectral lines of other elements is smaller and the values of half widths are determined with a larger error. Therefore, it is not recommended to use these lines for diagnostics.

Doppler shift. In macroscopic movement of plasma in relation to the observer the spectral lines carry out Doppler shift whose value is

$$\Delta\lambda = \lambda_0 \frac{v}{c}\cos\varphi. \qquad (1.48)$$

Here v is the plasma velocity, φ is the angle between the direction of plasma movement and the observation beam, λ_0 is the position of the centre of the non-shifted line (observation normal to the direction of plasma movement). Using equation (1.48), the plasma velocity can be determined from the measured shift of the spectral line. The error of the method

$$\frac{\delta v}{v} = \frac{\delta\Delta\lambda}{\Delta\lambda} \qquad (1.49)$$

is 5 10%. The modern methods make it possible to measure the minimum shift[4] $\Delta\lambda \sim 0.005$ Å which at $\lambda_0 = 5000$ Å and $\varphi = 0$ gives the lower restriction of the measured velocity $v_{min} = 300$ m/s.

1.4 Continuous spectrum

The continuous spectrum of low-temperature plasma forms as a result of photorecombination of the electron with the ion and deceleration of the electron in the ion field. The emission factor of the continuum formed during this process is written in the form[9]

$$\varepsilon_v = \frac{64\pi^{3/2}}{3\sqrt{6}}\frac{e^6}{m_e^{3/2}c^3}\xi(v)\frac{N_e^2}{\sqrt{kT}}, \quad v \leq v_g,$$

$$\varepsilon_v = \frac{64\pi^{3/2}}{3\sqrt{6}}\frac{e^6}{m_e^{3/2}}\xi(v)\frac{N_e^2}{\sqrt{kT}}\exp\left[\frac{(v-v_g)h}{kT}\right], \quad v \geq v_g. \qquad (1.50)$$

Function $\xi(v)$ and boundary frequency v_g take into account the specific features of the energetic levels of the atom. For example, for argon $v_g = 6.8\cdot10^{14}$ s^{-1} and the values of the function

ξ_v	1	0.73	0.88	1.15	1.47	1.7	1.75	1.9	2.1	2.05	1.85
$v\cdot10^{-5}$, s^{-1}	0	0.15	0.25	0.375	0.5	0.6	0.65	0.75	1.0	1.25	1.4

Equation (1.50) makes it possible to determine the electron concentration from the measured absolute intensity of continuum and the known plasma temperature. Varying the equation, we obtain the estimate of the error

$$\frac{\delta N_e}{N_e} = \frac{1}{2}\left(\frac{\delta\varepsilon_v}{\varepsilon_v} + \frac{\delta\xi}{\xi} + \frac{1}{2}\frac{\delta T}{T}\right), \tag{1.51}$$

which gives $\delta N_e/N_e \approx 15\%$ at $\delta\varepsilon_v/\varepsilon_v \approx 15\%$, $\delta\xi/\xi \approx 10\%$, $\Delta T/T \approx 10\%$. Relatively high accuracy is also obtained from the known temperature approximation. This method does not require that LTE exists in plasma.

Plasma temperature can be determined from the decrease of continuum beyond the boundary frequency. From equation (1.50) at $v > v_g$ we have

$$\varepsilon_{v_{rel}} = \exp\left(-\frac{(v - v_g)h}{kT}\right)$$

or

$$\ln\varepsilon_{v_{rel}} = \frac{hv_g}{kT} - \frac{hv}{kT}. \tag{1.52}$$

Plotting the dependence of $\ln\varepsilon_{v_{rel}}$ on hv, we obtain a fragment of the straight line whose slope gives the temperature. The accuracy of temperature measurement by this method depends on the frequency range $v_1 \div v_2$ in which the distribution continuum intensity is examined. According to (1.52), we can write

$$\ln\frac{\varepsilon_2}{\varepsilon_1} = \frac{hv_2 - hv_2}{kT} = \frac{\Delta hv}{kT},$$

and therefore

$$\frac{\delta T}{T} = \frac{kT}{\Delta h\nu}\,\frac{\delta(\varepsilon_2/\varepsilon_1)}{\varepsilon_2/\varepsilon_1}. \tag{1.53}$$

For example, at $\Delta h\nu/kT = 2$ and $\delta\varepsilon/\varepsilon \approx 10\%$, the error $\delta T/T \approx 5\%$. This method gives the electron temperature for non-equilibrium plasma.

At the known composition of LET plasma ε_ν from (1.50) can be represented as a function of temperature

$$\varepsilon_\nu = \varepsilon_\nu(T) \tag{1.54}$$

and the plasma temperature can then be determined from the absolute intensity of continuum. Figure 1.7 shows the dependence $\varepsilon_\nu(T)$ for argon plasma at atmospheric pressure. It also shows the transfer coefficient of the error $(\varepsilon_\nu/T)(\partial T/\partial\varepsilon_\nu)$ from equation (1.55) for estimating the error of temperature determination. Varying (1.53) we obtain

$$\frac{\delta T}{T} = \frac{\varepsilon_\nu}{T}\frac{\partial T}{\partial\varepsilon_\nu}\left[\left(\frac{\delta\varepsilon_\nu}{\varepsilon_\nu}\right)_{ran} + \left(\frac{\delta\varepsilon_\nu}{\varepsilon_\nu}\right)_{sys} + \frac{\delta\xi}{\xi}\right]. \tag{1.55}$$

Accepting that $\delta\varepsilon_\nu/\varepsilon_{\nu_{sys}} \approx 5\%$, $\delta\varepsilon_\nu/\varepsilon_{\nu sys} \approx 10\%$, $\delta\xi/\xi \approx 10\%$ we obtain that at $T < 130000$ K $\delta T/T \approx 3\%$, according to Fig.1.7.

1.5 Kirchhoff law

There is a relationship between the emission and absorption factors (Kirchhoff's law)

$$\frac{\varepsilon_\nu}{\kappa_\nu} = B_\nu, \quad B_\nu = \frac{2\pi h\nu^3/c^2}{\exp(h\nu/kT)-1}. \tag{1.56}$$

Here B_ν is the radiation of an absolutely black body at frequency ν and at temperature T. Measuring ε_ν and κ_ν, we can determine the plasma temperature from equation (1.56). If $\exp(h\nu/kT) \gg 1$ we have

$$\frac{\varepsilon_\nu}{\kappa_\nu} = (2\pi h\nu^3/c^2)\exp(-h\nu/kT), \tag{1.57}$$

and therefore

$$T = \frac{h\nu / k}{\ln(2\pi h\nu^3 / c^2) - \ln(\varepsilon_\nu / \kappa_\nu)}.$$
(1.58)

The error estimate follows from the variation of (1.57)

$$\frac{\delta T}{T} = \frac{kT}{h\nu}\left(\frac{\delta\varepsilon_\nu}{\varepsilon_\nu} + \frac{\delta\kappa_\nu}{\kappa_\nu}\right).$$
(1.59)

At $h\nu/kT = 4$ ($\lambda \approx 3000$ Å, $T = 10\ 000$ K), $\delta\varepsilon_\nu/\varepsilon_\nu \approx 15\%$, $\delta\kappa_\nu/\kappa_\nu \approx 5\%$, we have $\delta T/T \approx 5\%$. In non-equilibrium plasma, this method gives the electron temperature.

1.6 Spectral absorption lines

The atom concentration at the periphery of a plasma object outside the radiation zone can be determined from the spectral absorption lines. For this purpose, the object to be examined is irradiated with external radiation with a continuous spectrum. Absorption lines are obtained in a background of a continuous spectrum and are characterised by the total absorption value

$$A_\omega = \int_0^\infty \left(1 - \frac{I(\omega)}{I_0(\omega)}\right) d\omega,$$
(1.60)

where $I_0(\omega)$ is the intensity of the continuous spectrum outside the absorption line, A_ω is the area defined by the contour of the absorption line (Fig.1.11). The value of A_ω does not depend on broadening and the deformation of the absorption line by the apparatus function of the recording device. The absorption lines can be obtained on spectrographs with a relatively low resolving power. Only the linear dispersion of the device must be sufficiently high to ensure the required measurement accuracy. The value of the total absorption A_ω depends on the concentration of the absorbing atoms N, the length of the absorbing layer l, the power of oscillators of the spectral lines f and the parameters of the contour of the absorption line. For spectral lines with Voight's contour, i.e. when their broadening is determined by the collision processes and the Doppler effect, the contour of the line is characterised by the parameter

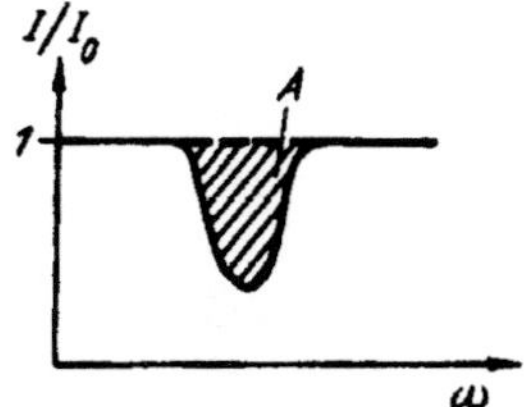

Fig.1.11 Contour of the absorption line.

$$a = \frac{\Delta\omega_L}{\Delta\omega_D}, \tag{1.61}$$

where $\Delta\omega_D = (\omega_0/c)(2RT)/\mu)^{1/2}$ is the Doppler e-halfwidth, $\Delta\omega_L = 4\sigma^2 N_j(2\pi RT\,(1/\mu + 1/\mu_j))^{1/2}$ Lorentz halfwidth, N_j, μ_j is the concentration and atomic weight of the perturbing particles, σ^2 is the effective cross section of 'optical' collisions of the atoms of the examined element with perturbing particles. Total absorption as a function of the number of absorbing atoms can be calculated for different parameters a (the so-called growth curves).[4] Figure 1.12 shows the growth curves for a homogeneous absorbing layer. These curves make it possible to determine the number of atoms from the measured total absorption of the line, if the gas temperature, parameter a and the power of oscillators f of the corresponding transition are known.

The spectrograms,obtained by the method of transverse photography in irradiating the heterogeneous object by radiation by a continuous spectrum (Fig.1.2a), record the absorption lines. The distribution of the total absorption A_ω along the line is determined. Photometric profiles are then constructed for all sections of the absorption line adjacent to each other. The characteristic curve of the photolayer is used for transition to relative intensities, and total absorption is determined by measuring the area defined by the photometric curve of the corresponding section of the absorption line (Fig.1.11). The parameter a (1.61) and the Doppler e-halfwidth are calculated from the known temperature and the concentration of the absorbing atoms is determined from the measured total absorption using the corresponding growth curve (Fig.1.12). A simplified method of taking into account the non-uniformity of the asymmetric object and determining the radial distribution of atom concentration in the absorption line was described in Ref.8. The cross section of the arc column is divided into annular zones of equal width and it is assumed that the gas temperature and the atom concentration are constant within each zone (Fig.1.13). The value $(A_\omega/2\Delta\omega_D)_1$ is determined from the

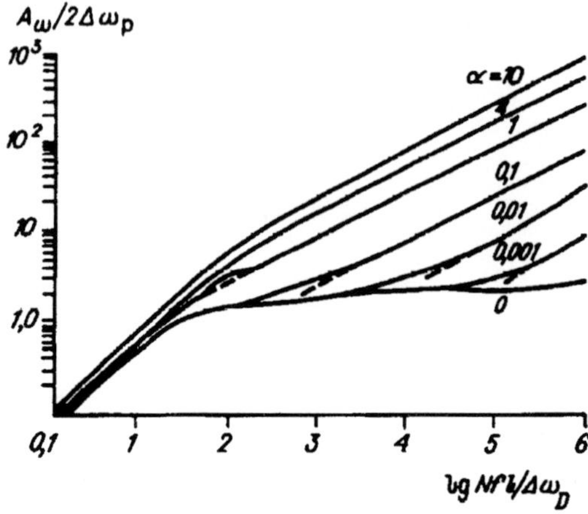

Fig.1.12 Total absorption A_α as a function of the number of absorbing ions N_i.

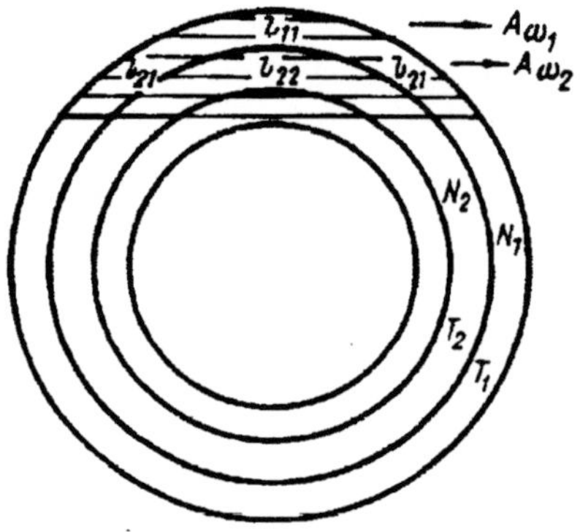

Fig.1.13 Calculation of the concentration of atoms at the absorption line.

measured total absorption for the outer zone $A_{\omega,1}$ and gas temperature T_1 and the value $(Nfl/\Delta\omega_D)_1$ is then taken from the growth curve corresponding to parameter a. This value is then used to determine the total number of absorbing atoms $(Nl)_1$ and the concentration $N_1 = (Nl_1)/l_{11}$, where l_{11} is the length of the chord of the first zone (Fig.1.13). In determining the number of absorbing atoms from $A_{\omega,2}$ it is assumed that the temperature along the absorbing beam is constantly equal to the mean temperature $\Delta \overline{T}_2 = (T_1 + T_2)/2$. The value $(Nfl/Dw_D)_2$ is determined for this temperature. This is followed by calculating the total number of absorbing atoms $(Nl)_2$ and the concentration for the second zone $N_2 = (Nl) - 2N_1 l_{21})/l_{22}$. Thus, the atom concentration is gradually calculated for all angular zones and each time it is assumed that the gas temperature along the absorbing beam is constant and equal to the mean value with respect to all intersecting zones.

The error of the method is evaluated using the equation on the right hand part of the growth curve

$$A_\omega = \left(\frac{4\pi^2 e^2}{m_e c} a \Delta\omega_D N f l \right)^{1/2} . \tag{1.62}$$

The variation of the equation (1.62), taking into account that parameter a is inversely proportional to temperature, gives

$$\frac{\delta N}{N} = 2\frac{\delta A}{A} + \frac{1}{2}\frac{\delta T}{T} + \frac{\delta f}{f} + \frac{\delta a}{a} . \tag{1.63}$$

Assuming in accordance with specific experiments[8] $\delta T/T \approx 20\%$, $\delta A/A \approx 7\%$, $\delta f/f \approx 10\%$, $\delta a/a \approx 10\%$, we obtain $\delta N/N \approx 50\%$. Figure 1.40 shows the results of determining the concentration of sodium and copper atoms from the absorption lines in the cross section of column of a carbon arc.[8] Plasma temperature was measured on the basis of the ratio of the intensities of copper lines CuI 5106 Å and CuI 5218/20 Å (crosses in Fig.1.14) and by the Lorentz method from the sodium line NaI 5890/96 Å (points in Fig.1.14), and the concentration of the sodium and copper atoms was measured from the absolute intensities of the lines NaI 4668 Å and CuI 5106 Å (crosses) and from the absorption lines of NaI 3303 Å and CuI 3243 Å (points). It may be seen that the absorption method greatly widens the spatial region of measurements.

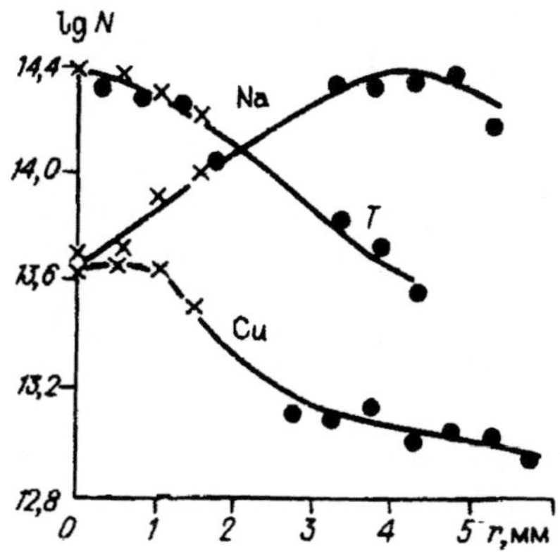

Fig.1.14 Distribution of temperature and concentration of sodium and copper atoms in the cross-section of the carbon arc.[8]

Chapter 2

KINETIC ENERGY, TRANSLATIONAL RELAXATION AND DOPPLER BROADENING OF LINES IN NON-EQUILIBRIUM PLASMA SPECTRA

2.1 Introduction

The nature of the translational motion of neutral particles in slightly ionised low-temperature plasma influences directly or indirectly all processes taking place in it and in many cases this effect is very strong. Therefore, any detailed examination of plasma objects includes the determination of the energies of translational motion of the atoms and molecules. In a large number of cases this is used to determine the temperature of the gas component. This is carried out taking into account that the translational relaxation is a fast process and leads to Maxwell's distribution with the temperature that is the same for all neutral components.

There are a relatively large number of methods of measuring the gas temperature. In the majority of cases, these methods are indirect in the sense that they are not related directly to determining the velocity distribution of the particles and use only the features of manifestation of different orders of the moments with respect to the distribution. They include calorimetric, thermal probe, pressure gauge and interferometric methods of measuring the particle density in different quantum states, etc.[1,2]

One of the widely used methods of measuring the gas temperature is based on examining the form of the line contours with Doppler broadening. It is well known that if the distribution of emitting particles with respect to the velocity v_z in the z observation direction is given by the function (v_z), then the contour of the radiation lines is[13]

$$\varphi(\nu) = \frac{c}{\nu_0} \Psi \left(c \frac{\nu - \nu_0}{\nu_0} \right).$$

(2.1)

Here ν is the observed radiation frequency, c is the velocity of light,

v_0 is the radiation frequency of the particle stationary in the examination system. The applicability of the initial equation (2.1) is restricted by the conditions of smallness of natural, impact and Stark broadening. In addition, it is implicitly assumed that the radiation frequency is greatly higher than the collision frequency, i.e. during radiation the particle does not change its velocity or, otherwise, $2\,l \gg \lambda$, where l is the free path, λ is the wavelength. For the visible region of the spectrum $\lambda \sim 10^{-4} \div 10^{-5}$ cm, and the last condition is justified up to the pressures of the order of atmospheric pressure. We shall assume that these assumptions are valid. The relationship between the form of the contour of the Doppler-broadened line and the distribution function of the particles with respect to the velocity modulus $P(v)$ is given in the general case by the integral equation[4,5]

$$\int_{\mu}^{\infty} K(v,\, \nu) P(v)\mathrm{d}v = \varphi(\nu), \quad \mu = \frac{|\nu - \nu_0|}{c}. \tag{2.2}$$

The structure of the kernel $K\,(v,\nu)$ is determined by the anisotropy of the particle velocity. Several simplest cases were examined in Ref.5. In the isotropic case, equation (2.2) assumes the form

$$\int_{\mu}^{\infty} \frac{P(v)}{v}\mathrm{d}v = \varphi(\nu) \tag{2.3}$$

It is relatively simple to calculate (analytically in many cases) the contour of the line for the derivative $P(v)$. Usually, the Doppler contour refers to the contour of the line corresponding to the Maxwell distribution of the thermal velocities of radiating particles

$$P_M(v) = \left(\frac{m}{2\pi kT}\right)^{3/2} v^2 \exp\left\{\frac{mv^2}{2kT}\right\}, \tag{2.4}$$

where m is the mass of the emitter, T is the gas temperature, k is Boltzmann constant. Integration of (2.3) with $P(v)$ in equation (2.4) gives the well known equation

$$\varphi(\nu) = \frac{c}{\nu_0}\left(\frac{m}{2\pi kT}\right)^{1/2} \exp\left\{-\frac{mc^2}{2kT}\frac{(\nu - \nu_0)^2}{\nu_0}\right\}. \tag{2.5}$$

The width of the contour at half height Δv_D

$$\Delta v_D = \frac{2v_0}{c}\sqrt{\frac{2\ln 2 \cdot kT}{m}}, \qquad (2.6)$$

Usually, (2.6) is also used to determine the gas temperature. However, the applicability of (2.6) both from the viewpoint of the Maxwell nature of $P(v)$ and with respect to the equality of the temperatures of the emitter and the main mass of the non-excited gases has been examined far less frequently. It should be noted that neither of these aspects is 'exotic'. In fact, as a result of the relatively low frequency of intermolecular collisions in reduced-density plasma, the velocity distributions of the short-lived electronically-excited and non-excited particles may differ. Therefore, the specific features of the nature of movement of the electronically excited particles are superimposed on the profile of the Doppler broadened lines. These special features are determined by both the rate of the relaxation processes of the special features of the excitation mechanism of the electronic states.

2.2 Determination of the type of velocity distribution of excited particles from spectral measurements

An inverse problem – determination of $P(v)$ from the measured contours of spectral lines, is in the mathematical sense an incorrect (according to Adamar) problem. Its solution consists of two stages. In the first stage, the experimentally recorded contour of the line $f(v)$ is used to determine its true form $\varphi(v)$, i.e. apparatus effects are excluded. This is reduced to deconvolution of the given convolution with the difference kernel

$$f(v) = \int_{-\infty}^{\infty} a(v - v')\varphi(v')dv' = f_0(v) + \varepsilon(v), \qquad (2.7)$$

where $a(v)$ is the apparatus function, $\varepsilon(v)$ is the noise with the zero mean. The second stage consists of determining $P(v)$ from (2.2) or (2.3) which requires differentiation of the experimentally determined $\varphi(v)$, i.e. the procedure also leading to instability of the solutions in relation to the experimental data errors. The progress in solving the problem examined in this section is determined to a large extent by the recently developed mathematical methods of solving such inverse problems using various regularisation methods (for example, see Ref.6–8). These methods and corresponding algorithms are also discussed to some extent in Refs.2,5,9) and in Chapter 20 of this book.

The experimental method of examining the form of the contours of the lines of emission spectra is relatively conventional and usually includes a high-resolution spectrometer linked with a Fabry–Perot interferometer. In photoelectric registration, the base of the interferometer can be changed by an electrical method or by varying the pressure.[2]

An important problem in these measurements is the reliable determination of the apparatus function $a(v)$ – a classic problem for which at least two new approaches have recently been proposed. Firstly, it is the use of laser radiation with a stable neutral frequency in a narrow line. Secondly, it is the application of the already mentioned mathematical method of deconvolution of the convolution (2.7) in which the type of $\varphi(v)$ is known with a high degree of reliability.[2,9] Of course, as previously, the task of the experimentator is to ensure the narrowest apparatus function so that the error in the final result can be reduced.

2.3 Velocity distribution of atoms and molecules in excited electronic states

If the process of excitation of the radiation state takes place in a reaction with the generation of a surplus energy

$$A + B \rightarrow C + D + \Delta E, \tag{2.8}$$

then the laws of conservation of the energy and the pulse show that the resultant reaction product should acquire an additional kinetic energy. If the radiating state has a short lifetime and the efficiency of collision processes is low, it should be expected that anomalous Doppler broadening of the spectral lines will take place. An example of such a phenomenon is shown in Fig.2.1. It shows the results of measurements of the form of the contour of the transition line[2] in the oxygen atom $O\ (3^3P{-}3^3S)$ in a discharge when the following mechanism takes place

$$\mathrm{Ar}\left(3p^5 4s\right) + \mathrm{O}\left(2^3\mathrm{P}\right) \rightarrow \mathrm{O}\left(3^3\mathrm{P}_{0,1,2}\right) + \mathrm{Ar}\left(^1\mathrm{S}_0\right) + 0.6\ \mathrm{eV}. \tag{2.9}$$

All curves are normalised with respect to the unit area. It may be seen that the width of the contour of the Doppler-broadened lines decreases with increasing gas pressure, indicating that relaxation collision processes take place. For comparison, the contour of the line of the same transition with normal Doppler broadening, corresponding to the gas temperature of $T = 330$ K, is shown. It should be noted that in most cases and under the same conditions the width of all three lines of the transition $3^3\mathrm{P}_{0,1,2}{-}3^3\mathrm{S}_1^0$ coincide.

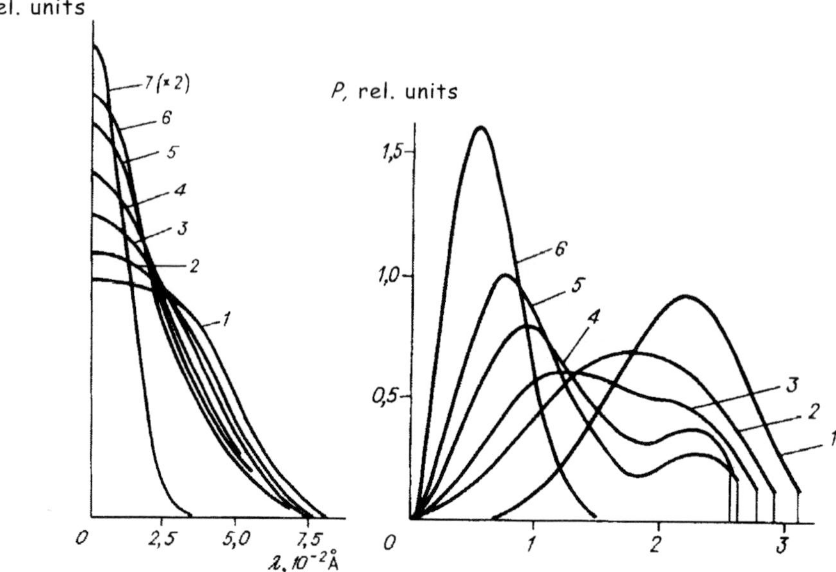

Fig.2.1. Contours of the spectral lines of radiation of atomic oxygen ($3^3P_{0,1,2} - 3^3S_1^0$) in a discharge in a mixture of O_2–Ar (1:36). Discharge current i = 20 mA, pressure of the mixture p = 0.2 torr (1); 0.5 (2); 3 (4); 10 (5); 15 (6); 7) Doppler contour of the radiation line of the oxygen atom for the Maxwell distribution at a gas temperature of T = 330 K; walls of the pipes with an inner diameter of 20 mm are cooled with running water.
Fig.2.2. (right) Velocity distribution functions of excited oxygen atoms under the conditions shown in Fig. 2.1. Pressure p = 0.2 torr (1), 3 (2), 5 (3), 10 (4) 15 (5); 6) Maxwell function at T = 330 K.

Figure 2.2 shows a set of velocity distribution functions of $O(^3P)$ atoms obtained in processing the contours shown in Fig.2.1. The Maxwell function of the distribution of oxygen atoms at T = 330 K is also shown. All curves are normalised for the unit area. The real distribution functions of the O (^3P) atoms are considerably wider than the Maxwell distribution. The width of the distribution gradually decreases with increasing pressure. Breaks of the function in Fig.2.2 correspond to the values of intensities at the wings of the line contours (see Fig.2.1), where the signal : noise ratio is ~1.

Thus, if the results of measurements of the Doppler width of the lines in the radiation spectra excited in the non-equilibrium plasma in the processes with non-resonance interactions of heavy particles are used to determine the neutral gas temperature, the results may be highly inaccurate. In this example, the widths of the lines, recorded at low densities, correspond to a temperature of ~2000 K, whereas the real gas temperature is considerably lower.

Similar investigations of the anomalous Doppler broadening of the lines in the spectra of molecular nitrogen and the examination of the velocity distributions of N_2 molecules ($C^3\Pi$) in a discharge were described in Ref.9, 10.

2.4 Measurement of gas temperature from Doppler broadening of spectral lines

A direct electron impact is one of the important channels of exciting the electronic states of the molecules and atoms. A collision between an electron and an atom is accompanied by a change of the velocities of the colliding particles. This change may influence the form of the spectral lines emitted by the short-lived excited particles for which the collision of relaxation is ineffective. We shall estimate the nature of this effect.

The pulse conservation law indicates that the velocity of the atom after collision with an electron is

$$\vec{v}' = \vec{v} - \frac{m_e}{M}\left(\vec{v}'_e - \vec{v}_e\right), \tag{2.10}$$

where $\vec{v}$ is the velocity of the atom prior to collision, M and m_e are the masses of the atom and the electron, $\vec{v}_e$ and $\vec{v}'_e$ are the velocities of the electrons prior to and after collision with the atom. The cases with $v'_e \ll v_e$ and $v'_e \approx v_e$ are most interesting for various applications. The first case is observed in low-temperature gas-discharge plasma where the mean electron energy is usually lower than the excitation threshold of the radiating electronic state of the molecules and atoms. The main role is played by the electrons with the energy slightly higher than the threshold energy because in this region the electron concentration rapidly decreases with increasing electron energy. The second case in which $v'_e \approx v_e$ is realised in exciting the molecules with a beam of fast electrons with the velocities $v'_e \gg v^o_e$ (v^o_e is the velocity of atomic electrons). These cases are encountered in, for example, probing the gas flow with an electron beam.

In the first case, equation (2.10) shows that the relative variation of the atom velocity $\Delta v/v$ in collision with an electron is

$$\frac{\Delta v}{v} = \frac{m_e}{M}\frac{v_e}{v}. \tag{2.11}$$

In estimates, value v will be represented by the most probable velocity of thermal motion $v_H = \sqrt{\dfrac{2kT}{M}}$, and v_e by the value of the electron velocity corresponding to the excitation threshold of the examined states v_e^{th}. In this case

$$\Delta w = \frac{\Delta v}{v_H} = \frac{m_e v_e^{th}}{\sqrt{2kTM}} \qquad (2.12)$$

The following table gives the values of Δw for some atoms and molecules at two gas temperatures T_1 = 100 K and T_2 = 300 K (v_e^{th} corresponds to the excitation pressure of the state emitting the spectra) often used for plasma diagnostics:

Atom, molecules, excited state	Excitation threshold, eV	$v_e^{th} \times 10^{-8}$, cm s^{-1}	Δw (%) at 100 K	Δw (%) at 300 K
H (n=4)	12.75	2.12	90	52
H_2 ($d^3\Pi$)	13.87	2.21	66	38
D_2 (d^3D)	13.88	2.21	47	27
He ($3d^1\Pi$)	23.07	2.85	60	35
N_2 ($C^3\Pi$)	11.03	1.97	16	9
CO ($B^1\Pi$)	10.77	1.88	15	9

It may be seen that, irrespective of the difference in the masses between the heavy particles and the electrons, in the process of excitation of the atoms and molecules the velocity of translational motion can greatly change. The value of Δw increases with a decrease of the mass of the atomic particles, with an increase of the excitation threshold of the state and with a decrease of gas temperature. For example, for a hydrogen atom at 100 K it is 90%, for a helium atom 60%, and for a heavier nitrogen molecule at T = 300 K it is does not exceed 10%.

The excitation of electronic states of atomic particles by fast electrons (this corresponds to the case with $v_e' \approx v_e$ has been studied quite sufficiently. The problem permits Born's approximation to be used, and its solution has been analysed in detail in Ref.11. The main role is played here by collisions causing scattering of the electrons through small angles with a small momentum transfer. The maximum value of the transferred momentum $P_{max} \approx m_e v_0$ (v_0 is the value of the order the velocity of atomic electrons). This result is physically evident. The process of collision with a transfer of momentum that is considerably greater than the momentum of atomic electrons, would lead simply to ionisation of the atom or molecule. In fact, under the examined conditions where the velocity of the beam electrons v_e is considerably higher than the velocity of the atomic electrons v_e^a, the atomic electrons can be regarded as free, and the collision with the atom as the elastic collision of the incident electron with the initially stationary atomic electrons. In collision with the transfer of a large momentum, both electrons (incident and atomic) acquire velocities of comparable magnitude and this leads to ionisation. Thus, to evaluate the relative variation of the velocity of the atoms excited by the beam of fast electrons, we can use equation (2.12) since $v_0 \approx v_e^{th}$.

We can now examine the behaviour of the Doppler contours of the spectral lines. For this purpose, it is necessary to find the distribution function of the excited atoms and molecules with respect to the component of the velocity of translational motion along the observation axis oz. We assume that the distribution prior to the interaction with the elec-

trons is of the Maxwell type $\Psi_M(v_z) = \dfrac{1}{v_H \sqrt{\pi}} \exp\left\{-\left(\dfrac{v_z}{v_H}\right)^2\right\}$.. As a re-

sult of interaction with electrons, the velocity of the excited atom changes. We also assume that the lifetime of the excited state of the atom is shorter than the characteristic free path time, i.e. the excited particles are not subjected to any collisions. It may easily be shown that the distribution of the radiating atoms and the molecules is described by the equation

$$\Psi(v_z') = \int_{-\infty}^{\infty} \Psi_M(v_z)\Psi_e(v_z, v_z')dv_z, \tag{2.13}$$

where the function $\Psi_e(v_z, v_z')$ is the distribution of the excited atoms with respect to the velocity component v_z' under the condition that the velocity of the atoms prior to collision with the electron was v_z. Using (2.10) for the isotropic velocity distribution of electrons, we obtain

$$\Psi_e(v_z, v_z') = \begin{cases} \dfrac{1}{2\Delta v}, & \text{if } \left|v_z' - v_z\right| \leq \Delta v, \\ 0, & \text{if } \left|v_z' - v_z\right| > \Delta v. \end{cases} \tag{2.14}$$

Substituting $\Psi_{\mathrm{M}}(v_z)$ and $\Psi(v'_z)$ into (2.13), we finally have

$$\Psi(v'_z) = \frac{1}{2\sqrt{\pi}\Delta v v_H} \int_{v'_z-\Delta v}^{v'_z+\Delta v} \exp\left\{-\left(\frac{v_z}{v_H}\right)\right\} \mathrm{d}v_z. \tag{2.15}$$

As a result of comparing $\Psi_{\mathrm{M}}(v_z)$ and $\Psi(v'_z)$, we can conclude that the velocity distribution of the excited atoms differs from the identical distribution of the non-excited atoms, and the distribution $\Psi(v'_z)$ is non-Maxwellian.

Equation (2.15) shows that the contour of the spectral line is described by the equation

$$\varphi(w)\mathrm{d}w = \frac{1}{4\Delta w}\left[\mathrm{erf}(w+\Delta w) - \mathrm{erf}(w-\Delta w)\right]\mathrm{d}w, \quad w = \frac{v'}{v_H}. \tag{2.16}$$

Function (w) is normalised for the unit area, i.e. $\int_{-\infty}^{\infty}\varphi(w)\mathrm{d}w = 1$. If the projection of the velocity onto the observation axis does not change during excitation, i.e. $v'_z = v_z$ (excitation by the electron beam), the line contour will be of the conventional Doppler type (Gaussian)

$$\varphi_D(w)\mathrm{d}w = \frac{1}{\sqrt{\pi}}e^{-w^2}\mathrm{d}w. \tag{2.17}$$

Figure 2.3 shows the contours of spectral lines normalised for the unit area. Curve 1 corresponds to the Gaussian contour (2.17), curves 2 and 3 to the contours (2.16). It may be seen that the contours described by equation (2.16) are wider than the Gaussian contour and the width of the contours increases with increasing Δw. For example, for $\Delta w = 0.3$ the difference from the Gaussian contour in respect of width is ~3%, and at $\Delta w = 0.6$ it is 13%.

Taking the actual measurement error into account, additional broadening should be considered at $w > 0.3$. Figure 2.4 shows the actual gas temperature dependence of the systematic error $\Delta T/T$ of determination of temperature from the width of the spectral lines of some atoms and molecules, associated with examining the above factor.

It can be seen that if measurements are carried out using the spectral lines of relatively heavy molecules (N_2, CO), the systematic error at the temperature found under the conditions of gas-discharge plasma

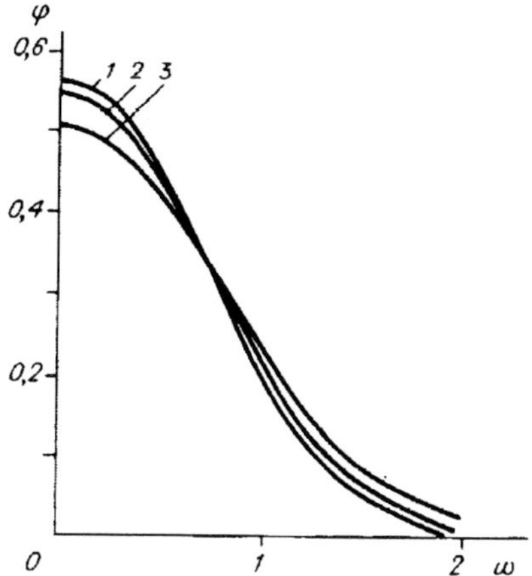

Fig.2.3 Contours of the spectral lines. 1) Gauss contour; 2,3) contours described by equation (2.16) at $\Delta w = 0.3$ and $\Delta w = 0.6$, respectively.

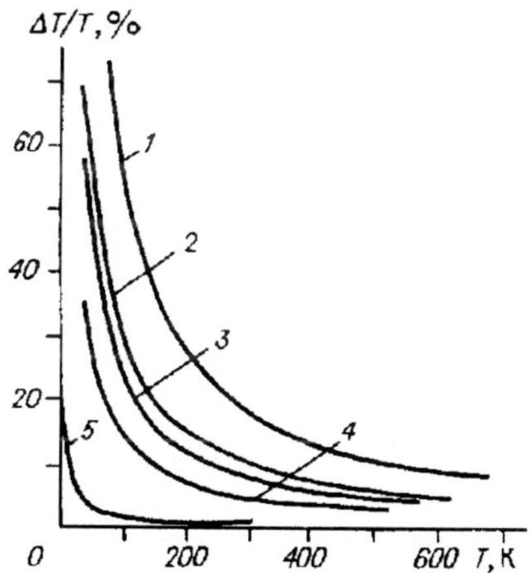

Fig.2.4 Dependence of the systematic error $\Delta T/T$ of determination of temperature on the basis of the width of the spectral lines H (1), H_2 (2), He (3), D_2 (4), N_2 and CO (5) on gas temperature.

in practice is negligible. A different situation exists for light particles at a low gas density. For example, $\Delta T/T > 10\%$ for the H atom at $T < 550$ K, for H_2 at $T < 330$ K, for He at $T < 250$ K, for D_2 at $T < 150$ K. In Ref.12 these conclusions were confirmed by experiments.

2.5 A simple model of relaxation of the mean kinetic energy. Effect of gas density

These considerations show that the processes of excitation of the electronic states of the atoms and molecules are accompanied by perturbations of the velocity distributions of the particles. The real distributions form under the effect of two factors. Firstly, excitation is accompanied by the formation of particles with specific velocity distributions. Secondly, during the lifetime of the excited particles these initial distributions relax as a result of collisions with the particles and transform to the form observed in experiments. In this case, especially in the processes of excitation in interaction of heavy particles, the initial deviations from Maxwellian can be large.

In Ref.13, the authors obtained general equations for the velocity distribution functions of the product of bimolecular reactions. The corresponding distributions for a number of modelling interaction potentials can be determined numerically or analytically. Not discussing this aspect, we shall only note that analysis carried out in, in particular, Ref.2, 14-16 on the basis of the Boltzmann equation shows that for the majority of cases interesting for practice the evolution of the mean energy can be satisfactorily described using the approximation in which the sequence of distributions in the relaxation process can be treated as a sequence of Maxwell distributions and the relaxation interactions are examined on the basis of the model of solid spheres.

Accepting these approximations, it is easy to construct a relatively simple model of relaxation of the mean molecular energy in the medium of a buffer gas at large deviations from equilibrium.

We examine the time dependence of the mean energy of translational motion $\Delta \overline{E}_N$ of molecules N during their movement in a buffer gas

$$\mathrm{d}\overline{E}_N = -n_M \tilde{v}_{N,M} \sigma_{N,M} \Delta \overline{E} \, \mathrm{d}t. \tag{2.18}$$

Here n_{M} is the concentration of the molecules of the buffer gas; $\tilde{v}_{N,M}$ is the mean velocity of the relative movement of N and M; $\Delta \overline{E}_N$ is the mean energy lost by the molecule during a single collision; $\sigma_{N,M}$ is the gas kinetic collision sections of N with M. We shall use the results of Ref.17 in which the authors also examined the model of solid spheres and it was assumed that both fast particles and the cold gas are described by Maxwell velocity distributions:

$$\tilde{v}_{N,M} = \overline{v}_N \left(\frac{m_N}{m_M} \frac{T_M}{T_N} + 1 \right)^{1/2} \tag{2.19}$$

(m_N, m_{M} are the mass of the particles N and M; T_{M}, T_N are the corresponding temperatures, $\overline{v}_N$ is the mean velocity of the molecules N),

$$\Delta\overline{E}_N = \overline{E}_N \mu\left(1-\frac{T_M}{T_N}\right), \quad \mu = \frac{8}{3}\frac{m_N m_M}{\left(m_N+m_M\right)^2}. \tag{2.20}$$

Taking into account (2.19) and (2.20), the relaxation equation is converted to the form

$$-\frac{d\overline{E}_N}{dt} = 4n_M \sigma_{N,M}\mu(3\pi m_N)^{-1/2}\left(\frac{m_N}{m_M}\overline{E}_M + \overline{E}_N\right)^{1/2}\left(\overline{E}_N - \overline{E}_M\right). \tag{2.21}$$

It should be noted that if deviations from equilibrium are small, i.e. $\overline{E}_N \approx \overline{E}_M$, equation (2.21) changes to the well-known relaxation equation with the constant relaxation time τ_{rel} (see, for example, Ref.11):

$$-\frac{d\overline{E}_N}{dt} = -\frac{1}{\tau_{rel}}\left(\overline{E}_N - \overline{E}_M\right), \tag{2.22}$$

where $\tau_{rel}^{-1} = 4n\sigma_{N,M}\mu\left(\overline{E}_M/3\pi m_N\right)^{1/2}\left(\frac{m_N}{m_M}+1\right)^{1/2}$.

Solution of equation (2.21) at the initial conditions $\overline{E}_N(t=0)=\overline{E}_N^0$ is written in the following form:

$$\overline{E}_N(t) = \overline{E}_M\left(1+\frac{m_N}{m_M}\right)\frac{\left[c'\exp\{t\tau_{rel}^{-1}\}+1\right]^2}{\left[c'\exp\{t\tau_{rel}^{-1}\}-1\right]^2} - \frac{m_N}{m_M}\overline{E}_M, \tag{2.23}$$

where

$$c' = \left[\left(\overline{E}_N^0 + \frac{m_N}{m_M}\overline{E}_M\right)^{1/2} + \overline{E}_M^{1/2}\left(1+\frac{m_N}{m_M}\right)^{1/2}\right] \times$$

$$\times\left[\left(\overline{E}_N^0 + \frac{m_N}{m_M}\overline{E}_M\right)^{1/2} - \overline{E}_M^{1/2}\left(1+\frac{m_N}{m_M}\right)^{1/2}\right]^{-1}. \tag{2.24}$$

To ensure determinacy, we assume that $N \equiv N_2$, $M \equiv Ar$. Figure 2.5 shows the time dependence of $\overline{E}_{N_2}(t)$ at $\overline{E}_{N_2}^0(t=0)=$ 2250 K, $\overline{E}_{Ar}=150\,K$. It can be seen that relaxation is very rapid: already af-

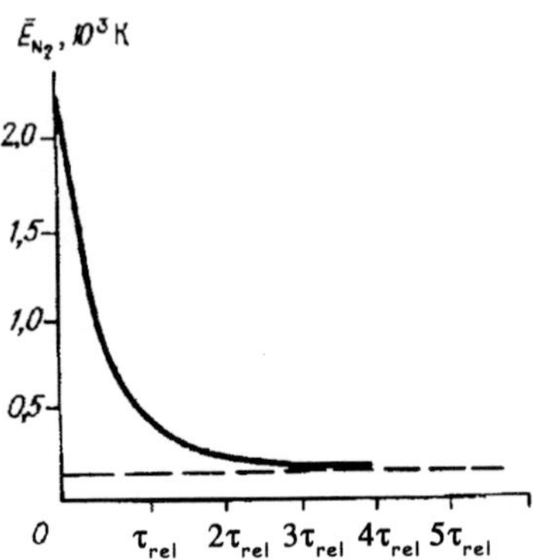

Fig.2.5 Time dependence of the mean energy of translational movement $\overline{E}_{N_2}$ of the molecules of N_2 $(C_3\Pi)$ in argon.

ter $t = 0.5\ \tau_{rel}$ the value of $\overline{E}_{N_2}$ decreases e times, and after $t = 3\tau_{rel}$ $\overline{E}_{N_2} \cong \overline{E}_{Ar}$.

In comparison with the experiments, carried out under the stationary conditions, it should be taken into account that the observed molecules formed, generally speaking, at different times in relation to the observation moment (scintillation) and the period during which they remain in the excited state differs. Therefore, when using equation (2.23) for analysis of the experimental data, it is necessary to carry out averaging on the basis of the lifetime of the molecules in the excited electronic state. The probability of an excited particle emitting a photon during the time t to $t + dt$ is given by the equation

$$dw(t) = \tau_p^{-1} e^{-t/\tau_p} dt, \tag{2.25}$$

where τ_p is the radiation lifetime of the excited state. For N_2 $(C^3\Pi)$ $\tau_p = (41 \pm 2.9)\cdot 10^{-9}$ s. Finally, averaging (2.23), we obtain

$$\left\langle \overline{E}_{N_2}\left(n_{Ar}, \sigma_{N_2, Ar}\right)\right\rangle = \tau_p^{-1} \int\limits_0^\infty \overline{E}_{N_2}\left(n_{Ar}, \sigma_{N_2, Ar}, t\right) e^{-t/\tau_p} dt. \tag{2.26}$$

Integral (2.26) can be calculated numerically. Comparing the dependence (2.26) with the experimental data[2,9] for discharges in the N_2–Ar (1:9) and N_2–He (1:10) mixtures, we can determine the effective sections of relaxation collisions for excited molecules of N_2 $(C^3\Pi)$ with argon atoms $\sigma_{N_2,Ar}$ and helium atoms $\sigma_{N_2,He}$. Figure 2.26 shows the

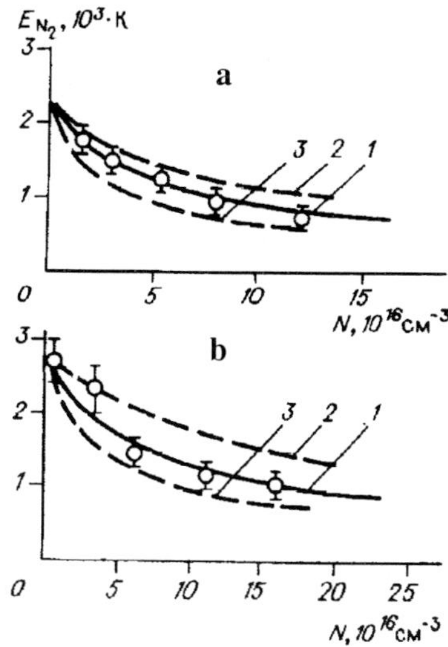

Fig.2.6 Dependence of the most probable values of energy for molecules of N_2 ($C_3\Pi$, $v' = 0$, $K' = 26$) on the concentration of He (a) and Ar (b). a) calculated from (2.26) for N_2–He with sections of 10^{15} cm²) $\sigma_{N_2,He}$ = 5.8 (1), 3 (2), 12 (3); b) calculated from (2.26) for N_2Ar with sections (10^{-15} cm²) $\sigma_{N_2,Ar}$ = 4.1 (1), 2 (2), 8 (3); the points indicate the experimental results.[29]

dependence of the most probable values of the energy for the N_2 molecules ($C^3\Pi$, $v' = 0$, $K' = 26$) (v', K' are the oscillatory and rotational quantum numbers) on the He and Ar concentration. The best agreements between the calculations and experimental values was obtained for $\sigma_{N_2,Ar}=(4.1\pm1.5)\times10^{-15}$ cm², = $(5.8\pm0.5)\times10^{-15}$ cm². It can be seen that the calculated dependence is sensitive to selection of the cross section. This makes it possible to determine the effective collision cross section at a relatively small error. It should be noted that if in the case of the N_2–Ar (1:9) mixture the value of $\sigma_{N_2,Ar}$ almost completely coincides with the gas kinetic collision section of the particles in the ground electronic states ($\sigma^0_{N_2,Ar}=4.3\times10^{-15}$ cm²), then for the N_2–He (1:10) mixture there is a difference ($\sigma^0_{N_2,He}=2.8\times10^{-15}$ cm²): the cross section obtained from the described comparison is approximately twice the $\sigma^0_{N_2}$ value. It should be noted that the cross sections for the electronically excited molecule and the molecule in the ground state may or may not coincide in collisions with another particle because the effective size of the molecule in the excited electronic state, generally speaking, increases.

2.6 Identification of excitation mechanisms

The proposed relaxation model can be used to determine the mean distribution energy at the known concentration of the buffer gas. However, an inverse problem can also be formulated: from the measured mean kinetic energy of the molecules we can determine the surplus of energy generated during excitation. This enables the process leading to molecular excitation to be identified.

Initially, we examine the process of transfer of excitation from the atom to the molecule. From the energy conservation law we obtain

$$E_e^a = E_{e,v,k}^{mole} - E_k^0 + E_{th}' - E_{th}^0, \qquad (2.27)$$

where E_e^a is the energy of the excited state of the atom, $E_{e,v,k}^{mol}$ is the energy of the examined electronic–oscillatory–rotational state of the molecule, E_k^0 is the rotation energy of the molecule prior to interaction, E_{th}^0 are the translational energies of the system of two particles prior to and after interaction, respectively. The translational energy of the system of two particles (with masses m_1 and m_2) can always be written in the form of the sum of the energy of the centre of the masses $E_{c.m}$ and the energy of the relative motion E_μ:[18]

$$E_{th} = E_{c.m} + E_\mu. \qquad (2.28)$$

Here $E_{c.m} = \dfrac{M_\Sigma v_{c.m}^2}{2}; \ M_\Sigma = m_1 + m_2; \ E_\mu = \mu v_{rel.}^2 / 2; \ \mu = \dfrac{m_1 m_2}{m_1 + m_2}.$ The velocity of the centre of the masses $\vec{v}_{c.m}$ and the velocity of the relative motion are determined as follows

$$\vec{v}_{c.m} = \frac{1}{M_\Sigma}(m_1 \vec{v}_1 + m_2 \vec{v}_2), \qquad (2.29)$$

$$\vec{v}_{rel} = \vec{v}_1 - \vec{v}_2. \qquad (2.30)$$

where $\vec{v}_1$ and $\vec{v}_2$ are the velocities of the first and second particles in the laboratory reference system.

In interaction of the particles $E_{c.m}$ does not change

$$E_{th}' - E_{th}^0 = E_\mu' - E_\mu^0, \qquad (2.31)$$

where $E_\mu^0 = (1/2)v_{rel}^{0^2}\mu$ and $E_\mu' = (1/2)v_{rel}'^2\mu$ are the energies of the relative motion of the particles prior and after interaction, respectively. From (2.29) and (2.30) we obtain

$$\vec{v}_{rel} = \left(\vec{v}_1 - \vec{v}_{c.m}\right)\frac{M_\Sigma}{m_2}. \tag{2.32}$$

Substituting the corresponding value of the relative velocity into equation (2.31) we obtain

$$E'_{th} - E^0_{th} = \frac{m_1}{m_2}\frac{M_\Sigma}{2}\left[\left(v'^2_1 - v^{0^2}_1\right) - 2\vec{v}_{c.m}\vec{q}_1\right], \tag{2.33}$$

where $\vec{q}_1 = \vec{v}'_1 - \vec{v}^0_1$.

Averaging over all possible directions $\vec{v}_{c.m}$ for the isotropic case leads to the disappearance of the term with $\vec{v}_{c.m}\vec{q}_1$, and, consequently

$$E'_{th} - E^0_{th} = \frac{M_\Sigma}{m_2}\left(\frac{m_1 v'^2_1}{2} - \frac{m_1 v^{0^2}_1}{2}\right). \tag{2.34}$$

Combining the equations (2.34) and (2.27) and setting $E^0_k = kT$, $(1/2)m_1 v^{0^2}_1 = (3/2)kT$ and $\dfrac{m_1 v'^2_1}{2} = \varepsilon^z_1$ we obtain that

$$E^{at}_e = E^{mole}_{e,v,k} - kT\left(1 + \frac{3}{2}\frac{M_\Sigma}{M_2}\right) + \frac{M_\Sigma}{m_2}\varepsilon^z_1. \tag{2.35}$$

Here ε^z_1 is the energy at the limit of zero gas concentration. To determine this, energy it is necessary to plot the dependence of the most probable energy on the concentration and find the asymptote of this dependence at the zero concentration limit. In interaction of nitrogen with argon for the N_2 ($C^3\Pi$, $v' = 0$, $K' = 26$) molecules, we obtain $\varepsilon^z_1 = 2860 \pm 300$ cm^{-1}, and at $T = 150$ K equation (2.35) gives $E^a_e = 93264 \pm 3600$ cm^{-1} which almost coincides with the energy of the state of Ar ($^3P_2$) equal to 93 144 cm^{-1}. Using other spectral lines of the second positive system of N_2 for measurements, in particular, those corresponding to the transitions from high rotational levels ($K' \sim 50$), we can determine the boundaries of the range of the states of argon from which excitation is transferred to the nitrogen molecules.

We shall now examine the process of transfer of excitation from molecule to molecule. In this case, the energy conservation law gives

$$E_{e,v}^{0} = E_{e,v}' - \varepsilon_k^{0}(1) - \varepsilon_k^{0}(2) + \left(E_{th}' - E_{th}^{0}\right) + \varepsilon_k'(1) + \varepsilon_k'(2), \tag{2.36}$$

where $E_{e,v}^{0}$ and $E_{e,v}'$ are the values of the electronic–oscillatory energy of the system prior to and after interaction respectively, $\varepsilon_k^{0}(1)$, $\varepsilon_k'(1)$, $\varepsilon_k^{0}(2)$, $\varepsilon_k'(2)$ are the rotational energies of the first and second molecule prior to and after interaction, E_t^{0} and E_t' is the translational energy of the system prior to and after interaction. Using equation (2.36) and set-

ting $\dfrac{1}{2}m_1 v_1^{0^2} = \dfrac{3}{2}kT$, $\dfrac{1}{2}m_2 v_1'^2 = \varepsilon_1^{\bar{z}}$, $\varepsilon_k^{0}(1) = \varepsilon_k^{0}(2) = kT$, we obtain

$$E_{e,v}^{0} = E_{e,v}' - kT\left(2 + \frac{3}{2}\frac{M_\Sigma}{m_2}\right) + \varepsilon_k'(1) + \varepsilon_k'(2) + \frac{M_\Sigma}{m_2}\varepsilon_1^{z}. \tag{2.37}$$

In transferring excitation from molecule to molecule, in addition to the translational energy whose value is easily determined from the conservation law, the second molecule can also transport the rotational energy $\varepsilon_k'(2)$ after interaction. In this case, to determine $E_{e,v}^{0}$ we must obtain additional information from the equation, in particular, we must know the maximum possible value of the translational energy of the second molecule max $[\varepsilon_k'(2)]$. Equation (2.37) shows that the maximum possible rotational energy of the second molecule is related to the mini-

mum value of the quantity $\dfrac{M_\Sigma}{m_2}\varepsilon_1^{\bar{z}} + \varepsilon_k'(1)$ for the first molecule, i.e.

$$E_{e,v}^{0} = E_{e,v}' - kT\left(2 + \frac{3}{2}\frac{M_\Sigma}{m_2}\right) + \min\left[\frac{M_\Sigma}{m_2}\varepsilon_1^{\bar{z}} + \varepsilon_k'(1)\right] +$$
$$+ \max\left[\varepsilon_k'(2)\right]. \tag{2.38}$$

We shall examine a specific example of excitation of the second positive system of N_2 in a discharge in pure N_2 or in N_2–He mixture. In the literature, there are at least two different viewpoints regarding the formation mechanism of the 'hot' group N_2 ($C^3\Pi$) in similar discharges: 1) as a result of de-excitation of the molecules which are in the state $N_2(E^3\Sigma)$;[19] 2) as a result of de-excitation of molecules present in the states $N_2(C'^3\Pi)$ or $N_2(D^3\Sigma)$.[20]

The products of these processes are the molecules N_2 ($X^1\Sigma$) and N_2 ($C^3\Pi$). The moments of inertia of the molecules N_2 ($X^1\Sigma$) and N_2 ($C^3\Pi$) are very close and it is therefore natural to assume that on the whole their rotational distributions coincide. The spectrum of the second positive system N_2 associated with the transitions of the molecules of 'hot' group contains lines up to $K' \sim 50$ and therefore max $[\varepsilon'_k(2)] = E_k$ ($K' = 50$). Investigations of the contours of the spectral lines showed that min $[2\varepsilon^z_1 + \varepsilon'_k(1)]$ corresponds to the molecules N_2^{hot} ($C^3\Pi$, $v' = 0$, $K' < 18$). In particular, for $K' = 18$ we obtain $2\varepsilon^z_1$ ($K' = 18$) + ε_k ($K' = 18$) = 2700 cm^{-1}. Finally, for the energy of the given state we obtain $E^0_{e,v}$ = 95 800 ± 700 cm^{-1}. This value is in good agreement with the energy of the state $E^3\Sigma^+_g$ equal to 95 772 cm^{-1} and even if we take into account the measurement error, it is lower than the energy of the state $C'^3\Pi$ ($E = 97\ 580$ cm^{-1}) and $D^3\Sigma$ ($E = 103\ 576$ cm^{-1}). Thus, the results show that only the state $E^3\Sigma^+_g$ is responsible in this case for the formation of the molecules of 'hot' group N_2 ($C^3\Pi$).

The agreement between the calculated and experimental results makes it possible to recommend the use of a simple model for analysing the processes of translational relaxation of the mean energy of the 'hot' particles. It should be noted that the examined cases are characterised by a high non-uniformity: the initial particle energy is more than an order of a magnitude higher than the equilibrium energy. As shown by com parison with the experiments, the simple analytical equations presented here make it possible to describe satisfactorily the evolution of the mean energy of 'hot' particles.

Chapter 3

OPTICAL ACTINOMETRY OF PLASMA

A common disadvantage of the methods of determining the concentration of molecules, atoms and radicals in plasma on the basis of the emission spectra is that these methods can be used to measure the particle concentration directly only in the excited state. The problem of determining the particle concentration in ground states (which is often more important) on the basis of the emission spectra is relatively complicated because it requires quantitative information on the mechanisms of excitation and deactivation of the particles of the examined type under the conditions of the studied gas discharge.

The optical actinometry method makes it possible in a number of cases to bypass these difficulties. The method is based on the following procedure.[1-3] A small amount of the actinometer – a gas with a known spectrum and the excitation and deactivation constants – is added to the examined plasma object. If the emitting states of the actinometer and the examined gas are excited by a direct electron impact from the ground state and have similar excitation potentials, the same group of electrons will take part in populating these states. Consequently, the rate constants of excitation of the emitting states should depend in the same manner on the plasma parameters.

In cases in which the radiation channel is the main deactivation channel of these states, the equations for the intensities of the emissions lines of the actinometer and the emitting gas have the same form:

$$I_{ik} = N_i A_{ik} h\nu_{ik} = N_0 n_e \langle \sigma v \rangle_{0i} A_{ik} \tau_i h\nu_{ik}, \qquad (3.1)$$

where I_{ik}, A_{ik}, ν_{ik} are the intensities of the lines, the probability of spontaneous emission and the frequencies of radiation transition $i \rightarrow k$, respectively; N_i, N_0 are the populations of the excited and ground electronic states; n_e is the electron concentration; $\langle \sigma v \rangle_{0i}$ are the rate constants of excitation of the emitting states by the electron impact from the ground electronic states; τ_i is the radiation lifetime of the emitting state.

50

In accordance with (3.1), the following equation can be written for the ratio of the intensities of the emission lines of the actinometer (the values with index 1) and the examined gas (values with index 2):

$$\frac{I_1}{I_2} = \frac{N_{01}\langle\sigma v\rangle_{01}A_1\tau_1 V_1}{N_{02}\langle\sigma v_0\rangle_{02}A_2\tau_2 V_2} =$$

$$= \frac{N_{01}}{N_{02}}\frac{\sigma_{01}^m}{\sigma_{02}^m}\frac{\displaystyle\int_{\Delta E_1}^{\infty}\varphi_{01}(\varepsilon)f(\varepsilon)\varepsilon^{1/2}d\varepsilon}{\displaystyle\int_{\Delta E_2}^{\infty}\varphi_{02}(\varepsilon)f(\varepsilon)\varepsilon^{1/2}d\varepsilon}\frac{A_1\tau_1 V_1}{A_2\tau_2 V_2}. \tag{3.2}$$

In (3.2), the rate constants of excitation are written in the explicit form[4]

$$\langle\sigma v\rangle_{0i} = \sigma_{0i}^m \int_{\Delta E_i}^{\infty}\varphi_{0i}(\varepsilon)f(\varepsilon)\varepsilon^{1/2}d\varepsilon, \tag{3.3}$$

where σ_{0i}^m is the size of the excitation cross section at the maximum; ΔE_i is the excitation threshold of the emitting state; $f(\varepsilon)$ is the energy distribution function of electrons in the plasma; $\varphi_{0i}(\varepsilon)$ is the normalised function describing the form of the dependence of the excitation cross sections of the emitting states on the electron energy.

The exact solution of equation (3.2) encounters principal difficulties associated mainly with insufficient knowledge of the energy distribution function of the electrons $f(\varepsilon)$. However, these difficulties can be overcome if the conditions $\varphi_{01}(\varepsilon)\simeq\varphi_{02}(\varepsilon)$ and $\Delta E_1 \simeq \Delta E_2$ are fulfilled. As shown in, for example, Refs.4 and 5, the first of these conditions is approximately fulfilled for a large number of atomic and molecular states. The second condition can be satisfied by selecting the corresponding excited emitting states of the actinometer and the examined gas. If both conditions are fulfilled, the following equation can be written

$$\frac{I_1}{I_2} = k\frac{N_{01}}{N_{02}}, \tag{3.4}$$

where k is a constant that depends only on the properties of specific atoms

$$k = \frac{\sigma_{01}^m A_1\tau_1 V_1}{\sigma_{02}^m A_2\tau_2 V_2}. \tag{3.5}$$

Thus, the absolute concentration of the particles of the examined gas in the ground electronic state is determined only by the ratio of the intensities of the corresponding spectral lines and by the absolute concentration of the particles of the gas–actinometer

$$N_{02} = k \frac{I_2}{I_1} N_{01}.$$
(3.6)

When examining the molecular states, the equation (3.2) should slightly be modified owing to the fact that it is necessary to take into account the excitation of the molecular emitting states by electron impact from the different vibration levels of the ground electronic states of the molecules. Taking into account the experimental data presented in Refs.6 and 7 according to which the relative cross sections of excitation of the electronic–vibration states by an electron impact are proportional to the Frank–Condon factors of the corresponding transitions, and taking into account the actual populations of the vibrational levels of the ground electronic states, we can write a relationship for the relative intensities of two molecular bands

$$\frac{I_1}{I_2} = \frac{N_{01}}{N_{02}} \frac{\sigma_{01}^m}{\sigma_{02}^m} \frac{\int\limits_{\Delta E_1}^{\infty} \varphi_{01}(\varepsilon) f(\varepsilon) \varepsilon^{1/2} d\varepsilon}{\int\limits_{\Delta E_2}^{\infty} \varphi_{02}(\varepsilon) f(\varepsilon) \varepsilon^{1/2} d\varepsilon} \frac{\sum\limits_{v^0=0}^{\infty} q_{v^0}^{(1)} \Psi^{(1)}(v^0)}{\sum\limits_{v^0=0}^{\infty} q_{v^0}^{(2)} \Psi^{(2)}(v^0)} \frac{A_1 \tau_1 v_1}{A_2 \tau_2 v_2}.$$
(3.7)

Equation (3.7) differs from the examined relationship between the intensities of two atomic spectral lines (3.2) by the fact that it contains

additional factors containing the sums $\sum\limits_{v^0=0}^{\infty} q_{v^0}^{(i)} \Psi^{(i)}(v^0)$, where $q I_{v^0 0}^{(i)}$ are

the Frank–Condon factors linking the vibrational levels of the ground electronic state v^0 with the zero vibrational level of the electronically excited state from which the emission transition starts; $\Psi^{(i)}(v^0)$ is the distribution function of populations of the vibrational levels of the ground electronic state of the examined molecules.

For the relative intensities of the atomic spectral line and the molecular band, we must write a slightly different relationship

$$\frac{I_1}{I_2} = \frac{N_{01}}{N_{02}} \frac{\sigma_{01}^m}{\sigma_{02}^m} \frac{\displaystyle\int_{\Delta E_1}^{\infty} \varphi_{01}(\varepsilon) f(\varepsilon) \varepsilon^{1/2} d\varepsilon}{\displaystyle\int_{\Delta E_2}^{\infty} \varphi_{02}(\varepsilon) f(\varepsilon) \varepsilon^{1/2} d\varepsilon} \frac{1}{\displaystyle\sum_{v^0=0}^{\infty} q_{v^0 0} \Psi\left(v^0\right)} \frac{A_1 \tau_1 \nu_1}{A_2 \tau_2 \nu_2}. \tag{3.8}$$

After appropriate transformations, the equations (3.7) and (3.8) can be reduced to the relationship (3.6) in which, however, constant k has a slightly different meaning. For example, for the case described by (3.7)

$$k = \frac{\sigma_{01}^m}{\sigma_{02}^m} \frac{\displaystyle\int_{\Delta E_1}^{\infty} \varphi_{01}(\varepsilon) f(\varepsilon) \varepsilon^{1/2} d\varepsilon}{\displaystyle\int_{\Delta E_2}^{\infty} \varphi_{02}(\varepsilon) f(\varepsilon) \varepsilon^{1/2} d\varepsilon} \frac{\displaystyle\sum_{v^0=0}^{\infty} q_{v^0 0}^{(1)} \Psi^{(1)}\left(v^0\right)}{\displaystyle\sum_{v^0=0}^{\infty} q_{v^0 0}^{(2)} \Psi^{(2)}\left(v^0\right)} \frac{A_1 \tau_1 \nu_1}{A_2 \tau_2 \nu_2}. \tag{3.9}$$

Evidently, the discussed method was proposed for the first time and applied in Ref.1 and 8 where the absolute concentrations of oxygen atoms were determined under the condition of glow discharge plasma in carbon dioxide.

The authors of Refs.1 and 8 compared the relative intensities of oxygen lines O ($3p^5P \to 3s^5S$) and the bands of the Angström system of carbon oxide CO ($B^1\Sigma$, $v' = 0 \to A^1\Pi$, v''). The upper levels of these emission transitions have similar excitation potentials (energy difference between them is $\Delta E \simeq 320$ cm^{-1} and, in addition to this, the excitation cross sections of these states are well known from the literature). In Refs.1, 8, the given concentration of carbon oxide molecules were used to determine the concentration of oxygen atoms in relation to the discharge conditions. The advantage of the proposed method is that the actinometric gas was represented by CO present naturally in the discharge.

Optical actinometry was reborn in studies of the diagnostic zone of non-equilibrium chemically active plasma generated in low-pressure high-frequency discharges used for etching silicon and silicon dioxide by active halogen atoms (fluorine, chlorine, bromine).[9-18] It was shown that the etching rate is proportional to the absolute concentration of the halogen atoms.

The authors of Ref.2 proposed to add gaseous argon to the plasma as an actinometer. The intensity of the argon line Ar λ = 7504 Å (excitation potential $\simeq$13.5 eV) was compared with the emission intensity of the line of the fluorine atom F λ = 7037 Å (excitation potential $\simeq$14.5 eV). The results show that the concentrations of the fluorine atoms, determined by the actinometer, in the high-frequency gas discharge (f = 13.56 MHz) with a power of W = 100 W in a mixture CF_4/O_2 at a pressure of $p \simeq$ 0.04 torr are in good agreement with the data obtained by other methods.

The authors of Ref.3 carried out detailed examination of the actinometer properties of argon (λ = 7504 Å) and molecular nitrogen ((0–2) band of the second positive system N_2 ($C^3\Pi \rightarrow B^3\Pi$)) for measuring the absolute concentration of F and O atoms and CO and CO_2 molecules in the high-frequency discharge (f = 27 MHz) with a power of W = 50 W in a mixture of CF_4/O_2 at a pressure of $p \simeq$ 1 torr. In particular, it was shown that although the excitation energies of the examined states vary over a wide range (8–20 eV), the concentrations of F, O, CO and CO_2 can nevertheless be calculated using equation (3.6), where Ar and N_2 are used as an an actinometer.

The accuracy of the optical actinometer method has been verified many times by comparison with the independent method of determining the particle concentration in the ground electronic states. For example, in addition to optical actinometry, the concentration of fluorine atoms in Refs.9 and 10 was determined independently by the titration method in which CO_2 was added to the initial gas mixture CF_4/O_2 and as a result of rapid substitution reaction F + $Cl_2 \rightarrow$ FCl + Cl the authors recorded recombination chemiluminescence of chlorine atoms in the range 600–700 nm. The glow intensity was used to determine the concentration of the fluorine atoms. Results show that the actinometry and titration methods give identical results.

In Ref.11, optical actinometry was used to determine the concentration of bromine atoms in etching GaAs crystals with bromine. The absolute concentration of the Br atoms was measured both by actinometry (argon was used as the actinometer) and by the linear absorption method. The results show that the ratio of the intensities of the bromine and argon lines is proportional to the absolute concentration of the Br atoms in a wide range of rearrangement of the discharge excitation frequency (f = 0.1 – 13 MHz). In accordance with (3.6) this shows that actinometry can be used to examine discharges of this type.

The accuracy of the optical actinometry method was verified in Ref.12 by laser-induced fluorescence enabling independent measurement of the concentration of CCl radicals in CCl_4 plasma (molecular nitrogen was used as an actinometer. In this case, actinometry could not be used because the excited CCl* radicals formed in the discharge not during direct electronic excitation but during dissociation of CCl_2, CCl_3 and CCl_4.[12]

The general conclusion relating to the results obtained in Refs.9–12 can be described as follows: under the conditions where it is possible to use the linear absorption methods or laser-induced fluorescence it may not be necessary to use optical actinometry. However, as a result of its availability and simple experimental set up, actinometry is often more attractive.

An important moment which forms the basis of the method of optical actinometry is the assumption on the excitation of emitting states by a direct electron impact. In Ref.13, this circumstance was verified by investigating the form of the contours of the spectral lines of fluorine, chlorine and argon atoms in the plasma of ac discharge ($f = 20$–50 kHz) in $CF_4/O_2/Ar$ and Cl_2/Ar mixtures. It is well known that in a low-pressure discharge the form of the contours of the spectral lines, excited by direct electron impact, should have the form of the Gauss function with the gas temperature used as the parameter.[19] In Ref.13 it was shown that the contours of the emission lines of the fluorine and argon atoms for the $CF_4/O_2/Ar$ mixture indeed have the Gaussian form with the temperature $T = 360 \pm 70$ K which corresponds to the gas temperature. At the same time, the Cl_2/Ar mixture was characterised by large differences in the contours of the emission lines of the chlorine and argon atoms in comparison with the Gaussian form at the gas temperature. It was concluded that the chlorine and argon atoms in the discharge in the Cl_2/Ar mixture are excited during dissociation in the processes of non-resonance collision of heavy particles but not in excitation with the electrons, i.e. the optical actinometry method cannot be used in this case.

Analysis of the strength of the possible effect of the gas–actinometer on the properties of the examined plasma was carried out in Ref.14 where the effect of the additions of argon, helium and molecular nitrogen on the properties of the plasma and the SF_6/O_2 mixture was examined theoretically and by experiments. The results show the effect of argon, up to 10% content in the working mixture, has almost no influence on the energy distribution function of the electrons, whereas the additions of molecular and nitrogen have a strong effect on the properties of SF_6/O_2 plasma as a result of effective dissipation of the electron energy in excitation of the vibrational levels of N_2.

Optical actinometry can also be used to identify the excitation channels of the quantum states of the molecules of chemically reacting gases. For example, the authors of Ref.15 examined the mechanism of excitation of the N_2 ($C^3\Pi$) state (argon was used as the actinometer) and of the N_2^+ ($B^2\Sigma$) state (actinometer – neon). The results show that the relationship (3.6) is fulfilled in a wide range of the variation of the discharge conditions for the corresponding states. Since the emitting states of argon and neon were excited with the direct electron impact under the given experimental conditions, it was concluded that the N_2 ($C^3\Pi$) and N_2^+ ($B^2\Pi$) states are also populated by the direct electron impact.

In Ref.20, the absolute concentration of the nitrogen atoms was measured under the low-pressure arc discharge conditions with a consumable cathode made of titanium. Argon was used as the actinometer. Taking into account the fact that the excitation potentials of the lines of nitrogen atoms NI λ = 4099.94 Å (E_1 = 13.7 eV), NI λ = 4109.98 Å (E_1 = 13.7 eV) and argon atoms ArI λ = 4158.96 Å (E_2 = 14.55 eV) are quite close, the nitrogen atoms concentrations can be measured in principle using equation (3.6). All assumptions made in deriving equation (3.6) were confirmed for the experimental conditions used in Ref.20. The experimental results of the measurement of concentration in relation to the conditions in the vacuum arc discharge plasma were explained in Ref.20 on the basis of a kinetic model which takes into account the formation of nitrogen atoms in dissociation of N_2 by a direct electron impact, dissociated recombination of N_2^+ with electrons in the recombination of N on the walls of the discharge chamber.

Comparison of the experimental results obtained for the dependences of the absolute concentration of nitrogen atoms and the growth rate of the nitride–titanium condensate on the substrate of gas pressure made it possible to propose an important channel of formation of titanium nitride in the vacuum arc discharge through recombination of the nitrogen and titanium atoms on the surface.

The excitation potentials of the line of the oxygen atom OI λ = 7771.9 Å, of the Angström system of the CO molecule and the second positive system of N_2 are relatively similar, and to measure the absolute concentrations of the oxygen atoms and the carbon oxide molecules in the ground electronic state it is convenient to use the optical actinometry method. The authors of Refs.22 and 23 carried out detailed measurements of the concentration of these components in chemically active plasma. Molecular nitrogen was used as the actinometer in all cases. Naturally, the initial stage was the confirmation of the assumptions of the method under the experimental conditions.[22,23] The experimental object in Ref.22 was a capillary glow discharge in the CO_2/N_2/He (1:1:8) mixture at medium pressure (p = 20÷100 Torr) used for pumping waveguide CO_2 lasers, and in Ref.23 it was the superhigh frequency (microwave) medium pressure discharge (p = 70–100 Torr) with transverse blowing of CO_2 used to dissociate the carbon dioxide. In the latter case, molecular nitrogen was added to the discharge in small amounts for diagnostic purposes.

The results for the spatial distribution of the chemically active molecules of CO and O atoms, presented in Refs.22 and 23, provided important information of the special features of plasma chemical transformations in the examined discharges so that it will be possible to optimise these systems.

When adding the actinometric gas M to the examined gas X, the processes of direct electron excitation may be accompanied by reactions in which heavy particles take part. For example, the reaction of quasi-resonant energy transfer in collisions should take place:

$$X + M^* \xrightarrow{k_1} X^* + M \pm \Delta E. \tag{3.10}$$

The rate of the process (3.10) is often very high because ΔE is small. As already mentioned, a low value of ΔE is one of the main prerequisites of the actinometry method.

Taking into account the fact that in addition to (3.10) the reactions of excitation of the emitting states by the direct electron are also effective:

$$M + e \xrightarrow{k_2} M^* + e, \tag{3.11}$$

$$X + e \xrightarrow{k_3} X^* + e, \tag{3.12}$$

the equations of the balance of excitation and deactivation of the actinometer and the examined gas assume the form

$$\left[M^* \right] (A + v)_M = [M] n_e k_2, \tag{3.13}$$

$$\left[X^* \right] (A + v)_X = [X] n_e k_3 + \frac{[M] n_e k_2}{(A + v)_M} [X] k_1, \tag{3.14}$$

where [ ] is the concentration of the corresponding particles; A_i and v_i are the probability of spontaneous emission and the quenching rate of the excited states M* and X* respectively, k_1, k_2, k_3 are the rate constants of the reactions (3.10), (3.11) and (3.12); n_e is the electron concentration.

Since the equations for the intensities of the lines (or bands) have the form

$$I_M = \left[M^*\right] A_M,$$
$$I_X = \left[X^*\right] A_X.$$

(3.15)

equations (3.13)–(3.15) show that

$$\frac{[X]}{[M]} = \frac{I_X}{I_M} \frac{(1+v/A)_X}{(1+v/A)_M} \left\{ \frac{k_3}{k_2} + k_1 \frac{[M]}{(A+v)_M} \right\}^{-1}.$$

(3.16)

It can be seen that in justifying the actinometry method it is important to take into account the processes of interaction of the heavy particles, especially the processes of quenching the emitting states and excitation transfer processes. This is especially important (as indicated by equation (3.16)) when examining the plasma of discharges of higher pressures and also the plasma with a higher concentration of the particles of the actinometric gas.

Chapter 4

LASER METHODS OF PLASMA DIAGNOSTICS

4.1 Introduction

The invention of the laser has offered investigators a light source which greatly expands the possibilities of the optical methods of plasma diagnostics. As a result of the unique properties of laser radiation – extremely high brightness, high coherence and the possibility of producing ultrashort laser pulses – lasers are not only used widely as light sources in classic diagnostic methods such as shadow, interference and schlieren methods, but have also been used to develop completely new diagnostic methods such as non-linear, resonance and laser interferometry, holographic diagnostics, scattering and fluorescence methods. In this chapter, we examine briefly the plasma diagnostic methods in which the use of lasers as light sources of principal importance.

More detailed information (of course, restricted by the publication dates) on this problem can be found in a number of monographs[1-3] and review articles.[5,6]

4.2 Interference examination of plasma using lasers

The advantages of lasers as light sources in classic two-beam interferometry with visualisation of field are obvious. It is the higher time and spatial coherence of radiation which reduces the requirements on straightening of the optical paths in two branches of the interferometer and on the accuracy of aligning the corresponding beams on the interferogram. The high intensity and monochromatic nature of laser radiation make it possible to avoid illuminating the interferogram with natural radiation of plasma when using corresponding optical filters and this greatly widens the range of plasma objects available for interference investigations. The possibility of generating ultrashort laser pulses greatly increases the time resolution.

Because of these obvious advantages, the lasers are used as light sources in the majority of interference methods, especially in investigations on large plasma systems and in diagnostics of dense pulsed plasma.

In this paragraph, we shall examine only the interference methods that can be realised only when using laser radiation. These methods include non-linear and resonance interferometry and also laser interferometry with photoelectric recording.

Non-linear interferometry

The development of laser technology has been accompanied by rapid advances in non-linear optics. In turn, the non-linear conversion of the frequencies of laser radiation (by doubling or displacing frequencies) has been used widely in interference plasma investigations. In particular, there have been a large number of studies in which plasma probing is carried out by simultaneous radiation of the basic frequency and the second harmonics of a ruby or neodymium laser. At the output of the interferometer these radiations are divided using light-dividing mirrors and selective filters and two interferograms, corresponding to two wavelengths, are recorded. As a result of combined treatment of these interferograms, it is possible to separate the contributions to the refraction of the plasma of electrons and atoms.

In all these studies the laser radiation frequency is converted to passage through the interferometer and the examined object. However, there is a whole group of non-linear interferometers[7] in which non-linear transformation of the waves takes place after the passage through the examined object. This offers a number of new possibilities to interference plasma diagnostics.

They include the possibility of probing plasma with the radiation of a single frequency situated in the spectral range and suitable for recording (for example, in the infrared range), followed by transformation of the radiation frequency to the visible range and recording the interferogram using conventional photographic material. To realise this possibility in plasma diagnostics, it is necessary to greatly increase the sensitivity of measuring the electron concentration. In fact, it is well known that the contributions of electrons and heavy particles (atoms and ions away from their absorption lines) to plasma refraction are described by the relationships

$$n_e - 1 = -\frac{C_e N_e}{\omega^2} = -4.49 \cdot 10^{-14} \lambda^2 N_e, \tag{4.1}$$

$$n_a - 1 = C_a N_a, \tag{4.2}$$

where $C_e = \dfrac{2\pi e^2}{m}$; e and m are the charge and electron, respectively;

ω and λ is the frequency and wavelength of probing radiation, respectively; $C_a = \dfrac{1}{N_L}\left(A + \dfrac{B}{\lambda^2}\right)$ is the refraction of the atoms of ions per single particle, cm³; N_e and N_a is the concentration of electrons and atoms; N_L is the Loschmidt number. The coefficients A and B are presented in tables in Ref.8.

In the visible and infrared ranges for the majority of atoms $B/\lambda^2 \ll A$ and, consequently, C_a is almost completely independent of radiation frequency. In accordance with (4.1) and (4.2), the phase shift, introduced into the wave by the electronic component

$$\varphi_e = \frac{\omega(n_e - 1)l}{c} = -\frac{C_e N_e l}{c\omega} \tag{4.3}$$

(l is the thickness of the irradiated layer, c is the speed of light), increases in transition to the IR range. At the same time, the phase shift determined by the atoms

$$\varphi_a = \frac{\omega(n_a - 1)l}{c} = \frac{\omega C_a N_a l}{c}, \tag{4.4}$$

increases in the IR region. Thus, the transition from the visible to the IR range is an efficient method of increasing the sensitivity of measuring the electron concentration in the plasma.

If plasma is probed with IR radiation with frequency ω_1 which is then mixed with the flat pumping wave with frequency ω_2 (Fig.4.1), the phase relief φ_3 of the wave of the total (or difference) frequency $\omega_3 = \omega_1 \pm \omega_2$ in the first approximation is equal to the phase relief φ_1 of the wave with frequency ω_1. Thus, the sensitivity of measuring the electron concentration corresponds to frequency ω_1 of IR radiation and the

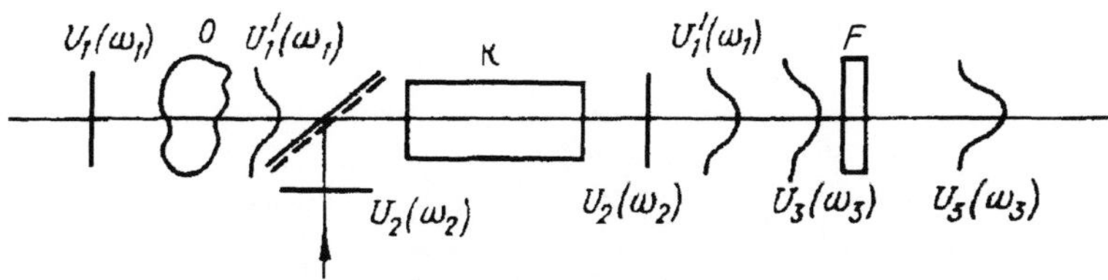

Fig.4.1 Diagram of non-linear transformation of the waves. The wave U_1 (ω_1) passes through the object O and is mixed with the flat wave U_2 (ω_2) in a non-linear crystal K; wave U_3 with a frequency $\omega_3 = \omega_1 \pm \omega_2$ is separated by the filter F.

interferogram can be recorded at frequency ω_3 in the spectrum range suitable for recording.

Another new possibility of non-linear interferometry is the direct comparison of the shape of the fronts of waves of different frequency.

The difference in the shape of the wave fronts of waves of different frequency, passing through a transparent object, is determined by differences in the refractive index of the object for radiation with these frequencies, i.e. by the dispersion of the object. The problem of direct determination of the difference of the refractive indices, i.e. comparison of the form of the wave fronts of waves of different frequency passing through the object, cannot be solved by classic interferometry methods because waves of different frequency do not form a stationary interference pattern. A non-linear interferometry makes it possible to solve this problem.

The principal diagram of a non-linear dispersion interferometer is shown in Fig.4.2. The object O (plasma) is situated between two non-linear elements K_1 and K_2. Two waves with frequency ω_1 and ω_2 are directed to the input of the interferometer. These waves are partially transformed to the wave with total frequency ω_3 in the first non-linear element. The object is then irradiated with three waves $U_1\,(\omega_1)$, $U_2\,(\omega_2)$ and $U_3\,(\omega_3)$, and the waves $U'_1\,(\omega_1)$, $U'_2\,(\omega_2)$ and $U'_3\,(\omega_3)$ leave the object. The phases of these waves are respectively

$$\varphi'_1 = \omega_1\left[n(\omega_1) - n_0(\omega_1)\right]\frac{l}{c}; \tag{4.5}$$

$$\varphi'_2 = \omega_2\left[n(\omega_2) - n_0(\omega_2)\right]\frac{l}{c}; \tag{4.6}$$

$$\varphi'_3 = \omega_3\left[n(\omega_3) - n_0(\omega_2)\right]\frac{l}{c}. \tag{4.7}$$

where n and n_0 are the refractive indices of the object and the environment.

In the second non-linear element, the frequencies of the waves $U'_1\,(\omega_1)$, $U'_2\,(\omega_2)$ are mixed and this leads to the formation of the wave $U''_3\,(\omega_3)$ whose phase is $\varphi''_3 = \varphi'_1 + \varphi'_2$. Interference of two waves U'_3

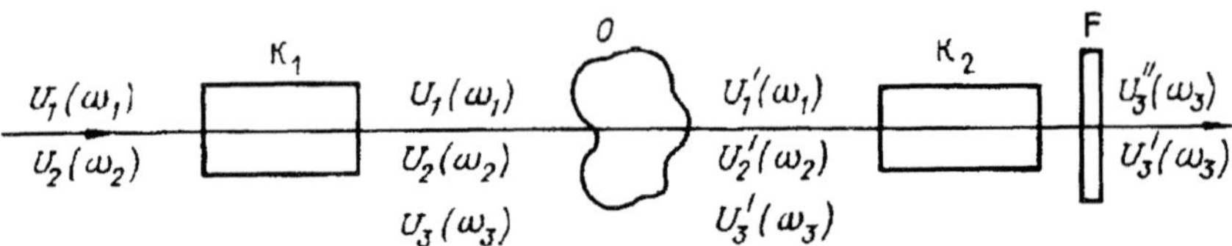

Fig.4.2 Diagram of a non-linear dispersion interferometer.

and U_3'', separated by the filter F, takes place at the outlet of the interferometer. The frequency of these waves is ω_3 and their distinguishing feature is that one of these waves was obtained as a result of mixing of the waves with the frequencies ω_1 and ω_2 prior to their passage through the object, and the second – after passage through the object. The phase difference of these waves

$$\Delta\varphi_D = \varphi_3'' - \varphi_3' = \varphi_1' + \varphi_2' - \varphi_3', \tag{4.8}$$

and the shape of their wave fronts differs only as a result of differences in the refractive indices of the object and the environment for radiation with the frequencies ω_1, ω_2 and ω_3. If the dispersion of the environment can be ignored, the shifts of the fringes on the interferogram will be determined only by the dispersion properties of the object. Substituting the values φ_1', φ_2' and φ_3' from (4.5), (4.6) and (4.7) into (4.8) and setting $n_0(\omega_1) = n_0(\omega_2) = n_0(\omega_3)$, we obtain

$$\Delta\varphi_D = \left\{\omega_1\left[n(\omega_1) - n(\omega_3)\right] + \omega_2\left[n(\omega_2) - n(\omega_3)\right]\right\}\frac{l}{c}. \tag{4.9}$$

A partial case of the dispersion interferometer is the non-linear interferometer based on frequency doubling proposed by the authors of this book[9] and almost simultaneously in Ref.10. In this case, the frequencies ω_1 and ω_2 are equal, $\omega_3 = 2\omega_1$ and

$$\Delta\varphi_D = \frac{2\omega_1 l}{c}\left[n(\omega_1) - n(\omega_3)\right], \tag{4.10}$$

and the shifts of the fringes on the interferogram are determined by the differences of the refractory indices of the object for radiations of the basic frequency and the second harmonics.

Representing plasma refraction as the sum of refractions of the electrons and the atoms (see equations (4.1) and (4.2)) gives

$$n - 1 = -\frac{C_e N_e}{\omega^2} + C_a N_a. \tag{4.11}$$

This shows that the dispersion of the refraction index of plasma is determined mainly by the electron concentration. Substituting n from (4.11) into (4.9) gives an equation for the shifts of interference fringes on the dispersion interferogram

$$k_D = \frac{\Delta\varphi_D}{2\pi} = \frac{C_e N_e l}{2\pi c}\left\{\frac{1}{\omega_1} + \frac{1}{\omega_2} - \frac{1}{\omega_3}\right\}. \tag{4.12}$$

For the case of the interferometer with frequency doubling $\omega_1 = \omega_2 = \omega_3/2$ and

$$k_D = \frac{3C_e N_e l}{4\pi c \omega_1}. \tag{4.13}$$

Therefore, the dispersion interferometer is not sensitive to the atomic component of plasma and makes it possible to take selective measurements of the electron concentration on the same dispersion interferogram.

In Ref.11, the non-linear dispersion interferometer based on frequency doubling was used to determine the electron concentration in the plasma of a laser spark in air. One of the interferograms, obtained in Ref.11, is shown in Fig.4.3a. For comparison, the photograph also shows an interferogram in the light of the basic frequency ω_1 (Fig.4.3) of the radiation of a ruby laser. The shifts of the fringes on the latter interferogram are considerable and determined mainly by the redistri-

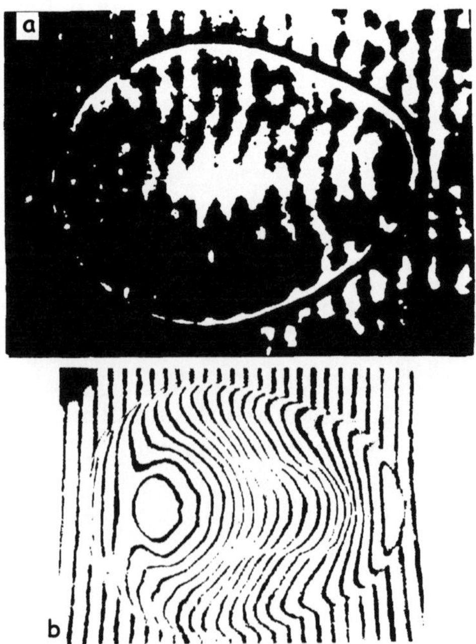

Fig.4.3 Interferograms of a laser spark. a) dispersion, b) in the light of the main frequency of radiation of a ruby laser.

bution of heavy particles (displacement from internal regions into the shock wave). Small shifts, determined by the electronic component, observed on the dispersion interferogram, are found only in the central highly ionised region, and in the region in the immediate vicinity of the front of the shock wave the fringes remain practically straight thus indicating that the interferometer is not sensitive to heterogeneities with no dispersion and associated with the distribution of heavy particles.

The accuracy of measuring N_e on the dispersion interferogram is considerably higher than when determining N_e using two interferograms obtained in the light with ω_1 and ω_3 because in the latter case the small shift, due to the electronic component $\Delta k_{1,3}$, is determined as a difference of large values ($\Delta k_{1,3} = 2k_1 - k_3$). Here k_1 and k_3 are the shifts of the fringes on the interferograms corresponding to ω_1 and ω_3.

Of special interest is the case in which one of the mixed frequencies, for example, ω_1, is situated in the long-range or medium IR range, and the second one, ω_2, in the visible range. Therefore, $\omega_1 << \omega_2$, ω_3 and

$$k_D \approx \frac{C_e N_e l}{2\pi c \omega_1}.$$ The sensitivity of measuring the electronic component of

plasma depends on frequency ω_1 and is ω_3/ω_1 higher than when probing plasma with the radiation with frequency ω_3 whose light is used to record the interferogram.

Resonance interferometry

The resonance interferometry methods are based on obtaining an interferogram in the light of radiation close to the absorption lines of one of the components of the examined plasma. Refraction ($n-1$) and the absorption factor χ (Fig.4.4) in the vicinity of the line with a dispersion contour are described by the relationships

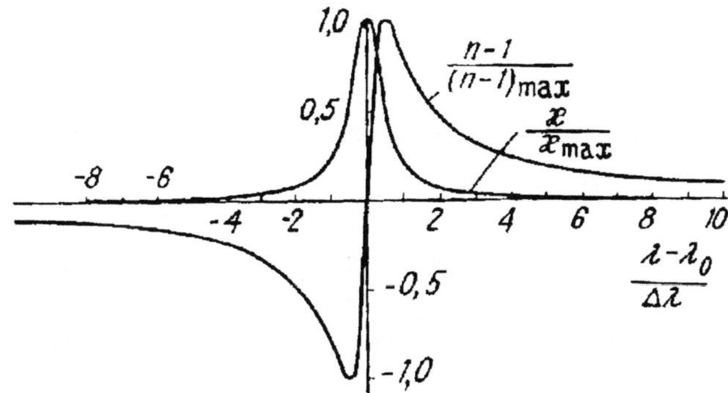

Fig.4.4 Dependence of refraction and the coefficient of absorption in the vicinity of the spectral line on the wavelength.

$$n - 1 = C\lambda_0^3 N_a f \, \frac{(\lambda - \lambda_0)}{(\lambda - \lambda_0)^2 + (\Delta\lambda/2)^2}, \tag{4.14}$$

$$\kappa = 2\pi C\lambda_0^2 N_a f \, \frac{\Delta\lambda}{(\lambda - \lambda_0)^2 + \left(\dfrac{\Delta\lambda}{2}\right)^2}. \tag{4.15}$$

Here λ_0 is the length of the wave corresponding to the maximum of the absorption line, $\Delta\lambda$ is the width of the line measured at half height, λ is the wavelength of radiation, f is the force of the line oscillator, N_a is the concentration of the atoms on the absorbing level, $C = \dfrac{e^2}{4\pi m c^2}$.

On approach to the absorption line, the refraction of the corresponding atoms rapidly increase and can exceed by several orders of magnitude the refraction of the same atom away from the absorption line. Thus, using the radiation with the wavelength close to the absorption line for obtaining interferograms, it is possible to increase greatly the sensitivity of measuring the concentration of appropriate atoms.

A proposal to produce the interference pattern in the light of radiation close to the absorption line of one of the components of the examined plasma was made prior to inventing lasers.[12] However, the invention of lasers facilitated the introduction of this method to the practice of plasma investigations. A radiation source with the lines close to the absorption lines of the atoms to be determined was specially selected in the first studies of resonance interferometry. For example, the authors of Ref.13 presented a table of laser radiation lines situated close to the absorption lines of different atoms.

Of special interest is the possibility of using tuneable dye lasers in resonance interferometry. This not only greatly widens the range of objects that can be studied by this method but also makes it possible to vary over a wide range the sensitivity of interference measurements displacing the position of the generation line in relation to the wavelength of the absorption line.

The sensitivity and ranges of application of resonance interferometry were calculated in Ref.14, 15. In Ref.14, calculations were carried out assuming monochromatic probing radiation for the case of the absorption line whose contour is determined by the combined effect of dispersion and Doppler broadening. In Ref.15, calculations were carried out for the absorption line with a dispersion contour, taking into account the finite width of the probing radiation line.

The relationship between the shift of the interference fringes on the resonance interferogram and the concentration of the absorbing atoms can be found using equation (4.14) for plasma refraction

$$k = \frac{(n-1)l}{\lambda} \approx \frac{C\lambda_0^2 N_a fl(\lambda - \lambda_0)}{(\lambda - \lambda_0)^2 + (\Delta\lambda/2)^2}. \tag{4.16}$$

Therefore, setting $k_{min} = 0.1$, it is easy to obtain the concentration determined by this method as a function of the distance between the wavelength of the line of probing radiation λ and the absorption line λ_0. Equation (4.16) shows that the minimum detectable concentration is obtained at a distance from the centre of the absorption line equal to its half width, i.e. at $\lambda = \lambda_0 \pm \dfrac{\Delta\lambda}{2}$:

$$N_{a,min} \approx \frac{k_{min}\Delta\lambda}{C\lambda_0^2 fl}. \tag{4.17}$$

Thus, the minimum detectable concentration of the atoms is directly proportional to the width of the absorption line.

The upper concentration limit available for investigations by laser interferometry is determined by the absorption of probing radiation by plasma and a corresponding reduction of the contrast of the interference fringes. Figure 4.5 shows the boundaries of applicability of the methods calculated in Ref.15 using these assumptions. The curves, shown in this graph, correspond to the probing radiation line whose width is considerably smaller than the width of the absorption line. For the case in which the width of the probing radiation line is comparable with that of the absorption line, the shifts of the fringes, corresponding to different parts of the contour of the probing line, differ and this reduces the contrast of the interference pattern. As shown by the calculations in Ref.15, when resonance interferograms are produced using a radiation with the line width exceeding the width of the absorption line, the sensitivity of the method greatly decreases and the range of the measured concentrations is smaller.

Laser interferometry with a dye laser as a light source was used to examine the plasma of the laser jet on a target made of an alloy of lithium with lead. The equipment circuit (Fig. 4.6) consisted of two lasers whose Q-factor was modulated by the same rotating prism P. Radiation of neodymium laser A_1 was focused onto the surface of the solid

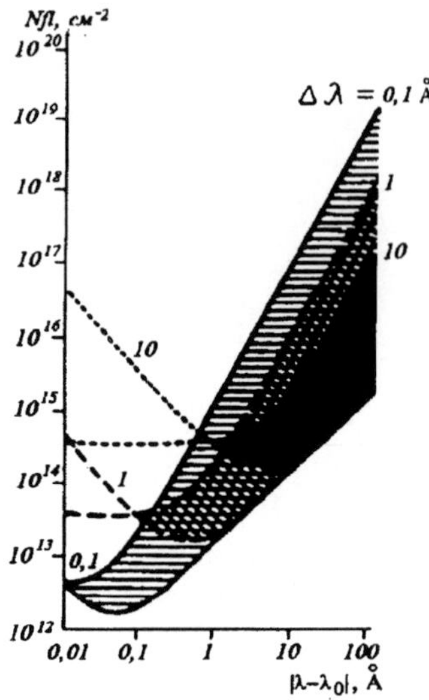

Fig.4.5 Boundaries of applicability of the method of resonance interferometry at different widths of the absorption line and the infinitely narrow line of the probing radiation.

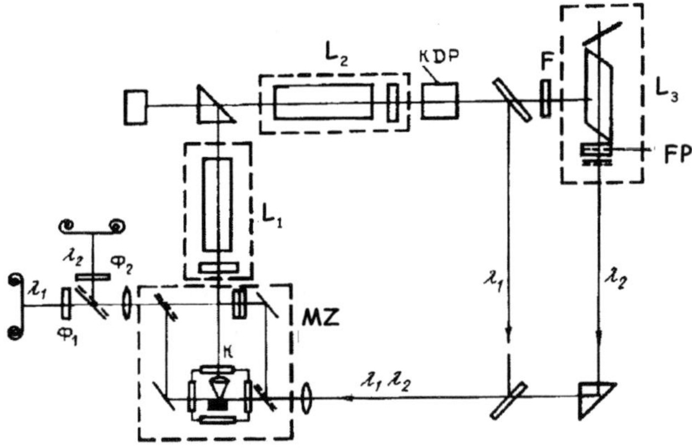

Fig.4.6 Diagram of equipment for interference examination of the laser jet using a dye laser.

target inside a vacuum chamber K placed in one of the arms of the Mach-Zender interferometer (MZ). The dye was pumped into the cuvette by radiation of the second harmonics of the ruby laser L_2. The cavity of the dye laser L_3 consisted of a semitransparent mirror and a diffraction grating which when rotated resulted in smooth readjusting of the length of the generation line in the vicinity of the resonance doublet of lithium ($\lambda_0 = 6707.8$ Å). A Fabry–Perot (FP) interferometer with the distance between the mirrors of 0.01 cm was placed inside the cavity of the dye laser to reduce the width of the generation line. The generation spectrum of the laser had the form of a single line 0.5 Å wide.

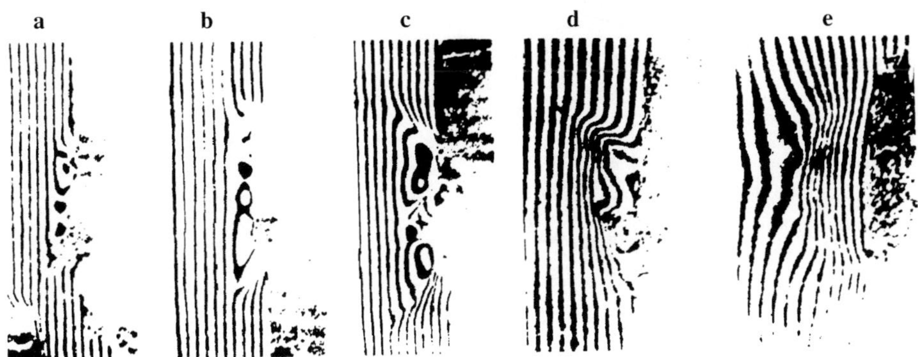

Fig.4.7 Interferograms of the laser jet on a target made of a lithium alloy with lead for different moments of time. t = 0.3 (a), 0.8 (b), 1.3 (c), 1.8 (d), 2.3 5s (e); λ-λ_0 = 30 (a,b), 8 (c,d), 4 Å (e).

The radiation of the ruby dye laser was directed onto the interferometer using glass prisms and sheets. Two interferograms were photographed at the outlet of the interferometer in the light of radiation of the ruby laser λ_1 and the dye laser λ_2.

Figure 4.7 shows interferograms obtained in the light of a dye laser and corresponding to different stages of jet acceleration (t = 0.3-2.3 µs). Optimum sensitivity was selected for each specific moment by varying the value $\lambda-\lambda_0$. For times of 0.3 and 0.8 µs when determining the concentration of lithium atoms, it was necessary to process interferograms obtained simultaneously in the light of radiation of the dye laser and the ruby laser. Starting from 1.3 µs, there are no shifts of the fringes of the interferograms obtained in the light of the ruby laser. Therefore, the shifts of the fringes on interferograms, corresponding to radiation of the dye laser, can be fully attributed to the refraction of lithium atoms.

The spatial distributions of the concentration of lithium atoms, shown in Fig.4.8, were obtained as a result of processing the interferograms for different stages of jet acceleration. As a result of modifying the wavelength of the radiation of the ruby laser, it was possible to take reliable measurements of the concentration of lithium atoms with the concentration varying in the range of approximately four orders of magnitude.

In Ref.17, resonance interferometry was used to determine the concentration of excited hydrogen atoms in the plasma of a laser spark. The laser spark was produced in a chamber filled with hydrogen with a pressure 2 to 10 atm. Interferograms were obtained using the radiation of a dye laser with a wavelength close to the line H_α. The measured

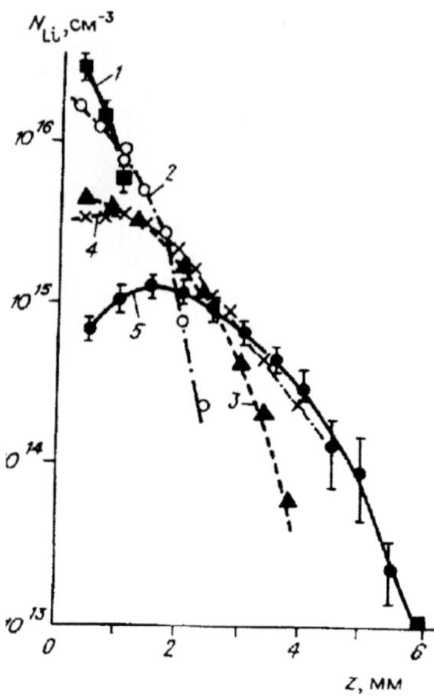

Fig.4.8 Distribution of lithium atoms along the laser beam axis for the moments of acceleration of the laser jet of 0.3 (1), 0.8 (2), 1.3 (3), 1.8 (4), 2.3 µs (5).

parameter in this case was the concentration of the atoms on the second energy level.

Laser interferometry

The term 'laser interferometry' usually refers to different interference methods in which the laser plays simultaneously the role of a light source and an interferometer, and the phase shifts are recorded by photoelectric methods.

If a diaphragm separating part of the interference fringe is placed at the output of an interferometer, and a photoelectric receiver is placed behind this diaphragm, then the variation of the photoflux can be used to examine the variation of the optical difference of the path with time. However, in this type of recording, it is not possible to obtain information on the spatial distribution of the refractive index that can be obtained in systems with visualisation of the field. The spatial resolution is determined by the diameter of the probing beam.

An advantage of the photoelectric recording method is the possibility of increasing the sensitivity of determining the electron concentration by using long wave radiation because a relatively large range of photoelectric receivers is available for operation in the IR range. In addition, photoelectric recording opens wide possibilities for using different radioelectronic systems employed in interference diagnostics of plasma in the microwave range (see Ref.18). Consequently, the accuracy of measuring the phase shift can be greatly increased.

The diagram of the simplest laser interferometer[19] is shown in Fig.4.9a. A flat mirror M_3 is added to the cavity of the helium–neon laser consisting of spherical M_1 and flat M_2 mirrors. The mirrors M_2 and M_3 form an additional Fabry–Perot cavity linked with the main laser cavity. The intensity of laser radiation depends strongly on the phase of the signal reflected from mirror M_3. The variation of the optical length of the M_2–M_3 cavity as a result of changes of the parameters of the plasma placed inside it caused periodic changes of the phase of the reflected signal and the modulation of laser radiation leaving through the mirror M_1. The signal modulated with respect to amplitude is recorded by the photoreceiver and oscillographically processed. Evidently, the variation of the optical length of the cavity corresponding to a single modulation period is $\lambda/2$.

It is well known that the helium–neon laser can emit simultaneously a number of lines in the visible and infrared spectrum ranges, including lines $\lambda = 0.63$; 3.39 µm. Both these lines have a common upper level and, therefore, the modulation of the intensity of laser radiation at a wavelength $\lambda = 3.39$ µm is also accompanied by the modulation of visible radiation $\lambda = 0.63$ µm. Placing a silicon filter between mirror M_2 and plasma it is possible to probe the plasma by IR radiation characterised by higher sensitivity to determination of the electronic density, and the modulation of the intensity of laser radiation in the visible range can be examined using a conventional photomultiplier.

Laser interferometry methods are used mainly for examining pulsed

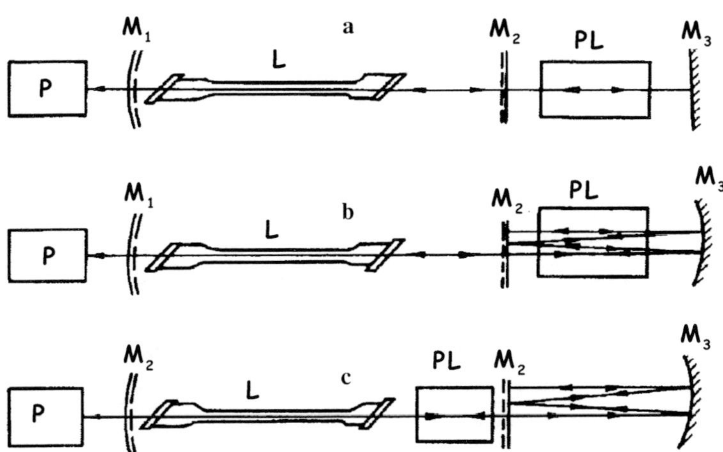

Fig.4.9 Diagrams of three-mirror laser interferometers. a) internal resonator of the Fabry–Perot type; b) resonator with a spherical mirror; c) plasma is situated inside the laser resonator; L - helium-neon laser; PL - plasma; P - photoreceiver.

plasma. If the number of maxima on the oscillogram from the moment of formation of plasma to the given moment of time t is $k(t)$ and it is known that refraction is caused mainly by the electronic component of the plasma and changes monotonically during this period of time, then taking into account the double passage of radiation through the plasma, we have

$$N_e(t) = \frac{1.12 \cdot 10^{13}}{\lambda l} k(t).$$
(4.18)

To determine the parameters of stationary plasma, it is necessary to disconnect rapidly the energy source supplying power to the plasma and count the modulation maxima from the moment of disconnection to complete collapse of plasma.

When the concentration changes non-monotonically, the form of the oscillogram does not make it always possible to determine whether the shift per fringe corresponds to the increase or decrease of N_e. Additional phase modulation is often used to remove ambiguity in interpreting the fringes. In the absence of plasma, laser radiation is modulated at a constant frequency. The introduction of plasma into the cavity disrupts the periodicity and the direction of variation of the refractive index of plasma can be determined by examining whether the distance between maxima becomes smaller or greater.

The accuracy and sensitivity of measuring $N_e l$ depend on the wavelength of probing radiation and the accuracy of measuring the shift of the fringes Δk:

$$\Delta(N_e l) = \frac{1.12 \cdot 10^{13}}{\lambda} \Delta k.$$
(4.19)

As indicated by (4.19), when using IR radiation of a helium–neon laser with $\lambda = 3.39$ μm a shift per fringe corresponds to $\Delta(N_e l) = 3 \cdot 10^{16}$ cm^2. In direct count of the maxima on the oscillogram, the accuracy of measurement is, according to the data of different authors, 1 to 0.25 cycles. The method of phase modulation using an acousto-optical cell introduced into the M_2–M_3 resonator, or vibrations of mirror M_3, make it possible to measure the phase difference with an accuracy of approximately 10^{-3} fringe.[20,21] Substituting $\lambda = 3.39$ μm and $\Delta k = 10^{-3}$ into (4.19), the resultant measurement error is $\Delta(N_e l) = 3 \cdot 10^{13}$ cm^2.

Sensitivity can be increased using an interferometer in which the flat mirror in the outer resonator is replaced by a spherical one[22] (Fig.4.9b).

Not only longitudinal but also transverse modes are excited in this resonator so that when measuring the length of the resonator, the intensity of laser radiation is modulated with a frequency S times higher than the laser interferometer with a flat mirror M_3. As shown in Ref.23, the quantity S has a simple geometrical meaning – it is the number of double passages of the light beam between the mirrors of the cavity after which the beam returns to the initial (entry) point. This geometrical interpretation is especially convincing when the laser beam is not directed along the axis of a reference resonator (Fig.4.9b). The increase of sensitivity in this case can be regarded as a result of multiple (S times) passage of radiation through plasma. In successive passages, the beam is displaced thus impairing the spatial resolution. This can be avoided by placing the plasma inside a laser resonator[24,25] (Fig.4.9c).

As already mentioned, the sensitivity of determining the electron concentration can be increased by increasing the wavelength of laser radiation. For example, when using the radiation of a CO_2 laser ($\lambda = 10.6$ μm) $\lambda(N_e l) \approx 10^{16}$ cm^{-2} corresponds to the shift per fringe, and when using a HCN laser ($\lambda = 337$ μm) it is $\lambda(N_e l) \approx 3 \cdot 10^{14}$ cm^{-2}.

When measuring the atom concentration, the sensitivity can be increased selectively using the radiation close to the absorption lines of these atoms (see equation (4.14)). For example, in Ref.26 the concentration of metastable helium atoms on the level 2^3S_1 was determined using the radiation of a helium–neon laser ($\lambda = 1.0798$ μm) whose distance from the absorption line $\lambda = 1.0830$ μm was 32 Å. A similar method was used in Ref.27 to determine the concentration of metastable neon atoms on the level $1s_5$. Plasma probing was carried out by the radiation of a helium–neon laser ($\lambda = 0.6401$ μm) 1 Å from the line formed in transition to the level $1s_5$ ($\lambda_0 = 0.6402$ μm). In this case, the concentration of the metastable neon atoms corresponding to the shift per fringe was $\Delta N_a \approx 1.1 \cdot 10^{11}$ cm^{-3} (column length $l = 42.5$ cm).

Laser heterodyne systems are characterised by a very high sensitivity to the changes of plasma refraction. The addition of plasma inside a laser resonator shifts the laser radiation frequency

$$\frac{\Delta\omega}{\omega} = \frac{l\Delta n}{d_r}, \quad \frac{\Delta\omega}{\omega} = \frac{l\Delta n}{d_r}, \tag{4.20}$$

where l is the thickness of the plasma layer, d_r is the laser resonator length. Frequency shift can be measured by optical heterodyning. According to the estimates in Ref.28, the minimum frequency shift measured by this method is around 10 Hz which corresponds to the $\Delta n = 2 \cdot 10^{-13}$ and an electron concentration of $3 \cdot 10^8$ cm^{-3} (in calculation it was assumed that l = 20 cm, d_r = 100 cm and λ = 1.15 μm).

A laser heterodyne system was used for plasma diagnostics in Ref.29. The radiation of two helium–neon lasers with plasma added to the resonator of one of these lasers and the second laser was a reference one, was directed onto a photoreceiver. The recorded signal showed pulsations at frequency Ω equal to the difference of the generation frequency of two lasers ($\Omega = \omega_1 - \omega_2$). The introduction of pulsed plasma into the laser resonator caused changes in the pulsation frequency. The minimum measured value of the electron concentration, obtained by the authors of Ref.29, was 10^{11} cm^{-3} (at a layer thickness of 24 cm).

A superheterodyne laser interferometer was described in Ref.30, 31. In this case, the radiation of a laser was divided into two channels in a Mach–Zender interferometer. An ultrasound cell was placed in one of the arms of the interferometer. The light signal at the output was recorded using a photomultiplier. Ultrasound frequency Ω was present in the frequency spectrum of the photoflux. The addition of plasma to the second arm of the interferometer changed the phase of the oscillations and this frequency situated in the radio range. The phase variation was recorded using an electronic phase meter. The sensitivity of determining the phase shift was $1°$,[31] which corresponded to $(N_e l)_{min} \approx 10^{15}$ cm^{-2} at λ = 0.63 μm.

Different circuits of laser interferometers with heterodyne and homodyne frequency conversion were described in Ref.32.

4.3 Holographic plasma diagnostic methods
Possibilities and special features of holographic plasma diagnostic methods

Holography is a method of recording and subsequent restoration of waves based on recording the interference pattern formed as a result of the interference of the subject wave and the reference wave coherent with it. The interference structure recorded on the light-sensitive material is referred to as the hologram. The diffraction of the reference beam on the hologram leads to restoration of the subject wave.

In examining the plasma by classic shadow, schlieren and interference methods, the phase distortions of the light wave which has passed through the plasma are visualised in the form of a shadow pattern or interferogram directly at the moment of passage of radiation through plasma. Holography makes it possible to record the wave passed through the plasma, restore it and examine it by various methods. As in the direct application of these methods, the measured quantities are the refractive index or its derivatives, and in measuring the amplitude distortions – the absorption factor.

Data on the plasma parameters are obtained through their link with the refractive index and the absorption factor, i.e. the holographic method offers nothing new in comparison with the conventional optical method. Nevertheless, in many cases the holographic methods have considerable advantages in comparison with the conventional optical methods and sometimes make it possible to solve problems which could not be solved by these conventional methods. The new possibilities of the holographic method include:

– recording and subsequent restoration of the wave passed through the pulsed plasma, diagnostics of the latter under the standard conditions by different optical methods;

– restoration by means of a single hologram of light waves passed through the plasma in different directions within the limits of the solid angle defined by the hologram makes it possible to obtain the spatial distribution of the parameters of the three-directional object in the absence of axial symmetry;

– interference examination of the plasma enclosed in a vessel with optically non-uniform windows;

– low requirements on the quality of the optical components of the interferometer and, consequently, the possibility of interferometric examination of plasma of almost unlimited dimensions;

– simultaneous or consecutive recording of holograms in the light of radiation passed through the plasma with several wavelengths and examination of the interference pattern formed by the waves restored by such a hologram;

– increase of the sensitivity of interference measurements;

– recording of the amplitude distortions of the transmitted wave and measurement of the absorption factor of the plasma without interference from the side of its natural radiation;

– the possibility of using various movie holographic systems for examining plasma dynamics.

The holographic methods were used for the first time for examining plasma in 1966.[33] At present, they are widely in plasma experiments. We shall now present the results of a number of investigations in which these special features were most evident together with the possibilities of holographic methods of plasma diagnostics. More detailed information can be found in a number of books and review articles.[2-5]

Examination of pulsed plasma under the stationary conditions

The light wave passed through the pulsed plasma is recorded on the hologram and then restored and examined by different methods which supplement each other.

The schlieren circuit was used in the restoration stage already in the initial investigations of holographic plasma diagnostics.[33,34] It was found

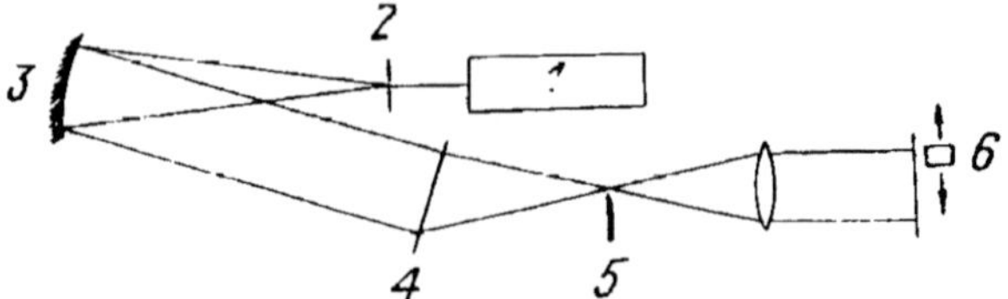

Fig.4.10 Schlieren-system for examining wave fronts restored by a hologram. 1) laser, 2) spot diaphragm, 3) bent mirror, 4) hologram, 5) blade, 6) scanning photoreceiver.

that the plasma of the laser spark is equivalent to the negative astigmatic lens, and the focusing distance of this lens was used to determine the refractive index of the plasma and estimate the electron concentration in it.

A schlieren circuit, shown schematically in Fig.4.10, was used to record a wave restored by the hologram in Ref.35 where a laser spark in argon induced by the radiation of a CO_2 laser was examined. Information on the spatial distribution of the gradient of the refractive index of the plasma was obtained using a scanning photodetector.

The authors of Ref.36 describes a holographic interferometer which enables the frequency and orientation of the fringes on the interferograms to be changed in the restoration stage. The circuit of the interferometer is shown in Fig.4.11. The hologram 1 was produced by the double exposure method, and the angle of incidence of the subject beam on the hologram was varied between exposures. The waves restored by this hologram and corresponding to each exposure propagate under an angle to each other and can be separated in the focus of the objective 2. Reflected from different faces of the prism 3, the waves propagate in two different arms of the interferometer and form an interferogram in plane 4 where the frequency and orientation of the fringes are meas-

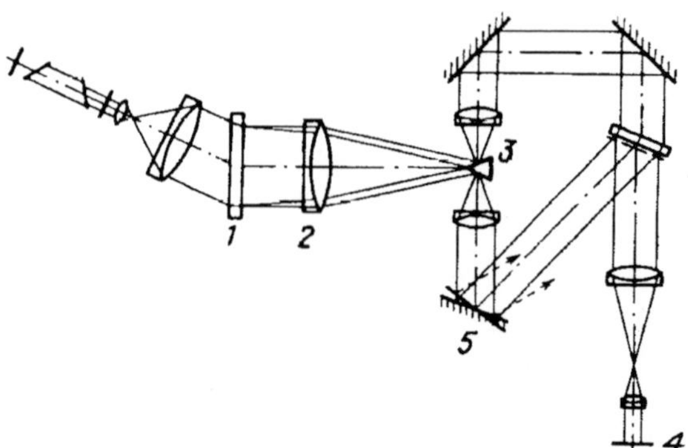

Fig.4.11 Diagram of a holographic interferometer.

ured by tilting one of the mirrors 5 of the interferometer. Overlapping one of the interference beams, the shadow images of the examined object are obtained. This interferometer was used by the authors of Ref.36 to examine Z- and θ-pinch plasma.

Examination of the spatial distribution of the plasma parameters using holographic interferometry data

As in the majority of optical methods (except the methods of laser scattering and resonance fluorescence, Fig.3), the holographic method makes it possible to determine only the plasma parameters integral with respect to the direction of the beam. In the case of objects with axial symmetry, the transition from the measured quantities to the radial distribution is usually carried out by solving the integral Abel equation.

In the absence of axial symmetry of plasma the information carried by the light wave, passed through the plasma in one direction, is insufficient to determine the local plasma parameters. If in producing a hologram the plasma was illuminated over a wide solid angle range, for example, using a diffusion scatterer, then using a single hologram it is possible to restore the light waves passed through the plasma under different angles so that the spatial distribution of the refractive index in the plasma can be determined.

In principle, some information can be obtained by classic interferometry. In this case, it is necessary to examine the plasma simultaneously using several interferometers or carry out consecutive examination of the interferograms of plasma with the variation of the mutual position of plasma equipment and the interferometer. From the technical viewpoint, both methods are highly cumbersome and difficult to apply and, in addition, when using the second method it is necessary to ensure exceptionally high stability or reproducibility of the examined object because random errors, associated with the non-reproducibility of the plasma, can lead to an unacceptable reduction of the accuracy of calculations of the local values of its parameters. Therefore, it is not thought that classic interferometry can determine t the local parameters of non-axis symmetric plasma with sufficient efficiency. At the same time, the holographic methods can be used to solve these problems.

The problems of examining the spatial distributions of the refractive index of three-dimensional objects were examined in Ref.37-40. The most detailed and consecutive examination of these methods was carried out in Ref.41-43. Without discussing the mathematical aspect of the problem, we shall examine only the main procedures of obtaining holographic interferograms corresponding to irradiation of the phase object under different angles.

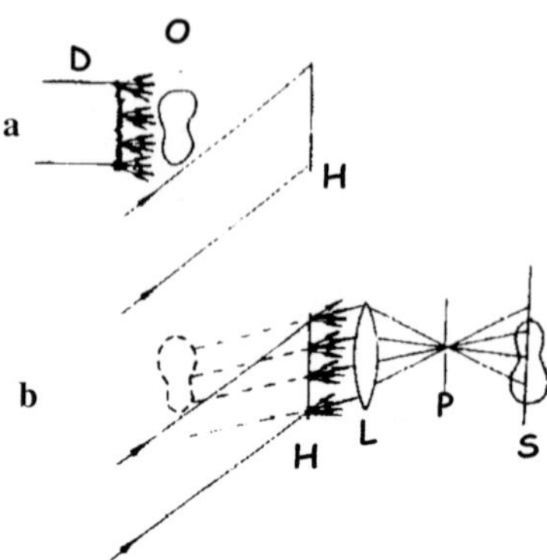

Fig.4.12 Diagram of recording the hologram of a transparent object with a scatterer (a) and restored waves corresponding to the specific direction of illumination (b).

The simplest method is to irradiate the object O using the scatterer D, for example, matt glass (Fig.4.12a). In the restoration stage, the spot diaphragm P is placed in the focal plane of the lens L (b) which projects the apparent image of the object on to the screen S; the diaphragm separates beams restored by the hologram H and the parallel beam corresponding to the specific irradiation direction. Changing the position of the diaphragm, it is possible to obtain a series of interferograms corresponding to different examination directions using a single double-exposure hologram.

It should be noted that when producing holograms by equipment with a scatterer, various parts of the initial subject would interfere with the reference beam. This imposes stringent requirements on the spatial coherence of the light used to produce the hologram. Therefore, the systems with the scatterer usually include single-mode lasers. An exception is a system proposed in Ref.44 in which the scatterer is projected on to a hologram so that it is possible to combine the mode structure of subject and reference beams and use a multimode laser as a radiation source.

The second disadvantage of the system with the scatterer is the spotty structure typical of images of diffusion objects illuminated with coherent radiation.

In certain cases, the matt scatterer can be efficiently replaced by a phase diffraction grating which enables the object to be irradiated under several discrete angles.[45,46] For example, in Ref.45 the holographic grating with 50 lines/mm generated bright beams of three diffraction orders on each side of the normal so that it was possible to obtain 7 interferograms with a vision angle of 11° (in 1.8° steps) from a single hologram. The restored interferograms do not have the spotted structure because in this case there are no diffusion scattering elements. The equipment with the diffraction grating was used in Ref.40 to obtain holographic interferograms corresponding to different irradiation directions. These interferograms were used to restore the field of gas temperatures above the heated wire.

Insensitivity of holographic diagnostics to the quality of optical elements

In a conventional two-beam interferometer, two light beams propagating along its two arms interact. Defects of mirrors and other optical elements lead to different distortions of the beams and this has a direct effect on the form of the interference pattern. However, in holographic interferometry (see, for example, Ref.3, 43) the waves passing along the same path interact but at different moments of time. This is one of the most important special features of the holographic method – its differentiality.

Recording on a hologram two waves corresponding to two states of the examined object and restoring them at the same time, it is possible to examine an interference pattern whose form is determined only by changes which took place in the object during the period between the two exposures. Thus, the comparison wave forms automatically and repeats in every detail the waves scattered by the object in the initial state. Consequently, the requirements on the quality of optical components of equipment becomes less stringent. Distortions of the wave fronts, associated with the imperfection of the optical elements of the holographic system are the same for two interference waves and have almost no effect on the form of the interference pattern.

Because of the insensitivity of holographic interferometry to imperfections of the optical elements it is possible to examine plasma in large plasma systems, for example, Z- and θ-pinch.[47,49] It is difficult to carry out these investigations by classic interferometry because devices with large mirrors are extremely difficult to produce and are very expensive. Because of this advantage of holographic interferometry it is also possible to diagnose the plasma enclosed in a vessel with non-uniform walls, especially plasma of pulsed lamps.[50-52]

If distortions in interference comparison of two waves distorted to the same extent by optical elements are excluded, then in examining the waved restored by the hologram using shadow or schlieren methods the distortions of the wave front are recorded to the same degree both as a result of the passage through the examined phase object and as a result of the imperfection of the optical systems, especially due to aberration of the lenses and spherical mirrors. However, holography makes it possible to exclude distortions of the wave front caused by the imperfection of the components of the optical system,[53,54] by recording two

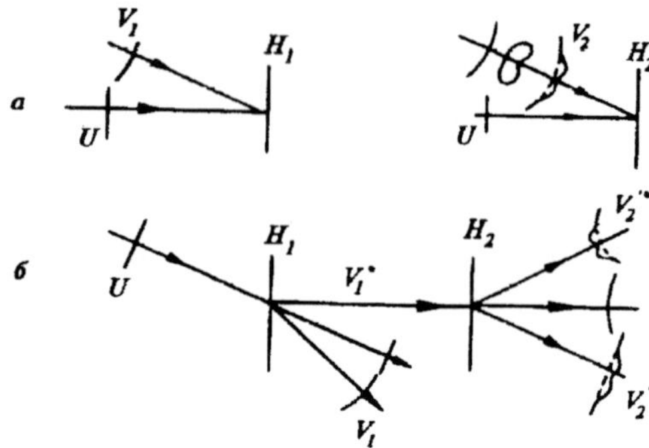

Fig.4.13 Repair of aberration of wave fronts. a) recording of the holograms H_1 and H_2; b) restoration of non-aberration wave front V_2'.

holograms on the same holographic equipment: H_1 in the absence of the object and H_2 in its presence. In the restoration stage the conjugate wave front 1, restored by the first hologram, with the phase relief inverse in relation to the wave V_1, is used to illuminate the second hologram. This results in the restoration of the aberration-free wave front V'_2 whose phase relief is determined only by the passage of the wave through the examined object (Fig.4.13).

Sensitivity of holographic plasma diagnostic methods and methods of measuring it

The sensitivity of double-exposure holographic interferometry is almost identical with that of the conventional method. The minimum observed variation of the optical thickness $\Delta(nl)_{min}$ is associated with the minimum measured shift of the fringes on the interferogram k_{min} by the relationship

$$\Delta(nl)_{min} = k_{min}\lambda, \tag{4.21}$$

where λ is the wavelength, l is the thickness of the plasma layer.

The error of measuring the shifts on the fringes on the holographic interferogram of medium quality, as in visual processing of conventional interferograms, is assumed to be usually equal to 0.1 fringe.

Minimum value of the electron and atom concentration can be determined substituting into (4.21) the values of the refraction of electrons (4.1) and atoms (4.2):

$$(N_e l)_{min} = 2.2 \cdot 10^{13} \frac{k_{min}}{\lambda}, \tag{4.22}$$

$$\left(N_a l\right)_{min} = \frac{k_{min}}{C_a \lambda}.$$ (4.23)

As in conventional interferometry, the transition to the IR range is the selective method of increasing the sensitivity of measuring the electron concentration in accordance with (4.22). At the same time, the selective increase of the sensitivity of measuring the concentration of a specific sort of atoms or ions can be achieved by displacing the wavelength of probing radiation to their absorption lines (see (4.14)), i.e. by resonance holography.[55-58]

In addition, there are a number of methods of increasing the sensitivity of interferometry. Some of these methods are used in both conventional and holographic interferometry, others are holographic only.

Sensitivity of both conventional and holographic interferometry can be increased by systems with multiple passage over the subject beam through plasma. The plasma is situated between two semitransparent mirrors A and B (Fig.4.14a). To separate the light beam, corresponding to the given number of passes, one of the mirrors is tilted in relation to the other one by small angle . The consecutive beams are inclined in relation to each other by angle 2α and can easily be separated using a spatial filter (a lens with a diaphragm positioned in its focal plane). A similar variant of the multipass method was used to include the sensitivity in examining shock waves in ballistic trajectories at reduced pressure.[59] Interferograms corresponding to 11 passages of radiation through the object were obtained.

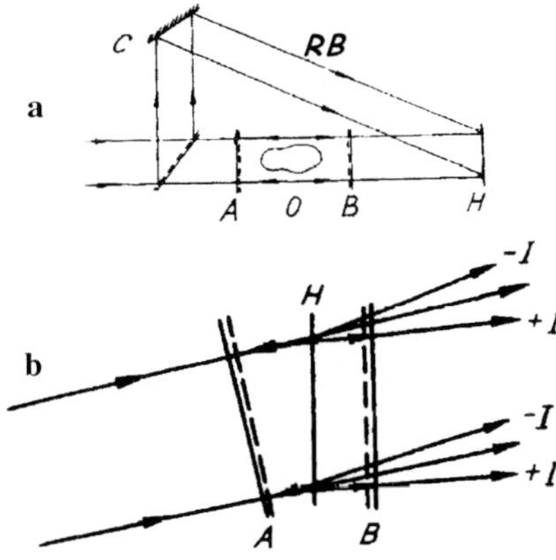

Fig.4.14 Holographic diagram with multiple passage of the probing light beam through the object (a) and the diagram of increasing sensitivity as a result of multiple passage of the light beam through the hologram in the restoration stage (b).

A considerable disadvantage of this variant of the multipass method is the displacement of the light beam in consecutive passages through the object thus reducing the spatial resolution of the method. This shortcoming can be eliminated by another variant of the multipass holographic method proposed in Refs.60, 61 in which the mirrors A and B between which the object O is situated are positioned strictly parallel in relation to each other and normal to the probing beam (Fig.4.14a). If the distance between the mirrors is greater than the coherence length of radiation of the lasers, then from the set of the beams with different numbers of passes through the plasma only the beam whose width differs from that of the reference beam by less than the coherence length of laser radiation will take place in the formation of the hologram. Moving the mirror C, it is possible to change the length of the optical path travelled by the reference beam RB and produce holograms corresponding to the given number of radiation passes through the plasma.

The sensitivity of holographic interferometry can be increased not only by the multiple passage of radiation through the object in the stage of producing the hologram but also as a result of multiple passage of radiation through the hologram in the restoration stage.[62,63] In this case (Fig.4.14b), one of the semitransparent mirrors is perpendicular to the beam restored by the hologram and directs the beam onto the hologram as a restoring beam. Since the thickness of the hologram is considerably smaller than the longitudinal dimensions of the examined plasma, the transverse displacement of the light beam during its multiple passage to the hologram is considerably smaller than in passage through the plasma. A disadvantage of this method is the faster (than in direct passage through the object) reduction of the intensity of consecutive beams determined not only by the transmission factor of the mirrors but also by the diffraction efficiency of the hologram.

The second specific holographic method of increasing the sensitivity is the use of non-linear effects and the possibility of producing an interference pattern in superimposition of conjugate waves (corresponding to the apparent and real images) with inverse phase reliefs.

If the hologram is recorded on a non-linear photomaterial, in addition to the zeroth order it also restores ± 1-st order of the wave of higher orders (Fig.4.15) whose phase relief is k times (k is the number of the order) higher than the waves of the first order. A similar method of increasing the sensitivity of holographic interferometry was proposed in Ref.54, 65.

Illuminating a hologram obtained by non-linear recording with two light beams results in the propagation of two waves of different orders in the given direction (Fig.4.16). In particular, if the restoring beams are directed in such a manner that the waves of $+k$-th and $-k$-th order

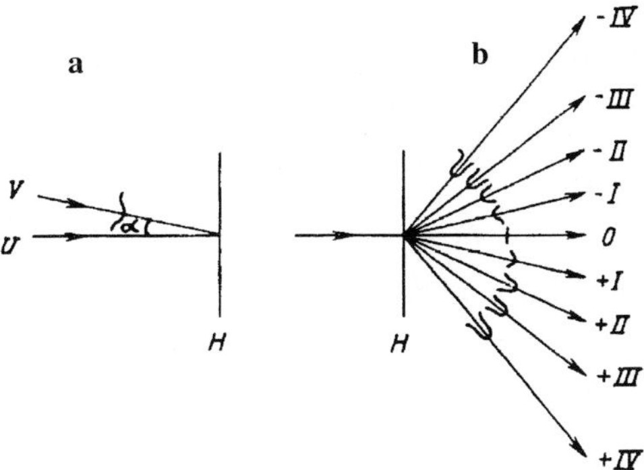

Fig.4.15 Diagrams of recording the hologram (a) and restoration of the wave fronts of a non-linearly recorded hologram (b).

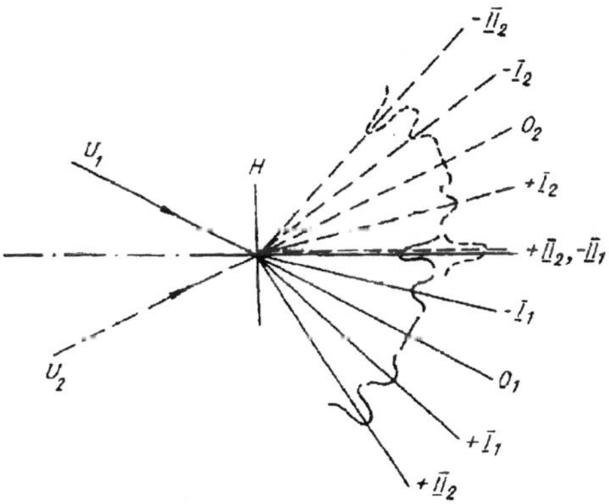

Fig.4.16 Diagram explaining the increase of the sensitivity of holographic interferometry by a factor of $2k$ in interference of the waves of the $+k$-th and k-th orders.

are restored in the direction normal to the hologram, then the gain in sensitivity as a result of their interference is $2k$.

However, the use of non-linear effects for increasing the sensitivity intensifies not only the distortions of the wave front caused by the examined object but also any distortion caused by the imperfection of the optical elements of holographic equipment. Therefore, either holographic systems designed for increasing sensitivity are produced on the basis of interferometers with high quality mirrors or it is necessary to compensate the distortions introduced by the optics in the restoration stage. The methods of this compensation of aberrations using waves of high orders were examined in Ref.66, 67.

Infrared and holographic interferometry

As already mentioned, the transition to the IR range leads to a selective increase of the sensitivity of determining the electron concentration. For example, in accordance with equation (4.22), the use of a CO_2 laser ($\lambda = 10.6$ μm) instead of a ruby laser ($\lambda = 0.69$ μm) leads to a 15 fold increase of the sensitivity of determining N_e, with other conditions being equal.

One of the main difficulties of transition to the IR range is that no high sensitivity and high-resolution media for recording IR holograms are available at the moment.

The upper limit of the spectral sensitivity or silver-halide photomaterials corresponds to a wavelength of the order of 1 μm. To record holograms and interferograms in the range of longer waves, it is necessary to use different materials (wax, gelatine, graphite, thermochromic materials, polymer films, metal films deposited on a glass substrate, etc.) whose optical properties undergo reversible or irreversible changes as a result of heating with laser radiation.

The common property of the majority of these materials associated with the thermal recording mechanism is the relatively low sensitivity in the range from 0.1 to 1 J/cm^2, whereas the sensitivity of silver halide materials for recording holograms in the visible range is 10^{-3}–10^{-7} J/cm^2.

A second special feature of the materials for recording IR holograms is the dependence of their resolution power on the exposure time determined by 'erosion' of the temperature relief as a result of heat conductivity. Therefore, continuous lasers can be used only for recording, with these materials, interferograms with a frequency of several lines per millimetre, whereas when using powerful pulse lasers with the same materials it is possible to record holograms with a frequency to 100 lines/mm.

A new promising material for recording holograms is a phase interference transformation reversed light reflector (PITRLR)[68] based on the semiconductor–metal phase transition in vanadium oxides. The sensitivity of this material to the radiation of a ruby laser is $5 \cdot 10^{-4}$ J/cm^2 but at longer wavelengths the sensitivity decreases as a result of a decrease of the absorption factor of vanadium oxides in the IR range. In recording all PITRLR holograms in the light of radiation of a CO_2 laser ($\lambda = 10.6$ μm) the layer of vanadium oxides is thicker as a result of thermal conduct with the substrate absorbing radiation with the absorbing layer of silicon oxide deposited on the surface of PITRLR. According to the data in Ref.69 the sensitivity in this case is $5 \cdot 10^{-3}$ J/cm^2.

High sensitivity is also typical of the method of recording IR holograms proposed in Ref.70–72 which utilises the temperature dependence of the sensitivity of conventional photomaterials. Infrared radiation generates a temperature relief on the surface of photographic material which is also illuminated with the actinic radiation of a pulsed lamp synchronised with the radiation of the IR laser. The sensitivity of the laser is 10^{-2} J/cm^2, resolution power 50 lines/min.

A more detailed analysis of the properties of different media and the methods of recording IR holograms can be found in Refs.73 and 74.

Infrared holographic interferometry was used to examine electric discharges,[75,76] exploding wires,[75] and plasma generated in focusing laser radiation in gases and on solid targets.[67,77,78]

The diagram of holographic equipment for diagnostics of a spark breakdown differs from conventional holographic systems in the visible range only by the presence of germanium optics (Fig.4.17). Holograms were recorded using a pulsed CO_2 laser on bismuth films deposited on a glass subject.

Holography in the light of a pulsed CO_2 laser was used in Refs.77, 78 for examining late stages of development of the laser spark in air and of the laser jet on a solid target. The plasma was induced by a second powerful CO_2 laser synchronised with the first one. Holograms were recorded on triacetate cellulose films.

The same material was used in holographic examination of an arc discharge.[76] The holographic equipment (Fig.4.18) included the germanium light-dividing wedge 2, and the electrodes of the arc 6, situated in the vicinity of the wedge, were projected by means of the spherical mirror 3 with ×2 reduction on the surface of triacetate cellulose film 5. The threshold properties of the material and the relief-phase nature of recording resulted in highly non-linear recording and the appearance of intensive waves of high orders in the restoration stage. Interferograms with improved sensitivity (Fig.4.19) were obtained as a result of interference of the restored waves of high orders (up to ± 7th order). The sensitivity of measuring the electron concentration was $(N_e l)_{\text{min}} =$

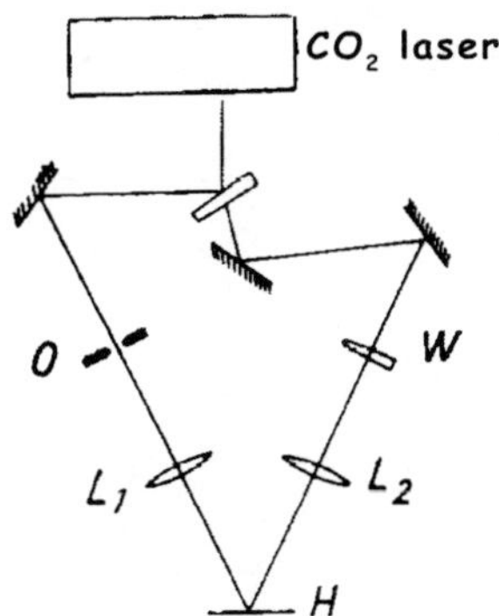

Fig.4.17 Equipment for examining the spark breakdown using IR holography. W is the germanium wedge, L_1 and L_2 are germanium lenses, H is the hologram, O are the electrodes of the spark.

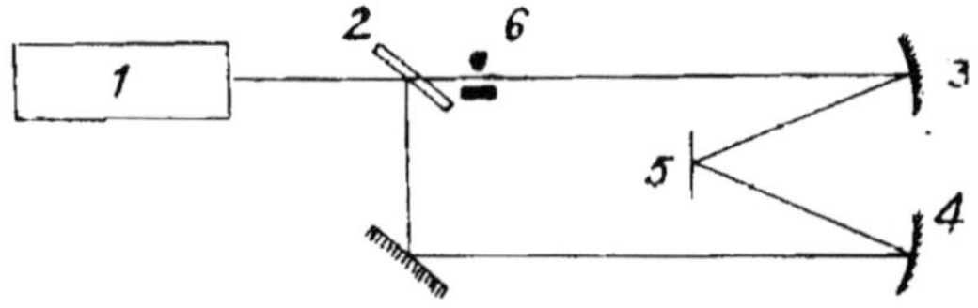

Fig.4.18 Diagram of equipment for producing IR holograms of arc discharge. 1) CO2 laser, 2) germanium wedge, 3,4) spherical mirrors, 5) hologram, 6) arc electrodes.

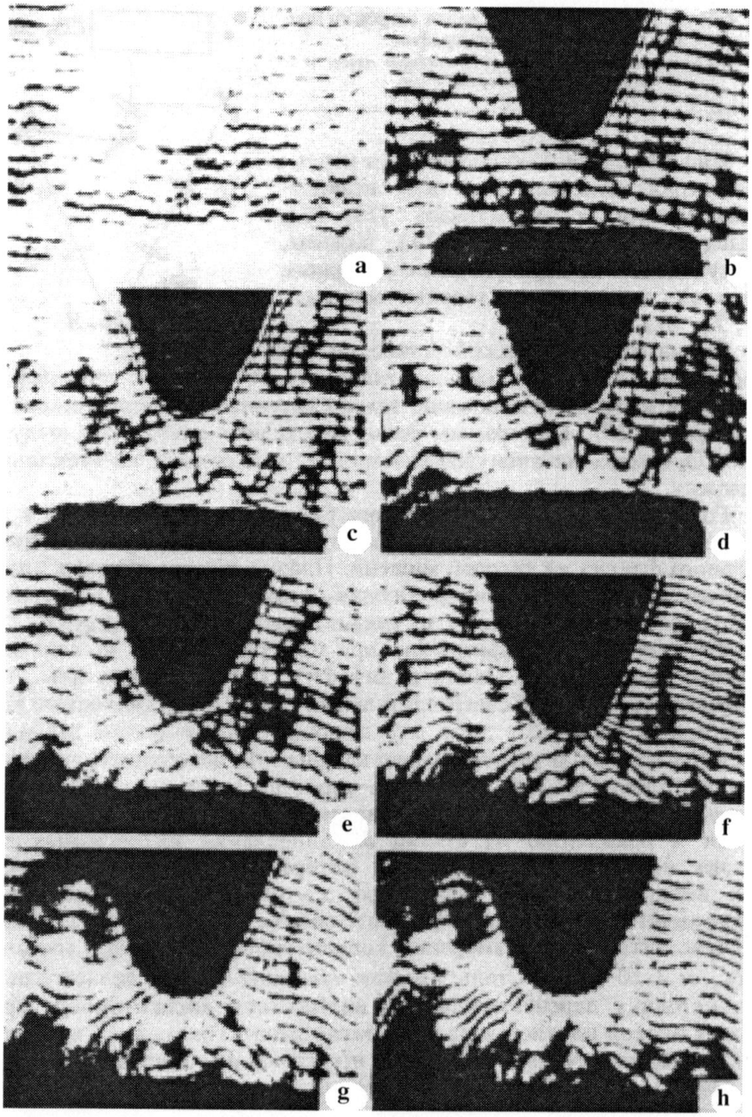

Fig.4.19 Holographic interferograms of the arc discharge plasma produced at interference of the waves of the orders of 0 and +1 (a), ± 1 (b), ± 2 (c), ± 3 (d), ± 4 (e), ± 5 (f), ± 6 (g) and ± 7 (h).

$2 \cdot 10^{14}$ cm^{-2}. This high sensitivity makes the described method promising for diagnostics of low-density high-temperature plasma in Tokamak-type systems. The equipment designed for these investigations was described in Ref.79; it was also used for analysis of late stages of acceleration of the laser jet on a carbon target.[67]

Two-wavelength, resonance and dispersion holography

To separate contributions to the refraction of electrons and atoms in both conventional and holographic interferometry, it is necessary to record at least two interferograms at two wavelengths. If the object and reference beams in the stage of producing the hologram consist of radiation with two wavelengths λ_1 and λ_2, the resultant hologram represents a superimposition of two interference (holographic) structures with different spatial frequencies ν_1 and ν_2 which depend on the wavelength

$(\nu_{1,2} = \dfrac{2\sin \alpha / 2}{\lambda_{1,2}}$ where α is the angle between the subject and reference beams).

In illuminating such a hologram with monochromatic radiation, the restored waves corresponding to each hologram, propagate under dif-

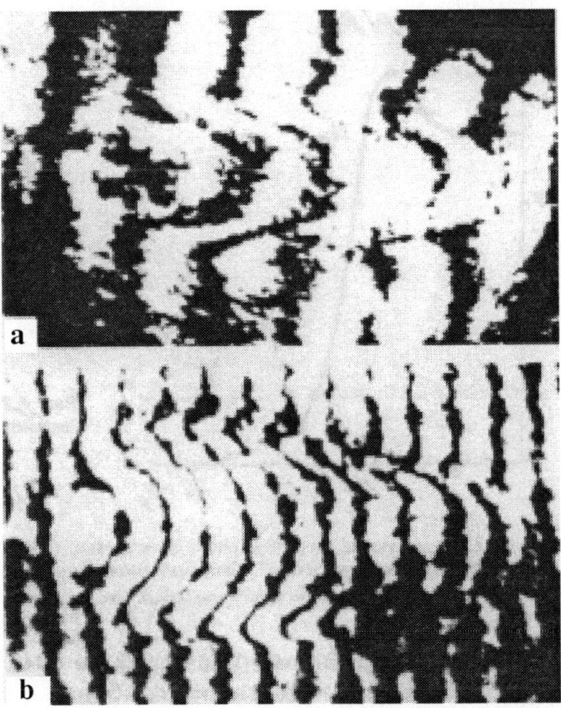

Fig.4.20 Holographic interferograms of the laser spark in helium at a pressure of 6 atm, corresponding to the radiations of the basic frequency (a) and second harmonics (b) of a ruby laser.

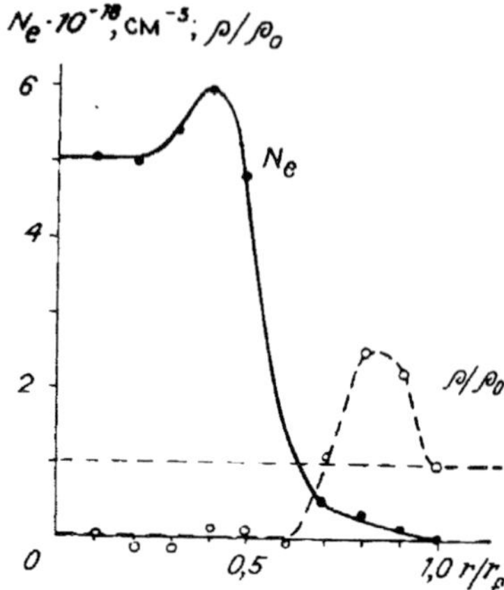

Fig.4.21 Variable distributions of the electronic concentration N_e and gas density ρ, obtained at combined processing of the interferograms shown in Fig.4.20; r_f is the radius of the front of the shockwave, ρ_0 is the initial density of the gas.

ferent angles. If the hologram is produced by the double exposure method, it can be used to restore two interferograms separated in space and corresponding to each wavelength. Two such interferograms of a laser spark in helium, corresponding to the radiation of the main frequency in the second harmonics of a ruby laser, are shown in Fig.4.20. Combined processing of these interferograms[80] yielded the spatial distribution of the concentration of electrons and gas density in the laser spark (Fig.4.21).

If in recording the hologram one of the wavelengths present in the radiation was close to the absorption line of the specific type of particles, the sensitivity of measuring their concentration increases in accordance with (4.14), i.e. we have a holographic variance of a resonance interferometry in this case.

Resonance holographic interferometry was used for the first time in Refs.55 and 56 for examining the laser jet on a potassium target. Plasma probing was carried out by radiation with three wavelengths: a ruby laser ($\lambda_1 = 6943$ Å), its second harmonics ($\lambda_2 = 3472$ Å) and forced combination scattering of the radiation of a ruby laser in nitrobenzol ($\lambda_3 = 7658$ Å). Wavelength λ_3 differs by only 7 Å from the wavelength of one of the components of the resonance doublet of potassium. As a result of processing three simultaneously produced interferograms corresponding to the three wavelengths, it was possible to separate the contributions to the refraction of electrons, air molecules and potassium atoms described by the equations (4.1), (4.2) and (4.14), respectively.

The method of resonance holographic interferometry with a dye laser used as a light source was used for examining the laser jet on a barium target[57] and cesium plasma of a pulsed arc discharge.[58]

In all the studies described previously interferograms corresponding to different wavelengths and separated in space were produced in the restoration state. However, holography makes it possible to examine directly the pattern of interference of the restored waves which are holographic copies of waves of different frequency.

As regards plasma diagnostics, of special interest is the method of dispersion holographic interferometry[81] in which the hologram on the phase object is recorded by means of the main frequency (λ_1) and the second harmonics ($\lambda_2 = \lambda_1/2$) of laser radiation on a non-linear photomaterial. In the process of restoration of waves using such a hologram H in relation to the waves of the ±1-st order waves of high orders were restored (Fig.4.22). The wave of the second order, corresponding to λ_1, propagates in the same direction as the wave of the first order corresponding to λ_2. Interference of these restored waves results in the formation of an interferogram where the shifts of the fringes are proportional to the dispersion of the object, as in the case as when using a non-linear dispersion interferometer (see section 4.2). In turn, the dispersion of the object is determined by the electron concentration in the plasma (see equations (4.10)–(4.11)).

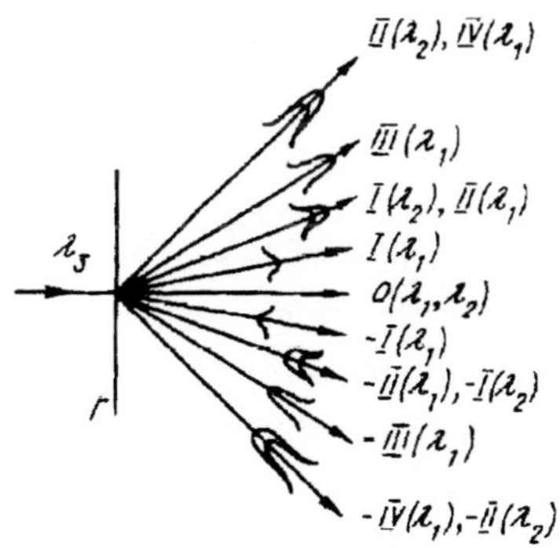

Fig.4.22 Diagram of producing a dispersion holographic interferogram.

Dispersion holographic interferometry was used in Ref.82, 83 to examine the laser spark in air. The resultant interferograms were used to determine the electron concentration in the laser spark without making any corrections for the redistribution of the concentration of air molecules associated with the formation of a shock wave.

Kinoholography

Holography offers considerable possibilities for constructing kinoholographic systems, i.e. equipment producing a series of holograms

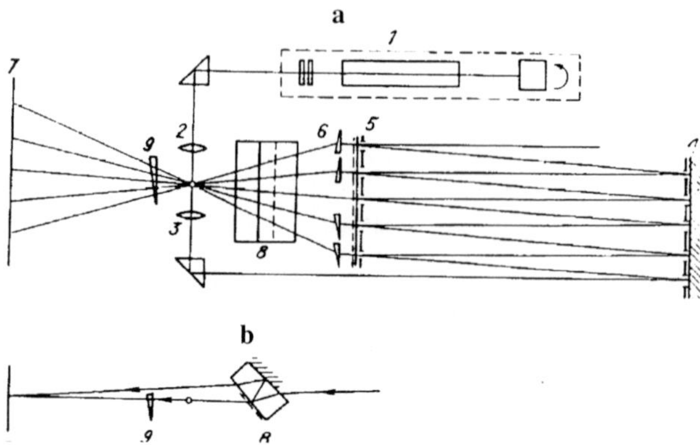

Fig.4.23 Kinoholographic equipment for examining the laser spark. a) top view, b) lateral view, 1) pulsed ruby laser, 2,3) co-focal lenses, 4,5) mirrors of the optical delay line, 6) wedges, 7) holograms, 8) light-dividing wedge, 9) wedge used to produce finite-width fringes.

of a high-speed process corresponding to successive stages of its development. These systems differ in the number of frames and time periods between them, the method of formation of light pulses displaced with time and the methods of holographic recording.

Figure 4.23 shows the circuit of one of the first kinoholographic systems used for examining the laser spark in air.[84,85] The optical delay line consists of opaque and semitransparent mirrors and is used to shape a series of light pulses shifted in space and with time. The distance between the mirrors is 6 m which corresponds to the time period between the frames of 40 ns. The light beams, passed through the semitransparent mirror, were directed onto a laser spark using glass wedges. Each of the beams was divided into subject and reference and then mode structures were combined in the plane of the hologram. This gave five holographic interferograms corresponding to five consecutive moments of development of the laser spark (Fig.4.24). A similar five-frame system with the period between the frames of 0.1 ns was described in Ref.86.

In the previously examined system, the object beams passed through plasma under different angles. However, in certain cases, for example, in examining Z- and θ-pinches and current layers, it is important to ensure that all subject beams illuminate the plasma strictly in one direction (along the discharge axis). The following holographic systems can be used in this case (Fig.4.25). In examining holograms using the system shown in Fig.4.25a, the subject beams, correspond to different times are combined in space whereas the reference beams illuminate the hologram under different angles. Thus, the same section of the photosensitive layer records several holograms with different frequency and orientation of the holographic structure. In the restoration stage, such

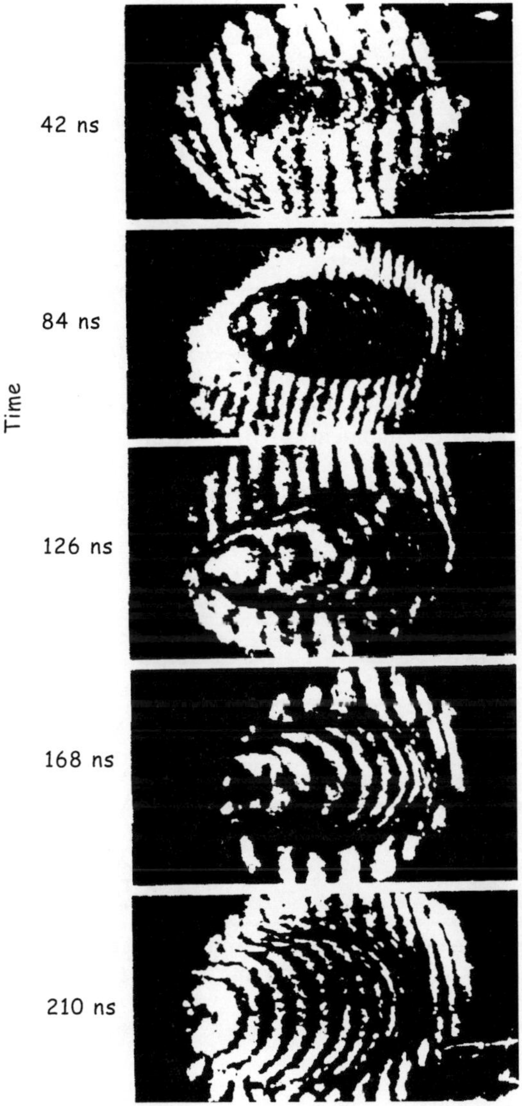

Fig.4.24 Five consecutive holographic interferograms of a laser spark.

a hologram makes it possible to examine the images of the objects separated in space and corresponding to different moments of time.

A shortcoming of this system is the low diffraction efficiency D_1 which is inversely proportional to the square of the number n of holograms recorded in the same area, and for the case of amplitude holograms is described by the equation

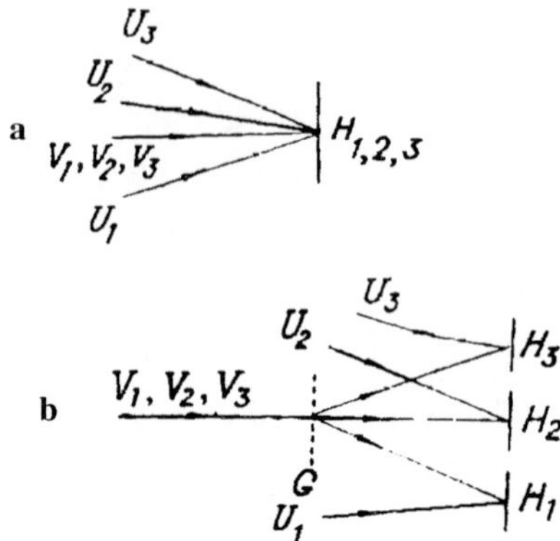

Fig.4.25 Diagrams of producing kinoholograms in spatial combination of the beams $(V_1$-$V_3)$, corresponding to various moments of time. a) three holograms are recorded on the same section of the photolayer, b) holograms corresponding to different moments of time are recorded in different sections of the photolayer, U_1, U_2, U_3 are reference beams.

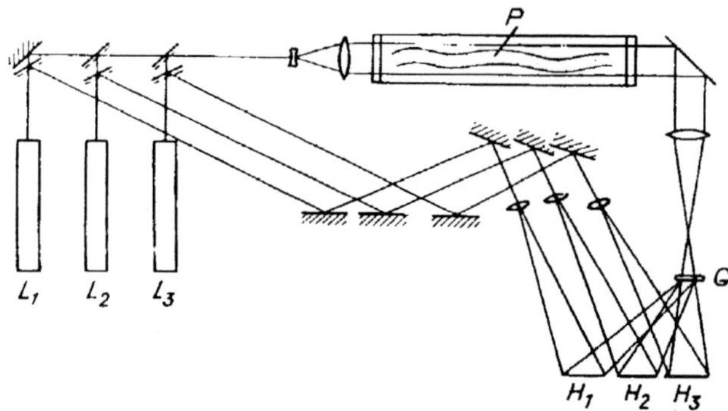

Fig.4.26 Diagram of equipment for kinoholographical examination of the θ-pinch. G is the diffraction grating, $H_1 - H_3$ are holograms, L_1, L_2, L_3 are lasers, P is plasma.

$$D_1 \leq \frac{1}{16n^2}. \tag{4.24}$$

In the system shown in Fig.4.25b, the subject beams combined in space are divided into several parts using diffraction grating G and are directed onto different areas of the photosensitive material. Each of these sections is illuminated with one reference beam. This system was used in Ref.48 to examine the θ-pinch. Three pulsed ruby lasers were used to produce light passes separated in time (Fig.4.26).

As shown in Ref.87, the diffraction efficiency of the hologram is described by the relationship

$$D_2 \leq \frac{1}{4\left(\sqrt{n+1}\right)^2}$$

(4.25)

and decreases at a considerably lower rate when the number of holograms increases. For example, at $n = 5$ D_2 is almost an order of magnitude higher than D_1.

The system with the diffraction grating was also used to examine the plasma of a current layer.[87-89] In this case, beams delayed in time were produced using an optical delay line (Fig.4.27). As an example, Fig.4.28 shows five holographic interferograms produced during a single pulse of plasma equipment and corresponding to the processes of failure of the current layer. As indicated by the photographs, in the initial uniform distribution of the electronic concentration the layer (frame 1) contains a region with reduced concentration in the centre of the layer (frames 2, 3) and this is followed by a decrease of the concentration in the entire layer (frames 4, 5). Figure 4.28*b* shows curves of equal electron concentration calculated as a result of processing a series of interferograms (see Fig.4.28*a*) and illustrating the process of failure of the layer.

The real time method can be used for kinoholographic plasma diagnostics. The circuit of one of these systems, designed for examining

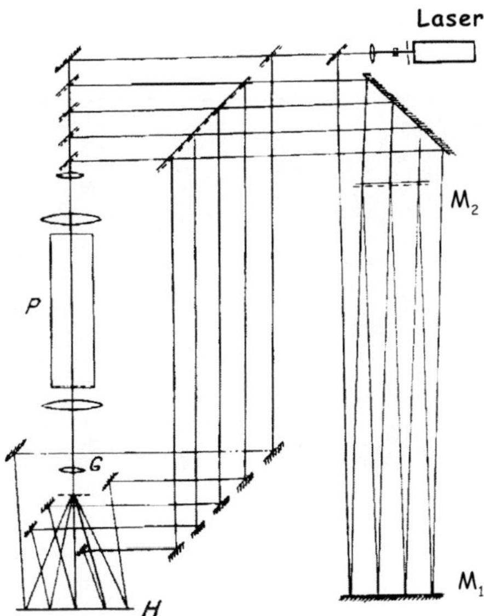

Fig.4.27 Diagram of kinoholographic equipment for examining current layers. *G* is the diffraction grating, M_1, M_2 are the mirrors of the optical delay line, *P* is the plasma chamber, *H* is the hologram.

a

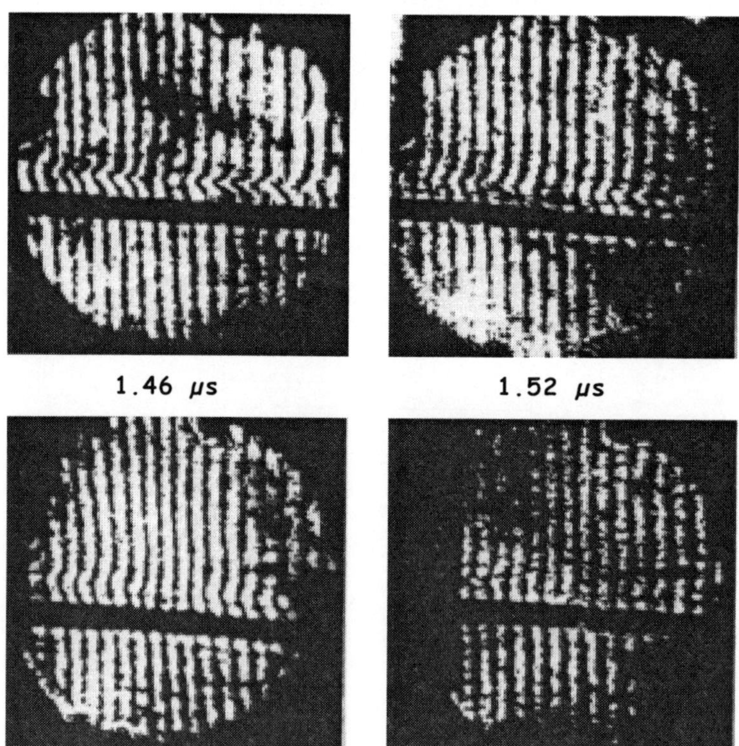

1.40 μs

1.46 μs   1.52 μs

1.58 μs   1.64 μs

Fig4.28 (continued on page 95).

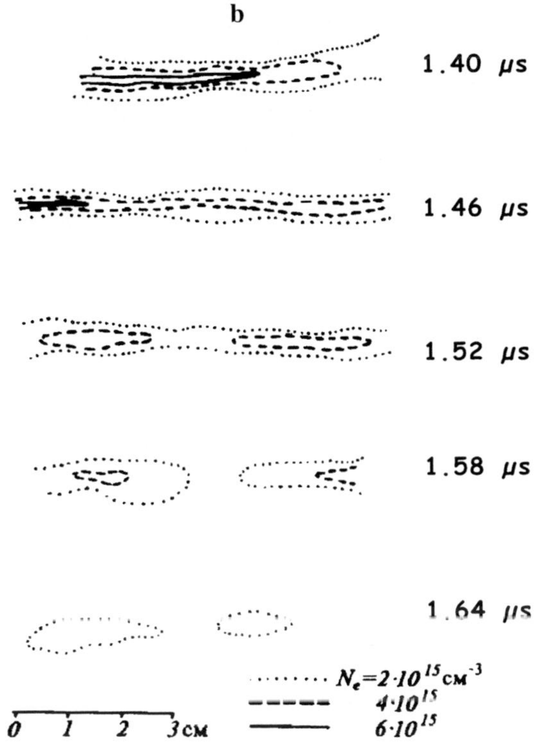

Fig.4.28 Five consecutive holographic interferograms of the current layer (a) and the lines of equal electronic density constructed from them (b).

the plasma jet,[90,91] is shown in Fig.4.29. Radiation was generated by ruby laser 1 operating in the free generation regime. Initially, the hologram 2 was exposed in the absence of plasma. The hologram was developed and processed on site using a special cuvette. The wave, restored by this hologram, then interfered with the wave passing directly through plasma 3. The interferograms were photographed using the superhigh speed camera 4. The exposure time of the individual frames was determined by the duration of the laser radiation beams and did not exceed 1 μs.

Identical equipment was used in Ref.92 to examine flames. Radiation was generated by a 50 mW continuous He–Ne laser. This restricted the time resolution of the method to 10^{-3} s.

4.4 Plasma diagnostics on the basis of scattering of laser radiation

In all previously described optical methods of plasma diagnostics, the

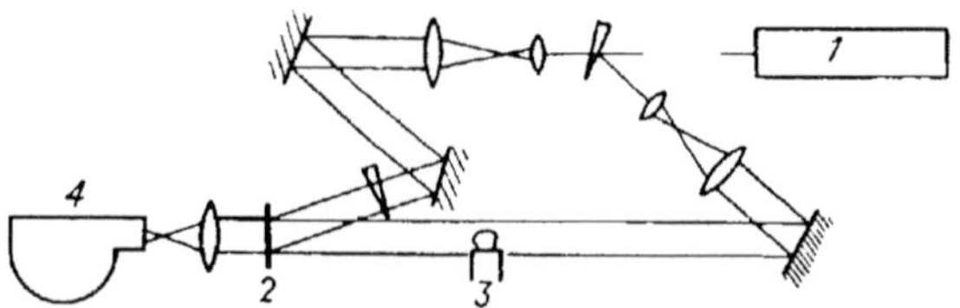

Fig.4.29 Equipment for kinoholographic examination of the plasma jet by the real time method.

information obtained directly from the experiment is integral with respect to the examination direction. The transition from integral to local values takes place by means of relatively complicated calculations and results in a large reduction of accuracy. Therefore, it is of special interest to examine the diagnostics method enabling direct measurements of the local plasma parameters. The method of scattering of laser radiation is one of these methods.

Scattered radiation is observed from a small volume formed in intersecting the light beam, probing the plasma, with a cone within which the radiation scattered by the plasma is recorded.

If the wavelength of the probing radiation is situated at a relatively large distance from the absorption line of the atoms and ions, the main contribution to light scattering by the plasma is provided by the electrons. The total, the so-called Thomson scattering cross section of radiation by an electron is equal to $6.65 \cdot 10^{-25}$ cm^2 which results in a weak signal of the scattered radiation measured in the experiments. Because of this weak signal, the scattering method could not be used for laboratory plasma diagnostics until lasers were invented.

The proposal to use the scattering of laser radiation for determining the plasma parameters was made almost simultaneously in Ref.93 and 94 in 1962. The theoretical fundamentals of the scattering method and the possibilities resulting from this method in plasma diagnostics were examined in a number of review articles[95–98] and also in monographs.[1,2,32,99-101]

The spectrum of the radiation scattered by plasma and its interpretation depend strongly on the plasma parameters. Scattering at free electrons (Thomson scattering) takes place in low-density plasma, whereas collective scattering takes place at high values of N_e.

Scattering on a free electron

Under the effect of an incident light wave $\vec{E} = \vec{E}_0 e^{i\omega_0 t}$, propagating along the x axis, the electron starts to vibrate and becomes at the same time

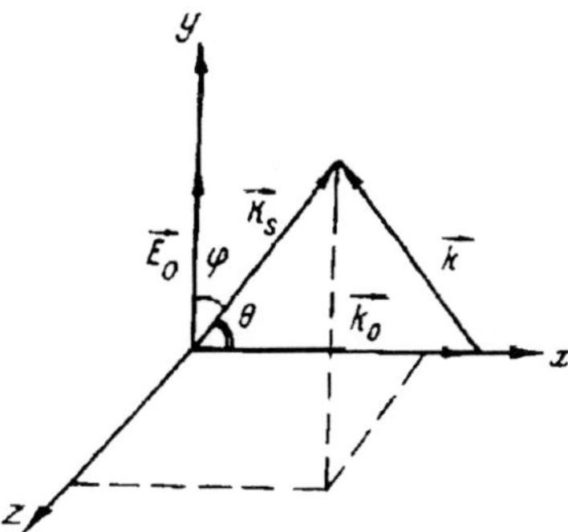

Fig.4.30 Scattering of the light wave by a free electron.

an emitting centre. Figure 4.30 shows the wave vectors of the incident $\vec{k}_0$ and scattered $\vec{k}_S$ waves whose absolute values are equal to

$$\left|\vec{k}_0\right| = \frac{2\pi}{\lambda_0} = \frac{\omega_0}{c} \quad \text{and} \quad \left|\vec{k}_S\right| = \frac{2\pi}{\lambda_s} = \frac{\omega_s}{c} \text{ , and the directions coincide with the}$$

directions of propagation of the incident wave and the examination direction.

The amplitude of the radiation scattered by an electron at distance r from it is

$$E_s = \frac{r_0}{r} E_0 \sin \varphi, \tag{4.26}$$

where $r_0 = \dfrac{e^2}{mc^2} = 2.8 \cdot 10^{-13}$ cm is the classic electron radius, φ is the

angle between the vector of the strength of the incident wave $\vec{E}_0$ and

the wave vector of scattered radiation $\vec{k}_s$.

The effective scattering cross section, representing the ratio of the light flux, scattered in the direction φ within the limits of the unit solid angle, to the intensity of incident radiation is

$$\sigma_e = \frac{E_s^2 r^2}{E_0^2} = r_0^2 \sin^2 \varphi. \tag{4.27}$$

Integrating (4.27) in all directions gives the total Thomson section

$$\sigma_{Th} = \frac{8\pi}{3} r_0^2 = 6.65 \cdot 10^{-25} \text{ cm}^2. \tag{4.28}$$

If the velocity of the electron on which scattering takes place is not equal to zero, in accordance with the Doppler effect the frequency of scattered radiation is displaced in relation to the incident radiation frequency by the value

$$\omega = \omega_s - \omega_0 = \vec{v} \cdot \left(\vec{k}_s - \vec{k}_0 \right) = \vec{v} \cdot \vec{k}, \tag{4.29}$$

where $\vec{v}$ is the vector of the electron velocity. Thus, the frequency shift in scattering is proportional to the projection of the velocity of the electron on the difference of the wave vectors $\vec{k}_s$ and $\vec{k}_0$ and not on the examination direction as in the case in studying the radiation of a moving source, for example, an atom.

Since $\left| \vec{k}_s \right| \approx \left| \vec{k}_0 \right|$, then

$$\left| \vec{k} \right| \approx 2 \frac{\omega_0}{c} \sin \frac{\theta}{2} \approx \frac{4\pi}{\lambda} \sin \frac{\theta}{2} \tag{4.30}$$

$$\omega \approx 2 \frac{\omega_0 v_k}{c} \sin \frac{\theta}{2}, \tag{4.31}$$

where θ is the angle between the direction of propagation of the incident wave and the examination direction, $v_k = v \cos \varphi$ is the projection of the electron velocity on the vector $\vec{k}$.

Scattering of radiation by plasma

The complex amplitude of the radiation scattered by plasma is determined by summing up the vectors of the complex amplitudes scattered by the individual electrons.

Thin layers within which all particles are scattered in the same phase are defined in the plasma. (In the case in which both the incident and scattered waves are flat, these layers are parallel to the bisectrix of the angle θ between directions of propagation of these waves (Fig.4.31).) At distances l, $2l$, $3l$ and so on from the first layer there are layers giving scattered waves whose phases differ from the phase of the waves scattered by the first layer by 2π, 4π, 6π, etc. From geometrical considerations

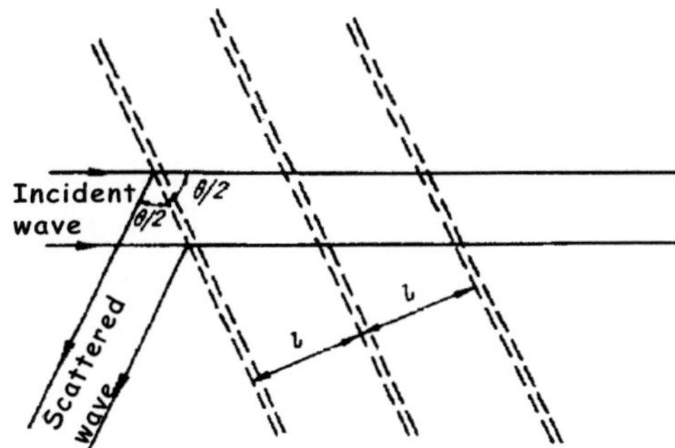

Fig.4.31 Scattering of a light wave by plasma.

$$l = \frac{\lambda}{2\sin \theta / 2} = \frac{2\pi}{|\vec{k}|}.$$

(4.32)

In the gap between these layers there are layers of the same thickness scattered in the antiphase in relation to the first layers. Consequently, if the number of electrons of each layer were the strictly the same, the scattered wave would be completely quenched. In reality, because of fluctuations of the electron density in the plasma, the complete quenching of the waves, scattered by the individual layers, does not take place.

The amplitude of the waves scattered by the k-th layer is $A_k = E_s n_k e^{ik}$, where $n_k = \bar{n}_e + \delta n_k$ is the number of the electrons in the k-th layer differing from the mean value $\bar{n}_e$ by the value n_k, and φ_k is the phase of the wave scattered by the given layer. The total amplitude of radiation, scattered by all layers, is

$$A = \sum_k A_k = E_s \sum_k \left(\bar{n}_e + \delta n_k \right) e^{i\varphi_k} = E_s \sum_k \delta n_k e^{i\varphi_k},$$

(4.33)

$\sum_k \bar{n} e^{i\varphi_k} = 0$ in accordance with the previous considerations.

The mean value of the intensity of the radiation scattered in the given direction is

$$I_s = \frac{c}{8\pi} \overline{AA}^* = \frac{c}{8\pi} E_s^2 \sum_{k,j} \overline{\delta n_k \delta n_j} e^{i(\varphi_k - \varphi)}.$$

(4.34)

Thus, the intensity of the radiation scattered by the plasma is directly linked with the correlations of the fluctuations of the electron density in plasma. The case of Thomson scattering (or scattering on free electrons) corresponds to the complete absence of correlation. The so-called collective scattering takes place in the presence of correlations.

Thomson scattering of radiation by plasma

In low-density plasma where the energy of charge interactions is negligible in comparison with the thermal energy of the particles, it may be assumed that individual electrons are separated in space randomly and the fluctuations of the number of particles within the limits of the elementary layers are independent. Statistical averaging leads in this case to the following expression for the intensity of scattered radiation:

$$I_s = \frac{c}{8\pi} E_s^2 \sum_k \overline{\delta n_k^2},\tag{4.35}$$

and since for the random distribution $\overline{\delta n_k^2} = n_k$, then

$$I_s = \frac{c}{8\pi} E_s^2 \sum_k n_k = \frac{c}{8\pi} E_s^2 n_e = \frac{c}{8\pi} E_s^2 N_e \Delta V.\tag{4.36}$$

Here $n_e = \sum_k n_k$ is the total number of electrons in the scattering volume V; $N_e = \dfrac{n_e}{\Delta V}$ is the electron concentration in plasma. It is evident that this equation can also be obtained by direct summation of the intensities of the radiation scattered by all electrons within limits of the volume V.

Using equation (4.27), equation (4.36) can be easily transformed to the form

$$I_s = \frac{c}{8\pi} \sigma_e \frac{E_0^2}{r^2} N_e \Delta V = \sigma_e \frac{I_0}{r^2} N_e \Delta V,\tag{4.37}$$

where $I_0 = \dfrac{cE_0^2}{8\pi} = \dfrac{W_0}{\Delta S}$ is the intensity of incident radiation, W_0 and S is the total power and area of the cross section of the light beam falling onto the plasma.

The total light flux of the scattered radiation, recorded by the detector within the limits of the solid angle $\Delta\Omega$, is determined by the expression

$$\Phi_s = I_s r^2 \Delta\Omega = \sigma_e N_e \Delta V \Delta\Omega I_0. \tag{4.38}$$

Since each electron in this case scatters radiation independently, the scattered radiation spectrum forms as a result of superimposition of the radiations of the individual electrons. The shift of the frequency of radiation scattered by an individual electron is related with its velocity by equation (4.29). If the velocity distribution of electrons is described by the function $f(v_x)$, the probability of the electron having the component of the velocity v_x in the range from $v_x + dv_x$ is

$$dw = f(v_x)dv_x, \tag{4.39}$$

The scattered radiation spectrum $S(\omega)$ is determined substituting the value of v_x from (4.31) into $f(v_x)$ (in this case it is assumed that the x axis coincides with the direction of the vector $\vec{k}$)

$$S(\omega_s - \omega_0) = \frac{S_0}{f_0} f\left(\frac{\omega_s - \omega_0}{|\vec{k}|} \right). \tag{4.40}$$

Thus, examining the spectrum of the radiation scattered by the plasma it is possible to determine directly the velocity distribution function of the electrons. For Maxwell's distribution

$$f(v_x) = \sqrt{\frac{m}{2kT_e\pi}}\, e^{-\frac{mv_x^2}{2kT_e}} \tag{4.41}$$

and the contour of the line of scattered radiation has the form of a Gauss curve

$$S(\omega_s - \omega_0) = S_0 \exp\left[-\frac{m}{2kT_e} \frac{(\omega_s - \omega_0)^2}{|\vec{k}|^2} \right] = S_0 \exp\left[-\left(\frac{\omega_s - \omega_0}{\omega_e} \right)^2 \right], \tag{4.42}$$

where $\omega_e \equiv \sqrt{\dfrac{2lT_e}{m}|\vec{k}|^2}$. The width of the scattering line, measured at half height, is

$$\Delta\omega = 2\omega_e(\ln 2)^{1/2} = 2\left[2(\ln 2)\frac{kT_e|\vec{k}|^2}{m}\right]^{1/2}. \qquad (4.43)$$

Transferring to the wavelength and substituting $|\vec{k}|$ from (4.30) gives

$$\Delta\lambda = 4\frac{\lambda_0}{c}\left(\sin\frac{\theta}{2}\right)\left(\frac{2kT_e\ln 2}{m}\right)^{1/2}. \qquad (4.44)$$

i.e. the electronic temperature of plasma can be determined from the width of the scattering line.

The electron concentration can be determined from the absolute intensity of scattered radiation (integral over the spectrum) in accordance with equation (4.38).

We shall now determine the plasma parameters at which the acts of radiation scattering by individual electrons can be regarded as independent.

As a result of the interaction of the charged particles in plasma the disruption of quasi-neutrality of plasma takes place only within the limits of the volume whose linear dimensions are considerably smaller than the so-called Debye radius

$$D = \left(\frac{kT_e}{4\pi N_e e^2}\right)^{1/2}. \qquad (4.45)$$

In the volume with the linear dimensions considerably smaller than D the fluctuations of the number of electrons can be regarded as independent. Thus, the criterion of applicability of the Thomson scattering model is the relationship between the Debye radius and the distance l (see equation (4.32)) between the plasma layers for which the scattered radiation phase changes by 2π. According to Salpet[102] such a criterion is the parameter

$$\alpha = \frac{1}{\left|\vec{k}\right|D} = \frac{1}{2\pi D} = \frac{\lambda}{4\pi \sin \dfrac{\theta}{2}} \sqrt{\frac{4\pi N_e e^2}{kT_e}}. \tag{4.46}$$

The case $\alpha \ll 1$ correspond to scattering on three electrons, and the case $\alpha \gg 1$ to collective scattering.

Collective scattering

At an arbitrary value of parameter α the intensity and spectrum of the radiation scattered by the plasma are determined by the time and spatial spectra of the fluctuations of the electronic density. The spectral flow of the radiation scattered by the plasma can be calculated from the equation

$$\frac{d\Phi_S}{d\omega} = \sigma_e s\left(\vec{k}_s - \vec{k}_0, \omega_s - \omega_0\right)\Delta V \Delta \Omega I_0, \tag{4.47}$$

where $s\left(\vec{k}, \omega\right)$ is the spatial–time Fourier transforma from the function of the paired correlations of the electronic density.[100] Integrating (4.47) with respect to frequencies gives

$$\Phi_s = S\left(\vec{k}_s - \vec{k}_0\right)\sigma_e N_e \Delta V \Delta \Omega I_0, \tag{4.48}$$

where $S\left(\vec{k}_s - \vec{k}_0\right) = \dfrac{1}{N_e} \displaystyle\int_{-\infty}^{\infty} s\left(\vec{k}, \omega\right)d\omega$ characterises the difference of the

collective scattering section from the scattering section on free electrons.

When $l \gg D$ ($\alpha \gg 1$) the thickness of the layers with coherent radiation is greater than the dimension of the regions within which the quasineutrality can be disrupted. The motion of each electron is strongly affected by the electrostatic field of ions and other electrons. The motion of the individual electrons can no longer be regarded as independent and the fluctuations of electronic density are random. When the electronic and ion temperatures are equal ($T_e \approx T_i$), the velocity of the electrons is considerably higher than that of the ions. Consequently, any displacement of the ion is accompanied by the collective displacement of the entire electron cloud screening the ion charge. On the other hand, the ions do not manage to follow the movement of the electrons and this leads to the disruption of quasineutrality for the period of the order of $\tau_e \approx \dfrac{D}{v_e} = \left(\dfrac{m}{4\pi N_e e^2}\right)^{1/2}$. The value reciprocal to this time is

$$\omega_p = \frac{1}{\tau_e} = \left(\frac{4\pi N_e e^2}{m}\right)^{1/2}, \tag{4.49}$$

and represents the so-called electronic plasma frequency.

According to these considerations, in the centre of the fluctuations of the electronic density we can separate the high-frequency (electronic) component determined by the movement of the free electrons, and the low-frequency (ion) component characterising the collective displacement of the electron cloud associated with the movement of the ions.

Consequently, as $s(\vec{k}, \omega)$ can be written with the form

$$s\!\left(\vec{k}, \omega\right) = s_e\!\left(\vec{k}, \omega\right) + s_i\!\left(\vec{k}, \omega\right). \tag{4.50}$$

A similar equality can also be written for the quantities integral over the spectrum

$$S\!\left(\vec{k}\right) = S_e\!\left(\vec{k}\right) + S_i\!\left(\vec{k}\right). \tag{4.51}$$

The spectrum of electronic density fluctuations was calculated by many authors.[102-106] Here we present (Fig.4.32a) the results of calculation carried out by Salpeter for the case of Maxwell's velocity distribution of electrons assuming that $N_e D \gg 1$, i.e. in a sphere whose radius is equal to Debye radius there are a large number of electrons. The value $\alpha = 0$ is related with the Gauss contour whose width is linked with T_e by the relationships (4.43) and (4.44). As the parameter α is increased the electronic component of the spectrum breaks up into two components whose distance from the probing radiation line is $\pm\Delta\omega$ and which satisfies the equation

$$\Delta\omega^2 = \omega_p^2 + \frac{3kT_e}{m}\left|\vec{k}\right|^2 = \omega_p^2\left(1 + \frac{3}{\alpha^2}\right). \tag{4.52}$$

At $\alpha \gg 1\,\Delta\omega$ tends to ω_p.

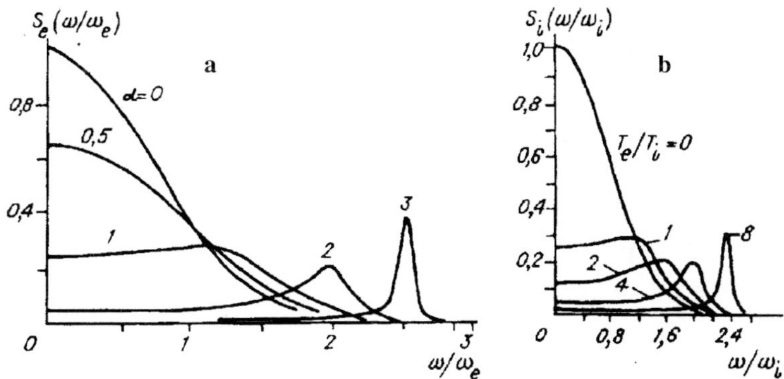

Fig.4.32 Scattering spectra at different plasma parameters. a) electronic component of the spectrum at different values of parameter α,[102] b) ionic component of the spectrum at $\alpha > 1$ and different ratios of the electronic and ionic temperatures;[103] the abscissa gives the distances from the laser line $\omega = \omega_s - \omega_0$ related to $\omega_e = (2kT_e/m)^{1/2}$ and $\omega_i = (2kT_i/M)^{1/2}$, where M is the mass of the ion.

The integral (with respect to frequency) contribution over the electronic component to the intensity of scattered radiation is determined by the equation

$$S_e\left(\vec{k}\right) = \frac{1}{1+\alpha^2} \tag{4.53}$$

and rapidly decreases with increasing α, whereas the contribution of the ion component to the scattering radiation intensity increases with increasing α:

$$S_i\left(\vec{k}\right) = \frac{z\alpha^4}{\left(1+\alpha^2\right)\left(1+\alpha^2 x \dfrac{T_e}{T_i}\alpha^2\right)}. \tag{4.54}$$

The width of the ion component of the scattering spectrum is determined by the velocity of the ions and is considerably smaller (approximately $\sqrt{M/m}$ times) than the width of the electronic component. The shape of the ion line depends on the ratio of the electron and ion temperatures and on the ion charge. At $T_e \ll T_i$ it consists of two components shifted in relation to ω_0 by the frequency of ion acoustic vibrations, and at $T_e \ll T_i$ the ion line has the Gaussian contour (Fig.4.32b).

Thus, at the values $\alpha \gg 1$ the scattered radiation spectrum consists of a narrow central peak whose width is determined by the thermal velocities of the ions, and two weak electronic components displaced in relation to the centre by the value of the electronic plasma frequency ω_p. Consequently, at $\alpha \gg 1$ the width of the ion peak is used to calculate the ion temperature and the electron concentration is calculated from its intensity integral with respect to the spectrum. In addition, the electron concentration can be found from the distance from the central maximum to the side maxima, in accordance with equation (4.49).

When $\alpha \approx 1$, the plasma parameters are estimated by comparing the observed contour of the scattering line with the calculated theoretical contours corresponding to different temperatures and electron concentrations in the plasma.

As indicated by equation (4.46), the value α depends not only on the plasma parameters but also the examination angle and the wavelength of scattered radiation. For example, for the same plasma in scattering under large angles where $\sin(\theta/2) \cong 1$, the value of α can be lower than 1 and the scattering close to Thomson scattering. In scattering under very small angles where $\sin(\theta/2) \ll 1$, there may be cases in which the value of parameter α exceeds the unity and collective effects appear in the scattered radiation spectrum. The appearance of the collective effect is also supported by the increase of the wavelength. The appearance of the collective effect is also supported by the increase of wavelength. For example, for the same examination angle, the transition from the radiation of a ruby laser to that of a CO_2 laser increases the value of α up to 15 times.

Experimental procedure and apparatus for examining plasma-scattered radiation

Experimental equipment for examining the plasma by the scattering method should ensure the introduction of laser radiation into a plasma chamber, collection of radiation, scattered within the limits of the given angle, and recording the spectrum of scattered radiation.

Figure 4.43 shows the circuits of two typical systems for examining scattering. In the first system, (Fig.4.33a) scattered radiation is observed under a high angle θ to the direction of incident radiation (in the majority of cases $\theta = 90°$). Figure 4.33b shows equipment for examining radiation under small angles. Both systems consist of lasers 1, focusing optics 2, plasma chamber 3, optical elements 4, collecting the scattered radiation and directing on to the spectral device 5 whose output contains the radiation detector 6, and the device for recording the scattering signal 7.

As already mentioned, the main problem in measuring the intensity of plasma-scattered radiation is the weak used signal. The light flux, leaving the spectral device (monochromator), is determined by the relationship[110] within the limits of the spectral range

$$\Phi_{reg} = b_\lambda L \, (\delta\lambda)^2, \qquad (4.55)$$

where b_λ is the spectral brightness of the light source, $L = \tau\beta S\lambda_\varphi$ is the light force of the monochromator, τ is the transmission factor which takes into account the losses of light in the device, β is the angular height of the input slit, S is the effective area of the dispersing element, D_φ is angular dispersion.

In turn, the spectral brightness is represented by the light flux emitted by the unit visible surface of the source in the unit solid angle and the unit spectral range

$$b_\lambda = \frac{D_S}{l d \Delta\Omega\Delta\lambda}, \qquad (4.56)$$

where l is the length of the scattering volume, d is its diameter. Substituting Φ_S for (4.38) into (4.56) and taking into account that $\Delta V = l\Delta S$, and $I_0 = W_0/\Delta S$, and gives

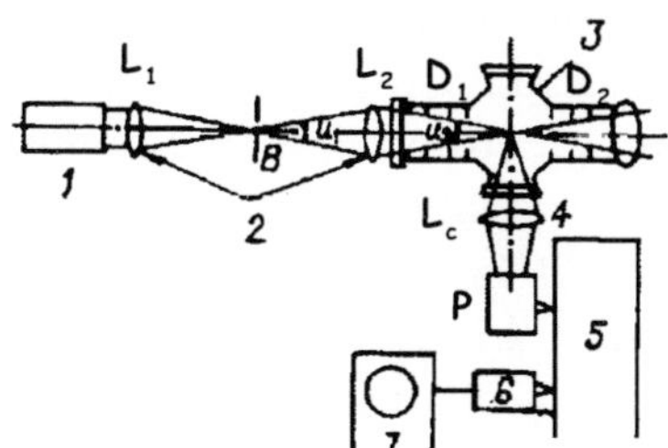

DFS-12 oscilloscope

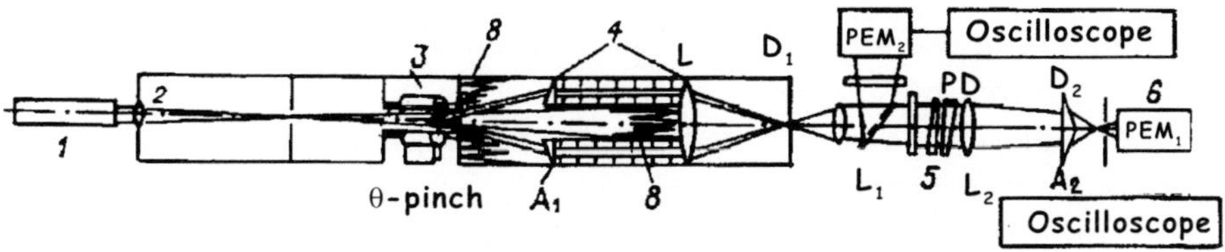

Fig.4.33 Typical diagrams of equipment for plasma diagnostics by the scattering method. a) equipment for examining scattering under an angle of $\theta = 90°$,[107,108] b) the same, under a small angle θ (in the forward direction).[109]

$$b_\lambda = \frac{\sigma_e N_e W_0}{d\Delta\lambda}.$$

(4.57)

To construct the contour of the scattered radiation line, the width of the separated spectral range $\delta\lambda$ must be p times smaller than the width of the spectrum of scattered radiation $\Delta\lambda$ (p is the number of points used to construct the scattered radiation spectrum), i.e. $p = \Delta\lambda/\delta\lambda$. Consequently, substituting b from (4.57) into (4.55) we obtain

$$\Phi_{reg} = \frac{\sigma_e N_e W_0 L\Delta\lambda}{p^2 d}.$$

(4.58)

The useful signal recorded by the photodetector can be determined as the number of electrons n_e knocked out by the recorded radiation from its photocathode

$$n_e = \frac{\eta\Phi_{reg}\Delta t}{h\nu} = \frac{\eta\sigma_e N_e L\lambda\Delta\lambda E_0}{hcp^2 d}.$$

(4.59)

Here η is the quantum yield of the photocathode, $h\nu = h\dfrac{c}{\lambda}$ is the quantum energy, Δt is the laser pulse time, $E_0 = W_0\Delta t$ is the total laser pulse energy.

In accordance with (4.59), the strength of the useful signal is determined directly by the laser pulse energy, the light force of the spectral device and the sensitivity of the radiation detector. Therefore, general requirements of the elements of equipment for examining scattering are reduced to using lasers that emit a sufficiently large amount of energy in the pulse and to the use of apparatus with a high light force and the most sensitive radiation detectors (with a high quantum yield).

At the same time, the useful scattering signal is recorded on the background of various types of interference associated with the intrinsic glow of plasma and also the parasitic scattering of laser radiation at optical elements and components of plasma equipment. Therefore, to increase the signal/noise ratio, it is necessary, on the one hand, to increase the laser radiation power and, on the other hand, take measures to reduce the level of parasitic scattering.

The requirements of apparatus used for examining scattering were examined in detail in Ref.2. Here, we shall confine ourselves only to more detailed examination of two typical experimental systems described previously.

Equipment shown in Fig.4.33a was used for examining the scattering of the light by the plasma of dc arcs in a magnetic field.[107,108] The radiation source was a ruby laser in the free generation regime (E_0 = 25 J). Laser radiation was focused in the centre of the plasma column using a system consisting of two lenses L_1 and L_2, the first of which focuses radiation on diaphragm B and the second constructs the image of the diaphragm in the centre of the chamber. The diameter d_1 of diaphragm B is determined from the condition $d_1 = \xi f_1$, where ξ is the divergence of laser radiation, f_1 is the focusing distance of length L_1. A large part of laser radiation passes through the diaphragm and, at the same time, the cone of the beams directed onto the plasma is restricted. This is very important for attempts to reduce the intensity of parasitically scattered radiation.

Two nozzles with glass windows are used for input and output of radiation from the chamber. The output is positioned under the Brewster angle to the incident plane-polarised laser beam thus preventing the propagation of the radiation reflected from it into the chamber. Inside the nozzles there is a system of diaphragms D_1 and D_2 used to reduce the radiation scattered by the windows and walls of the chambers.

The radiation scattered by the plasma under angle $\theta = 90°$ is collected by the condenser lens L_c from the volume 7 mm long and 0.6 mm in diameter in the solid angle $\Delta\Omega = 1/32$ mean and after passing through the prism P, which rotates the image by 90°, is directed onto the slit of the DSS-12 double monochromator whose parameters, determining its light force, are $\tau = 0.1$, $S = 180$ cm^2, $D = 2.5 \cdot 10^4$ cm^{-1}, $\beta = 0.05$.

Radiation is recorded using the FEU-38 photodetector and an oscilloscope. The signal corresponding to scattering in plasma is determined from equation

$$I_S = I - I_{pl} - I_{par} \qquad (4.60)$$

where I is the signal recorded in the presence of the laser pulse and in the presence of plasma, I_{pl} is the signal determined by the natural glow of plasma in the absence of the laser pulse, I_{par} is the parasitic signal observed in the presence of the laser pulse and in the absence of plasma. The contour of the scattering line is recorded from individual points. Regardless of the fact that the parasitic signal I_{par} at the wavelength of laser radiation exceeded 30 times the useful signal I_S, the 'wings' of the scattering line were recorded with sufficient reliability.

Absolute calibration of the entire measuring system was carried out on the basis of Rayleigh scattering in helium with which the plasma chamber was filled. This made it possible to eliminate errors associated with measuring quantities such as the absolute values of the laser energy, the light force of the spectral device, the sensitivity of photodetector, etc.

The measured width of the scattered radiation contour $\Delta\lambda = 43$ Å was used to determine the electron temperature $T_e = 1.8$ eV, and the electron concentration $N_e = 2.5\cdot10^{13}$ cm^{-3} was determined from the absolute intensity of scattered radiation. The parameter calculated for these values of T_e and N_e was $\alpha = 0.04$, i.e. scattering in this case took place in free electrons.

The equipment shown in Fig.4.33b is typical for examination of scattering and small angles.[109] Scattering took place in the plasma of θ-pinch in the 'forward' direction. The direct laser beam, passed through the plasma, was absorbed by special traps 8. The radiation, scattered under small angles, was outputted using a conical length (axicone) A$_1$ with a orifice in the central pass through which the direct laser beam passed. The axicone deflects all beams falling on it under the same angle $\varphi \approx (n - 1)\beta$ (β is the angle at the base of the cone). Thus, all the beams, leaving the scattering volume under the angle $\theta = \varphi$ to the axis, propagate parallel to the axis of the system after passing through the axicone and are separated by the diaphragm D_1 positioned in the focus of the lengths L. Moving the diaphragm along the optical axis it is possible to observe the radiation scattered by the plasma under different angles.

The spectrum of scattered radiation was examined by a Fabry-Perot (FP) interferometer and the second axicone A$_2$ directing the beams, corresponding to one of the interference rings, to diaphragm D$_2$ positioned in the focus of the lens L$_2$. The diameter of the diaphragm determined in this case the width of the separated spectral range and the spectrum was scanned by the moving axicone along the optical axis.

In initial studies in this area[107-109,111,112] the scattering signal was recorded using a photodetector and an oscilloscope and the spectrum was obtained from the individual lens. Subsequently, to record the entire spectrum of scattered radiation in a single pulse it was necessary to develop different multichannel systems[103,104] and also electron–optical converters (EOC). For example, in Ref.115,116 EOC was used to examine the spectrum of collective scattering in examining a laser spark. The region of the spectrum 10 mm long with the centre on the line of the ruby laser and separated by the spectral device was projected on to the cathode of the EOC. To avoid superimposition of the natural glow of the plasma on the scattered radiation, the time sweep of the spectrum was carried out. Figure 4.34 shows microphotographs of the spectrum photographed from the screen of the EOC and corresponding to moments of time 3.6

(*a*), 10.4 (*b*) and 21 μs (*c*) after formation on the laser spark. The spectrum shows clearly the central ion maximum and satellites at the plasma frequency.

Further advances in experimental techniques used in the laser scattering method were associated with using this method of diagnostics of the high-temperature plasma in large toroidal systems of the Tokamak type.[6] To increase the information content of the method, investigators used multichannel recording systems based on polychromators of different types. The focal plane of a spectral device is separated into a number of spectral regions using a package of light guides so that the entire spectrum of scattered radiation can be recorded in a single plasma pulse.

The radial distribution of the plasma parameters is determined using a set of packets of light guides where each light guide shows an image of a specific region of the slit of the spectral device in the direction of height, or a set of modules (polychromators) of the same type is used where each records the radiation spectrum scattered from some point of the plasma cord.

Simultaneous recording of scattering spectra in a large number of spatial points in TFTR Tokamak equipment[117] was carried out, using as a detector a two-coordinate television system matched with a polychromator. Measurements were taken in 76 points in space and 25 channels with respect to spectrum.

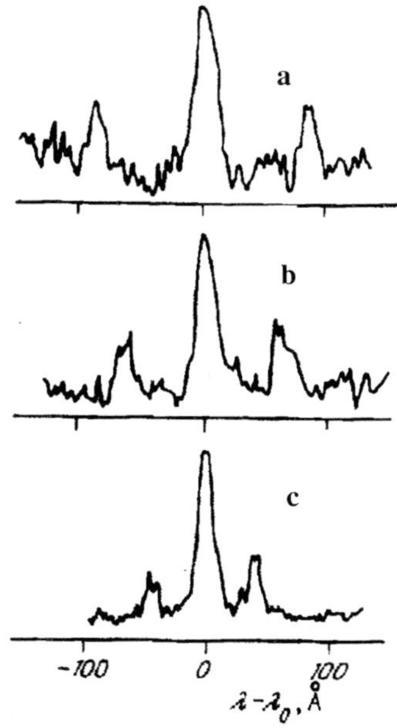

Fig.4.34 Microphotographs of spectra of collective scattering photographs from the screen of the EOC.[116]

The information possibilities of the scattering method are indicated most convincingly by the diagnostic complex used in ASDEX Tokamak equipment.[118,119] It includes a laser ($\lambda = 1.06$ µm) operating in the frequency regime (the energy in each pulse is 0.8 J), silicon avalanche detectors and a set of 16 modules of three-channel polychromators. The system makes it possible to measure, on the basis of the scattering spectrum, to measure the temperature and electron concentration simultaneously at 16 points on radius. The frequency laser generation regime makes it possible to find the radial distributions T_e (r) and N_e (r) every 16 µs. Up to 400 temperature and electronic density profiles can be recorded in a single discharge.

Determination of plasma parameters from scattered radiation spectra
Since it is not possible to review the large number of investigations in which the diagnostics of various plasma objects was carried out using the scattering method, we shall demonstrate here the possibilities of evaluating various plasma parameters on the basis of scattered radiation spectra using several typical studies.

As already mentioned, the interpretation of the scattering spectra depends on the value of parameter α. In scattering of radiation on low-density plasma ($\Delta\alpha \ll 1$) with the Maxwell distribution of electron velocity, the scattering line has the Gaussian contour (see (4.42)), and the width of this contour can be used to determine the electronic temperature of plasma T_e using equation (4.44). As an example, Fig.4.35 shows the spectra of scattered radiation recorded in Tuman-2 toroidal equipment.[120] The Gaussian contours were constructed on the basis of experimental points using a computer. The spectra shown in Fig.4.35 were obtained prior to (a) and (b) adiabatic compression of the plasma by the magnetic field and indicate a large increase of the electron temperature in the compression stage.

If the velocity distribution of the electrons differs from Maxwellian, the contour of the scattering line can be used to draw conclusions on the nature of these differences. In particular, the authors of Ref.121 found an anisotropy in the velocity distribution of the electrons in the direction along and across the magnetic field, and the form of the spectrum examined in Ref.122 indicated the presence in plasma of two electron groups each of which had Maxwellian velocity distribution corresponding to two different temperatures.

In the presence of directional movement of the electrons the scattering line is shifted in relation to the wavelength of laser radiation. For example, in scattering on an electron beam with an energy of 2 kV (see Ref.123), the Doppler shift of the scattering line examination under an angle of 65° in relation to the direction of electron motion reached

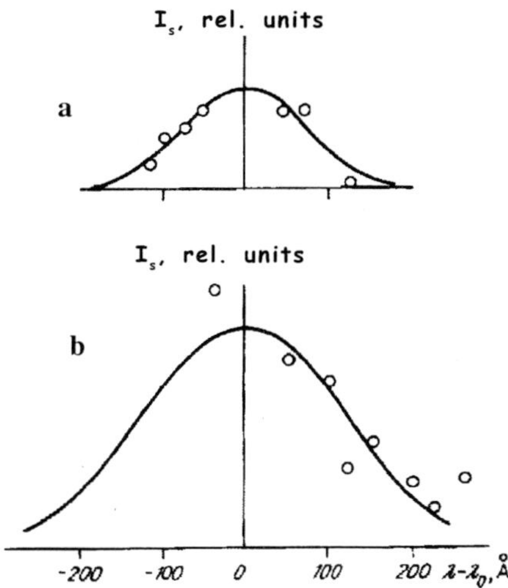

Fig.4.35 Spectra of scattered radiation obtained in equipment Tuman-2.[120] Points are experimental data, the curves are the calculated Gauss contours.

259 Å. The Doppler shift of the scattering line was also observed in investigations of the laser spark[124-126] so that it was possible to measure the travel velocity of the plasma boundary in the direction against the laser radiation which equalled around 100 km/s.

In accordance with (4.38), measuring the scattered light flux Φ_s, we can determine the electron concentration N_e in plasma. However, the absolute energy measurements are always associated with large experimental errors. It is far easier and more accurate to calibrate apparatus on the basis of Rayleigh scattering. For this purpose, the plasma chamber is filled with a gas with the known scattering section and the Rayleigh scattering signal is measured at the same intensity of the laser pulse, the geometry of optical equipment and the sensitivity of the measuring system. Consequently, all absolute measurements are eliminated and only the ratio of two signals – scattering on plasma and Rayleigh scattering – is measured:

$$\frac{\Phi_S}{\Phi_R} = \frac{\sigma_e}{\sigma_\varphi}\frac{N_e}{N_a} = \frac{\sigma_{Th}N_e}{\sigma_R N_a}.$$

(4.61)

Here σ_φ is the Rayleigh scattering section in the direction φ in the unit solid angle calculated for a single atom in the unit volume, and σ_R is the Rayleigh scattering section integral with respect to angles. This section can be calculated from Rayleigh's equation[127]

$$\sigma_R = \frac{8\pi}{3}\left(\frac{\omega}{c}\right)^4 \alpha_a^2,$$

(4.62)

where α_a is the polarisability of the atoms.

Equation (4.61) can be used to determine N_e when $\alpha \ll 1$. In the opposite case, σ_{Th} in equation (4.61) should be replaced by $\sigma_{Th}S(\vec{k})$ where $S(\vec{k}) = S_e(\vec{k}) + S_e(\vec{k})$. At $\alpha \gg 1$, the main contribution to the intensity of scattered radiation, integral with respect to spectrum, is provided by the central ion component and, in accordance with (4.54), at $z = 1$ and $T_e = T_i$ $S_i \cong 1/2$, i.e. the scattering section is half the Thomson section. The intensity of the ion peak was used to examine the spatial distribution of the spectrum concentration in a laser spark plasma in air.[112] Figure 4.36 shows the distribution of N_e in the laser spark obtained in Ref.112.

When $\alpha \approx 1$, the main contribution to the intensity of the scattering spectrum is provided by the electronic component $S_e(\vec{k})$. The plasma parameters are determined comparing the contour of the scattering line with the calculated contours corresponding to different α. The observed and theoretical contours can be matched only by selecting both the accurate value of α and a specific scale on the frequency axis. Since the theoretical curves (Fig.4.32b) are formed by plotting ω/ω_e on the ab-

scissa, then we determine not only α but also $\omega_e = |\vec{k}|\left(\frac{2kT_e}{m}\right)^{1/2}$ and this

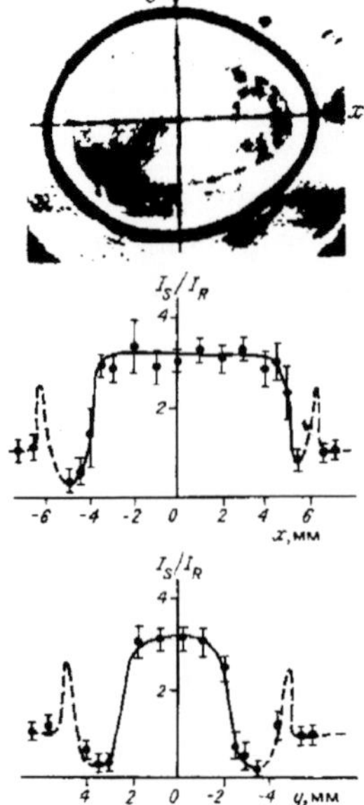

Fig.4.36 Spatial distribution of the concentration of electrons in the plasma of the laser spark in the stage $t = 3$ 5s.[112]

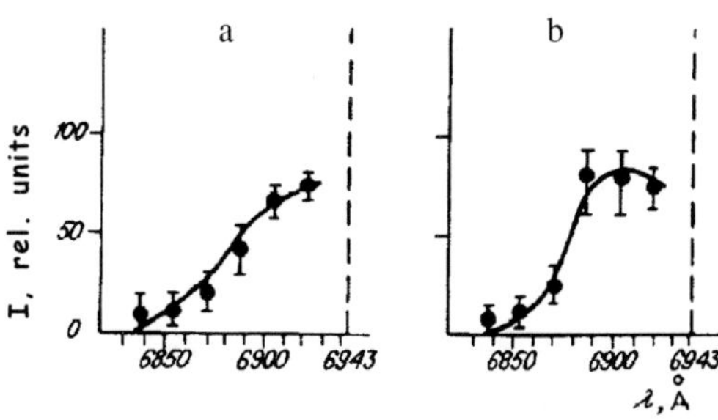

Fig.4.37 Spectra of radiation scattered by the plasma of θ-pinch.[128] Points are the experimental values, solid lines are the experimental contours of the scattering line: a) T_e = 90 000 K, N_e = 2·10^{16} cm^{-3}, α = 0.53; b) T_e = 55 000 K, N_e = 4·10^{16} cm^3, α = 0.97; broken lines are the positions of the laser lines.

is used to determine T_e. Substituting T_e into (4.46), N_e is calculated from the known value of α.

As an example, Fig.4.37 shows the scattering spectra on the plasma of θ-pinch obtained in Ref.128.

When α >> 1, the scattering spectrum consists of a narrow central ion peak and weak satellites whose distance from the ion peak is approximately equal to the distance of the electronic plasma frequency ω_p so that equation (4.49) can be used to calculate N_e. If the width of the range defined by the spectral device is considerably greater than the width of the ion and electron peaks, the ration of their intensities is proportional to $2S_i(\vec{k})/S_e(\vec{k})$ and in accordance with (4.53) and (4.54) for z = 1 and $T_e = T_i$ is equal to

$$\frac{2S_i(\vec{k})}{S_e(\vec{k})} = \frac{2\alpha^4}{1+2\alpha^2},$$

(4.63)

so that the value of α can be determined. Consequently, substituting the values of α and N_e into (4.46) we can determine T_e.

Of special interest for experiments is the plasma for which α > 1 in examining scattering at large angles, and in examination at small angles α > 1. An example of such a plasma is the θ-pinch characterised by the values $N_e \approx 10^{15} \div 10^{16}$ cm^{-3}. Figure 4.38 shows the scattering spectra on the plasma of θ-pinch obtained in the postglow stage in examination under angles θ = 13.5° (a) and 90° (b).[129] The values α = 3, N_e = 2.4·10^{15} cm^{-3} and T_e = 1.1 eV were determined from the ratios of the intensities of the central and lateral maxima and from the distance between them (Fig.38a). Similar values, N_e = 2.4·10^{15} cm^{-3} and T_e =1.0 eV, were determined from the intensity integral with respect to the spectrum and from the width of the contour of the scattering line

(Fig.4.38b). The value of α in this case was ≈ 0.5.

The information on the plasma parameter is also obtained by examining the shape of the ion line and the plasma satellite.[130] In particular, the width of the ion peak is used to determine the ion temperature T_i and the ratio T_e/T_i is determined from the shape of this peak (see Fig.4.32b).

It is also important to note the possibility of measuring the strength of the magnetic field in the plasma from the scattering spectrum. In the presence of a magnetic field, the electrons move in the plasma along a spiral around force lines with the rotation frequency $\omega_B = \dfrac{eB}{mc}$ (electronic cyclotron frequency). Consequently, the contour of the scattering line is modulated by the frequency ω_B. The modulation depth is maximum if the angle γ between vector $\vec{k}$ and the vector of the strength of magnetic field $\vec{B}$ is equal to 90°. If γ differs from 90°, the modulation depth rapidly decreases and this restricts the value of the solid angle $\Delta\Omega$ within which the radiation scattered by the plasma is collected. Regardless of considerable experimental difficulties, in a number of studies[131,132] it was possible to record the modulation of the scattering spectrum with respect to frequency ω_B (Fig.4.39). The strength of the magnetic field (125 kHz) was determined from the distance between the adjacent components (≈ 5.7 Å).

Plasma diagnostics based on radiation scattering on atoms

As mentioned at the beginning of this section, if the wavelength of laser

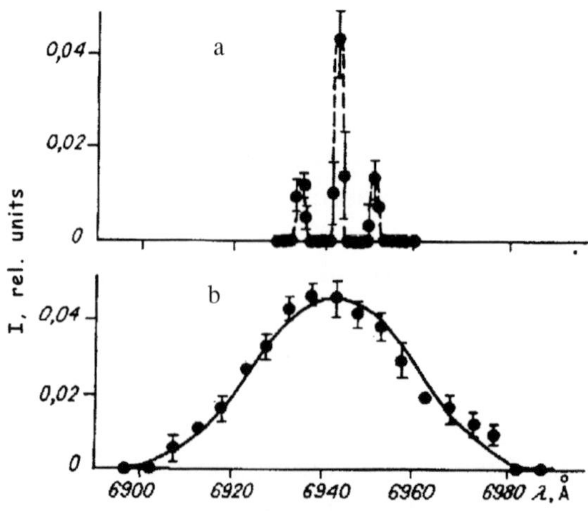

Fig.4.38 Spectra of radiation scattered by θ-pinch plasma in the forward direction (a) and under a right angle to the laser beam (b).[129]

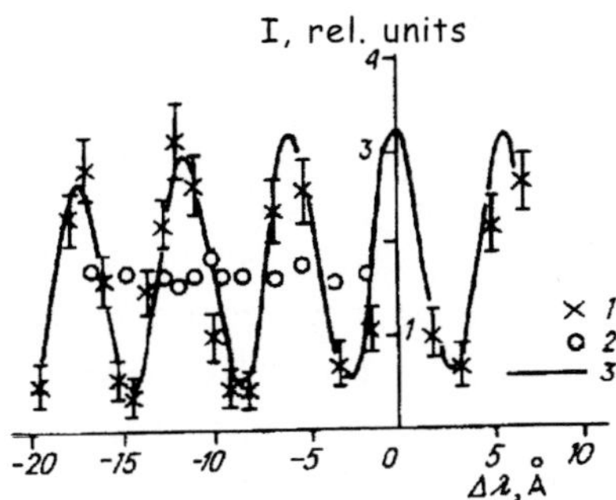

Fig.4.39 Spectra of radiation scattered by the plasma in the magnetic field.[131] $B = 125$ kilogauss (1), 0 (2); 3) theoretical curve.

radiation is at a sufficiently large distance from the absorption lines of the atoms, the main contribution to the scattering signal comes from the electronic component of the plasma. Rayleigh scattering on atoms or molecules is used (because of its small section ($\sigma_R \ll \sigma_{Th}$)) mainly for calibrating the measuring system. To ensure that the ratio of the scattering signals with respect to the electronic component F_s to the Rayleigh scattering signal F_R is of the order of unity (see equation (4.61)), the atom concentration N_a should be several orders of magnitude higher than the measured electron concentration N_e.

However, when approaching the absorption line, the scattering section on the atoms rapidly increases so that this phenomenon can be used for plasma diagnostics. The scattering process consists of two acts: absorption of the photon and subsequent radiation. The scattering section on the atoms σ_a can be determined from equation (4.15) transferring from wavelength to frequencies

$$\sigma_a = \frac{\kappa}{n_a} = r_0 c f_{ki} \frac{(\Delta v / 2)}{(v_{ki} - v)^2 + (\Delta v / 2)^2}, \tag{4.64}$$

where r_0 is the electron radius, f_{ki}, v_{ki} is the force of the oscillator and the frequency corresponding to the transition between the levels k and i, Δv is the width of the absorption line which in the case of transition between the ground and excited levels is determined by the radiation lifetime of the upper level.

When approaching the absorption line, both Rayleigh scattering without frequency change, i.e. at frequency v, and fluorescence at frequency v_{ki}, induced by collisions (Fig.4.40a), can take place simultaneously. The laser scattering lines are narrow and the width is determined by the temperature of the atoms and ions, whereas the lines of induced fluorescence broaden as a result of collisions and their contents provide information on the processes of inelastic collisions of excited atoms. The ratio of the intensities of Rayleigh scattering and induced fluorescence depends on frequency difference $\nu_{ki}-\nu$, the laser radiation intensity and the parameters of the examined plasma, especially its optical thickness.[133]

Resonance fluorescence appears when the frequency of the scattering radiation coincides with the natural frequencies of the atoms. This name was used due to tradition because in the initial studies, carried out prior to invention of the lasers, investigators used resonance radiation corresponding to transition between the ground and first excited levels of the atom. However, in principle, radiation can coincide with any sufficiently intense absorption line, including the one which starts on the excited level. As a result of the absorption of the light quanta with frequency ν_{ik} corresponding to the transition between the levels k and i (Fig.4.40c), part of the atoms is transferred from the lower state k to the upper state i. Spontaneous transitions of the atoms from the level i to k results in radiation with the initial frequency ν_{ik}. If transitions from state i are possible not only to level k but also other underlying levels (n, m, etc.), the fluorescence spectrum also contains frequencies ν_{in} and ν_{im}, etc.

The fluorescence method has greater historical merit in comparison with spectroscopy because it was one of the main methods of examining the distribution of energy levels of the atoms. Recently, as a result of extensive use of tuneable dye lasers, this method has been used on an increasing scale in fluorescence analysis[134] and also plasma diagnostics.[6,135]

The fluorescence intensity is determined by the number of absorbing atoms and its measurement can be used to determine their concentration N_k. The lifetimes of the atoms of a given level are measured examining the time dependence of the fluorescence signal. The form of the fluorescence lines is determined by the same processes as the form

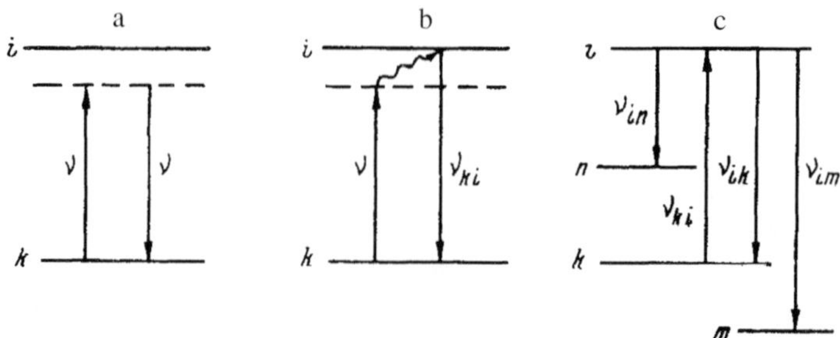

Fig.4.40 Scattering of radiation on atoms. a) Rayleigh scattering, b) induced, c) resonance fluorescence.

of the radiation lines of plasma particles so that the contour of the fluorescence line can be used to determine the electron concentration if the Stark effect in the electric fields of charged particles is the main broadening mechanism, or temperature if the line broadens as a result of the Doppler effect.

An advantage of the resonance fluorescence method in comparison with the conventional emission diagnostic method is its localisation. As in the laser scattering method, the examined radiation is observed from a small plasma volume in which the laser beam and the cone of the beams recorded by the detector intersect and, consequently, provides information on the plasma parameters within the limits of this volume. Strictly speaking, this holds for optically thin plasma. In examination by the fluorescence method of the plasma with a large optical thickness it is important to take into account the absorption of resonance radiation in the vicinity of the scattering region where the effective volume of this region may increase and the glow brightness may change.

The strength of the recorded fluorescence signal is

$$\Phi_{fl} = j_{ik}\Delta V \Delta \Omega, \tag{4.65}$$

where j_{ik} is the light flux emitted by the unit volume in the unit solid angle, ΔV is the plasma volume from which fluorescence is observed, $\Delta \Omega$ is the solid angle within which the radiation is recorded. In turn,

$$j_{ik} = \frac{1}{4\pi} A_{ik} h\nu_{ik} \Delta N_i, \tag{4.66}$$

where A_{ik} is the probability of a radiation transition, $\Delta N_i = N_i - N_i^0$ is the change of the population of the upper level as a result of absorption of probing radiation. Substituting (4.66) into (4.65) gives

$$\Phi_{fl} = \frac{1}{4\pi} A_{ik} h\nu_{ik} \Delta N_i \Delta V \Delta \Omega. \tag{4.67}$$

The population of level N_i^0 in the absence of laser radiation is determined by the combined effect of different processes (collisions with electrons and atoms, spontaneous and forced radiation, absorption, cascade transitions from higher levels, etc.). Under LTE conditions, the combined effect of these processes leads to the Boltzmann distribution

$$\frac{N_i^0}{N_k^0} = \frac{g_i}{g_k} e^{-\frac{h\nu_{ik}}{kT_e}} ,$$
(4.68)

where g_i and g_k are the statistical weights of the levels.

In passage of a laser pulse through the plasma the equilibrium distribution of the population is disrupted. For accurate calculation of the populations of the level, it is important to take into account all the processes leading to an increase or decrease of the population, and it is also essential to know the probabilities of the corresponding impact and radiation transitions. However, in a number of partial cases, the problem can be greatly simplified.

As an example of such a simplification, we shall examine a two-level system in an electromagnetic field. It is assumed that the initial population of the upper level prior to applying the field is close to zero ($N_i^0 \approx 0$ and $\Delta N_i \approx N_i$). This corresponds to the case of low temperatures ($kT_e << h\nu_{ik}$). Further population of the level takes place as a result of absorption of radiation, and its depletion – as a result of forced and spontaneous radiation transitions and also quenching collisions. Consequently, the differential equation, describing the pumping process, has the form[136]

$$\Delta \frac{dN_i}{dt} = N_{ki}\rho_\nu(t)N_k(t) - \left[B_{ik}\rho_\nu(t) + A\right]N_i(t),$$
(4.69)

where $A = A_{ik} + A_q$; B_{ki}, A_{ik} are the Einstein coefficients; A_q is the probability of quenching collisions; $\rho_\nu(t)$ is the spectral density of the pumping radiation flux. Assuming that the total number of the atoms on both levels $N_0 = N_k + N_i$ does not change, and the laser pulse is right-angled, i.e. $\rho_\nu(t) = \rho_\nu$ at $0 \leq t \leq t$ and $\rho_\nu(t) = 0$ outside this range, equation (4.69) is transformed to the form

$$\frac{dN_i}{dt} = B_{ki}\rho_\nu N_0 - \left[(B_{ki} + B_{ik})\rho_\nu + A\right]N_i.$$
(4.70)

If a laser pulse is sufficiently long, a stationary state is established in which $\frac{dN_i}{dt} = 0$. Therefore, from (4.70) we determine the stationary value

$$N_i^{st} = \frac{N_o \rho_v B_{ki}}{\left(B_{ki} + B_{ik}\right)\rho_v + A}.$$

(4.71)

We shall now examine two extreme cases – high and low values of ρ_v. If $\rho_v \ll \dfrac{A}{B_{ki} + B_{ik}}$, then

$$N_i^{st} \approx \frac{N_0 N_{ki} \rho_v}{A},$$

(4.72)

i.e. the population of the upper level in the stationary state is directly proportional to the radiation density. Therefore, the fluorescence signal also increases linearly with increasing ρ_v. Substituting N_i^{st} from (4.72) into (4.67) instead of ΔN_i and replacing A by A_{ik} (this corresponds to the absence of quenching collisions) we obtain

$$\Phi_{fl} = \frac{1}{4\pi} h\nu_{ik} \Delta V \Delta\Omega N_0 \rho_v.$$

(4.73)

In the second extreme case, where $\rho_v \gg \dfrac{A}{B_{ki} + B_{ik}}$, as indicated by (4.71)

$$N_i^{st} \approx \frac{N_0 B_{ki}}{B_{ki} + B_{ik}} = \frac{N_0 g_i}{g_i + g_k}$$

(4.74)

(it is taken into account here that $B_{ki} g_k = B_{ik} g_i$ and does not depend on ρ_v). Correspondingly, the fluorescence signal is also independent of ρ_v and in this case is equal to

$$\Phi_{fl} = \frac{1}{4\pi} A_{ik} h\nu_{ik} \frac{N_0 g_i}{g_i + g_k} \Delta V \Delta\Omega.$$

(4.75)

This case corresponds to absorption saturation and fluorescence. Strictly speaking, saturation starts at infinitely high values of ρ_v ($\rho_v \rightarrow \infty$), but high non-linearity of the dependence of Φ_{sl} on ρ_v is already observed at

$$\rho_v = \rho_v^{sat} \equiv \frac{A}{B_{ki} + B_{ik}}. \tag{4.76}$$

This value is usually used to estimate the saturating flux density. In the absence of quenching collisions, taking into account that

$$A = A_{ik} = \frac{8\pi h}{\lambda^3} \frac{g_k}{g_i} B_{ki}, \text{ and transferring to } \rho_\lambda = \rho_v \frac{c}{\lambda^2}, \text{ we obtain}$$

$$\rho_\lambda^{sat} = \frac{8\pi h c g_k}{\lambda^5 (g_i + g_k)}. \tag{4.77}$$

A characteristic feature is the strong dependence of ρ_λ^{sat} on wavelength. This leads to a rapid increase of the values of ρ_λ^{sat} in the ultraviolet region of the spectrum. The table of calculated values of the saturating flux density for different atomic transitions was presented in Ref.134.

From the equations (4.73) and (4.75) for the light fluorescence signal we can transfer to the total number of the photoelectrons recorded by the detector during the laser pulse time Δt

$$n_e = \Phi_{fl} T \eta \Delta t, \tag{4.78}$$

where T is the transmission of the optical system, η is the quantum yield of the photocathode. Consequently, after specifying the minimum measured value of n_e, this makes it possible to estimate the minimum atom concentration N_0 detected by the examined method.

Sensitivity of the fluorescence method is so high that individual atoms can be recorded in some cases.[137] This is possible due to multiple absorption and repeated radiation of the photons in an ideal two-level system. When using a three-level system, sensitivity is lower. For example, the sensitivity of detecting Pb atoms in Ref.138 was 250 at/cm^3.

The equations (4.73) and (4.75) were derived ignoring the population of the upper level in the absence of laser radiation. At sufficiently high temperatures when the initial population of the upper level cannot be ignored, N_0 in equations (4.73) and (4.75) should be replaced by $N_k \left(1 - \frac{g_k}{g_i} \frac{N_i}{N_k}\right)$. Here the expression in the brackets takes into account the depletion of the upper level as a result of induced emission. The fluorescent signal is observed in the background of natural plasma radiation.

The natural radiation flux of the plasma, recorded by the detector, originates from the volume V and is determined by the relationship

$$\Phi_{pl} = \frac{1}{4\pi} A_{ik} h\nu_{ik} N_i^0 V \Delta\Omega. \tag{4.79}$$

From (4.67) and (4.79) we obtain that

$$\frac{\Phi_{fl}}{\Phi_{pl}} = \frac{\Delta N_i \Delta V}{N_i^0 V} = \left(\frac{N_i}{N_i^0} - 1\right)\frac{\Delta V}{V}. \tag{4.80}$$

The ratio of the volumes $\Delta V/V$ in the first approximation is equal to the ratio of the diameter of the focusing spot of the laser beam to the linear dimension of the plasma and is usually far smaller than unity. Substituting in (4.80) the value N_i^0, corresponding to LTE conditions (see (4.68)), and N_i which corresponds to the saturation conditions, and assuming that $N_I^0 + N_k^0 = N_i + N_k = N_0$, we obtain

$$\frac{\Phi_{fl}}{\Phi_{pl}} = \frac{g_k\left[\exp\left(\dfrac{h\nu_{ik}}{kT_e}\right) - 1\right]}{g_i + g_k}\frac{\Delta V}{V}. \tag{4.81}$$

Equation (4.81) shows that at the values $kT_e > h\nu_{ik}$, the ratio $\Delta\dfrac{\Phi_{fl}}{\Phi_{pl}} \ll 1$. This reduces the sensitivity of measuring the atom concentration.

Thus, natural glow is in fact the main interference when examining the fluorescence signal. The interference from the side of parasitic scattering of the optical elements of the system and the walls of the plasma chamber in this case play a considerably less significant role than in the laser scattering method, owing to the fact that the fluorescence cross section is considerably greater than the Thomson scattering cross section. In addition, the corresponding signals can be separated on the basis of their different time dependence (parasitic scattering takes place without inertia, and the duration of the fluorescence signal is determined by the lifetime of the upper level). In cases in which the observed signal does not correspond to the transition whose emission causes pumping of the upper level, the wavelengths of the fluorescence and parasitic scattering signals differ and they are quite easy to separate. This also relates to examining the induced fluorescence signal displaced with respect to frequency in relation to the laser radiation frequency by the value $\nu_{ki} - \nu$.

In initial studies concerned with the application of the resonance fluorescence method for plasma diagnostics,[139,140] investigations were carried out using barium plasma and the light source was a high-pressure arc. This was followed by using this method in a series of investigations into the diagnostics of the plasma containing alkali metals, using tuneable lasers as a radiation source. In Ref.141, the concentration of barium atoms in the flame was determined by the fluorescence of the resonance barium line excited using a dye laser.

In Ref.142, the fluorescence method was used for diagnostics of potassium plasma. Fluorescence was excited by radiation of a ruby laser whose wavelength coincided with the potassium line $\lambda = 6939$ Å $(4P_{3/2} \to 6S_{1/2})$ as a result of cooling the ruby crystal to a temperature of $-45°$C. The fluorescence signal was observed under an angle of $90°$ to the incident laser beam and was recorded using a monochromator, a photomultiplier and an oscilloscope. When the wavelength of laser radiation was displaced by 0.06 Å from the absorption line, the fluorescence signal disappeared. Fluorescence was observed at two lines $\lambda = 6939, 6911$ Å. The decay time of fluorescence in measuring the plasma parameters was examined. In this case, the decay time depended on the electron concentration so that these measurements could be used to determine N_e. The measurements of the ratio of the fluorescence signal to the intensity of the emission line enabled the authors to determine T_e (see equation (4.81)). According to the estimates of the authors, the method is suitable for potassium plasma diagnostics with the parameters $N_e = 10^{11} \div 10^{11}$ cm^{-3} and $T_e = 2000 \div 3000$ K.

In Ref.143, the resonance fluorescence method was used to measure the concentration of sodium atoms in the octupole plasma and in Ref.144 – for determining the spatial distribution of the atoms and ions of barium in an arc discharge in a magnetic field. The spatial resolution power obtained in Ref.144 was 0.2 mm^3.

In a number of studies, the fluorescence method was used to examine the distribution of impurities in the plasma evaporating from the walls of the plasma chamber and separated from metallic electrodes. The concentration of Fe, Cr, Ni, Mo, Ti, W in both cold[145,146] and hot[147,151] plasma was determined.

The fluorescence method was used to measure the concentration of hydrogen,[152,153] helium,[154] argon[155] and other gases.

In Ref.154, a dye laser set to a wavelength of $\lambda = 4471$ Å was used for selected excitation of the $4d^3D$ level of a helium atom. The fluorescence signal was detected at both the line $\lambda = 4471$ Å, corresponding to the transition $4d^3D \rightarrow 2p^3P$ and in a number of other helium levels, including those originating from levels to which radiation transitions from the level $4d^3D$ are forbidden. The population of these levels is determined only by collisions with atoms and electrons. Examining the time dependence of the fluorescence signal for the corresponding lines, the authors of Ref.154 estimated the probabilities of radiation and impact transitions.

In low-temperature hydrogen plasma, the fluorescence method in the visible range was used to measure the concentration on the second[152] (in excitation of fluorescence of the line H_α) and on higher levels.[153]

Fluorescence on the line H_α in Ref.156 and H_β in Ref.156 was used for determining the concentration of excited hydrogen atoms in high-temperature plasma in Tokamak-type equipment. The atom concentration was measured directly on levels with the main quantum number $n = 2$; 3, and the concentration of normal atoms ($n = 1$) was calculated on the basis of balance of populations of the excited state in accordance with the physical model of the plasma. For Tokamak plasma, the physical model can be represented by the coronal model. Figure 4.41 shows the spatial distribution of the concentration of neutral atoms in the plasma of ST-1 Tokamak in two regimes differing in the electron concentrations.[156]

The direct measurement of the concentration of neutral hydrogen atoms from the fluorescence on the resonance line is very interesting. However, the resonance line of hydrogen L_α ($\lambda = 1216$ Å), unlike the resonance lines of the majority of gases, is situated in the vacuum ultraviolet range. As already noted, in movement into the ultraviolet region of the spectrum the spectral density of the flux at which fluorescence is saturated ($\rho_\lambda^{sat} \sim \lambda^{-5}$) greatly increases. For example, if in examining fluorescence on H_α, ρ_λ^{sat} is several kW/(cm^2·nm), in the vicinity of L represents several thousands of kW/(cm^2·nm). In addition, tuneable lasers are not available for this region of the spectrum.

To vary the wavelength of radiation in vacuum ultraviolet, it is necessary to use the methods of non-linear optics, especially third harmonic generation. The initial radiation is the radiation of ruby, neodymium, excimer lasers and dye lasers and the non-linear medium is a mixture of inert gases with metal vapours. For example, radiation at a wavelength of L_α with a record power of hundreds of watts was produced.[158,159] Regardless of the fact that this power can be considerably lower than that required for fluorescent saturation, this radiation source was used in ASDEX Tokamak for examining the interaction of plasma with the chamber wall in 'cleaning' discharges.[158]

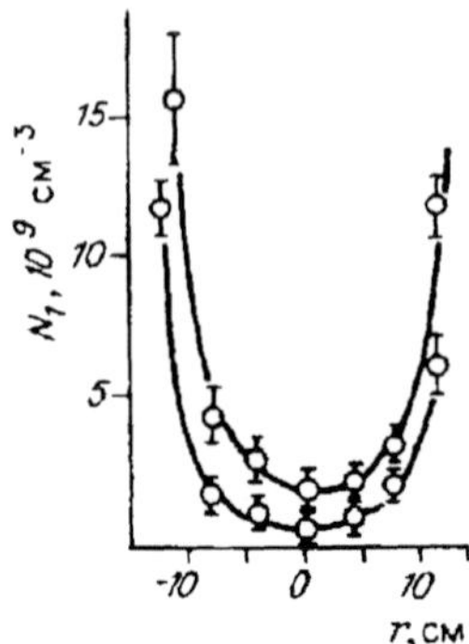

Fig.4.41 Distribution of the concentration of hydrogen atoms along the diameter of the plasma column in a FT-1 Tokamak produced by the fluorescence method on the line H_α.[156]

With the existing technical possibilities, the minimum concentration of neutral hydrogen, measurable by the examined method, is $10^9 \div 10^{10}$ cm^{-3}. Thus, the sensitivity of measuring N_1 from the fluorescence L_α is approximately an order of magnitude lower than when measuring on the basis of the H_α line with subsequent conversion in accordance with the accepted model.[156,157]

Another possibility of direct examination of the concentration of non-excited atoms is to utilise the two-photon excitation. The fluorescence signal in two-photon excitation is proportional to the square of intensity, so that at higher flux densities two-photon excitation may prove to be more promising than a single-photon one. In addition, the wavelength of pumping in this case is twice the wavelength of the resonance line. In particular, for the two-photon excitation of the fluorescence of the line L_α it is necessary to apply radiation with the wavelength $\lambda = 2434$ Å which is far easier than in the region $\lambda = 1216$ Å. The first experimental results were obtained in Ref.160 and 161.

As shown by the results, the resonance fluorescence method is used at present mainly for determining the local values of the concentration of normal and excited atoms and ions in the plasma. In individual studies, the temperatures of the atoms and ions were determined from the contours of the fluorescence line.[143,162]

The resonance fluorescence method can also be used to examine the dynamics of processes leading to population and depletion of the levels. Therefore, in addition to the previously mentioned study in Ref.157 in which the kinetics of population of the levels in the helium plasma was investigated, it is also important to note the Ref.134, 164 concerned with the analysis of the processes of population of the hydrogen levels.

There is another possibility of the resonance fluorescence method - examination of the diffusion processes in plasma.[165,167]

Chapter 5

SPECTROSCOPY OF GROUND ELECTRONIC STATES OF MOLECULES IN PLASMA USING TUNABLE LASERS

The advantages of using tunable lasers in comparison with conventional light sources in quantitative plasma spectroscopy are often caused by the following parameters: high monochromatism, high intensity, coherence and directionality of radiation.[1] If these properties are efficiently utilised, it is possible to change qualitatively many conventional spectroscopy methods. The possibilities of absorption spectroscopy,[2] interferometry,[3] the spectroscopy of Rayleigh,[4] Thomson,[5,6] Raman[7] light scattering have been greatly expanded. These possibilities are used efficiently in spectral and optical plasma diagnostics. Methods based on recording changes of the plasma properties in resonance absorption of laser radiation by plasma particles are being developed.[8,10] In this book, we shall discuss two methods of spectroscopy of molecular plasma using the frequency-tunable lasers: diode spectroscopy and CALS (coherent anti-Stokes spectroscopy of Raman light scattering).

These methods are attractive because of the following reason. The molecular plasma conditions are characterised by a low ($\lesssim 10^{-6}$) degree of ionisation and the fraction of electronically excited emitting particles ($\lesssim 10^{-5}–10^{-7}$). Therefore, in the majority of cases special interest is attracted by the processes in which vibrationally-rotationally-excited molecules in ground electronic states take part. These methods make it possible to determine directly the behaviour of the molecules under these conditions. The second reason is that these methods often suitably supplement each other.

5.1 Diode spectroscopy
Laser diodes
The name of this method was derived from the fact that the light source in adsorption measurements is represented by laser diodes made of semiconductor materials generating radiation in passage of current through them in the region of the $(p–n)$-junction. The characteristic dimension of the diode in each measurement are less than 1 mm thick

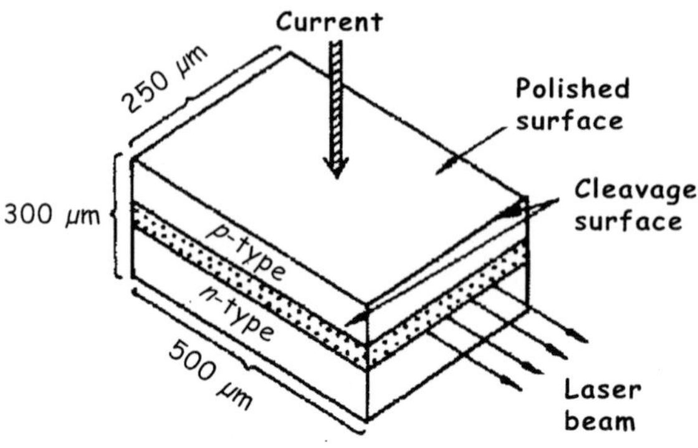

Fig.5.1 Laser diodes.

(Fig.5.1). The cleaning surfaces are used as the reflecting elements of the resonator.

The generation frequency can be tuned by many methods. The most useful methods are the variation of the pumping current and the specimen temperature. The typical region of frequency tuning for the diode of the given composition is ~100–300 cm^{-1}. Pulsed and continuous generation regimes are possible. In the pulsed regime, frequency tuning forms in non-stationary heating of the (p–n)-junction by a current pulse. The tuning characteristics are piecewise-continuous characteristics with zones of smooth tuning Of several centimetres to the minus first degree. This is associated with the mechanisms of formation of laser modes. In the continuous regime, the width of these bands is usually 1–2 cm^{-1}, in the pulsed regime to 10 cm^{-1} and more.[11] The spectrum range is selected by selecting the diode materials and can be varied from 0.7 to 50 μm.12 Figure 5.2 shows schematically the data on the types of laser used and the characteristic frequencies in the spectra of several relatively simple molecules.

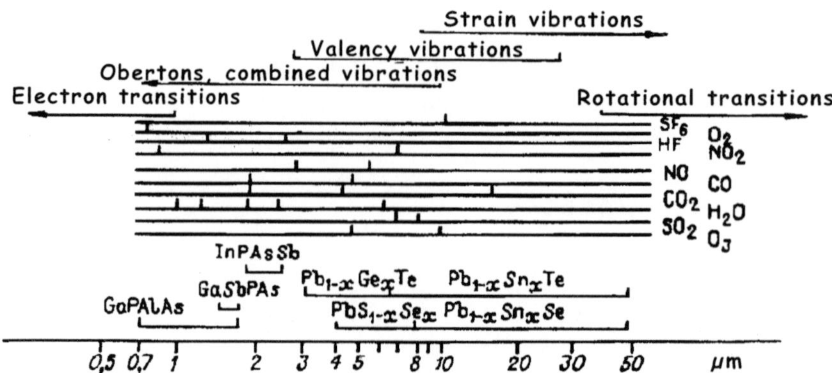

Fig.5.2 Characteristic types of laser diodes used in various areas of the spectrum and the frequency in spectra of certain molecules.

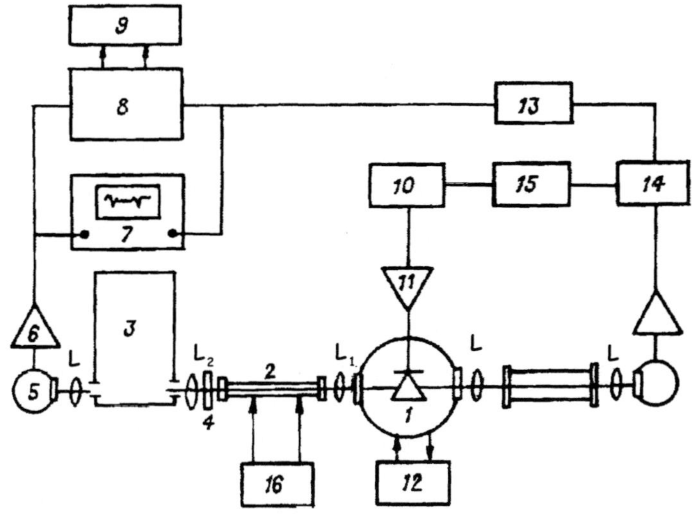

Fig.5.3 Circuit of the pulsed–periodic diode spectrometer.

Spectrometer

The diagram of a diode laser spectrometer[11] operating in the pulsed–periodic regime, is shown in Fig.5.3. The radiation of the diode laser 1, shaped by the lenses $L1$ and $L2$ into the beam with a transverse dimension of ~1 mm, takes place through the examined object 2 (discharge tube with a high voltage current source 16). The monochromator 3 is used for rough determination of the frequency of generation and selection of the laser modes. The relative frequency calibration is specified by the Fabry-Perot interferometer 4. The signal from the photodiode 5 passes through the amplifier 6 to the oscilloscope 7 and the stroboscopic integrator 8 followed by recording on the automatic recording device 9. PbSe diodes, used in Ref.11, with the generation in the range 4–6 µm, are excited by rectangular current pulses with a frequency of ~1 A, duration 200 µs and a repetition frequency of 10–100 Hz. The generation power is 10–100 µW. The pulses are shared by the generator 10 and magnified with the magnifier 11. The generation frequency is tuned by passing a current pulse through the laser. The position of the tuning zone is varied by varying the temperature (in the range 30–80 K) of the cooling line on which the laser is secured. The sensitivity of the generation frequency to the variations of temperature for the laser diode is ~1 cm^{-1} K^{-1}. Therefore, to obtain a high spectral resolution, it is sufficient carry out accurate regulation of temperature. The long-term stability of temperature is not lower than 10–2 K and is achieved using the electronic system 12 as a feedback. The moment of the start of recording the spectrum with a stroboscopic integrator is set by the shaper of the time delay 13. The temperature fluctuations in the period between two consecutive pulses of the pumping current are compensated by the optical triggering system 14 with the delay shaper 15.

This system makes it possible to ensure a resolution power of $\sim 10^{-4}$ cm^{-1}, the width of the zone of continuous tuning to 11 cm^{-1}, frequency tuning rate 10^4–10^6 cm$^{-1}\cdot$s^{-1}, operating speed of the recording system $\sim 10^{-8}$ s. At these powers and widths of the radiation spectrum the effective brightness temperature of the emitter can be estimated at $\sim 10^5$ K which enables quantitative absorption measurements to be carried out even at very high excitation levels of the examined system. At the same time, the relatively low power of the semiconductor lasers enables measurements to be taken in the linear absorption regime.

Special features associated with higher resolution under the conditions of piecewise-continuous tuning characteristics

The presence of breaks in the frequency-tuned characteristic of the laser creates certain difficulties in investigations in practice. However, the latter is partially compensated by a higher spectral resolution. This shall be explained on the following example of examination of distribution of CO_2 molecules on the vibrational–rotational levels in the active medium of a gas discharge CO_2 laser.[11,13,14]

The absorption bands of CO_2 corresponding to its four normal vibrations v_1, v_2 (twice degenerate), v_3, are situated in the spectrum regions 8, 15 and 4 μm. In conventional approaches utilising the methods of spectroscopy with moderate resolution, to examine the overall pattern of distribution of vibrational energy investigation must be carried out especially in these regions, and v_2 and v_3 are reflected in the absolute spectra and the symmetric vibration v_1 in the Raman scattering spectrum. The application of the high resolution technique, restricted by the width of the line like, for example, in diode spectrometry, permits a different approach. It is sufficient to work in the region of one of the frequencies, for example, v_3, and use the superimposed rotational structure of sequential transitions $v_1 v_2' v_3 \rightarrow v_1 v'_2 (v_3 + 1)$, where v and l are the quantum numbers in the generally accepted notations.

Figure 5.4 shows a fragment of the absorption spectrum of CO_2 in a band approximately 1 cm^{-1} wide in the region 4.3 μm in a discharge in the CO_2–N_2–He mixture and in the absence of discharge. In the discharge the spectrum is enriched. This corresponds to absorption from excited vibrational levels. For the example discussed here, it is important to note that certain vibrational states of the molecules are found already in such a narrow (in the conventional sense of the word) spectral range, including the states belonging to isotopically substituted molecules. In

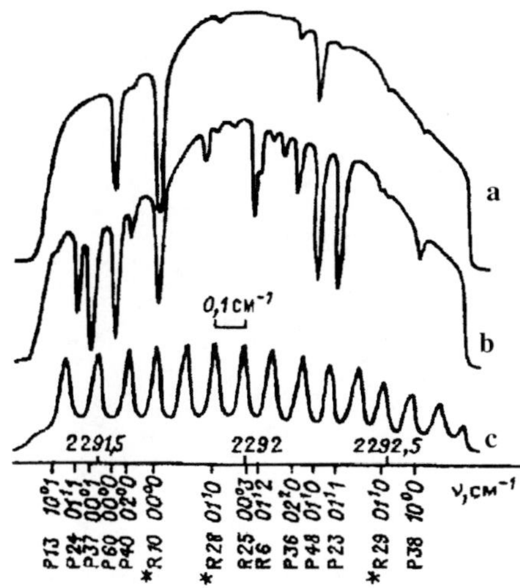

Fig.5.4 A fragment of absorption spectra of CO_2: a) in the CO_2–H_2–He mixture (1:1:8) in the absence of discharge, pressure 36 torr; b) in the discharge, the mixture and the pressure are the same, current density 1.5 mA/cm^2; c) transmittance of the Fabry-Perot resonator.

identifying the transition shown in Fig.5.4, we note a branch and also rotational and vibrational states being lower states for the transitions. The asterix indicate the transitions in the $C^{12}O_2^{18}$. Other transitions are in the normal isotope $C^{12}O_2^{16}$.

Using the intensities of absorption and the literature data on the values of the matrix elements of the dipole moment for the vibrational–rotational transitions, it is possible to determine the distributions of populations of the rotational and vibrational levels.[11,14] Figure 5.5 shows the distribution of CO_2 molecules on rotational levels for several vibrational levels, and Fig.5.6 is the distribution on the vibrational levels. It may be seen that the rotational distributions are of the Boltzmann type, and since they exist in long-life ground electronic states, the rotational temperature corresponds to the kinetic temperature of the neutral particle.[15] The vibrational distributions also demonstrate on the whole the Boltzmann-type distributions within the limits of each vibrational mode. Only the block with anti-symmetric vibrations v_3 shows a weak tendency for Trinor-type deviations.[16] The vibrational temperature of the anti-symmetric mode T_3 = 2650 ± 30 K, and of the symmetric and strain modes $T_1 = T_2$ = 560 ± 15 K and does not coincide with the gas kinetic temperature T_g = 470 ± 10 K.

We shall discuss here physical processes that determine the formation of corresponding distributions because we are interested in this case in the procedural aspect of the matter. It is only important to note that, firstly, the relative error of temperature determination is sufficiently

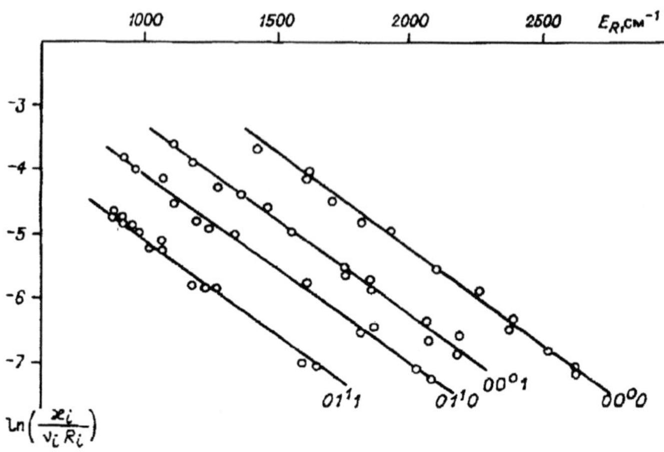

Fig.5.5 Distribution of CO_2 molecules on the basis of the rotational levels for several vibrational levels. Discharge in a capilliary 2 mm in diameter, 50 mm long, current $i = 9$ mA, gas mixture CO_2–N_2–He (1:1:8:), pressure 30 torr.

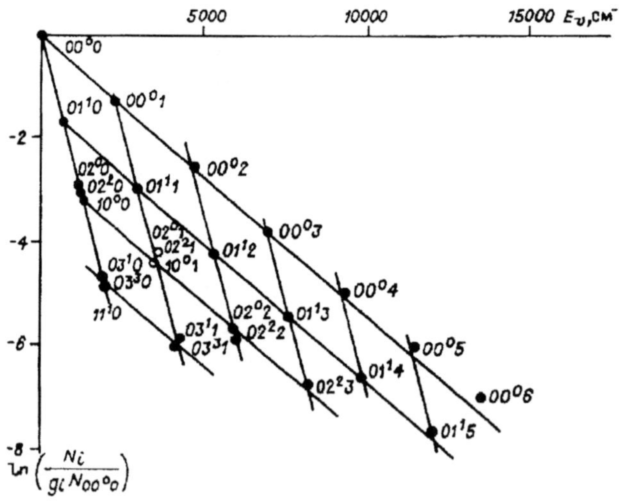

Fig.5.6 Distribution of CO_2 molecules on vibrational levels. Symbols are the same as in Fig.5.5.

small, being $\Delta T/T \lesssim 2\%$. This is due to the fact that a large number (~1000) of vibrational–rotational transitions was used in the measurements. Secondly, the information on the vibrational distributions in Fig.5.6 was obtained in the energy range of vibrational states ~15000 cm^{-1}, but within the limits of a very narrow spectral range ~75 cm^{-1} since the differences in the frequencies of vibrational transitions are associated only with anharmonism. Naturally, this approach can be used only when using high spectral resolution. This is also possible in diode spectroscopy.

Recording low-concentration particles

Due to high spectral resolution, diode spectroscopy has high sensitivity. When used for plasma diagnostics, this makes it possible to examine excited states and measure low particle concentrations.

In the simplest case of direct linear absorption, the reliability of measurements on the level of recording the changes of intensity of laser radiation $\Delta I/I \sim 10^{-2}$ is sufficiently reliable. For the example of examination of the energy distribution of CO_2 molecules, described in the previous section, this corresponds to a molecule concentration of $\sim 10^8$ cm^{-3} on one vibration-rotational level and an optical path of 50 mm. The use of the methods of two-beam spectroscopy enables measurements to be taken at $\Delta I/I \sim 10^{-5}$. Approximately the same possibilities are offered by the use of autocorrelation methods in recording complicated spectra.[17] Modulation methods can be used to increase sensitivity: the frequency modulation of laser radiation, Stark and Zeeman modulation.

In recording the ions in plasma in discharges excited by an alternating field or DC discharges with a specially applied alternating field, it is also possible to carry out the modulation of the speed of motion leading to the modulation of the Doppler shift, combined the method of phase-sensitive detection of radiation.[18] Figure 5.7 shows a fragment of the modulation spectrum of absorption of the three-atom negative ion NCS^- in ac discharge (25 kHz) in a gas mixture of 1.7 torr NH_3 +0.5 torr CS_2. Figure 5.8 show the radial distribution of density of ArH^+ ions in DC discharge in an H_2 + Ar mixture.[20]

The radicals, ions and electronically-excited particles in the plasma recorded up to the year 1988 by diode spectroscopy, are given below. In many cases, investigations were not restricted to concentration meas-

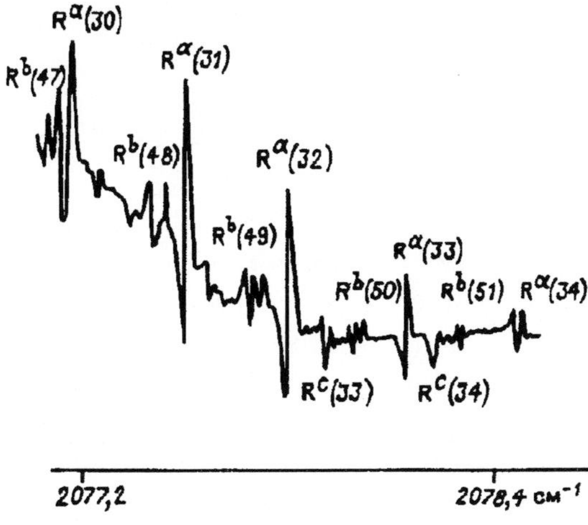

Fig.5.7 Fragment of the absorption spectrum of NCS^-.[119] a) band v_1, b) $v_1 + v_2 - v_2$, c) $v_1 + v_3 - v_3$.

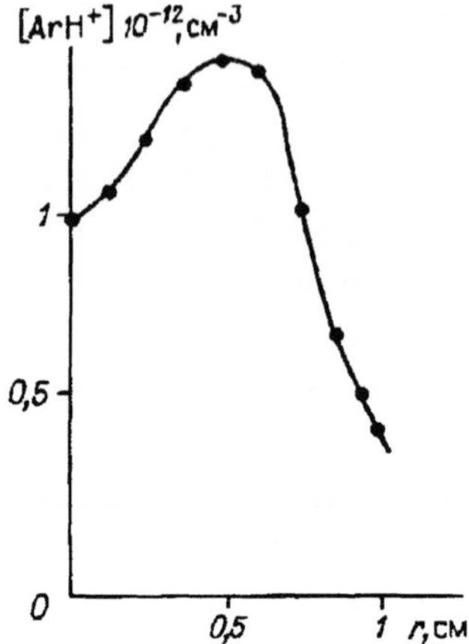

Fig.5.8 Radial profile of ArH⁺ ions in a discharge tube.[20]

urements but also included determination of the distribution of particles on the vibrational–rotational states, analysis of the shape of the contours of spectral lines and obtaining information on molecular constants:

Radicals, unstable particles
AsH, BF, CCl, CF, CN, CS, FO, NCl, NF, NS, OH, OD, PH, PO, PCl, PN, PS, PF, SF, SCl, SiH, SO, SiN, BO_2, C_2D, CD_2, CH_2, CF_2, C_2O, C_2H, FCO, FO_2, HCO, HO_2, DO_2, NF_2, CD_3, CF_3, CH_3, CH_2F, NO_3, ClBO, FBO, HBO, S_2O, HBNH

Positive ions
Ar^+, ArH^+, CF^+, ClC^+, HCl^+, NO^+, NeH^+, OH^+, CO_2^+, ClH_2^+, D_3^+, DCO^+, DN_2^+, H_3, HBF^+, H_2D^+, HD_2^+, HCO^+, HCS^+, HN_2^+, H_2O^+, $HCNH^+$, H_3O^+, SH_3^+, HeH^+, SH^+

Negative ions
C_2^-, OH^-, SH^-, FHF^-, $ClHCl^-$, FDF^-, N_3^-, NCO^-, NCS^-

Electronically-excited particles
I, Kr, D_2, N_2

Non-stationary processes
It has already been noted that the pulsed-periodic regime of operation of the laser diode is characterised by a very high frequency tuning rate. It is thus possible to record the spectra of non-stationary objects. For example, Fig.5.9 shows the evolution of amplification in a pulsed periodic CO_2 laser at atmospheric pressure on the P(20) line of the 00°1–10°0 generation transition. In contrast to the conventional methods of measuring the amplification factor in the centre of the line, it is possible to examine the change of the entire profile of amplification with high (less than 1 µs) time resolution.[18a]

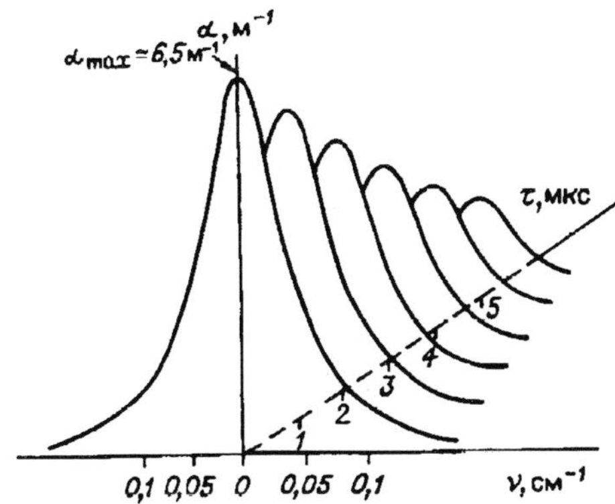

Fig.5.9 Evolution of amplification in a CO_2 laser.

It should be noted that as a result of obtaining extremely high rates of frequency scanning in diode spectroscopy combined with the high-speed radiation recording system, the rate of recording the absorption spectra can be limited no longer by technical but purely physical factors. If the width of the absorption line is $\Delta\omega$, and the duration of recording the contour of the line Δt, it is evident that if the indeterminacy relation $\Delta\omega\Delta t \gtrsim 1$ is not fulfilled, the contour is recorded in the distorted form. In fact, the coherent radiation passing through it polarises this contour. The phases of relaxation processes during periods longer than $\tau_{rel} \sim \Delta\omega^{-1}$ leads to dephasing of dipoles and induced polarisation disappears. In scanning the contour during the periods shorter than or comparable with τ_{rel}, after the effect of laser radiation (i.e. when $\omega(t) > \omega_{ik} + \Delta\omega/2$, where $\omega(t)$ is the tuned frequency, ω_{ik} is the frequency corresponding to the centre of the contour) the medium retains the molecules with the induced dipole moment. The molecular dipoles continue to vibrate at frequency ω_{ik}, they are phased, and this leads to emission of coherent radiation by the medium in the direction of incident light. This radiation is combined with probing radiation, and depending on the ratio of the phases of these fields the resultant signal may both increase or decrease.

This situation was demonstrated by experiments in Ref.21 in recording the contour of the line from the vibrational–rotational spectrum of CO_2 ($\lambda \approx 4.2$ µm) with a diode spectrometer with a frequency scanning speed of up to 10^6 cm$^{-1}\cdot$s^{-1}.

Figure 5.10 shows the oscillograms of the radiation passed through a gas. These oscillograms were recorded at tuning rates of $\mu = 10^5$; 10^6 cm$^{-1}\cdot$s^{-1} and the width of the line $\Delta\omega = 4.4\cdot10^{-3}$ cm^{-1}. Case *a* corresponds to 'slow' recording $\Delta\omega^2/\mu = 36.6$, and the oscillogram reflects

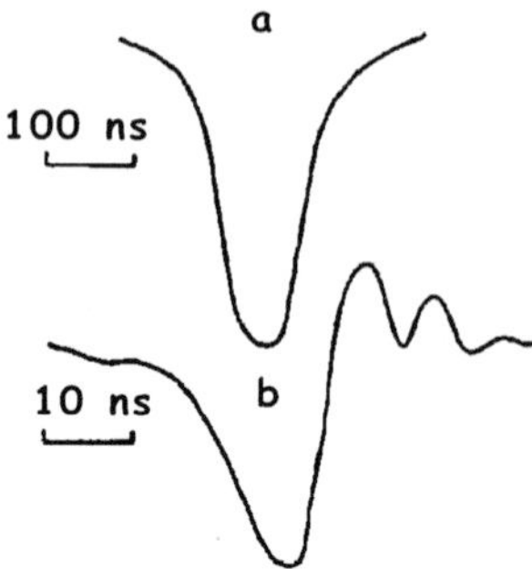

Fig.5.10 Oscillograms of intensity in recording the contour of the absorption line of CO_2 at scanning speeds of $\mu = 10^5$ (a), 10^6 cm^{-1}·s^{-1} (b).

the actual line contour. In the case b $\Delta\omega^2/\mu = 36.6$ and the effects described previously are observed.

Restrictions in the application of diode spectroscopy are associated mainly with two circumstances: 1) absence in a number of cases of adsorption vibration IR spectra (in dipole approximation), for example homonuclear diatomic molecules; 2) no localised measurements in the direction of propagation along the laser beam.

CALS spectroscopy has no such restriction.

5.2 CALS spectroscopy
Introduction
Coherent anti-Stokes spectroscopy of Raman light scattering (CALS) is a non-linear analogue of the classic Raman scattering method.[22] The effect of two power coherent sources with frequencies ω_1 and ω_2 on the radiation medium generates new coherent radiation at frequency ω_3. From the microscopic viewpoint, CALS is a four-photon process in which the molecule absorbs two photons with frequency ω_1 and emits two photons with frequencies ω_2 and $\omega_3 = 2\omega_1 - \omega_2$, and the quantum state of the molecule does not change (Fig.5.11). The radiation at frequency ω_2 is forced in the field of incident radiation with frequency ω_2. The radiation with ω_3 is coherent and characterised by the wave vector

$\vec{k}_3 = 2\vec{k}_1 - \vec{k}_2$. In principle, this process takes place in any medium, but the radiation intensity at frequency ω_3 rapidly increases if the difference $\omega_1 - \omega_2$ is close to the frequency of the Raman-active transition. Varying the difference $\omega_1 - \omega_2$ we obtain a spectrum containing the same information as the spectrum of spontaneous Raman scattering (SRS). The two main advantages of CALS in comparison with SRS are the rapid increase of the signal amplitude in the directional light beam (this is important when examining objects with a low density), and the possi-

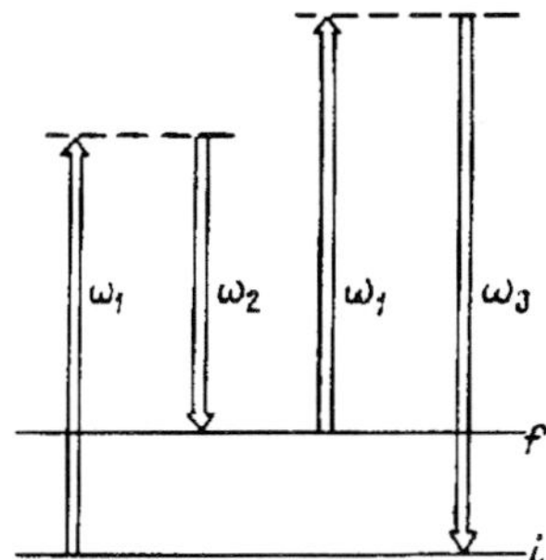

Fig.5.11 Generation of CALS signal with a frequency of $\omega_3 = 2\omega_1 - \omega_2$ under the effect on the molecule of two laser beams with frequencies of ω_1 and ω_2 (i, f are the quantum states of the molecules).

bility of localising the region of generation of the CALS signal, due to the non-linear nature of non-interaction of the light with the medium.

A similar process also takes place in the Stokes region where $\omega' = 2\omega_2 - \omega_1$ although the anti-Stokes region is more suitable for practice because of the absence of undesirable radiations in it: luminescence of optics, scattered laser radiation, etc.

To determine the intensity of CALS spectra and their relationship with the densities of the quantum states of the molecules, it is important to use the conclusions of quantitative theory, similar to that described in a number of reviews and monographs (see, for example, Ref.22).

Under the effect of the field of the light wave the medium is polarised $\vec{P}$. This polarisation can be described by a series with respect to the exponents of the total strength E of the electric field

$$\vec{P} = \chi^{(1)}\vec{E} + \chi^{(2)}\vec{E}^2 + \chi^{(3)}\vec{E}^3 + ..., \tag{5.1}$$

where $\chi^{(i)}$ is the dielectric susceptibility tensor. The first term with $\chi^{(1)}$ describes refraction and light reflection phenomena. Susceptibility $\chi^{(1)}$ is linked with permittivity by the relationship $\varepsilon = 1 + 4\pi\chi^{(1)}$. The term with $\chi^{(2)}$ describes the phenomena of doubling the frequency and optical detection. For isotropic media, such as gases relevant to this work, this and all subsequent even terms in (5.1) are equal to zero because the tensor χ is symmetric. Thus, the term with $\chi^{(3)}$ is a term of a lower order describing the non-linear effect in isotropic media, including all coherent defects of Raman scattering, including CALS.

Calculations within the framework of conventional electromagnetic theory which take into account the symmetry properties $\chi^{(3)}$ for a medium with magnetic permittivity $\mu = 1$, give the following expression for the intensity of the CALS signal:

$$I_3 = \frac{25\pi^4\omega_3^2}{n(\omega_3)n(\omega_2)n(\omega_1^2)c^4}\left|\chi^{(3)}(-\omega_3, \omega_1, \omega_1, -\omega_2)\right|^2 I_1^2 I_2 l^2 \sin c\left(\frac{\Delta kl}{2}\right). \qquad (5.2)$$

Here I_j is the intensity of the wave with frequency ω_j; l is the length of interaction; $k = k_3 - 2k_1 + k_2$; c is the speed of light; n is the refraction index. In accurate phase matching ($\Delta k = 0$) $I_3 \sim l^2$. At $k \neq 0$, I_3 is the periodic function of l which reaches the first maximum at a length $l_c = \pi/\Delta k$ referred to as the coherence length. Generally speaking, the condition $\omega_3 = 2\omega_1 - \omega_2$ does not ensure that $\Delta k = 0$ is fulfilled, but in gases due to low dispersion (weak dependence $n\ (\omega_j)$) in the case of colinear propagation of the beams the phase synchronism is fulfilled over a large length. For the majority of gases $l_c \sim 1$ m, and in the majority of real systems $l < l_c$ so that it can be assumed that $I_3 \sim l^2$.

Examination of the susceptibility tensor $\chi^{(3)}$ on the microscopic level and determination of the relationship with the density of particles and molecular constants shows that its component, corresponding to resonance with the Raman-active transition, can be expressed as follows:

$$\chi_p^{(3)} = \frac{(N_f - N_i)n_1 c^4}{2\hbar n_2 \omega_2^4}\left(\frac{d\sigma}{\sigma\Omega}\right)_{if}\left(\omega_{if} - \omega_1 + \omega_2 - i\Gamma_{if}\right)^{-1}. \qquad (5.3)$$

Here N_i and N_f are the densities of the molecules in the states i and f; σ is the cross section of spontaneous Raman scattering; Ω is the solid angle; Γ_{if} is the half-width at half height of the line of spontaneous Raman scattering. Cross sections SRS σ and $\dfrac{d\sigma}{d\Omega}$ are well known for a large number of molecules.[23]

Thus, in a relatively narrow spectral range

$$I_3 \sim \left(N_f - N_i\right)^2 I_1^2 I_2.$$

The direct relationship of $\chi^{(3)}$ with SRS cross sections shows that the rules of selection of CALS are identical with those for SRS. In particular, for diatomic molecules they have the form

$$\Delta v = 1; \quad \Delta J = \begin{cases} 2 \text{ is } S-\text{branch} \\ 0 \text{ is } Q-\text{branch} \\ -2 \text{ is } O-\text{branch} \end{cases}$$

$$\Delta v = 0; \quad \Delta J = \pm 2 \text{ is } S-, O-\text{branch,} \tag{5.4}$$

where v, J are the vibrational and rotational quantum numbers. At the same time, since CALS is a parametric process in which the amplitude of the scattered wave represents the sum of contributions from all molecules in the interaction region, in calculating I_3 it is necessary to carry out averaging with respect to molecular orientation. Consequently, differences in the matrix elements of the polarisability tensor form in the microscopic sense in the SRS and CALS cases. However, this does not affect the selection rules (5.4),[24] and the matrix elements of polarisability α_{if} will have the following dependence on v and J (in models of a harmonic oscillator and a rigid rotator):

$$Q-\text{branch:} \quad \left|\langle v, J|\alpha_{if}|v+1, J\rangle\right|^2 = \frac{\hbar}{2M\omega_v}\left[\alpha'^2 + \frac{4}{45}b_{J,J}\gamma'^2\right](v+1),,$$

$$O-, S-\text{branches:} \quad \left|\langle v, J|\alpha_{if}|v+1, J\pm 2\rangle\right|^2 = \frac{\hbar}{2M\omega_v}\left[\frac{1}{15}b_{J,J\pm 2}\gamma'^2\right](v+1). \tag{5.5}$$

Here M is the reduced molecular mass, ω_v is the frequency of vibrations, b is the Plachek–Teller coefficient

$$b_{J,J} = \frac{J(J+1)}{(2J-1)(2J+3)}, \quad b_{J,J-2} = \frac{3J(J-1)}{2(2J-1)(2J+1)},$$

$$b_{J,J+2} = \frac{3(J+1)(J+2)}{2(2J+1)(2J+3)}, \quad \alpha_{if} = \left(\frac{c}{\omega_2}\right)^4\left(\frac{d\sigma}{d\Omega}\right)_{if},$$

α' and γ' describe the mean polarisability and its anisotropy.

For the majority of molecules, the anisotropy is relatively small.[25] To record the Q-branch with high intensity (in comparison with O- and S-branches) in the CALS spectrum, the correction for the intensities of the vibrational–rotational lines is small (this correction is the largest in the region of small J).

For the H_2 and N_2 molecules it is given below, %:

J	0	1	2	3
H_2	−8	5	1	0.3
N_2	−6.5	4	0.8	0.3

This holds if the resonance component of the tensor $\chi^{(3)}$ is considerably higher than the non-resonant one. In practice, this can be fulfilled in cases in which the examined molecules represent a small fraction of the total number of particles in the gas. In these cases, the interpretation of intensities in CALS spectra is greatly complicated and, most cases, steps are taken to exclude the non-resonance component in formulating an experiment. In particular, the authors of Ref.26 proposed an efficient method of suppressing it based on the fact that when the polarisation rates of pumping rates do not coincide with the polarisations of the resonant and non-resonant signals of the CALS do not coincide.

Spectrometer. Figure 5.12 shows the typical circuit of a CALS spectrometer.[27,28] Since the particle concentration in the gas and plasma is usually relatively low, to use non-linear-optical methods it is necessary to use radiation of a relatively high power. The master oscillator in the described circuit is the pulsed-periodic aluminium–garnet laser 1 (Nd-YAG). Radiation of with a wavelength of $\lambda = 1.06$ µm, a repetition frequency of 20 Hz and a pulsed time of ~10 ns is converted to the radiation with a doubled frequency (second harmonics, $\lambda = 532$ nm) in DKDP crystal 2 (the efficiency of energy conversion is ≈30%) which is used as a beam with reference frequency ω_1. The radiation pulse energy in the second harmonics is ~30–50 mJ, the spectral line with ~0.1–0.2 cm⁻¹. Residual radiation with $\lambda = 1.06$ µm, not converted in the doubler 2, is doubled with respect to frequency in the second DKDP

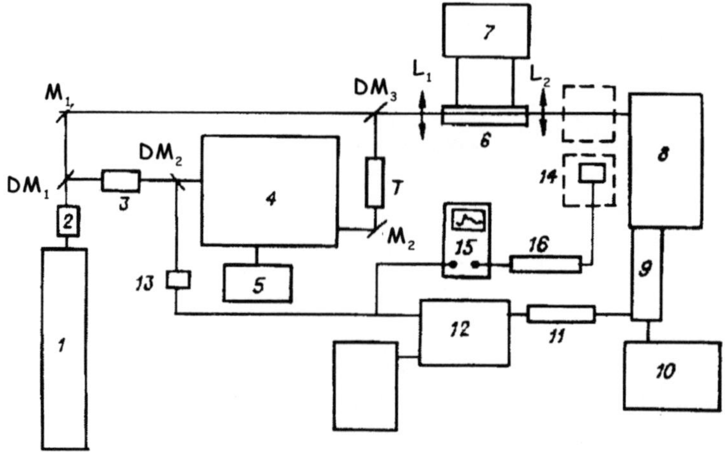

Fig.5.12 Diagram of CALS spectrometer.

crystal 3 (~20 mJ with λ = 532 nm) and is used to pump the dye laser 4. To separate the radiation beams of the first and second harmonics of the (Nd-YAG) laser, the authors used the DM_1 and DM_2 dichroic mirrors. The width of the dye laser radiation line was 0.1-0.2 cm^{-1}, the pulsed energy 1–5 mJ. The frequency of the laser 4 is tuned by the microprocessor 5. Telescope T is used for optimum matching the diameters d_1 and d_2 of the beams ω_1 and ω_2. After passage through the telescope, radiation ω_2 is co-linearly combined with the beam ω_1 using the DM_3 mirror. Both beams are focused in the examined object 6 by the power unit 7 using the lens L_1. The CALS signal is separated from laser radiation using the symmetric four-prism monochromator 8,[29] which enables operation in a relatively wide spectrum range at the constant position of the prisms. The radiation detector was the photodetector 9 with the power unit 10. The signal from the photoelectric multiplier travels through the cable delay line 11 to the stroboscopic integrator 12 operating in the synchronous detector regime. The integrator is activated from the avalanche photodiode 13 illuminated with laser radiation 1. The delay line is selected in such a manner as to ensure the transfer of the signal from the photoelectric multiplier in the working range of the time delays of the integrator. The beam passage path ω_1 and ω_2 are selected such as to ensure the simultaneous arrival of the light pulses with the given frequencies to the examined object. This is controlled by the avalanche photodiode 14 whose signal travels to the oscilloscope 15 through the delay line 16.

Typical parameters of this spectrometer are: spectral resolution 0.1-0.2 cm^{-1}; the range of combination vibrational frequency 1000–4200 cm^{-1}, sensitivity with respect to concentration $N_f - N_i \sim 10^{10}$–10^{14} cm^{-3} for H_2, N_2, CO molecules...; localisation of the region of measurements across the beams ~10–100 µm, along the beams 1–20 mm.

This spectrometer was constructed on the basis of the so-called 'narrow-band' 'co-linear' systems. Other variants can also be used, depending on the requirements imposed by the specific features of the examined object. For example, if necessary, the localisation of the measurements along the probing beams is improved using a non-co-linear system in which the conditions of phase synchronism of the beams are satisfied at the large angles from their convergence selected by specific procedure.[30] Additional advantages of this system are due to simple recording of CALS spectra with small combination shifts, for example, purely rotational spectra.

The 'narrow-band' system in which both lasers operate with a high monochromaticity is preferred when examining objects operating under stationary or pulsed-periodic conditions. In the case of objects with a monopulse regime, it is recommended to use the 'wide-band' variant of the spectrometer.[31] in which a dye laser generates a wide spectrum and the CALS spectrum is recorded in a single laser pulse under the condition of photography, microchannel or matrix (instead of photoelectric multiplier) detection. However, as shown by practice, the sensitivity in this case is approximately an order of magnitude lower than in the 'narrow-band' system.

In recent years, the method of using CALS spectrometers has been successfully developed; we shall mention the use of lasers with the picosecond pulse duration,[32] systems with increased spectral resolution power which include, as the master laser, a high-stability gas laser with subsequent light amplification in the (Nd-YAG) medium,[33] and a number of other interesting proposals.

Examples of using CALS for investigations of vibrational–rotational distribution of molecules in gas discharge plasma

The majority of applications of CALS in low-temperature plasma diagnostics are associated with examination of vibrational–rotational distributions of simple diatomic molecules.

The authors of Ref.34 examined the distribution of nitrogen molecules on vibrational–rotational levels of the ground $X^1\Sigma$ state in a low-pressure discharge ($p = 2 \div 4$ torr, $i = 80$ mA). The sensitivity of the method enabled molecules to be recorded on vibrational levels with $v \leq 14$. The measured rotational temperatures in the vibrational states $v = 0 \div 10$ at $p = 2$ torr were within the error range equal to $T_{rot} = 530 \pm 30$ K. The vibrational distribution was of the non-Boltzmann type with the vibrational temperature describing the relative population of lower levels with $v = 0$ and $v = 1$ equal to $T_{10} = 5300 \pm 350$ K. However, the vibrational distribution at a pressure of $p = 2$ torr was not described within the framework of the well-known Trinor model (Fig.5.13). This

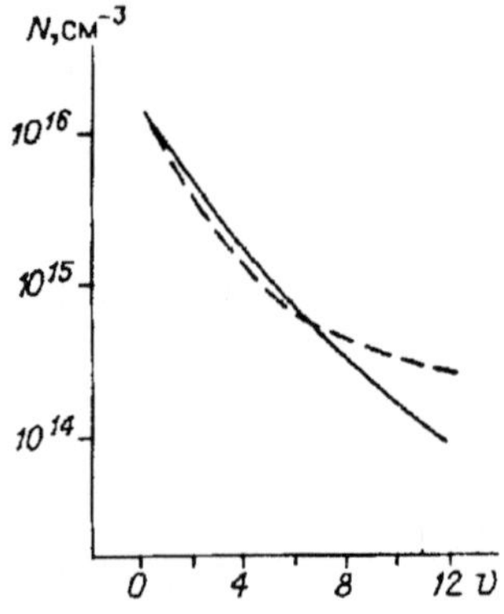

Fig.5.13 Populations of the vibrational levels of N_2 in the discharge. Pressure $p = 2$ torr, discharge current $i = 80$ mA;[34] solid line - experiments, broken line are the calculated data.

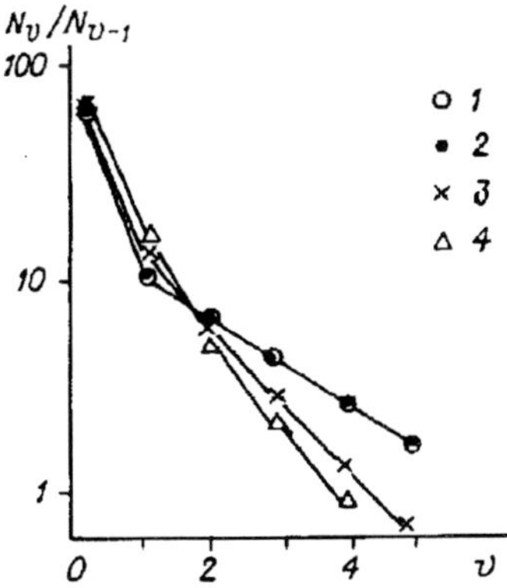

Fig.5.14 Distribution of the concentration of N_2 molecules on the vibrational states in a pulsed (duration 40 ns) discharge (current 1.4 A) after a pulse with different time delay. $t = 50$ (1), 100 (2), 6 (3) and 20 μs (4).

is attributed to the effect of vibrational relaxation on the walls. An additional argument, according to the authors, was that at high pressure ($p = 4$ torr) the agreement between theory and experiment improved.

The authors of Ref.35 examined the distribution of N_2 molecules in the vibrational–rotational states in the condition of a pulsed ($\tau = 40$ ns) high-current ($i = 1.4$ A) discharge. The discharge zone was probed both during the current pulse and after its completion over a period of 20 μm. Results are shown in Fig.5.14. The non-Boltzmann type of distribution is again found. Comparison with the theory of vibrational relaxation enabled the authors to determine the constants of the ($V–V$) exchange. Similar measurements were taken in Ref.36 at higher energy inputs ($\tau = 200$ ns, $i = 260$ A), at delay times after the start of the current pulses of 400 ns–1 ms. In this case, to explain the experimental results, the authors had to assume that in addition to the direct electron impact and ($V–V$) processes, there are additional channels of formation of vibration distributions, especially the population of the vibrational levels of the ground electronic state in the processes of quenching of metastable electronic levels.

Detailed investigations of the distribution of hydrogen molecules on the rotational levels in a gas discharge were carried out in Refs.27 and 28. The higher value of the rotational quantum of the H_2 molecule and the presence of modifications with different nuclear spin caused that at high gas temperatures the rotational distribution was non-Boltzmann, even under the stationary conditions. Figure 5.15 shows an example of such a non-equilibrium distribution in H_2 ($X^1\Sigma$, $v = 0$) under the conditions of a gas discharge in a H_2–He (1:3) mixture at a pressure of 0.5 torr and a current of 30 mA in a discharge tube 14 mm in diam-

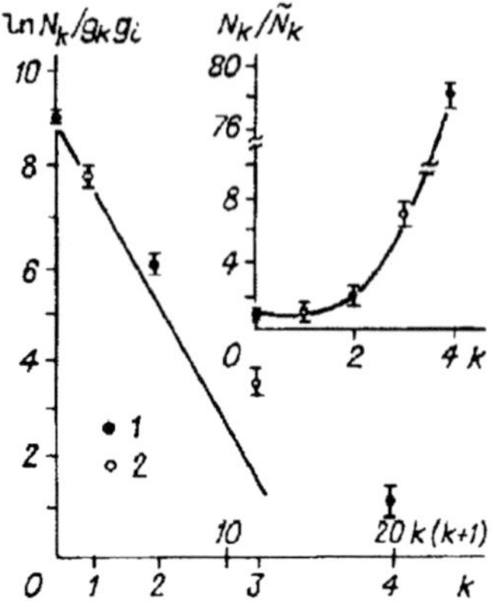

Fig.5.15 Distribution of the concentration of H_2 molecules ($X^1\Sigma$; $v = 0$) on the rotational levels at the discharge axis. The straight line is the Boltzmann distribution at a gas temperature of $T_g = 145$ K; 1) vapour - H_2; 2) ortho-H_2.

eter with the walls cooled with liquid nitrogen. The population of the rotational levels is related to the unit statistical weight taking into account the nuclear statistical weight so that we can examine a single dependence of the number k of the levels for ortho- and paramodifications of the molecules. The slope of the straight line corresponds to the gas temperature measured by independent methods (thermocouple, the rotational structure of the spectrum of small nitrogen impurities, the width of Doppler contours) which in this case is $T_g = 145$ K at the discharge axis. To provide further information, the upper part of the figure shows, on the linear scale, the ratio of the measured population N_k of the level to the calculated population $\tilde{N}_k$, corresponding to the Boltzmann distribution at the rotational temperature equal to the gas temperature. The quantitative interpretation of the deviations was provided by the authors of Ref.28 on the basis of examining the balance of the excitation rates of the rotational levels by the electronic impact and rotational relaxation.

Some other possibilities of CALS spectroscopy as a local measurement method are indicated by Figs.5.16 and 5.17 which show the spatial distributions of the H_2 molecules ($X^1\Sigma$, $v = 0$) on individual rotational levels under the conditions of a discharge in hydrogen with standing layers and cooled with liquid nitrogen.

Figure 5.16 shows the radial distributions in the relative units of the concentration of H_2 molecules ($X^1\Sigma$, $v = 0$, $k = 2$; 3) in a discharge. For comparison, the graph also shows the graph of the Bessel function describing the radial distribution of a concentration under the diffusion discharge regime.

Figure 5.17 shows the distribution of the H_2 molecules ($X^1\Sigma$, $v = 0$, $k = 2$; 3) along the stationary layer of the discharge under the experiment

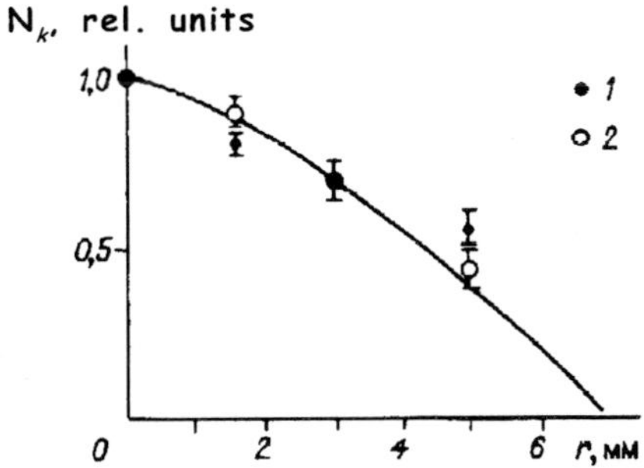

Fig.5.16 Radial distribution of the concentration of H_2 molecule ($X^1\Sigma$; $v = 0$; $k = 2$ (1), 3 (2)) in a H_2 discharge in a tube cooled with liquid nitrogen. Pressure $P = 0.5$ torr, discharge current $i = 40$ mA, solid curves is the Bessel function $J_0\left(2.4\dfrac{L^3}{R}\right)$.

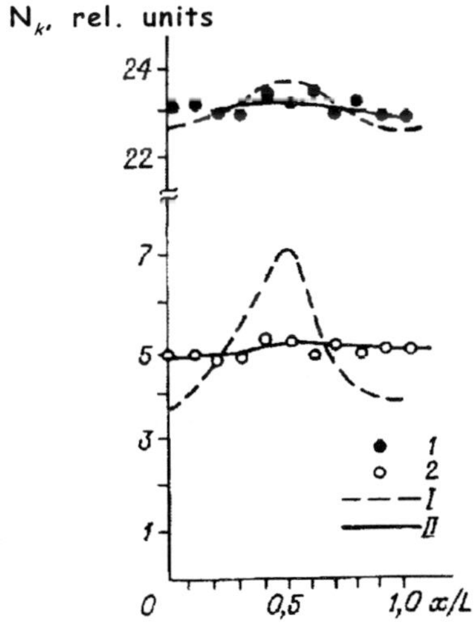

Fig.5.17 Distribution of the concentration of H_2 molecule ($X^1\Sigma$; $v = 0$; $k = 2$ (1), 3 (2)) in a discharge. Calculations were carried out disregarding (I) and taking diffusion into account (II).

conditions corresponding to Fig.5.16. The directions of propagation of the laser beams coincide with the axis of the discharge tube (Fig.5.12). The spatial resolution along the beams is 1 mm, the layer length $L = 1$ cm. As a result of a large change of the parameters of the electronic component along the layer one could also expect changes in the concentration of the molecule in different rotational states. However, measurements show that no such changes take place. The result is explained by the authors of Ref.28 by the fact that during the rotational relaxation time which is relatively short for the hydrogen molecules the axial profiles are smoothed out by diffusion. The results of the corresponding calculations are shown by solid curves in Fig.5.17.

Concluding this section, it may be noted that the described methods of diode and CALS spectroscopy enable us to transfer to a relatively new level of investigation of the processes in a non-equilibrium systems in which vibrationally–rotationally excited molecules take part in comparison with conventional methods of classic spectroscopy. This, however, does not mean that the classic spectral methods of diagnostics have lost their importance. On the contrary, the combination of various approaches opens new wider possibilities of detailed investigations of these systems.

Chapter 6

DETERMINATION OF THE CONCENTRATION AND TEMPERATURE OF HEAVY PARTICLES FROM THE SPECTRA OF RAYLEIGH-SCATTERED LIGHT

6.1 Introduction

The method of diagnostics of low-temperature plasma, based on examining the spectra of Rayleigh-scattered radiation probing the plasma, has been attracting special attention recently. This is associated not only with the fact that the method is relatively simple but mainly with the possibility of direct measurement of the concentration of heavy particles without using complicated and expensive methods of measuring light absorption in the vacuum ultraviolet range or the methods of multiphoton absorption of laser radiation which are still difficult to apply. In a number of cases, the Rayleigh scattering method can be used to measure independently also the translational temperature of the particles. This is of special interest in investigations of non-equilibrium plasma.

As is well known, Rayleigh scattering occurs when the size of the scatterers is considerably smaller than the wavelength of probing radiation. Light scattering by gas molecules can be regarded as a consequence of fluctuations of the concentration of molecules causing fluctuations of permittivity. The latter is directly included in the equation that determines the intensity of Rayleigh-scattered light.[1] The spectrum of this radiation (the structure of the Rayleigh line) depends on the nature of damping with time of concentration fluctuations which in turn is associated with the nature of the physical processes determining the propagation of concentration fluctuations in the gas.[2]

6.2 Principles of the method

Depending on whether the Rayleigh scattering is a microscopic or macroscopic process, there are two limiting scattering conditions: kinetic and hydrodynamic whose boundaries are determined by the value of the scattering factor

$$\alpha = 1/Kl, \tag{6.1}$$

where l is the mean free path of gas molecules, K is the modulus of the wave vectors $\vec{K} = \vec{K}_s - \vec{K}_0$, where $\vec{K}_0, \vec{K}_s$ are the wave vectors of the probing and scattered waves ($K_0 = 2\pi/\lambda_0$; $K_s = 2\pi/\lambda_s$, λ_0, λ_s are the wave lengths of the probing and scattered light). Since $K_s \approx K_0$, the value $K = 2K_0 \cdot \sin(\theta/2)$, where θ is the angle between the vector of the probing and scattered waves. Setting $K = 2\pi/\Lambda$, we obtain

$$\Lambda = \lambda_0/2 \cdot \sin(\theta/2) \tag{6.2}$$

It may be concluded that the scattering in the examination direction forming the angle θ with the direction of the probing beam is determined by concentration fluctuations with a wavelength of $\Lambda/2\pi$. For the scattering parameter we obtain the following equation, taking into account (6.2)

$$\alpha = \Lambda/2\pi l = (\lambda_0/l)\,(1/4\pi)\,\sin(\theta/2) \tag{6.3}$$

which shows that α can be treated as the reciprocal value of the Knudsen criterion for the scattering process.

The inequality $\alpha \gg 1$ determines the boundary of the hydrodynamic scattering regime in which the scattering spectrum consists of the central line λ_0 and two Brillouin components λ_B symmetric in relation to the central line. These components are caused by thermal fluctuations of the concentration propagating as a sound wave through the gas and are therefore shifted with respect to frequency in accordance with the Doppler effect.[3]

At $\alpha \sim 1$, the wavelength of fluctuations is close to the mean free path of the molecules. In this case, the scattering regime is kinetic and the scattering spectrum can be determined only by solving the system of kinetic equations;[4,5] The Brillouin components in the spectrum merge almost completely with essential line.

The inequality $\alpha \ll 1$ corresponds to the collisionless scattering regime. Concentration fluctuations are non-correlated, the scattering acts in every molecule of the probed volume become independent, and the frequency of the light scattered by the individual molecules is shifted in accordance with the Doppler effect by the value which depends on the velocity of the molecule and the direction of its movement in relation to the examination direction. In other words, under conditions in which the molecule velocities have some distribution (for example, Maxwell distribution), the line of probing radiation simply assumes Doppler broadening in the scattering spectrum.

In calculating the scattering parameter for the specific plasma conditions it is important to know the mean free path of the molecules which can be evaluated using relationships obtained in Ref.6 assuming that the model of collisions of solid spheres and the ideal gas flaw are valid:

$$l = \frac{8\mu}{5p}\left(\frac{2kT}{\pi m}\right)^{1/2},$$

(6.4)

where μ, p, T is the viscosity, pressure and temperature of the gas, respectively, k is Boltzmann's constant, m is the mass of the gas molecule. The accuracy of estimating l from equation (6.4) is fully acceptable for the majority of cases relevant for practice, whereas the data from Ref.7 can be used for more accurate estimates.

For the large majority of plasma and plasma chemical systems, with the exception of cases with very small scattering angles of very high pressures, the parameter $\alpha < 1$, i.e., the scattering regime is close to collisionless. The scattering spectrum consists of the components determined by the Doppler shift of the line of probing radiation as a result of kinetic movement of the light-scattering molecules. If the distribution of the molecules of every component i of the plasma with respect to velocities v is of the Maxwell type

$$f_i(v) = B_i^{1/2} \exp(-B_i),$$

(6.5)

where $B_i = m_i/2\pi kT$; m_i is the mass of the molecule of component i; T is temperature, then the scattering spectrum is

$$S_i(v) = B_i^{1/2} \exp\left(-\pi B_i\left[\frac{(v-v_0)c}{2v_0 \sin\theta/2}\right]^2\right),$$

(6.6)

where v is the frequency of scattered light; $v_0 = c/\lambda_0$, c is light velocity; θ is the scattering angle. Thus, measuring the width of the scattering line, we determine the translational temperature of scattering molecules.

The concentration of scatterers can be determined in principle by measuring the integral intensity $I(v)$ of the scattering line on the basis of the equation

$$I(v) = AI_0 N \sum_{i=1}^{z} x_i \sigma_{R_i} S_i(v),$$

(6.7)

where A is some constant, I_0 is the intensity of probing radiation, N is the total gas concentration, x_i is the molar fraction of the particles of the i-th thought, σ_{Ri} is the Rayleigh scattering cross section, z is the number of components. Equation (6.7) shows that for single component plasma it is quite easy to measure the molecule concentration. Obviously, in the case of multicomponent plasma it is necessary to know the plasma composition. In some cases, this difficulty can be bypassed by taking relative measurements of the intensity of scattered light, i.e. comparing the value measured in the plasma with that in the gas under known conditions and the same composition as the plasma. Evidently, in the presence of chemical reactions leading to a large change of the initial composition of the plasma gas, correct measurements of concentration are relatively difficult. Preliminary calculations of the composition of the reacting gas make it possible to bypass this problem in a number of cases, at least, estimate the error of concentration determination. For example, according to the data of the authors of Ref.9 who use the calculation model, the error of determination of concentration by the Rayleigh scattering method does not exceed 35% even for gas mixtures of some flames of rather complicated composition.

To estimate the contributions of each plasma component to the scattering signal, it is essential to know the Rayleigh scattering section σ_{Ri} of the molecules of this component; the section is usually computed from the equation[9]

$$\sigma_{R_i} = \frac{4\pi^2}{\lambda^4}\left(\frac{n_i - 1}{N_L}\right)\sin^2\theta, \qquad (6.8)$$

Gas	H	O	C	N	Ar	He	Ne
$(n_i - 1) \times 10^4$ (273 K, 10^5 Pa)	0.72	1.36	1.78	1.50	2.82	0.34	0.67
$\sigma_{Ri} \times 10^{28}$, cm^2 ($\theta = 90°$, $\lambda = 488$ nm)	0.47	1.78	3.05	2.16	76.7	0.11	0.41

Gas	Cl	H_2	O_2	N_2	OH	CO	NO
$(n_i - 1) \times 10^4$ (273 K, 10^5 Pa)	3.81	1.44	2.73	3.00	2.06	3.40	2.97
$\sigma_{Ri} \times 10^{28}$, cm^2 ($\theta = 90°$, $\lambda = 488$ nm)	13.9	1.89	7.18	8.66	4.08	11.12	8.50

Gas	HCl	Air (without CO_2)	H_2O	HO_2	CO_2	N_2O	H_2
$(n_i - 1) \times 10^4$ (273 K, 10^5 Pa)	4.47	2.93	2.55	3.43	4.50	5.08	6.40
$\sigma_{Ri} \times 10^{28}$, cm^2 ($\theta = 90°$, $\lambda = 4$ 88 nm)	19.2	8.16	6.26	11.3	19.5	24.8	39.4

Gas	O_2	NH_3	C_2H_2	CH_4	C_2H_4	C_2H_6	C_3H_8
$(n_i - 1) \times 10^4$ (273 K, 10^5 Pa)	6.61	3.76	5.80	4.44	6.36	7.76	10.94
$\sigma_{Ri} \times 10^{28}$, cm^2 ($\theta = 90°$, $\lambda = 488$ nm)	42.1	13.6	32.4	18.96	38.9	53.0	104.1

where n_i is the refraction index of the component i; N_L is the Lochsmidt number. The data on the refractive indices and the Rayleigh scattering sections for some gases are given below:

6.3 Concentration and temperature of gas in flames and electric arc

Possibilities, advantages and shortcomings of the Rayleigh scattering method were demonstrated most extensively in studies of the well-known investigators of turbulent flames L. Talbot and his colleagues.[6,9-13] The physical–chemical conditions in turbulent flames are relatively similar to those in plasma chemical objects (with the exception of evidently the temperature level); therefore, the results of these studies (or at least, the procedure) can be efficiently used in the diagnostics of plasma objects.

Figure 6.1 shows the diagram of experimental equipment[6] for examining the Rayleigh scattering spectra of laser radiation in flames. The source of probing radiation was an argon ion laser operating in the single-mode regime with frequency stabilisation (using an intraresonator reference) at a radiation power of approximately 1 W with at a wavelength of 488 mm. The laser beam was focused in the volume of the examined flame (constriction diameter around 0.03 mm). Diaphragms and traps were used to reduce the intensity of parasitically scattered light. The scattered light spectrum was produced using a piezoelectrically tuned Fabry–Perot interferometer (free spectrum range 0.41 cm^{-1}); the tuning range and speed were set using a computer. Radiation intensity at a hundred points uniformly distributed throughout the spectrum was measured using a photon counter with a corresponding discriminator at the input. The measured data were transferred to the same compu-

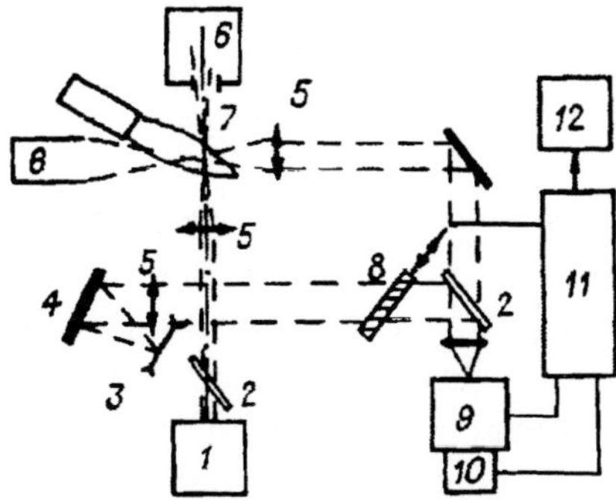

Fig.6.1 Diagram of equipment for examining Rayleigh scattering of laser radiation in a flame.[6] 1) laser, 2) mirror with 5% reflection, 3) scattering lens, 4) black screen, 5) collecting lens, 6) optical trap, 7) hydrogen flame in air, 8) screen, 9) Fabry-Perot interferometer, 10) photoelectric multiplier, 11) computer, 12) oscilloscope.

ter where they were processed using an algorithm enabling determination of the maximum intensity of the scattering line and the half-width of the profile of this line. In addition to the scattering line, the position and shape of the initial radiation line were inspected. This radiation was directed on to the input of the interferometer using the screen controlled by the same computer. An interference light filter was used to weaken the natural emission of the flame and increase the signal:noise ratio.

For the flame conditions[6] the scattering parameter did not exceed 0.1 so that the authors of Ref.6 could determine the gas temperature in the flame (~2 kK) from the half-width of the scattering line. With the measurements repeated ten times, the error of temperature measurement did not exceed 3%. In addition to the errors determined by a low signal:noise ratio, the total error was determined to a large extent by the instability of the spatial position of the flame. For the conditions outside the flame the scattering parameter was around 0.7 (kinetic scattering regime) so that it was necessary to calculate the contributions of Brillouin components to the scattering spectrum.

The intensity of scattered radiation was measured by numerical integration of the resultant spectra. Since the error of these measurements proved to be[6] relatively high (~20%), the authors of Ref.6 took only relative measurements of the density distribution in the flame. Evidently, the use of a spectral device with a smaller resolution (a narrow-band filter in the limit) made it possible to measure also the absolute values of the gas density. In fact, the majority of investigators took this approach in subsequent investigations; for example, in Ref.9–13 it was possible to measure the spatial distribution in the flame of not only the absolute values of gas density but also of its turbulent pulsations.

Two-dimensional temperature and concentration distributions of the components in turbulent flames were obtained in Ref.14 on the basis of measurements of the intensity of Rayleigh and Raman scattering of laser radiation (a dye pulsed laser with a radiation energy of 200 mJ in a pulse with a duration of 1.8 μs at a wavelength of 440 nm). The authors of Ref.15 measured the two-dimensional distribution of temperature in the diffusion turbulent flame of a mixture of methane with hydrogen where this distribution was obtained during a single pulse of laser radiation (duration 10 ns) using a special optical system of illumination of the flame by the laser radiation – the second harmonics of a neodymium laser (the energy in the pulsed 200 mJ). To select the scattered radiation they used an interference filter and the effective increase of the signal:noise ratio was achieved by using a vidicon with pulse control as a registering device. The spatial resolution of the optical system was $0.12 \times 0.12 \times 0.3$ mm^3, the error of temperature measurement in the range 1800–2300 K was around 7%.

Without discussing the experiments concerned with Rayleigh scattering in 'cold' gas flows (see, for example, Refs.16 and 17) in which measurements were taken of both the molecule concentration and its pulsations, attention will be given to the Ref.18 where the Rayleigh scattering method was used to examine the plasma of a dc arc freely burning in an argon atmosphere (purity 99.999%) at a current of 100 A (arc length 5 mm, argon flow rate 10 l/min). The plane-polarised radiation of the argon ion laser (wavelength 514.5 nm), modulated with respect to the intensity by a rotating perforated disc with a frequency of 1 kHz, was focused in the arc plasma. The laser beam leaving the arc chamber was absorbed in a trap. Scattered light was examined under an angle of 90° to the direction of the probing beam in the solid angle of 0.01 steradian. To separate the line of scattered radiation at the background of arc plasma radiation, the authors used gradually the Fabry–Perot resonator (free spectral range 0.25 nm) and a monochromator (focusing distance 1 m). The spatial resolution was 0.05×0.005 mm². To reduce the intensity of parasitically scattered light, they diaphragmed the laser beam and the laser itself was placed in a light impermeable jacket. The level of parasitically scattered light was determined by measuring the intensity of light scattered in helium and argon which filled the chamber (the ratio of the Rayleigh scattering sections of the gases was 66).[18] The results show that this level equals ~3% of the intensity of the light scattered by argon under the normal conditions so that sufficiently reliable measurements could be taken in plasma up to a plasma temperature of approximately 9 kK where the signal:noise ratio decreased to unity.

Measurements were taken of the radial distributions of the argon concentration in the arc column in different sections of the column; these data were then used to compute the distribution of plasma temperature using the equation of state of the ideal gas. Plasma temperature was also measured by an independent method on the basis of the absolute

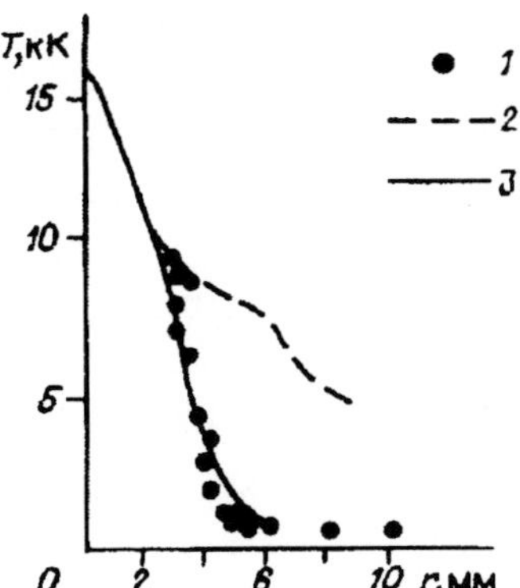

Fig.6.2 Radial distributions of temperature in the arc column in the cross-section 2 mm from the cathode. Total length of the arc column 5 mm, current intensity 100 A, argon; 1) results of measurements by the Rayleigh scattering method, 2) results of spectroscopic measurements, 3) calculations in the LTE approximation.

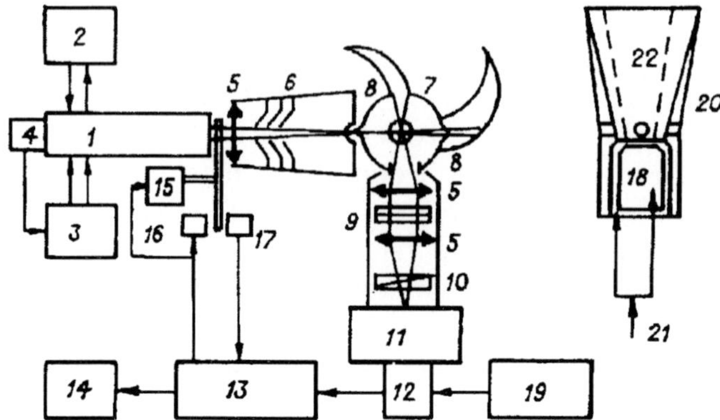

Fig.6.3 Diagram of equipment for investigating Rayleigh scattering or laser radiation in a plasma jet. 1) laser, 2) system for pumping and filling the laser tube, 3) light flux stabiliser, 4) sensor of laser radiation intensity, 5) collecting lens, 6) a set of diaphragms, 7) camera, 8) optical trap, 9) interference light filter, 10) polariser, 11) monochromator, 12) photoelectric multiplier, 13) unit for amplifying and recording signals and controlling equipment, 14) automatic recording device, 15) drive of the modulator, 16) light diode, 17) photoresistance, 18) plasma torch, 19) unit for cooling the photoelectric multiplier, 20) distribution of the plasma torch in the chamber (axial section), 21) feed of purified argon, 22) plasma jet.

intensity of ArI line at 696.5 nm. Figure 6.2 show some of the results.[18] It may be seen that the temperatures measured spectroscopically and by the scattering method greatly differ in the peripheral regions of the arc column. At the same time, the results of measurements by the scattering method are in good agreement with the results of calculations[19] carried out assuming the validity of the LTE model for the arc plasma. The calculated results are also in good agreement with spectroscopic data obtained for the near-axial regions of the arc. Thus, the peripheral regions of the arc were characterised by large deviations of the population of the excited levels from those corresponding to the LTE model. These deviations are explained by the excitation of argon atoms at the arc periphery by resonant radiation of the argon plasma of the high-temperature (~15 kK) core of the arc, as confirmed by the results of calculations published in Ref.20 in which the process of excitation and deactivation of argon atoms in the peripheral regions of the arc column was taken into account.

Thus, experiments with the method of Rayleigh scattering of light made it possible to determine the reasons for the deviation of the state of plasma from that described by the LTE model generally accepted for the examined object.

6.4 Plasma jet diagnostics

In our experiments, the method of Rayleigh scattering of light was used for the diagnostics of the argon plasma jet. The diagram of equipment is shown in Fig.6.3. We used a LG-106M argon ion laser produced in Russia, with a radiation power of 0.5–0.7 W at a wavelength of 488 nm. Since the operating life of lasers of this type is very short, the laser tube was connected to the argon pumping and inlet system so that in addition to replacing the working gas it was possible to maximise the radiation power by varying the argon pressure in the tube. Instead of the standard laser power unit, we used a special voltage stabiliser controlled by the sensor of the laser radiation intensity; it was thus possible to reduce the fluctuations of radiation intensity from 20 to 1.5%. The radiation of the laser output was modulated by a rotating perforated disc at a frequency of 729 Hz. Using a lens (focusing distance 207 mm), the laser beam was focused inside a chamber at its axial line which coincided with the axial line of the plasma jet travelling into the chamber from the plasma torch nozzle (diameter 3 mm); the plasma torch operated at frequencies of 90–120 A, arc voltage 22–24 V, argon flow rate 0.15–0.25 g/s. At the exit from the chamber, the laser beam was absorbed in an optical trap. Scattered radiation was examined under an angle of 90° to the axial line of the beam in the solid angle of ~0.03 steradian using a lens with a focusing distance of 195 mm. An interference filter was placed in the quasi-parallel light beam behind this lens (the position of the maximum of transmittance and the width of the band were 486.9 and 18 mm respectively). The second lens of the receiving optical system produced the image of the cross section of the probing beam at the input slit of the monochromator in the focus constriction area (the constriction diameter was around 60 µm); the slit of the spectrometer cut out from the image a section approximately 15 µm wide. These dimensions characterised the spatial resolution of equipment. A FEU-79 multiplier was positioned at the output of the monochromator. Its cathode was cooled using a thermoelectric cooler to a temperature of around −15°C. The signal from the output of the photoelectric multiplier was transferred to a selective amplifier with a synchronised detector at the output. The signal was recorded in an automatic recording device.

The main task in measuring the intensity of Rayleigh-scattered light was to ensure an acceptable signal:noise ratio. Noise sources were the parasitically-scattered light, natural plasma radiation, the noise of the photoelectric detector and the appearance of macroscopic particles in the probed volume of the plasma flow. To reduce the intensity of the parasitically scattered light, the internal walls of the chamber were ground and bevelled in relation to the axial line under an angle of 8°;

the conicity of the chamber walls also made it possible to prevent return flows in the chamber bringing macroscopic particles from the outside into the examined jet. The windows positioned initially under the Brewster angle in the chamber walls for transmitting the laser beam in examining scattering were dismantled so that the level of parasitically scattered light was greatly reduced. The beam was transmitted through the hole in the wall with a diameter of around 3 mm; a series of diaphragms with a gradually decreasing aperture were placed along the path of the beam from the length to inlet into the chamber. Optical systems of illumination and examination were placed in duralumin thin-walled tubes with black walls. The optical traps for absorbing probing radiation at the exit from the chamber and for absorbing parasitically scattered light (opposite the aperture of the receiving optical system) were made of glass with the black external surface. The natural radiation of plasma was suppressed quite efficiently using an interference filter, a monochromator and a polariser. Cooling of the cathode of the photo-electric multiplier made it possible to reduce the level of natural noise of the multiplier to a level at which the contribution of this noise to the noise signal was insignificant. To reduce the frequency of appearance of macroscopic particles in the probing volume, the argon fed into the plasma torch and for circular blowing of the plasma jet in the chamber was cleaned to remove dust in three consecutive filters FV-1.6; argon was dried prior to passing through the filters. All these measures and also careful adjustment of the optical system caused that the signal:noise ratio approached unity in measurements of the intensity of scattered radiation in the plasma with a temperature of around 9700 K.

In addition to determination of the intensity of Rayleigh-scattered light, we also measured the pulsations of the intensity (in the peripheral regions of the jet); the laser beam was not modulated in this case.

It is also important to note that the appearance of macroscopic particles in the probed volume could not be completely prevented; the frequency of appearance of the particles was on average 12 min^{-1}. Since the intensity of the light scattered by the particles exceeds the intensity of Rayleigh scattered light by 6–7 orders of magnitude, the particles greatly complicated the measurements. To suppress the noise signals (including those from the particles) whose amplitude was higher than the amplitude of the measured signal of Rayleigh scattering, we developed a special electronic circuit including a photoelectric multiplier for the duration of the effect of a strong noise; the level of signal discrimination was selected during measurements. More efficient suppression of the noise may be ensured by using the method of digital filtration of the signal.[21]

This equipment was also used to measure the temperature distribution in the plasma jet on the basis of the absolute intensity of lines ArI 430 and 696.5 nm. The radiation standard was SI-10/300 tungsten strip lamp. In the measurements were used the same monochromator and the amplifying-recording part of equipment.

Figure 6.4 shows the radial distribution of temperature in the plasma jet in the vicinity of the outlet of the plasma torch nozzle for two different mean mass enthalpies of the plasma. Temperature T was determined from the results of measurements of the outlet concentration by the scattering method using the equation of state of the ideal gas $p = NkT$, where p is pressure (here 10^5 Pa), N is the partial concentration, k is Boltzmann's constant. The same graph shows the results of spectroscopic measurement. In contrast to the experiment discussed previously,[18] there were no large differences between the results of individual measurements. It is possible that this is caused by a considerably lower level of temperatures in the jet core in comparison with that in the arc,[18] if we take into account the results of calculations carried out in Ref.20. Figure 6.5 shows the radial distribution of mean quadratic pulsations of the intensity of scattered light, i.e. pulsations of the argon atom concentration. It may be seen that the pulsation intensity at the periphery of the jet where rapid turbulent mixing of the plasma jet with the surrounding cold air takes place, the pulsation intensity reaches a high value. The presence of these pulsations leads to a large increase of the measurement error of the mean concentration and, consequently, the temperature in the corresponding regions of the jet.

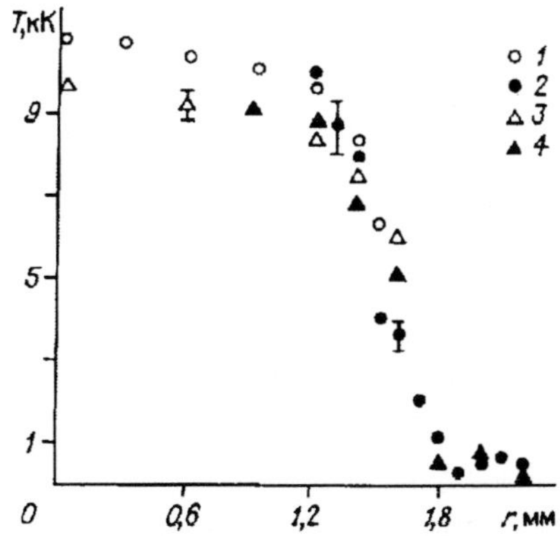

Fig.6.4 Radial distributions of the temperature in the cross-section of the plasma jet of argon situated 1.2 mm from the outlet from the plasma torch nozzle. 1,3) results of spectroscopic measurements, 2,4) results of measurements by the method of Rayleigh scattering; specific enthalpy of plasma 2.4 (1,2) and 1.4 MJ/kg (3,4).

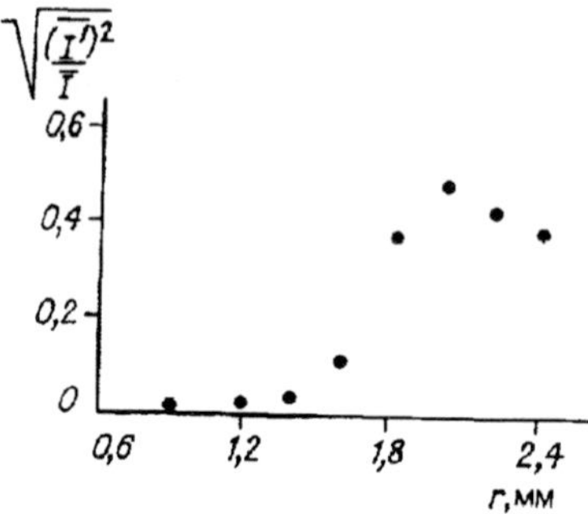

Fig.6.5 Radial distribution of the relative rms pulsations I' of intensity I of Rayleigh scattered light in the plasma jet in the section situated at a distance of 1.2 mm from the outlet of the plasma torch nozzle (specific enthalpy of the plasma 2.4 MJ/kg).

6.5 Gas temperature in glow discharge

This parameter is one of the important parameters that determine the course of many physical–chemical processes in the discharge plasma and on substrates placed in the plasma. Gas temperature T_g is measured on the basis of the rotational structure of the spectrum of diatomic molecules assuming that T_g coincides with the rotational temperature measured directly in this case, or by different contact methods on the basis of assumptions whose validity cannot always be verified. Interferometric methods of measuring the gas concentration in the discharge are complicated and the corresponding optical systems are very difficult to use. It was therefore interesting to measure the gas concentration in the discharge and, consequently, determine T_g by the Rayleigh scattering of light method.

Under the glow discharge conditions at $p \sim 3 \cdot 10^2$ Pa the gas concentration is 10^{16}–10^{17} cm^{-3}, and in comparison with the previously examined objects at the atmospheric pressure, the intensity of the scattering signal (with other conditions being equal) should decrease by almost three orders of magnitude. To increase the signal:noise ratio, the equipment whose circuit is shown in Fig.6.3, was modified. Instead of an argon laser, an LT-401 laser fitted with a radiation frequency converter was used; radiation power at a wavelength of 543 nm was 1.2-1.5 W. The conventional design of the glow discharge tube (internal diameter 48 mm) was modified in such a manner as to reduce the intensity of parasitically scattered light. The recording system used the Fabry-Perot interferometer, a polariser and a NDR-6 monochromator; signals were recorded in the memory of a D3-28 computer. The main interference in this case was the natural radiation of the discharge plasma. Since the signal:noise ratio at working pressures of the gas in the discharge of 0.1–1 kPa was 0.6–3, the scattering line was recorded many

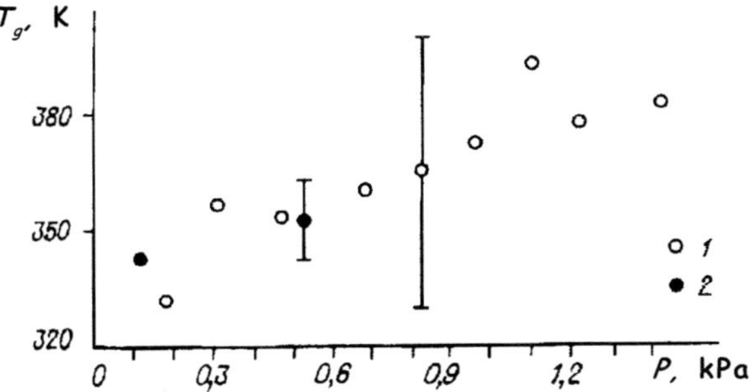

Fig.6.6 Results of measurements of the argon temperature in glow discharge at different pressures. The axial line of the cylindrical discharge tube with a diameter of 48 mm, discharge 50 mA.

times (usually 900–2000 times) in the presence and absence of plasma; the natural radiation of the plasma was recorded with the overlapped laser beam. The recording cycles were synchronised using a computer with the cycle of movement of the screen overlapping the beam. The algorithm of subsequent data processing made it possible to determine the average value of the intensity of the radiation scattered in the plasma normalised to that measured in the absence of plasma. The rms error of measurements was computed for each measurement cycle and its value varied with respect to temperature from ~17% at lower pressures to ~10% at higher pressures in the pressure range given previously.

Figure 6.6 shows the data on the gas temperature at different pressures in the discharge 1 obtained by the scattering method. The temperatures measured by Yu.A. Ivanov and V.A. Timakin by the thermal probe method 2, described in Chapter 13, are also presented here. Regardless of the relatively good agreement of the results of the individual measurements, it should be noted that the Rayleigh scattering method in this form is characterised by a very high error. On the basis of the experimental material, it appears that in Rayleigh diagnostics of stationary plasma it is necessary to use pulsed lasers with the pulsed time in the range from one to several microseconds with radiation energy in the pulse of several microJoules permitting retuning with respect to the wavelength (to eliminate the interfering lines of plasma radiation). It is thus possible to increase greatly the accuracy of measuring the gas concentration at low gas pressures in the discharge. It should only be remembered that for specific discharge conditions it is important to evaluate the degree of perturbation of the plasma by laser radiation at high laser radiation energy levels.

Chapter 7

REFRACTOMETRIC PLASMA DIAGNOSTICS METHODS

Refractometric examination of plasma is based on measuring its refractive index. Heterogeneous media are examined by interferometric and shadow methods. In the interferometric method, the refractive index is computed from the experimentally measured values of the optical difference of the path. In the shadow method, the refractive index is determined from the angle of deviation of the light beam from its initial direction as a result of deformation of the light wave during passage through the examined object. In the shadow method, we examine the shadow on the screen of some non-transparent barrier along the path of light propagation, and this is also reflected in the name of the method. Subsequently, using the relationship of the refractive index with the internal plasma parameters (temperature, electron and atom concentration), the values of these parameters can be determined. Refractometric devices together with the measurement of plasma temperature and the concentration of atoms and electrons can be used to examine the nature of plasma flows.[1] In the laminar gas discharge flow (plasma) there are no local density fluctuations. The interference pattern has fringes stable with time. Large-scale density pulsations leads to erosion of the interference pattern when the period of large-scale plasma flow fluctuations is considerably smaller than the exposure time. The large-scale pulsations are not visible on the interferograms with a short holding time. Local heterogeneities with high-density gradients, characteristic of turbulent flows, can be seen on interferograms as dark and light spots.

7.1 Refractive index of plasma

The group and phase velocities are defined in propagation of light in a medium. The group velocity is associated with energy transport and is always lower than the light velocity in vacuum. Phase velocity c_p is associated with the displacement of the phase of the light wave and can be both lower and higher than the light velocity in vacuum. The refractive index is determined by the difference of the phase velocity from the light velocity in vacuum c and is equal to the ratio $n = c/c_p$ and the values

$n > 1$ and $n < 1$ are possible. The value of the phase velocity is determined by the interaction of an electromagnetic wave with the electrons of the medium, both bonded and free.

The refractive index n and the absorption factor of the medium κ are linked by the Kramers–Kronig relationship[2]

$$n(\omega) - 1 = \frac{c}{4\pi} \int_{-\infty}^{\infty} \frac{\kappa(\omega')d\omega'}{\omega'(\omega' - \omega)}. \tag{7.1}$$

The refractive index at frequency ω depends on the absorption factor in the entire spectrum range.

The dependence (7.1) holds for any processes leading to light absorption and can be used to calculate the refractive index from the available spectral absorption factor. For the spectral lines, the absorption factor is:

$$\kappa(\omega) = \frac{\pi e^2}{m_e c} N f P(\omega), \tag{7.2}$$

where N is the concentration of absorbing atoms, f is the force of the oscillators in absorption.

Substituting (7.2) into (7.1) and computing the integral, we obtain for the Lorentz contour $P_L(\omega)$

$$n(\omega) - 1 = -\frac{\pi e^2 N f}{m_e \omega_0} \frac{\omega - \omega_0}{(\omega - \omega_0) + (\gamma/2)^2}, \tag{7.3}$$

where ω_0 is the centre of the line, γ is its half width.

Figure 7.1 shows the course of the function $n(\omega) - 1$ in the vicinity of the centre of the spectral line. The range of rapid variation of the refractive index in the vicinity of the spectral line is referred to as the anomalous dispersion zone.

Measurements of the refractive index in this zone are usually taken by the Rozhdestvenskii 'hook' method which enables, together with equation (7.3) to determine either the concentration of absorbing atoms or the force of the oscillators. The method is highly accurate in operation with homogeneous objects (for example, plasma in a shock tube). In the case of heterogeneous plasma, the 'hooks' are eroded and the method is difficult to use in this application.

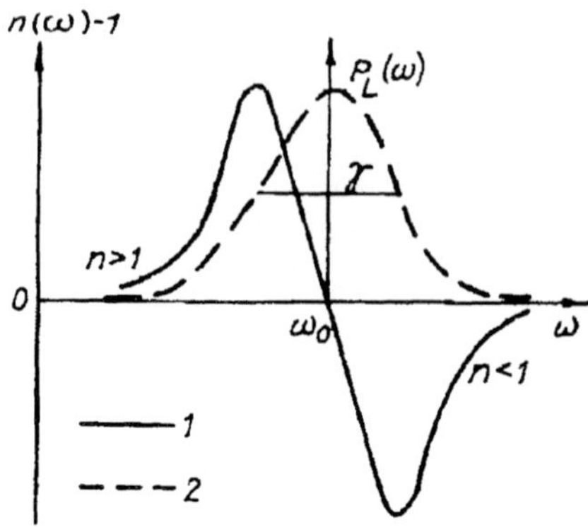

Fig.7.1 Refractive index for the Lorentz contour of the spectral line. 1) $n\omega$, 2) $P_L(\omega)$.

Away from the centre of the spectral line at frequency ω, satisfying the conditions $\omega \ll \omega_0$, $\omega_0 - \omega \gg \gamma/2$, equation (7.3) assumes the form

$$n(\omega) - 1 = \frac{2\pi e^2}{m_e \omega_0^2} Nf. \tag{7.4}$$

It may be seen that in this case the refractive index does not depend on the probing radiation frequency ω. Equation (7.4) holds for any contour of the spectral line (with the exception of the purely Doppler contour) because they have the same asymptotics in the distribution of the absorption factor. In calculating the refractive index of plasma at frequencies away from the spectral lines, it is sufficient to take into account the contribution of ground states of the atoms (ions) because their value Nf is considerably higher than for the excited states. The condition $\omega \ll \omega_0$ is fulfilled in examining gases (nitrogen, oxygen, argon, etc.) in the visible region of the spectrum because their resonance lines are situated in the vacuum ultraviolet region of the spectrum.

The contribution of the free plasma electrons to the refractive index can be explained by the interaction of the probing electromagnetic wave with a wave in plasma at the plasma frequency. The latter is determined by the superimposition of the ordering effect of Coulomb interactions in the Debye sphere on the chaotic motion of the electrons and is equal to

$$\omega_p = 2\pi \left(\frac{e^2 Ne}{\pi m_e} \right)^{1/2}.$$

Under laboratory plasma conditions, the plasma frequency is usually low and situated in the infrared range of radiowave frequency range. At the probing frequency $\omega \gg \omega_p$ the refractive index, determined by the free electrons, is lower than unity:[2]

$$\left[n(\omega) - 1 \right]_e = - \frac{2\pi e^2 N_e}{m_e \omega^2}. \tag{7.5}$$

When calculating the refractive index at a given frequency, it is important to take into account the contribution of the spectral lines and free electrons. For example, for argon plasma:

$$(n - 1) = (n - 1)_{ArI}, \; (n - 1)_{Ar*I} + (n - 1)_{ArII} + (n - 1)_e.$$

The components of this sum at $\lambda = 5000$ Å:[3]

$$(n - 1)_{ArI} = 1.04 \cdot 10^{-23} N_{ArI}, \; (n - 1)_{Ar*I} = 13.2 \cdot 10^{-23} N_{Ar*I},$$
$$(n - 1)_{ArII} = 0.715 \cdot 10^{-23} N_{ArII}, \; (n - 1)_e = -13.4 \cdot 10^{-23} N_e.$$

Here ArI, Ar*I, ArII denote the neutral and excited argon atoms and argon ions, respectively. Since $N_{Ar*I} \ll N_{ArI}$, then $(n - 1)_{Ar*I} \ll (n - 1)_{ArI}$ and the contribution of the excited states to the refractive index can be ignored, as mentioned previously. The contribution of argon ions is also negligible: it is considerably smaller than the contribution of free electrons ($N_{ArII} \approx N_e$). Thus, it is sufficient to take into account the contribution of the argon atoms in the ground state and the contribution of the free plasma electrons. As an example, Fig.7.2 shows the dependence of the refractive index of the argon plasma at atmospheric pressure at a wavelength of $\lambda = 6328$ Å. The refractive index of the plasma decreases with increasing temperature as a result of a decrease of the atom concentration due to thermal expansion of the gas at constant pressure and atom ionisation. The free electrons provide a negative contribution to the value $n - 1$ and determine mainly the refractive index at an argon ionisation degree higher than 10%. For most gases, the refractive index can be calculated quite accurately because it is determined by the spectral lines of the main series (the forces of oscillators of these lines are available with a high degree of reliability) and by free

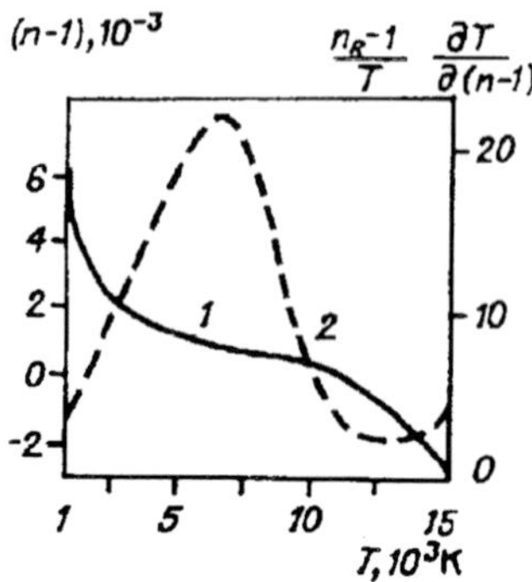

Fig.7.2 Refractive index of argon plasma at a wavelength of $\lambda = 6328$ Å at $P = 1$ atm (1) and the coefficient of transfer of the error (2).

electrons whose contribution is computed with high accuracy from equation (7.5).

In the case of LTE plasma, the dependence $n\,(T)$, identical with that shown in Fig.7.2, can be used to determine the plasma temperature from the measured refractive index. If the plasma is in the partial thermodynamic equilibrium state, the measurements of the refractive index at two relatively different frequencies ω_1 and ω_2 make it possible to determine the concentration of atoms and ions. In fact, the refractive index, determined by the atoms, does not depend on frequency (see equation (7.4)), whereas the refractive index, determined by the free electrons, is inversely proportional to the frequency (see equation (7.5)). We have two equations

$$n\,(\omega_1) - 1 = (2\ e^2/m_e\omega^2_0)\ fN_a - (2\pi e^2/m_e\omega^2_1)\ N_e,$$
$$n\,(\omega_2) - 1 = (2\ e^2/m_e\omega^2_0)\ fN_a - (2\pi e^2/m_e\omega^2_e)\ N_e$$

for determining the concentration of atoms N_a and the electrons N_e. Using Sah's equation, it is possible to calculate the electron temperature T_e and compare it with the temperature determined by a different method assuming LTE and evaluate at the same time the deviation of the state of plasma from local thermodynamic equilibrium.

The common feature of any measurement system of the refractive index is that the value $\zeta = n - n_R$ is determined directly; here n_R is the refractive index of the medium outside the examined object. In calculating temperature $n = \zeta + n_R$ is used. It is therefore necessary to know the refractive index n_R of the surrounding environment. Since n_R is usually considerably higher than the refractive index of plasma, this parameter can be determined with high accuracy. The value n_R can be measured directly by the interferometry method using, as a reference, a cuvette with a thoroughly scrubbed gas whose refractive index is known accurately (hydrogen). n_R can also be determined indirectly, as is the case in the combined spectroscopic–refractometric method of temperature detemination.[1] From the temperature T_0 at the axis of the plasma object determined by the spectroscopic method it is possible to find, using the dependence $n(T)$ (Fig.7.2), the refractive index at the axis n_0 from which $n_R = n_0 - \zeta_0$, where ζ_0 is the value measured by the refractometric method at the axis of the object.

The relationships between the errors of measurement of the refractive index and temperature in the case of LTE plasma will now determined. Varying $n = n(T)$, gives $\delta n = (\delta n/\delta T)\delta T$, and consequently $\delta T/T = (\delta T/T\delta n)\delta n$.

In turn, $\delta n = \delta \zeta + \delta n_R$, since $\delta T/T = (\delta T/T\delta n)(\delta \zeta + \delta n_R)$. The last equation will be written in the form $\delta T/T = (\delta T/T\delta(n-1))[\delta \zeta + \delta(n_R - 1)$ since in practice it is more convenient to operate with the value $n - 1$ because of the fact that n differs only slightly from unity. Approximately, it can be written that $\zeta = n_R - n \approx n_R - 1$ and $\delta \zeta/(n_R - 1) \sim \delta \zeta/\zeta$ and, consequently,

$$\frac{\delta T}{T} \approx \frac{n_R - 1}{T} \frac{\partial T}{\partial (n-1)} \left[\frac{\delta \zeta}{\zeta} + \frac{\delta(n_R - 1)}{n_R - 1} \right]. \tag{7.6}$$

The dependence of the transfer factor of the error $[(n_R - 1)/T] [\partial T/\partial(n-1)]$ on the argon plasma temperature is shown in Fig.7.2.

The accuracy of the refractometric method is considerably lower than that of the spectroscopic method. The refractometric method gives reliable results only for a highly stable object at a sufficiently high frequency of the measurements and detailed processing of the results. As shown in Ref.1, it is possible to measure the temperature with an error of 5-10%. An important advantage of the refractometric method of the measurement is that is enables the latter to determined at the periphery of the plasma object outside the radiation zone where the spectrometric method cannot be used. The combined application of the spectroscopic and refractometric methods enables a complete radial temperature distribution to be plotted.

7.2 Plasma interferometry

Optical systems[4,5] for determining the refractive index in interferometric experiments are oriented at measurements of the optical difference of the path of two beams, one of which passes through the examined object with thickness l with a refractive index n, the other one passes outside

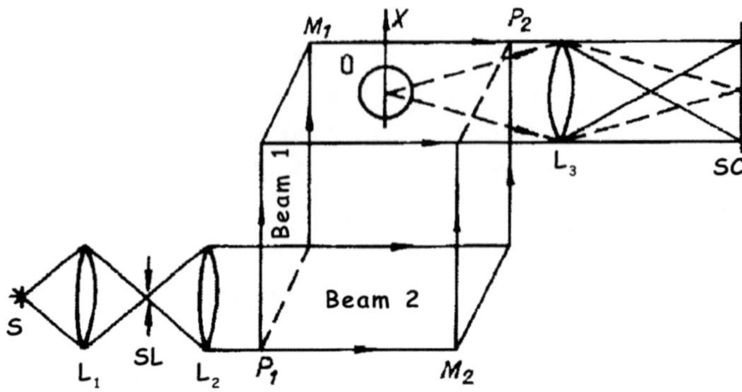

Fig.7.3 Mach–Zender interferometer.

the object where the refractive index is equal to n_R. The measured optical path difference is

$$\mu = l \, (n - n_R) \tag{7.7}$$

The most efficient device for measuring the optical path difference of gas or plasma objects is the Mach–Zender interferometer. However, this equipment is of complicated design and is expensive. An interferometer based on a modification of a standard IAB-451 shadow equipment is used more widely. It is also less sensitive to vibrations and is simple to set up.

Mach–Zender interferometer
The circuit of the Mach–Zender interferometer is shown in Fig.7.3. It is a two-beam interferometer consisting of two mirrors M_1 and M_2 and two plain-parallel semi-transparent sheets P_1 and P_2. The distances $M_1 P_1 = M_2 P_2$ and $M_1 P_2 = P_1 M_2$ are equal. The mirrors M_1, M_2 and the sheets P_1, P_2 are parallel and ranged in pairs. The light source S is reflected by the lens L_1 on the slit SL of the device situated in the focus of the collimator L_2. The beam light from the source of illuminating radiation is divided by the sheet P_1 into two beams which after reflection from M_1 and M_2 and passage through P_2 are projected onto the screen SC. The examined object in placed in the beam bundle 1. The lens L_3 forms the image of the object O on the screen. The beam bundles, the object 1 and reference 2, are coherent in relation to each other and generate an interference pattern on the screen. If all the mirrors and sheets are parallel, then in the absence of the object the width of the interference fringes, formed on the screen, is infinite (the interference field is uniformly illuminated). The introduction of an object leads to the appearance of fringes whose shape corresponds to the curves of equal

values of the optical path difference. The interference fringes are localised in the planes of the object and its image on the screen so that it is possible to determine the correspondence between the position of the fringe on the screen and the line equal to the optical path difference in the objects. If mirrors M and sheets P are positioned under a small angle in relation to each other, an additional path difference forms between the beams 1 and 2 and finite-width fringes appear on the screen. The fringes are lines of the equal path difference which differ for the adjacent lines by the light wavelength. The fringe width depends on the angle between M and P and increases with its decrease. In the presence of an optically heterogeneous medium (examined object) the interference fringes on the screen change their shape in accordance with the distribution of the optical path difference in the object. The distance between the branches of the Mach–Zender interferometer is sufficiently large so that very large objects can be examined. Interferograms can be recorded using a camera placed instead of the screen. Lasers are usually used as the source of probing radiation in plasma diagnostics. As a result of the high brightness of laser radiation and the use of a narrow-band light filter, it is possible to prevent illumination of the interferogram by plasma radiation.

Figures 7.4 and 7.5 show interferograms of the flame of a spirit torch in setting the interferometer in relation to the infinite-width and finite-width fringe. The fringes of the equal optical path difference (Fig.7.4) correspond to the additional path difference $\mu = k\lambda$, $k = 1, 2, ...$, with counting carried out from the periphery to the centre of the object. In processing the interferogram, the position of the fringes on the coordinate axis X with the start of counting in the centre of the object is measured in the given cross section of the object (axisymmetric in this case) and the dependence $\mu = \mu(x)$ is plotted. When processing the interferogram of finite-width fringes (Fig.7.5a), the additional path dif-

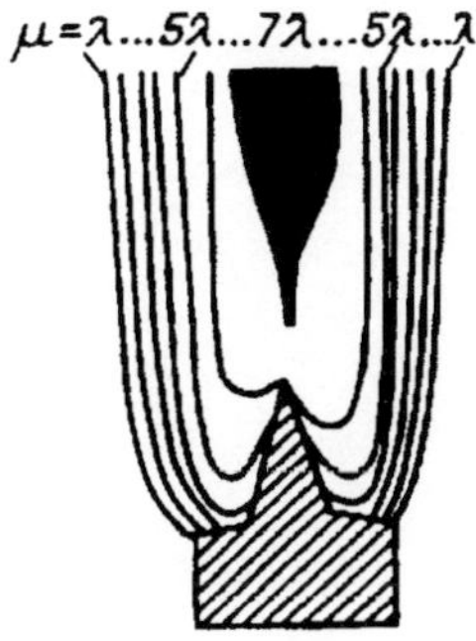

Fig.7.4 Interferogram of the flame of a spirit torch. The fringes of equal optical path difference.

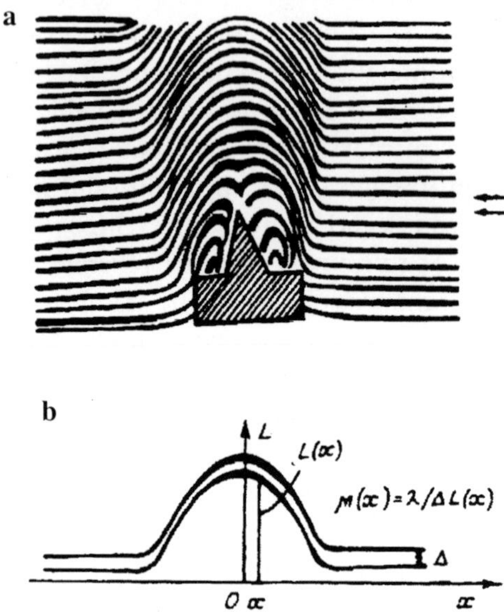

Fig.7.5 Interferogram of the flame of a spirit torch. Finite-width fringes (in the absence of the object the interference fringes are horizontal over the entire plane).

ference introduced by the examined object, is determined using the following procedure (Fig.7.5b). In the given cross section measurements are taken, for different values of the co-ordinate x, of the shift of the fringe in relation to the initial position in fractions of the distance between the pair of unperturbed fringes whose optical path difference is λ. The relationship between the measured additional path difference $\mu(x)$ and the required distribution of the value $\zeta(r) = n(r) - n_R$ ($n(r)$ is the refractive index at the circumference of the radius r, n_R is the refractive index outside the object) is given by the Able integral equation

$$\mu(x) = 2\int_{x}^{R} \frac{\left[n(r) - n_R\right] r\, dr}{\sqrt{r^2 - x^2}}, \tag{7.8}$$

which is converted to

$$\zeta(r) = n(r) - n_R = -\frac{1}{\pi}\int_{r}^{R} \frac{d\mu(x)}{dx} \frac{dx}{\sqrt{x^2 - r^2}} \tag{7.9}$$

Equations (7.8) and (7.9) are valid for small angles of deflection of the light beam in passage through the object. If the deflection angle is large, the interference fringe and its processing are greatly complicated. In this case, it is convenient to transfer to shadow examination methods.

The 'slit-filament in the focus' interferometer

The main simplified circuit of this interferometer is shown in Fig.7.6. The light source S is reflected by the lens L_1 onto the slit SL of the device which is situated in the focus of the collimator objective L_2. The parallel light beam, leaving the objective L_2, is focused by the camera objective L_3 in the focal spot F and then projected on to the screen SC. The non-transparent filament H is placed in the focus of the camera objective and overlaps the image of the slit. The thickness of the filament is slightly larger than the image width of the slit (filament thickness ~0.1 mm, the width of the slit image ~0.08 mm). The light carrying out diffraction on the filament uniformly illuminates the screen in the absence of the heterogeneous object. The examined object O is placed in the light beam between the objectives L_2 and L_3 and reflected by the objective L_3 on to the screen. Because of the presence of the gradient of the refractive index in this object, the light beams are deflected from the initial direction and pass away from the filament onto the screen. The deflected light beam is displayed on the screen at the same point as the unperturbed beam because the planes of the object and the screen are connected by the lens L_3. The rays of the light of the parallel beam, passing outside the object, are overlapped by the filament, but part of the light falls on to the screen as a result of diffraction on the filament. The light rays formed in this manner and passing through the object and outside it interfere and the screen shows the fringes of equal optical path difference, similar to those shown in Fig.7.4. At specific ratios between the size of the object and the width of the slit and the filament the interference fringe in the described interferometer is identical with that in the Mach–Zender interferometer with setting to the infinite-width fringe.

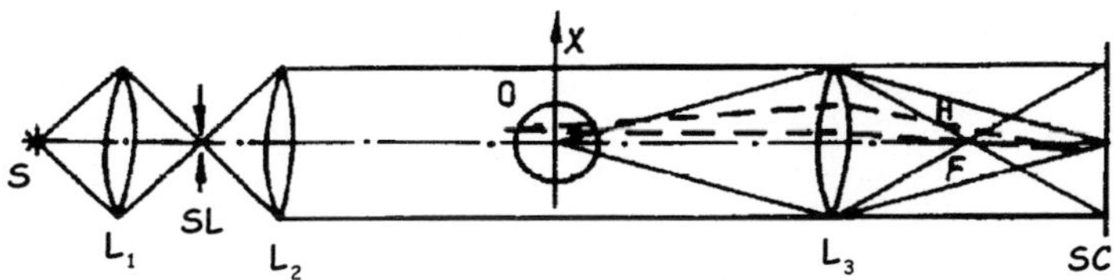

Fig.7.6 Diagram of the "slit-filament in focus" interferometer.

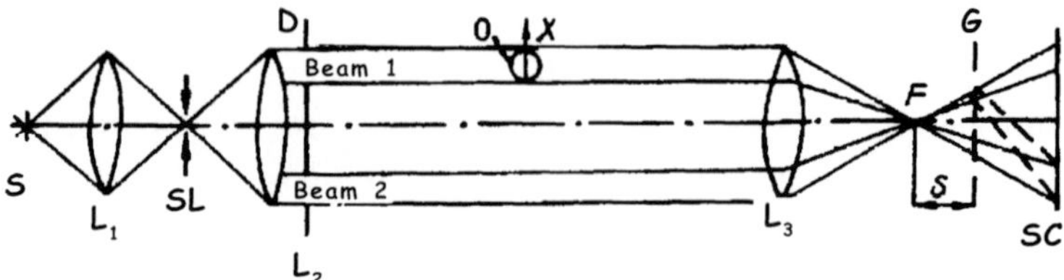

Fig.7.7 Diagram of an interferometer with a diffraction grating.

An interferometer with diffraction grating

A simplified circuit of the interferometer with a diffraction grating is shown in Fig.7.7. The parallel light beam is divided into two beams using diaphragm D. The examined object O is projected by the objective L_3 on the screen SC and is placed in one of these beams. The transparent diffraction grating DG (50–100 lines/mm) is positioned in the vicinity of the focus F of the camera object L_3. Each of the probing beams is split by the diffraction grating into a number of beams. The rays from the first and second bundles that overlap on the screen interfere together. In the absence of the object, the screen shows interference fringes of finite width with the spacing

$$\Delta = \frac{m\,b\,f}{\left(k_1 - k_2\right)s},\qquad(7.10)$$

where m is the scale of reflection of the object on the screen; b is the grating constant; s is the distance of the grating from the focus F; f is the focusing distance of the objective L_3; k_1, k_2 is the order of interference diffraction maxima of the first and second beams. When the examined object is placed in the beam 1 the interference fringes are distorted. This gives an interferogram identical with that shown in Fig.7.5a for the Mach–Zender interferometer. If the diffraction grating is placed in the focus F then in accordance with the equation (7.10) $s = 0$ and $\Delta \rightarrow \infty$ and the screen shows the fringe of infinite width. For this setting of the interferometer and placing the examined object in the beam 1 the screen shows fringes of the equal optical path difference of the type shown in Fig.7.4. Thus, the interferometer with a diffraction grating makes it possible to realise two setting schemes: finite width fringes at $s > 0$ and an infinite width fringe at $s = 0$.

Shearing interferometer

The diagram of the shearing interferometer differs from those described previously by the fact that a light-dividing device (a system of mirrors,

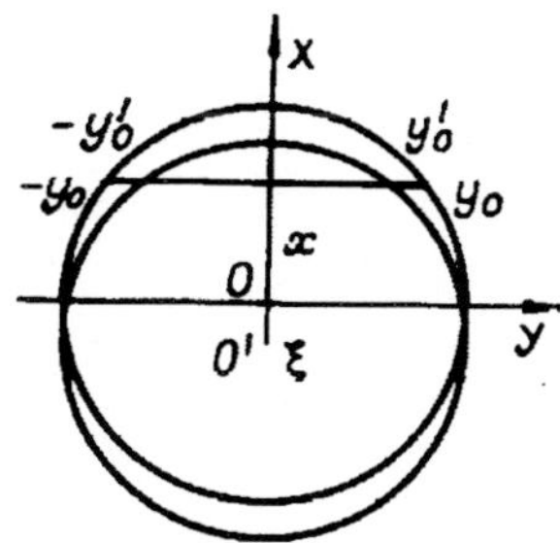

Fig.7.8 Displaced images of the object on the screen of a shearing interferometer.

Wollaston prism) is positioned in the vicinity of the focal plane of the camera objective and splits the light beam into two beams in such a manner that the screen shows two images of the examined object shifted in relation to each other (Fig.7.8). In this method, the light beams which pass through the object along the chords are separated by the shift distance ξ interfere. The corresponding optical path difference

$$\mu(x) = \int_{-y_0}^{y_0} \left[n(r, x) - n_R \right] dy - \int_{-y_0'}^{y_0'} \left[n(r, x + \xi) - n_R \right] dy, \qquad (7.11)$$

where the first integral corresponds to the optical path difference for the chord whose distance from the centre is x (initial image), the second integral for the chord $(x + \xi)$ (shifted image). For the most interesting case – small shift – we can write

$$n(r, x) - n(r, x + \xi) \approx \frac{\partial n(r, x)}{\partial x} dx = \frac{\partial n(r, x)}{\partial x} \xi$$

and then

$$\mu(x) = \int_{-y_0}^{y_0} \left[(r, x) \right] - n(r, x + \xi) dy = \xi \int_{-y_0}^{y_0} \frac{\partial n(r, x)}{\partial x} dy. \qquad (7.12)$$

Taking into account the rotational symmetry $r^2 = x^2 + y^2$

$$\frac{\partial n}{\partial x} = \frac{\partial n}{\partial r} \frac{\partial r}{\partial x}, \quad \frac{dr}{dx} = \frac{x}{r}, \quad dy = \frac{r dr}{\sqrt{r^2 - x^2}}.$$

Consequently, equation (7.12) takes the form

$$\mu(x) = 2\xi x \int_x^R \frac{dn}{dr} \frac{dr}{\sqrt{r^2 - x^2}}$$

or

$$\frac{\mu(x)}{\xi x} = 2 \int_x^R \frac{dn}{rdr} \frac{rdr}{\sqrt{r^2 - x^2}} \tag{7.13}$$

Equation (7.13) describes the relationship of the optical path difference with the gradient of the refractive index in the shearing interferometry method at a small image shift. This is an integral Able equation that is transformed to

$$\frac{dn}{rdr} = -\frac{1}{\xi\pi} \frac{d}{rdr} \int_r^R \frac{\mu(x)dx}{\sqrt{x^2 - r^2}}. \tag{7.14}$$

Multiplying both parts of (7.14) by rdr and integrating over the region from r to R gives

$$-\int_r^R dn \equiv n(r) - n_R = \frac{1}{\xi\pi} \int_r^R \frac{\mu(x)dx}{\sqrt{x^2 - r^2}}. \tag{7.15}$$

For comparison, the equation of the normal interference method (7.9) has the form

$$n(r) - n_R = -\frac{1}{\pi} \int_r^R \frac{d\mu(x)}{dx} \frac{dx}{\sqrt{x^2 - r^2}}.$$

It may be seen that equation (7.15) does not contain the operation of differentiation of the experimental function $\mu(x)$. This must increase the accuracy of determining $n(r)$ from the shearing interferograms in comparison with normal ones.

Figure 7.9 gives the shearing interferogram for a cylindrical object. The shearing interferograms are processed by the same procedure as normal ones. The shift of the fringe (7.9b) in relation to the initial

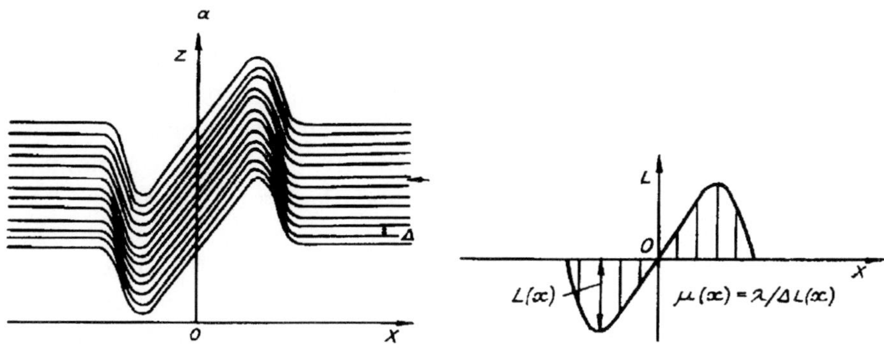

Fig.7.9 View of a shear interferogram (a), b) processing of shearing interferograms, $L(x)$ is the geometrical shift of the fringe, Δ is the distance between the maxima of the non-perturbed fringes.

position is measured for difference values of co-ordinate x in fractions of the distance between the pair of unperturbed fringes whose optical path difference is λ. This gives the dependence $\mu(x)$.

7.3 Shadow method

This method is based on widening the angle of deflection of the light beam from its initial direction in passage through an optically heterogeneous object. In experiments, the shadow from some barrier in the path of the light is observed on the screen. The displacement of this shadow from the initial position is measured. The more suitable method is the shadow method of the defocused inclined filament[6] (Fig.7.10). The following symbols are used in Fig.7.10: a is the distance from the object to the lens L_3 with a focusing distance f; s is the distance of the barrier (filament) to focus F (defocusing); A is the barrier (inclined filament); b) is the distance between the screen S and the barrier. The examined object is illuminated with a parallel light beam, B is one of the points of the object. Length L_3 reflects the object onto the screen, B' is the image of the point B.

Two beams, one of which (beam 1) passes through the object (point

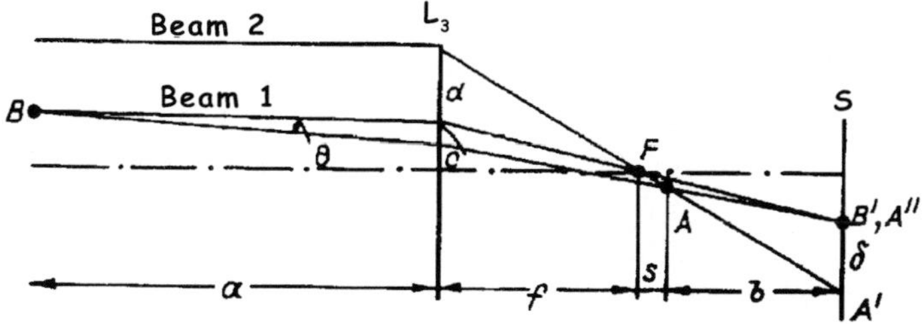

Fig.7.10 Diagram of a shadow device (chamber path) of a defocused inclined filament.

B), and the other one (beam 2), outside the object will now be examined. In the absence of the object, all beams pass through the focus *F* and the shadow from the barrier *A* by the beam 2 is projected to point *A′*. In the presence of the beam 1 is deflected by the angle θ, passes away from the focus and forms a shadow from barrier *A* at point *A″* whose distance from *A′* is δ. The beam 1 deflected by the object is displayed on the screen at the same point as the unperturbed one because the planes of the object and the screen are connected by the lens L_3. The relationship of the measured value of displacement of the shadow δ with the required deflection angle θ will be determined. In accordance with Fig.7.10, $c = a \tan \theta \approx a\theta$ for small angles. From the similarity of the triangles with common tips at the points *A* and *F* we have $\delta/b = (c + d)/(f + s)$, $d/f = \delta(b + s)$. Using the equation of the lens $1/f = 1/a + 1(f + b + s)$ and the derived relationships finally gives

$$\theta = (a - f)s\delta / f\left[f^2 - s(a - f)\right],$$

Usually $s \ll f$ and $a = 2f$ (the object is displayed on the screen in the natural size) giving

$$\theta = \frac{s}{f^2}\delta. \tag{7.16}$$

At the deflection angle θ the displacement of the shadow on the screen δ increases with decreasing defocusing of the filament *s*. However, the slit width also increases proportionately (Fig.7.11). In the beam optics approximation $h'/h = (b + s)/s$. From Fig.7.10 at $a = 2f$ we obtain $b + s = f$ and, using equation (7.16)

$$\frac{\delta}{h'} = \theta\frac{f}{h}. \tag{7.17}$$

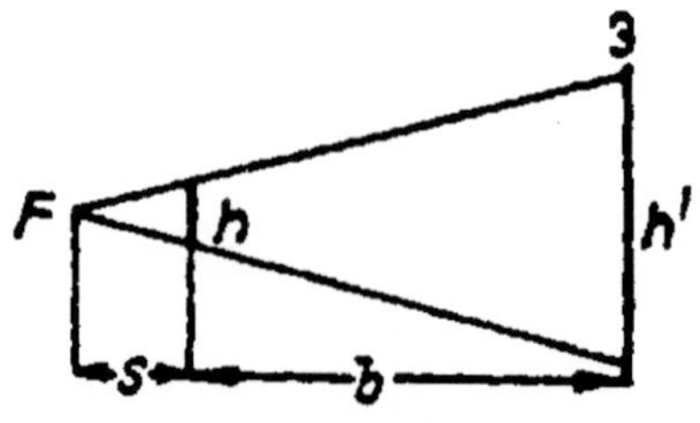

Fig.7.11 Projection of a filament onto a screen.

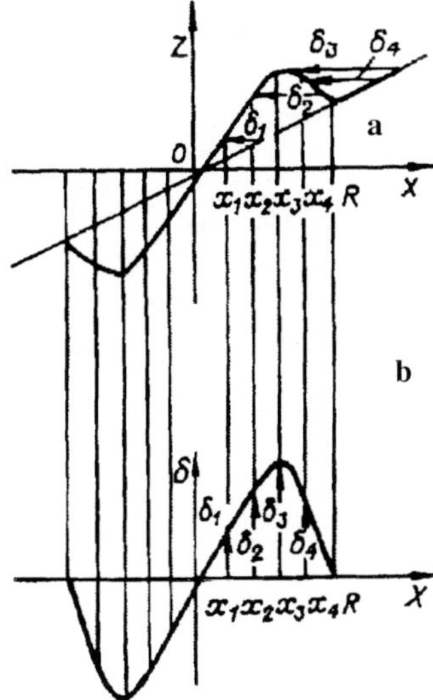

Fig.7.12 A shadow pattern of the method of the inclined defocused filament for a cylindrical non uniform object (a) and the function δ (x) (for processing of the shadow pattern) (b).

It can be seen that the fraction of the displacement of the shadow in relation to its width δ/h', which also determines the accuracy and sensitivity of the method, does not depend on the value s. The width of the shadow can be decreased using, as a barrier/filament, a phase sheet at half the wavelength.

The form of the shadow fringe in the method using an inclined defocused filament shown in Fig.7.12. To simplify considerations, it is assumed that the heterogeneous object is cylindrically symmetric with the gradient of the refractive index only in the radial direction. The initial shadow of the filament – the direct inclined line – is transformed when the examined object is placed in the path of the beams. The displacement of the shadow for several characteristic points is clearly visible. The lower part of the figure shows the dependence $\delta(x)$ obtained on the basis of the shadow pattern. Measuring $\delta(x)$ equation (7.16) can be used to determine $\theta(x)$.

The relationship of the deflection angle with the gradient of the refractive index and the radial distribution of the refractive index is determined using the following procedure. For a small deflection angle of the light beam (Fig.7.13)

$$d\theta = \frac{\partial \ln n}{\partial x} dy. \tag{7.18}$$

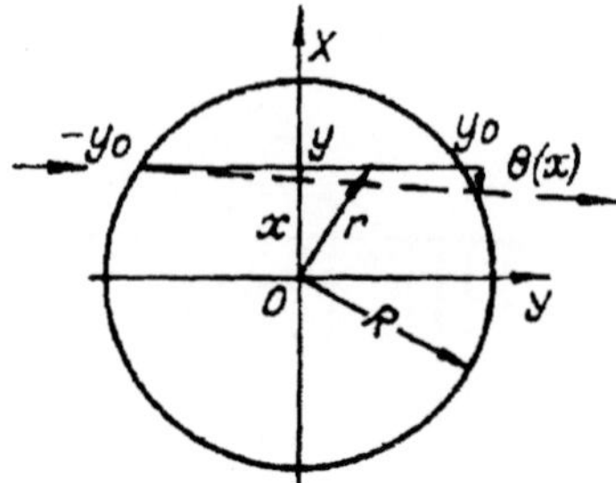

Fig.7.13 Diagram of passage of the beam through a non-uniform cylindrical object.

Integrating (7.18) along the chord $[-y_0, y_0]$, whose distance from the axis of the object is x, gives

$$\theta(x) = \int_{-y_0}^{y_0} \frac{\partial \ln n}{\partial x}\, dy. \tag{7.19}$$

Taking into account the rotational symmetry $r^2 = x^2 + y^2$

$$\frac{\partial \ln n}{\partial x} = \frac{\partial \ln n}{\partial r}\frac{dr}{dx}, \quad \frac{dr}{dx} = \frac{x}{r}, \quad dy = \frac{r\,dr}{\sqrt{r^2 - x^2}}$$

and equation (7.19) takes on the form

$$\frac{\theta(x)}{x} = 2\int_{x}^{R} \frac{d \ln n}{dr}\frac{dr}{\sqrt{r^2 - x^2}}. \tag{7.20}$$

The integral Able equation (7.20) gives the relationship of the angle of deflection of the light beam with the radial gradient of the refractive index. To determine the radial distribution of the refractive index, equation (7.20) is transformed to

$$\frac{d \ln n}{r\,dr} = -\frac{1}{\pi}\frac{d}{r\,dr}\int_{r}^{R} \frac{\theta(x)\,dx}{\sqrt{x^2 - r^2}}. \tag{7.21}$$

Multiplying (7.21) by $r\,dr$ and integrating it from r to R gives

$$-\int\limits_{r}^{R} d\ln n \equiv -\ln\frac{n_R}{n(r)} = n(r) - n_R = -\frac{1}{\pi}\int\limits_{r}^{R}\frac{\theta(x)dx}{\sqrt{x^2 - r^2}}. \tag{7.22}$$

Equation (7.22) is used to calculate the radial distribution of the refractive index from the measured deflection angle of the light beam. It should be noted that the equation (7.22) for the shadow method is identical with (7.15) for the shearing interferometry method and does not contain the derivative of the measured value. The latter determines the high stability of the calculations of $n\,(r)$ in relation to random measurement errors. It should be noted that the values $\theta(x)$ in the shadow method and $\mu(x)/\xi$ in the shearing interferometry method are identical, although they were obtained using different optical systems.

As mentioned, the equations (7.15) and (7.22) are valid for small angles of deflection of the probing beam light when the integral (7.19) is taken along the chord $[-y_0, y_0]$. Integration along the beam with an arbitrary trajectory leads to the expression[7]

$$n(r(s)) = n_R \exp\left[-\Phi(r(s))\right]. \tag{7.23}$$

Here

$$\Phi(s) = \frac{1}{\pi}\int\limits_{s}^{R}\frac{\theta(x)dx}{\sqrt{x^2 - s^2}}, \tag{7.24}$$

$$r \approx s\cdot\exp\left[\Phi(s)\right]. \tag{7.25}$$

From the measured deflection angle $\theta(x)$ we calculate the value $\Phi(s)$ from (7.24) and then determine $n\,(r(s))$ from (7.23) where the relationship of r with s is described by equation (7.25). The small angle approximation is present in these equations as a limiting case: $\Phi(s) \ll 1$. Since $\Phi(s)$ is determined from the experimental data on $\theta(x)$, we can also verify the degree of the reliability of this approximation. At $\Phi(s) \ll 1$, we have $r \approx s$ and $n(r) \approx n_R(1 - \Phi\,(r)) \approx n_R - \Phi(r)$, which is in agreement with equation (7.22).

7.4 Diffraction interferometer based on IAB-451 shadow equipment for plasma investigations

The circuit of IAB-451 equipment was proposed and developed by D.D. Maksutov. An advantage of this equipment is the possibility of easy separation of the collimator and camera parts from the main objectives so that it is possible to use different optical and mechanical attachments designed for operation with a specific method. The universality of this equipment in carrying out shadow and interference investigations was described by S.A. Abrukol and L.A. Vasil'ev.[5,8] The results[5,9] were used efficiently for the rapid development and construction of a diffraction interferometer on the basis of IAB-451 shadow equipment. IAB-541 was converted to a diffraction interferometer by introducing transparent optical gratings into the optical system of the equipment.

The main circuit of IAB-451 equipment used to examine the nature of flows is shown in Fig.7.14. Equipment consists of two identical systems: the collimator system, generating a parallel light beam for illuminating the examined heterogeneity, and the chamber system used for examination and photography. The collimator part consists of a mirror–meniscus objective with a diameter of 230 mm and a focusing distance of 1917 mm and a flat mirror for changing the direction of the beams. The condenser produces an intermediate image of the light source in the focal plane in the mirror–meniscus objective. The camera part has the same structure. Instead of the slit and the blade, transparent diffraction gratings P_1 and P_2 are placed in the focal planes of the objectives O_1 and O_2. To produce two beams (working and reference), the authors of Ref.9 proposed to place a field diaphragm with two windows in the plane of the object P. The grating P_1 is used only for increasing the intensity of illumination of the interference field from the source with a continuous or linear radiation spectrum. The application of the laser as a light source together with the grating P_1 does not require field diaphragm to be positioned in the object plane of the device because the required two light beams can be produced directly behind the grating P_1. The light beam from a helium-neon laser, falling through the objective O on the transparent diffraction grating P_1, is divided by the grating to a number of beams. Each beam represents

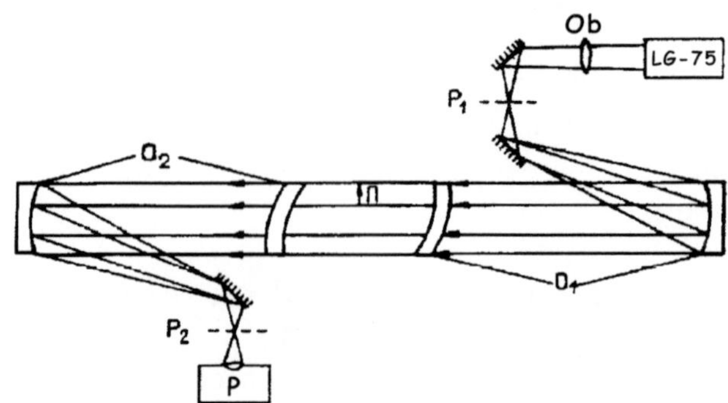

Fig.7.14 Diagram of IAB-451 device, *P* is the photorecording device.

a specific order of diffracted light, two of which – zero and first orders – pass through the objective of the collimator O_1 and are shaped into beams with flat wave fronts. One of the beams passes through the examined plasma, the other one is a reference beam. The remaining beams leave the grating P_1 under diffraction angles which are such that the beams are cut off by the mounting of the objective of the collimator and do not arrive in the reception part. The diameter of the beams a is given by the focusing distance F of the objective O placed between the laser and the diffraction grating P_1

$$a = a' \frac{f + F}{F} \approx a' \frac{f}{F},$$

where $f = 1917$ mm is the focusing distance of the collimator objective, $a' = 3.5$ mm is the diameter of the beam at the exit from the laser.

In the reception part of the device, these two beams are directed on to the diffraction grating P_2. The transparent grating P_2, which is a compulsory element of the circuit of the diffraction interferometer, ensures the formation of an interference field which coincides with the image plane. The distance between the centres of the light beams b satisfies the condition of producing an interference pattern: $bd = \lambda f$, since the collimator and reception sections of IAB-451 device are identical (λ is the wavelength of probing radiation, d is the spacing of the grating P_2). The diffraction grating represents the object plane in the form of the maxima of zeroth, $\pm 1^{st}$ and so on orders which are differently readjusted[9] in relation to the constant of the grating P_2 forming the interference field. Interference is of the two-beam nature because only two beams from two superimposed orders meet at each point of the interference field. If the spacing of the grating P_1 is equal to that of P_2, the interference pattern forms in superimposition of the zeroth and $+1^{st}$ or zeroth and -1st orders of the diffracted light of the reference and working beams. The even orders of light diffraction ($\pm 2^{nd}$, $\pm 4^{th}$) do not appear because the ratio of the width of the transparent groove to the non-transparent one in the diffraction gratings is equal to unity. The spacing of the gratings P_1 and P_2 is selected in accordance with the recommendations in Ref.5. The diffraction gratings P_1 and P_2 have the spacing $d = 0.01$ mm. The source of radiation in the device is a continuous gas laser LG-75, $\lambda = 6328$ Å. A pattern formed by the interference of the zero and $\pm 1^{st}$ orders of the working beam is used for the investigations. The grating grooves were parallel to each other. The infinite width fringes are used when the diffraction grating P_2 is situated exactly at the focal point of the main objective of the reception part of the device. In this case, the light waves from different points of the working field (and the reference field) arrive in the image plane in the same phase, and the intensity of illumination of the interference pattern is the same for all points of the image and depends on the difference of the phases between the waves propagating from the working section and their comparison section.

The working and the reference beam are positioned in space alongside each other in the horizontal plane or one above the other, depending on the system used.

The objects characterised by strong intrinsic glow are usually difficult to record in interferometric diagnostics because additional measures must be taken to weaken this glow. Although the addition of narrowband interference filters to the optical system of the device leads to the required weakening of the radiation of the object, it nevertheless greatly impairs the quality of interferograms: the monochromatic radiation of a laser generates additional interference fringes on the interferogram after passing through the light filter and these fringes influence the quality of the main interference pattern. The use of conventional absorbing light filters leads to smaller distortions but is not always suitable because such filters are characterised by a wide transmission band.

The required coefficient of weakening of radiation of the plasma jet is obtained in this case using several simultaneous methods. Firstly, attenuation is carried out using a KS-14 light filter placed in the reception part of the device in front of the photorecording device. Secondly, plasma radiation is weakened by a diaphragm positioned on the mirror objective of the reception pattern device. The dimensions and configuration of the holes in the diaphragm correspond to those of probing beams. The light from the plasma jet is weakened by the value determined from the ratios of the entire area of the mirror objective to the beam area. For example, at an objective diameter of 230 mm and diameters of two identical orifices in the diaphragm of approximately 60 mm, radiation is weakened by approximately an order of magnitude. The third factor weakening the parasitic plasma radiation is the diffraction of this radiation on the grating P_2. The grating divides 'wide' radiation of the plasma jet into a spectrum, the image of the jet is stretched in the direction normal to the groove of the grating. The degree of attenuation increases with increasing frequency of the grating. This property of the diffraction interferometer is an advantage in comparison with other types of interferometers. As a result of combined use of this method it is possible to produce an interference pattern that is not 'burdened' by the presence of natural plasma radiation.

The interferograms are recorded by a mirror-type photographic camera highly suitable for visual examination. The gain factor of the system in recording the interferograms is selected to ensure the photograph of the standard size shows the examined region. The high monochromaticity of probing radiation and overexposure enables interferograms of good quality to be obtained.

Both interfering beams are produced by the same components of the device and, consequently, this device is far less sensitive to vibrations than Mach–Zender-type interferometers. Special measures are taken to protect IAB-451 device against vibrations duration operation: the device is positioned on a foundation and there is a rigid link between the collimator and camera parts. The two-beam diffraction interferometer results in the error of measurement of the fringe shift of ~0.05λ.

7.5 Examination of the nature of the plasma flow
Two-jet plasma torch

The interferometric method described above was used to examine jets and the main plasma flow of a two-jet dc plasma torch.[10] Plasma jets were discharged into air. The main operating regime of the plasma torch: arc current 105 A, the consumption of the working gas (argon) 0.12-0.36 g/s, supplied power 16 W, efficiency up to 0.9. IAB-451 device was constructed for infinite width fringes. Exposure time was 2 ms. Typical interferograms of the plasma flow for gas flow rates of 0.12; 0.24; 0.36 g/s are presented in Fig.7.15a-c. The initial angle of convergence of the jets was 60°. For intermediate gas flow rates, the shadow pattern of the plasma flow does not differ qualitatively from those presented here and enables regions of the plasma where the flow is still laminar to be defined. The presence of stable interference fringes, detected visually, indicates the absence of large scale pulsations of density. Examination under different angles shows a direct relationship between the angle of convergence of the jets and the flow regime. Increasing initial angle between the plasma jets decreases the flow rate of the plasma forming gas at which turbulent pulsations appear.

Analysis of the interferograms shows that at argon flow rates less than 0.15 g/s and the angles of convergence of the jets of less than 60° the plasma flow is laminar over a large length. The length of the stable plasma flow is approximately 50 cm. At a gas flow rate of 0.24 g/s, the main plasma flow is laminar at a distance of up to 30 mm from the area of merger of the jets. This is followed by the appearance of pulsations. At a gas flow rate higher than 0.36 g/s these pulsations 'spread' the interference fringes downwards along the plasma flow up to the area of convergence of the electrode jets. A further increase of the gas flow rate results in unstable operating regime of the plasma torch,

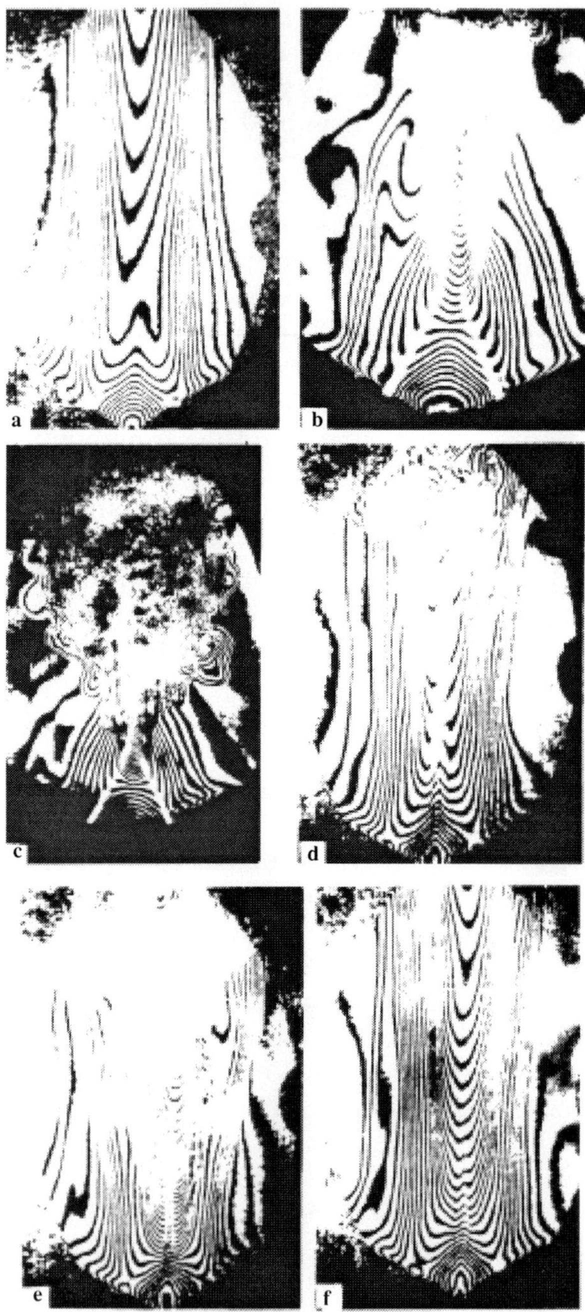

Fig.7.15 Interferograms of a plasma flow ($I = 105$ A) in the regime of infinite width fringes. At a flowrate the plasma forming gas of 0.12 (a), 0.24 (b), 0.36 g/s (c); d-f) at a flowrate of the plasma forming gas of 0.12 g/s and introducing a substance into the zone of convergence of the jet; argon with a flowrate of 0.03 g/s (d), aerosol (e) and suspended mixture (f) with a flowrate of the transport gas of 0.03 g/s.

formation of breaks in the jets and arc extinction. Arc extinction can be explained by the insufficient reserve of the voltage of the power sources; if the voltage is increased, a higher gas flow rate is required to extinguish the arc. However, the electrode jets of the plasma remain laminar even under these limiting conditions. The passage of electric current prevent stabilisation of the current-conducting jets.

Analysis of the experimental data on the nature of plasma flow in arc channels show that at Reynolds numbers Re_{in} ~3000–5000, estimated on the basis of the parameters of the cold gas at entry into the channel, the flow is still laminar. When converted to plasma temperatures (~10 000 K), the Reynolds number is ~1000. The estimates, obtained on the basis of the parameters of the plasma jets at an exit from the nozzle of the two-jet plasma torches in the equation

$$Re_{in} = 4G/\pi d\mu$$

at $G = 0.4$ g/s, $d = 5$ mm, gave Re $= 300 \div 400$, i.e. lower than critical (here d is the nozzle diameter, μ is plasma viscosity). During the passage of electric current, the temperature in plasma is maintained sufficiently high as a result of Joule heat generation and the Reynolds number is lower than critical. In the case of currentless plasma flows, rapid cooling of these flows is the result of heat exchange with the environment results in a rapid increase of the Reynolds number due to a rapid reduction of viscosity. This creates suitable conditions for the transition from the laminar to turbulent flow.

Analysis of the interference patterns shows that no reversed flows were observed at initial angles of convergence of the jets of less than 60° and a gas flow rate lower than 0.15 g/s.

It should be noted that the information interferograms, shown in Fig.7.15, corresponding to the laminar plasma flow, can also be used to determine the temperature and density field in examination from different directions. Figure 7.16 shows interferograms in finite width fringes.

For spectral analysis, plasma chemistry, the technology of spheroidisation of microparticles and coating, it is essential to know the special features of the processes of entry of the solutions and powders into the plasma and their effect on the flow regime. Qualitative examination of these processes was carried out by the interferometric method in IAB-451 equipment.

Analysis of the literature data shows that the laminar regime is suitable for a number of technological plasma processes. Therefore, the effect of added substances was investigated for the laminar flow (Fig.7.15a). This gave interferograms of the plasma flow when adding argon, aerosol

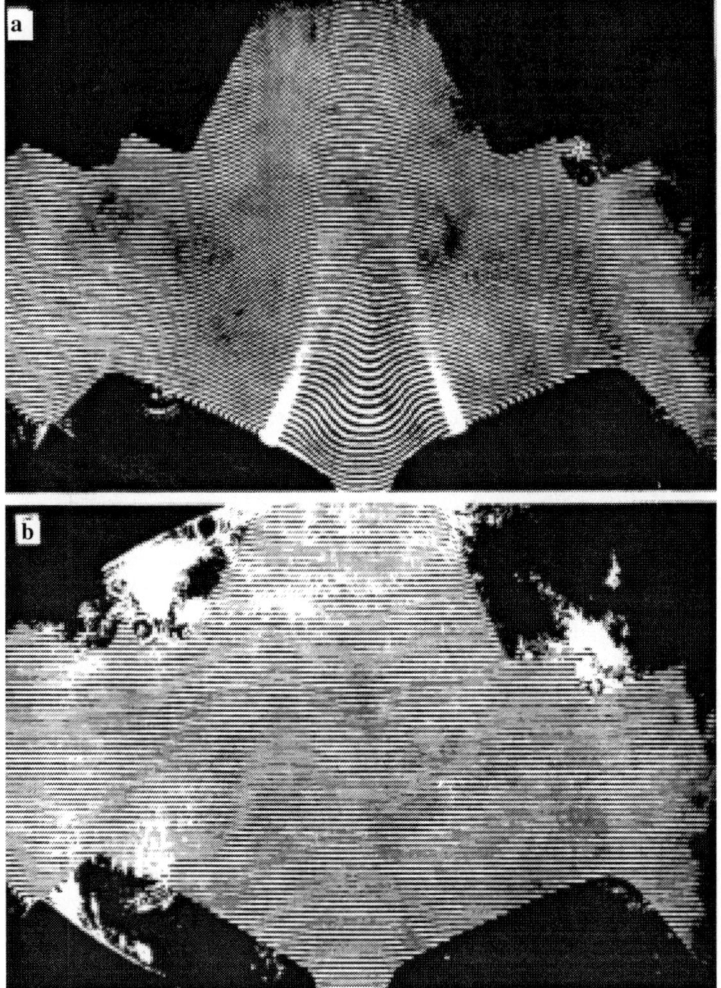

Fig.7.16 Interferograms of plasma jets (a) and the view of interference fringes in the absence of plasma (b) in the regime of finite-width fringes.

and aerosol suspension into the area of convergence of the plasma jets. The results are presented in Fig.7.15*d-f*. Figure 7.15*d* shows that blowing cold gas (0.03 g/s of argon) greatly changes the nature of the flow. The gas velocity in blowing through a pipe 0.08 cm in diameter and a gas flow rate of 0.03 g/s is on average 35–40 m/s. Disruption of the laminar nature of the flow takes place as a result of rapid cooling of the main plasma flow by the cold gas. This is due to an increase of the Reynolds number above the critical level due to a reduction of viscosity. The addition of aerosols no longer leads to such large changes in the nature of the flow (Fig.7.15*e*). In this case, pulsations evidently form as a result of explosion-like evaporation of the droplets of the solution in the plasma. When adding the powders (powder flow rate 0.3 g/s) with the flow of the transport gas (0.03 g/s of argon) even a strongly pulsating plasma flow becomes laminar (Fig.7.15*f*). The Reynolds number for powder particles with a diameter of 100 μm, estimated taking into account the temperature of the surrounding plasma, is ~1. At this Reynolds number, a laminar boundary layer forms at the surface of each particle. The presence of a large number of such particles stabilises the general flow. Evidently, this effect of laminarisation of the plasma flow by adding finely dispersed materials is common to all plasma jet devices and indicates the directions of controlling the flow regime not only in open discharges but also in plasma reactors where the turbulence greatly determines the losses on the walls.

'Laminar' plasma torch

The flow regime of a plasma jet of an 'laminar' plasma torch[1] was investigated in IAB-451 shadow equipment by the slit and filament method. The slit and the filament were placed parallel to the axis of the plasma jet. The slit was 0.07 mm wide, with the filament 0.08 mm thick. The shadow pattern of the plasma jet was recorded from a SVDSh1000 mercury lamp through a ZhS-18 light filter on RF-3 photofilm, exposure time 0.002 s. Typical photographs for different gas flow rates are shown in Fig.7.17. Each photograph shows two sections of the jet greatly

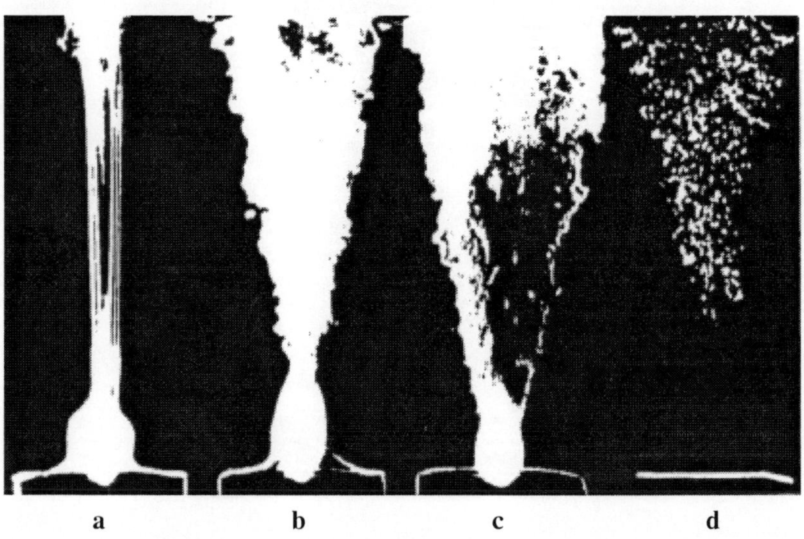

Fig.7.17 Shadow patterns of the plasma jet at a gas flowrate of 0.09 (a), 0.18 (b), 0.36 (c) and 0.72 g/s (d).

differing in the type of flow. Interference fringes appear in the section at the base of the jet. These fringes form as a result of diffraction, at the filament, of the light passed through the plasma jet. The presence of stable interference fringes, clearly observed by visual examination, indicates that the flow is laminar. Consequently, at an argon flow rate of 0.09 g/s, the plasma jet is laminar almost along its entire length, and at 0.18 g/s the laminar flow is clearly visible at a distance of 14 mm, i.e. along the entire bright part of the jet. At a flow rate of 0.36 g/s there is no laminar section, although an essential part of the plasma jet can again be laminar as a result of high viscosity.

Welding arc

Investigations were carried out into a welding arc in Ar and CO_2 for currents $I = 50{\div}90$ A, a flow rate of the plasma gas of 0.1–0.6 g/s, at different geometry and configuration of the nozzles and different distances from the outlet of the nozzle to the component (anode). Typical interferograms are presented in Fig.7.18. Similar interferograms were

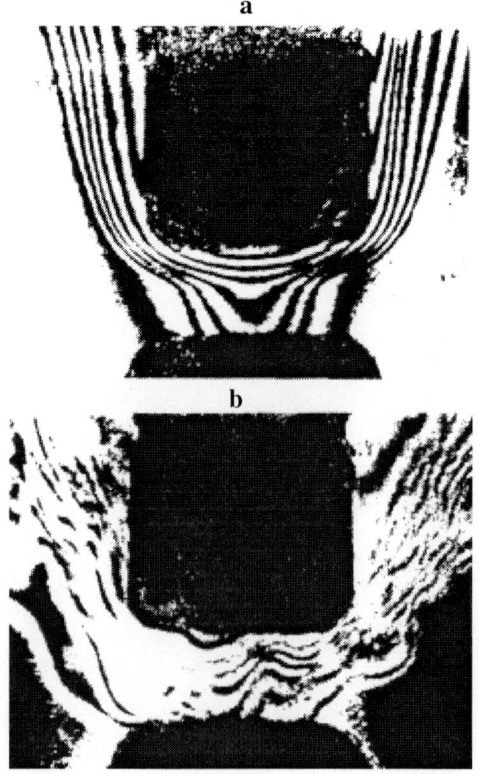

Fig.7.18 Interferograms of a welding arc in argon at a current of 60 A, distance from the outlet of the nozzle to the component 6 mm, flowrate of the plasma forming gas 0.1 (a), 0.25 g/s (b).

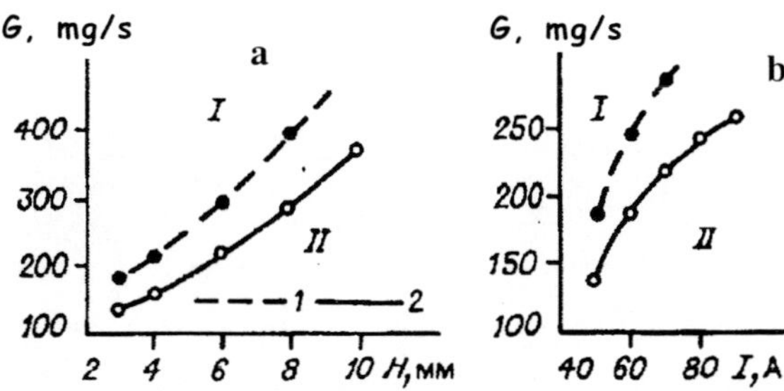

Fig.7.19 Effect of the flowrate of the plasma forming gas, current intensity and distance H from the component to the nozzle on the boundary of transition of the flow to the turbulent regime at a nozzle diameter of 3 mm. a) $I = 60$ A, b) $H = 5$ mm, I is the turbulent, II the laminar region; 1) argon, 2) CO_2.

obtained at all examined conditions. The photographs show clearly the laminar and turbulent flows. These investigations were used to determine the boundary of transition from the laminar to turbulent flow for different arc parameters (Fig.7.19). The results of these investigations were used as a basis for recommendations for selecting the optimum parameters of the welding conditions.

Alternating current arc

Figure 7.20 shows interferograms of an ac arc (current 10 A, carbon electrode) in infinite width fringes. The dynamic interferograms were recorded using an SKS-1 camera. At the same time, time marks from the generator of rectangular pulses were recorded on the film. The signal from the generator was also fed to the input of N-115 loop oscilloscope where arc current oscillograms were also recorded. This method makes it possible to identify each frame of the interferogram with the arc burning phase. Current oscillograms show that the arc burns for 5 ms and the remaining 15 ms is used up by the current break. This is confirmed by visual examination of the interferograms. The presence of axial symmetry enables the interferograms to be processed by a corresponding procedure (see section 7.1) and obtain quantitative data on the plasma arc temperature. At the initial moment of arc ignition, the frame 1 shows the region of strong turbulence between the electrodes. The temperature at this moment increases rapidly. In the following frame 2 ($\tau \sim 0.75$ ms), the fringes start to assume the form of concentric rings, their number increases and the temperature at the axis is close to 1000 K. In the following frame 3 (after 0.75 ms), the number of interference fringes increases and the temperature in the arc increases to 4000 K. The size of the arc cloud (frame 4) reaches 20 mm with the temperature at the axis being 6500 K. This is followed by a current break and the temperature of the axis slowly decreases to 3000–1000 K and the arc cloud

Fig.7.20 Interferograms of ac arc with half-wave current.

starts to travel upwards (frames 5 and 6). Examination of the interferograms shows that the radial velocity of the front of the thermal wave in arcing reaches ~1.8 m/s, the rate of convection in the current break is ~0.2–0.3 m/s.

7.6 Temperature measurements by the interferometric method

A two-beam diffraction interferometer, constructed on the basis of the shadow equipment, was used to measure the temperature distribution of the argon plasma in the jet of the 'laminar' plasma jet. A typical interferogram of the plasma jet, obtained during the experiments, is shown in Fig.7.21. The procedure of temperature measurements in the plasma jet by the interferometric method is reduced to determining the radial distribution of the refractive index $n\,(r)$, followed by the transition to temperature $T\,(r)$ using the well-known dependence $n\,(T)$.

Processing the interferograms

Plasma jets are examined by lateral examination with interference fringes of finite width oriented normal to the axis of symmetry of the jet. In a general case, the shifts of the fringes $\varphi(x)$ are proportional to the variation of the optical path in the plasma in relation to the surrounding medium. For the axisymmetric plasma with non-uniformly distributed parameters, the local values of the refractive index $n\,(r)$ were determined by the integral Able equation written on the basis of the results of lateral examination of the shift of the fringes at a wavelength of

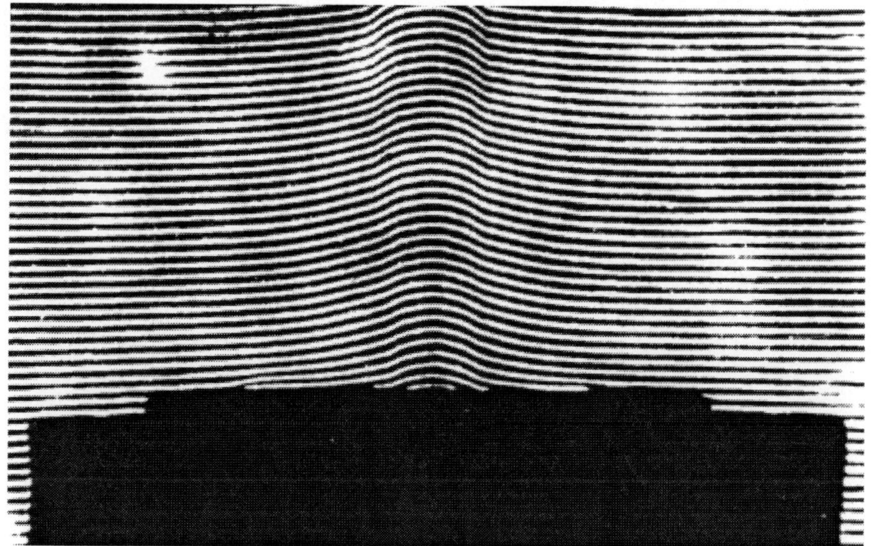

Fig.7.21 Interferogram of a jet.

$\lambda = 6328$ Å:

$$\varphi(x) = \frac{\mu(x)}{\lambda} = \frac{2}{\lambda}\int\limits_{x}^{R}\left[n(r) - n_R\right]\frac{r\,dr}{\sqrt{r^2 - x^2}},$$

where R is the radius of the plasma jet, n_R (r) is the refractive index of the medium surrounding the jet. To determine n (r), we used the transformation of the Able integral (7.9)

$$n(r) = n_R - \frac{\lambda}{\pi}\int\limits_{r}^{R}\frac{d\varphi(x)}{dx}\frac{dx}{\sqrt{x^2 - r^2}}. \tag{7.26}$$

The accuracy of determination of the refractive index n (r) from the data on the shift of the fringes $\varphi(x)$ depends on observation in the experiment of the expected axial symmetry and also the accuracy of measuring the dimensions of the zone of perturbation of the jet R and the refractive index n_R of the unperturbed region.

The deflection of the fringes φ can be measured from the coordinates of the shift for the whole or half numbers of the fringes (position of the maxima and minima of the interference field). For this purpose, in the examined section AA (Fig.7.22) we draw a line and measure the co-ordinates of the points of intersection of the line with the interference fringes. In this method, the experimentator obtains limited information on $\varphi(x)$ because the number of experimental points is small and the error in determining the position of at least one fringe leads to large errors in n (r).

The shift of the fringes was determined in the form $\varphi(x) = L(x)/\Delta$, where L (x) and Δ is respectively the linear deflection of the fringe from the initial position and the distance between the light (or dark) interference fringes. This method of construction of $\varphi(x)$ makes it possible

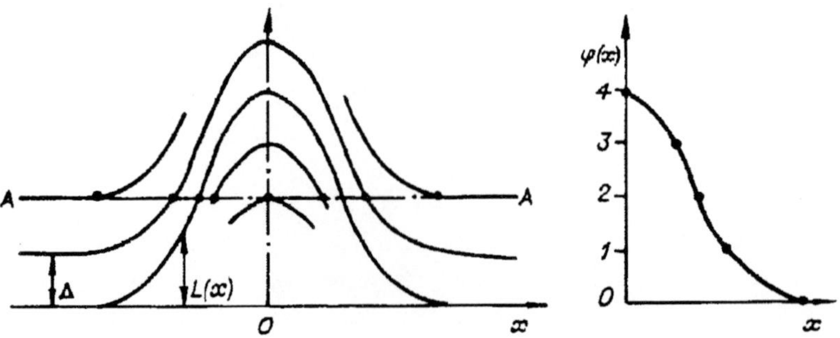

Fig.7.22 Processing of interferograms.

to obtain a considerably larger number of experimental points (their number depends only on the resolution power of the interferometer) and, consequently, use the available information more efficiently. However, it should be remembered that for real objects (arcs, jets, flames) processing must be carried out taking into account the axial variation of the parameters of the object. In addition, the accuracy of the measurement of $\varphi(x)$ depends on the non-ideal nature of the specific circuit of the interferometer causing apparatus distortions.

Taking apparatus distortions into account

In an ideal device, the interference fringes should be strictly straight and mutually parallel, but production defects result in distortion of the interference pattern in an actual device. Apparatus distortions were taken into account by counting $L(x)$ from the position of the corresponding initial interference fringes and not from the horizontal straight line normal to the jet axis. For this purpose, we measured the co-ordinates of the initial interference fringes in relation to the basic straight line on the interference patterns obtained without plasma. Measurements of the interferograms were taken on an x-y comparator with an accuracy of 0.01 mm.

Figure 7.23 shows the effect of apparatus distortions on the determination of the deflection of the interference fringe in the section $z = 25$ mm. In this case, ignorance of the apparatus distortions results in systematically too high values of $L(x)$ for the plasma axis jet (by ~10%). If necessary, a correction can be made not only for the deviation of L but also for the co-ordinate x (taking appropriate measures). When distortions are taken into account, the accuracy and reliability of measurements increases.

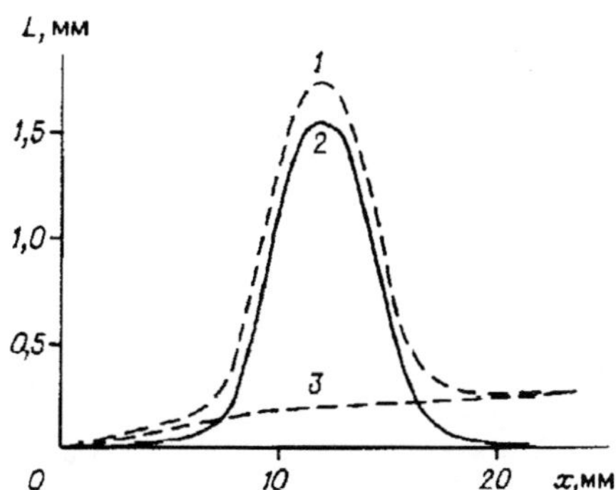

Fig.7.23 Diagram of taking into account apparatus distortions. 1) measured shift of the fringes, 2) shift of the fringes with apparatus distortions taken into account 3) magnitude of distortions.

Taking axial gradients into account

The correction for the axial gradients was introduced owing to the fact that in measuring $L(x)$ we determined not the shift of the phase of the wave front of the given fixed section but of the section positioned at a distance of $L(x)$ from it. For a flow with a cylindrical symmetry, these corrections are not required because the interference fringes, representing a boundary of the same delay of the front of the light wave, are identical in any section. For the real plasma jet, with steep axial power of density gradients, the temperature error may be 500–1500 K if appropriate corrections are not made.

To take axial gradients into account, it is necessary to reduce the deviation of the interference fringes to the fixed section. The following procedure is used. In the region of the plasma jet containing the relevant section the deviations of 3–4 adjacent interference fringes are checked. Subsequently, the graphs of deviation $L(x)$ of fringes is constructed for the fixed values of the x co-ordinate as a function of the distance z from the outlet of the plasma nozzle. The corresponding points for the adjacent fringes are connected by a straight line from which we can determine the deviation at a given fixed value x in the examined cross section of the jet (for a cylindrically symmetric source this curve degenerates into a horizontal straight line). Figure 7.24 shows the directly measured distribution of the deviation of the interference fringe at the deviation reduced on the basis of the described procedure to a single section. The graph reflects indirectly the processes of cooling the central parts of the jet and of heating the peripheral parts of the jet and the associated corrections.

Efficient processing and consideration of even small axial gradients are essential because the resultant curves represent integral quantities with respect to the examination beam, and the accuracy of subsequent

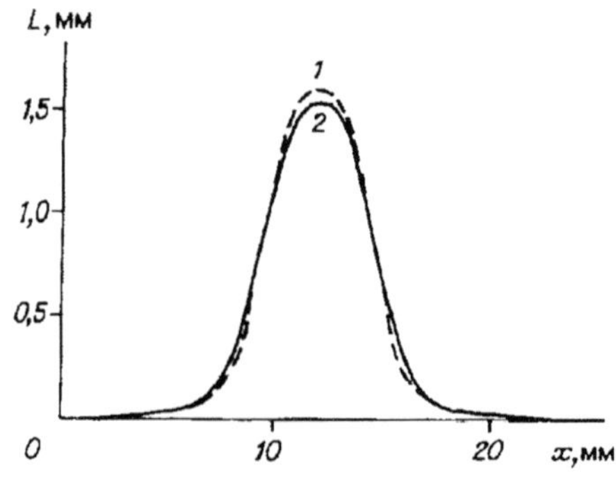

Fig.7.24 Deviation of the interference fringe taking into account (1) and disregarding (2) axial gradients.

determination of the local data is especially sensitive to the deformation of the initial integral contour.

The distribution $L(x)$ was constructed from 80–100 points which corresponds to a measurement range of ≈ 0.3 mm and is consistent with the resolving power of the shadow devices of ~10 lines/mm. Subsequently, equation (7.26) was used to determine $\zeta(r)$ in the form

$$\zeta(r) = n(r) - n_R = -\frac{\lambda}{\pi\Delta} \int_r^R \frac{dL(x)}{dx} \frac{dx}{\sqrt{x^2 - r^2}} \tag{7.27}$$

Numerical integration in (7.27) was carried out by dividing the experimental curve $L(x)$ into 40 zones. Statistical processing of the measurements was carried out by averaging the data for $\zeta(r)$. To obtain reliable results and evaluate random errors of the method, it was necessary to process 20 interferograms. The resultant values of $\overline{\zeta}(x)$ were used to determine the distribution of the refractive index in different sections of the jet.

Refractive index of the environment
The value of n_R was determined by means of mutually supplementing methods: analytical and experimental.

The analytical method is based on the functional dependence of refraction of air on pressure, temperature and moisture content

$$(n-1)_R = (n-1)_0 \frac{273}{T} \frac{P}{760} - \frac{4.1 \cdot 10^{-5} m}{760}$$

where P, T is the partial pressure and temperature of air, n is the partial pressure of water vapour, $(n-1)_0$ is the refraction of air under normal conditions. The error of calculations of n_R by this method is $\approx 0.5\%$, but the method is applicable only in cases in which the nature and magnitude of deviation of the composition of the atmosphere from the normal conditions are known.

In the atmosphere with unknown impurities, the value of n_R can be determined with high accuracy by the experimental method based on comparing the values with the refractive index of the reference gas (helium, hydrogen). Experiments were carried out on a two-beam diffraction interferometer where a gas cuvette with transparent plane-parallel glass sheets was placed in one of the arms of the interferometer. The required refractive index $n_R = n_e + k\lambda/l$, where n_e is the refractive index of the reference gas at the experiment temperature and pressure, l is the length of the cuvette, i.e. the path travelled by the light beam in the reference gas, k is the number of fringes by which the interference pattern was displaced when letting the reference gas into the cuvette, λ is the wavelength of probing radiation. According to the estimates, the error or measurement is 1.5% at a cuvette length of 5 cm and 0.9% and $l = 20$ cm.

Similar measurements of n_R were taken after each experiment series with a plasma jet. The resultant absolute values of the refraction of the atmosphere were in good agreement with the analytical determination data. It may therefore be concluded that anomalous impurities were not present during the experiment. The analytical method is more convenient, but in single measurements or if only insufficient statistical data are available, it can result in large errors of measuring the plasma temperature.

From the resultant average distributions of the refractive index of argon plasma in the jet of a laminar plasma torch, we determine the distribution of plasma temperature $T(r, z)$ using the standard dependence of plasma refraction on plasma temperature. The refractive index is an additive quantity and for a mixture of particles it is expressed as the sum of components forming this mixture:

$$(n-1) = \sum_j \left(n_j - 1\right) = \sum_j K_j N_j,$$

where K_j and N_j is the specific refraction and concentration of type j particles, respectively (atoms, ions, electrons). We used the following values of specific refraction: $K_a = 1.042 \times 10^{-23}$, $K_i = 0.715 \times 10^{-23}$, $K_e = -1.798 \times 10^{-22}$. The plasma composition was computed using the standard procedure for thermally equilibrium argon plasma at the characteristic atmospheric pressure in the experiment of $P = 0.9$ atm. The calculated values of $n\,(T)$ were used to plot the distribution $T\,(r, z)$ and estimate the experiment error.

Figure 7.25 shows the radial temperature distribution in the section $z = 5$ mm (from the outlet of the plasma torch). Vertical sections indicate the measurement errors. The rms error of measuring the temperature from a series of 20 measurements is ~5% for the entire temperature distribution. However, the error of a single measurement reaches 20%. Figure 7.25 also shows the distribution $T\,(r)$ obtained by the spectroscopic method.

Since plasma refraction at high temperatures is caused in most cases

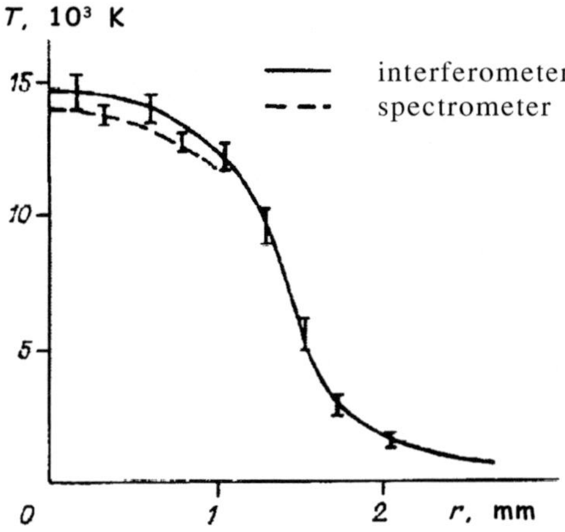

Fig.7.25 Temperature distribution in a 5 mm cross-section of the plasma jet of argon.

by the contribution of the electrons, in the temperature range 13000-15000 deg these two methods determine almost the same quantity – electron temperature. Therefore, in principle, the spectral data can be used to determine n_R by the normalisation method:

$$n_R = n\big|_{T=T_0} - \zeta(0).$$

Combined with the spectral data (or independently), interferometric measurements make it possible to construct the temperature distribution in the plasma jet from the maximum temperature at the axis to room temperature at the periphery.

Chapter 8

DIAGNOSTICS OF PLASMA FLOWS WITH A DISPERSED PHASE

8.1 Introduction

Heterophase plasma flows (in particular, those with a dispersed phase) are found most widely in plasma and plasma chemical technologies. The examination of the characteristics of these flows is important in processing powders of metals and non-metallic materials, processing coal and other carbon-containing solid substances, waste from various productions, in plasma chemical production of commercial carbon (soot), pigment titania, ultradispersed powders of a large number of metals and inorganic materials, etc.[1]

The two-phase plasma system examined up to now can be classified in two groups. The first group includes the processes in which the second phase is the initial one and is to be processed to the target product under the plasma flow conditions. The second group includes systems (initially homophase) in which a phase transition takes place under the plasma flow conditions. From the viewpoint of diagnostics this division is principally of secondary importance although the examination methods used can slightly differ.

It is well known that tasks of the diagnostics of homophase plasma flows include the measurement of the concentration of atoms and molecules, determination of the form of the distribution of energy between these particles (temperature in quasi-equilibrium plasma) and measurement of the flow velocity. Plasma flows are usually systems open in the thermodynamic sense and they contain gradients (often steep) of the concentration and temperature of the particles and the flow velocity. In the majority of cases interesting for practice the examined flows are turbulent. Because of these special features it is necessary to measure local values of these parameters and also ensure a sufficiently high resolution power in measurements of turbulent pulsations of the concentration, temperature and velocity.

When examining two-phase plasma flows, in addition to these tasks it is also necessary to solve a number of other problems associated with the appearance (or consumption) of the second phase in the flow. They include the measurement of the size of the second phase particles, their concentration, velocity and temperature. Until recently, the main method of examining the two-phase flows was the method of taking samples using probes of various type. For many low-temperature systems, this method gives fully satisfactory results even at present. However, when examining high-temperature (2.5–10 kK) plasma two-phase flows using probes, the results of probe measurements of the parameter of the second phase may be unreliable. This is associated mainly with the fact that the processes of condensation from the gas phase taking place in the channel of the probe in cooling the sample may greatly distort the initial size distribution of the particles (SDP). Distortion of SDP also forms as a result of the losses of particles in transport lines of the probe (these losses depend on the particle size). Finally, under normal steep radial gradients of the parameter of the plasma flow it is difficult to ensure the conditions of isokinetic sampling of second phase particles from the flow using finite size probes.[2] This also distorts the initial SDP.

Various optical methods are at present more promising for examining two-phase plasma flows. These methods are of two types. The first type includes the methods based on obtaining images of the individual particles or of their ensembles. The methods of the second type are based on measuring the characteristics of the light scattered by an individual particle or an ensemble of particles.

Measurement of the size of single particles in a flow without counting the number of particles of different sizes (i.e. without size classification of the particles) is of no significance for practice because it does not make it possible to determine the most often required mean temperature of the examined high-temperature aerosol. Therefore, one of the main types of diagnostics of two-phase plasma flows is the determination of SDP. In most cases, when examining these flows it is required to determine the kinetic characteristics of the system, especially the SDP kinetics. It is necessary to measure SDP at different times or in individual points of the flow which appears to rotate in space the corresponding time processes of formation of SDP. To relate the spatial co-ordinate of the flow and time it is necessary to know the velocity of the particles. It is well known that particles not larger than 1–2 μm are dragged away almost completely by the plasma flow, whereas as the particle size increases 'slipping' becomes more extensive. It is important to measure the velocity of distribution of the particles (VDP) correlated with the size distribution of the particles.

The currently available optical methods combine laser Doppler velocity measurement devices (LDVMD) and devices for analysis of the amplitude and duration of pulses of the light scattered by particles (see, for example, Ref.3 and 4). These methods are unsuitable for examining the kinetics of SDP under the conditions of the initially homophase plasma flows with a phase transition (at least, at the most interesting initial stage of nucleation of a new phase). This is due to the fact that the dimensions of nuclei of the phase are distributed in this case in the so-called Rayleigh region restricted by the condition $\pi\delta/\lambda \lesssim 0.3$, where δ is the size of the particles scattering the light, λ is the wavelength of probing radiation.

The restriction in this and of a number of other methods of measuring the SDP as regards the range of possible dimensions is not the only one. Almost all methods of light scattering are efficient under the conditions in which the concentration of scattering particles does not exceed some threshold value starting from which it is necessary to take into account secondary scattering of probing radiation. Optical methods require relatively complicated equipment and equipment itself is difficult to adjust. These methods do not make it possible to determine simultaneously the chemical composition of the reacting plasma flow. All these factors indicate that it is necessary to combine efficiently the optical and probe methods of diagnostics of two-phase plasma flows. The extent of perturbations caused by the probes in the flow (these perturbations are often the main obstacle in the application of the probes) is quite easy to control using corresponding optical methods. At the same time, taking samples with probes makes it possible to control the results of measurement by the optical method using the methods of mass concentration of the second phase particles and enables information to be obtained on the chemical composition of the flow.

Analysis of the problem of mutually correlated measurement of the size and velocity distribution of the particles transported by the plasma flow and, as shown later, also of the temperature distribution shows that in many cases the solution of this problem can be made easier by the use of the methods of mathematical modelling for describing the movement and heat exchange of the particles and also the kinetics of chemical reactions in the particle–plasma flow system. The simplified models of movement and heating of the particles in the plasma flow without chemical reactions were examined in Ref.5–7. Even this very simplified approach yields conclusions that may prove to be useful from the viewpoint of diagnostics. For example, as a result of losses through radiation the particle in plasma cannot, under certain conditions, be heated above a specific temperature which depends on the temperature and composition of the plasma.[7] Radiation effects are important for the relatively large particles whose material is characterised by a high density (blackness) coefficient and evaporation (sublimation) temperature, and also under the conditions with a small difference of the temperature of the plasma and particle surface. The mathematical model makes it possible to describe, on the basis of a relatively small volume of the experimental data, the kinetics of heating particles as a whole (and not only of its surface) and physical–chemical processes taking place in the plasma volume and on the particle surface.[8] It is evident that this information cannot be obtained using only contactless diagnostics methods.

8.2 Methods of measuring particle size

The main requirements on the method of measuring the size of dispersed particles in the plasma flows are usually reduced to the following: contactless nature, localisation, real scale time, the range of measured dimensions 1–10 μm, only weak dependence of the measurement results on the particle form and their relative refractive index.

We shall examine the currently available laser–optical methods of measuring the particle size and evaluate their suitability for the diagnostics of plasma jets with a dispersed phase in the context of the requirements described previously. Firstly, it is a group of methods that uses different parameters of the output signal of the LDVMD: visibility, phase, amplitude.

Visibility of the LDVMD signal (depth of modulation of the photoflux) depends on the particle diameter and the period of the interference field in the measured volume.[9] However, in order to use this dependence to determine the particle size it is necessary to satisfy a number of requirements[10] which in practice is possible only in a small number of cases. In addition, the method has shortcomings which complicate its use for the diagnostics of plasma flows: dependence on the measured results on the refractive index of the particles (in the case of extra-axial reception of the scattered radiation); sensitivity of the method to mutual coherence, polarisation and relative intensity of the laser probing beams; insufficient dynamic range ($\delta_{max}/\delta_{min} < 10$, where δ is the particle size); very stringent requirements on the quality of setting the optical system.

The phase method of determining the particle size is based on measuring the phase difference between LDVMD signals from two or more photodetectors.[11] The application of this method for a number of tested objects and non-plasma flows gave the following results. The method is characterised by a linear relationship between the measured phase shift and the particle size; its dynamic range $\delta_{max}/\delta_{min} \simeq 40$ without readjusting the optical system; the range of the measured parameters 1–10 µm; the size and velocity of particles can be measured simultaneously; only slight sensitivity to optical perturbations, weakening of the probing beams and detuning of the optical system; relatively high spatial resolution. The LDVMD manufactured by Dantec company[12] is fitted with a two-channel processor of the counting type which enables the Doppler period and phase shift to be measured; the range of the measured parameter is 1–10 µm; the error of determination of the dimension ±3%; the range of the measured Doppler frequency is 2–8 MHz; the maximum measured velocity 250 m/s; the error of velocity determination ±2%; maximum measured particle concentration 10^3 mm^{-3}; the error of measuring particle concentration ±30%.

Experience with the application of this diagnostic apparatus to plasma objects is not available but the following can be noted: the device has an insufficiently high upper limit of measured velocity; the LDVMD signal in the diagnostics of plasma flows is usually measured on the background of intensive noise, and the possibility of reliable analysis of the signal using a processor of the counting type is problematic (analysis of LDVMD signals in the diagnostics of plasma flows is usually carried out in spectrum analysers and photon correlators).[10,13]

The absolute intensity of radiation scattered on a spherical particle is proportional to the square of its diameter and can be calculated from the equations of the Mi series.[14] Consequently, the amplitude values of the base component of the LDVMD signals[15] can be used to determine the particle size by calibrating the measuring system in advance using particles with the known optical properties and the dimensions.[16–24] The velocity of the particles is measured at the same time using standard methods of laser Doppler anemometry.

The same principle of determination of the particle size from the intensity of radiation scattered on the particles is used in the so-called laser particle counters.[20,21] However, the Gauss distribution of the intensity and the indeterminacy of the particle path in the measuring volume cause that the measurements are ambiguous: a small particle in the centre of the measured volume can give a signal of the same amplitude as a large particle at the periphery of the volume. There are methods of eliminating this ambiguity that can be classified in two groups.

They include methods[17–21] in which the size distribution of particles is determined from the distribution of the amplitudes of the signals of a photodetector using a corresponding mathematical procedure for calculating the probability of intersection by the particle of specific points of the measuring volume for the known form of distribution of the intensity of the probing beam in this volume. In this case, it is not possible to take direct and simultaneous measurements of the particle size and velocity. The SDP can be determined only from a large number of recorded signals by computer processing. In addition, it is necessary to make several assumptions: the form of the SDP should be specified, the velocity of the particles in the measuring volume should be the same irrespective of their dimensions, etc. The error of measurements depends to a large extent on the accuracy of statistical processing and the accuracy of computer calculations. The range of the measured dimensions varies from fractions of a micron to 100 µm at $\delta_{max}/\delta_{min} \approx 10 \div 15$ and an error of ±10–15%. A shortcoming of the methods in the examined group is the complicated form of the algorithms of mathematical processing and the corresponding programmes.[22]

In the methods of the second group, a measuring volume with the known and homogeneous illumination is formed within the limits of the examined object. This is carried out by stopping down the probing beams, using additional detectors with matching circuits and other measures for separating the given region. The method proposed in Ref.23 and 24 is the simplest and most reliable. In the given optical circuit two beams of identical dimensions intersect in the centre of a large laser beam of a different colour or polarisation, forming a Doppler measuring volume which also represents the region with homogeneous illumination in the larger beam. The signal of the LDVMP from 'narrow' beams is used to measure the particle velocity, and the size of the particles is determined from the amplitude of the signal from 'large' beam. Theoretical investigation of the possibilities of the method and its experimental verification on a number of modelling objects have shown that it can be used efficiently for simultaneous measurements of the particle size and velocity, and the error of determination of dimensions is ±10%, with a dynamic range of $\delta_{max}/\delta_{min} \approx 30$.

Disadvantages of the methods of measuring the particle size on the basis of the intensity of scattered radiation include the dependence of the accuracy of measurements on the level of background radiation of heated gas and particles, fluctuations of the relative refractive index of the particles, absorption of radiation in components of the optical system, absorption and multiple scattering of radiation and the particles outside the measuring volume in operation in dense flows, and also from the level of the noise of the laser, photodetector and radioelectronic equipment.

Regardless of these shortcomings, the method of determining the particle size from the intensity of the radiation scattered by them is used widely for measuring the SDP in the flows of heated gases of various types.[18,20,21,23,24] In Ref.19, the amplitude of the base component of the

LDVMP signal was used to determine the size distribution of Al_2O_3 particles in the plasma jet conditions.

The small-angle scattering method[25-31] makes it possible to plot the SDP under the conditions of two-phase flows in the particle size range 3–150 µm. Two approaches can be used here. In the first approach,[25-28] the SDP is determined from the indicatrix of scattering of probing light at particles in the small-angle range solving the incorrect problem of light scattering. There are various methods of solving the inverse problem but they all lead to a large error (±10–15%), especially at relatively narrow SDPs.

The second approach[29,30] is based on the method of measuring the mean (with respect to the beam of sight) particle size in an ensemble on the basis of the integral characteristics of scattered light without solving the incorrect inverse problem. An interesting variant of the method of estimating the size of individual particles was proposed in Ref.31 where the integral characteristics of the radiation, scattered by each particle flying through a flat focused beam, was measured under small angles. Scattered radiation is recorded simultaneously through slit and sector diaphragms. The ratio of the amplitude of the corresponding signals at the output of the photodetectors is linked unambiguously with the particle size. Knowing the parameters of the receiving optical system and the dimension of the diaphragm, a calibrating curve can be plotted from the corresponding functional dependence. The range of the measured dimensions is 3–350 µm. In testing with lycopodium particles, the error of the method was ±2%.

An important advantage of the small angle method is its insensitivity to the refractive index of the particle material and slight sensitivity to the shape of these particles. It is promising to examine the possibility of combining the local small angle method[31] with a time-of-flight anemometer. It is also important to note the original and relatively simple method of measuring the dimensions of glowing particles[31] (see below the examination of the problem of measuring the surface temperature of particles). Here the image of the glowing particle intersects consecutively (during displacement of the particle itself) two slit diaphragms. The width of the first diaphragm is selected smaller than and that of the second one larger than the particle size. The ratio of the intensities of the light passed through these diaphragms is linked unambiguously with the particle size.

Optical methods of diagnostics of two-phase plasma flows in the Rayleigh region. The recently developed photon correlation methods[33,34] make it possible to measure quite reliably the mean particle size in a laminar flow without taking additional measurements of the light absorption factor. Until recently, these additional measurements until recently the only methods (combined with measurements of the intensity of scattered light) of estimating the mean size of Rayleigh particles.[35,36] However, the photon correlation methods have not as yet been used in the diagnostics of two-phase plasma flows. Evidently, these methods are promising for the diagnostics of laminar plasma flows with ultrafine particles.

When using certain methods of diagnostics of two-phase systems based on measuring the light scattering, it is usually necessary to know the particle shape and the refractive index of their material. For small and, in particular, ultrafine particles, the refractive index is a relatively indeterminate value. In the formation of particles as a result of plasma chemical reactions, it is often necessary to face the situation in which precursors of the solid phase particles are represented by ultrafine droplets of an almost unknown chemical composition. The problem of reliable determination of the shape and refractive index of the particles has not as yet been completely solved. It is possible that the concepts proposed in Ref.37,38 will help the development of appropriate methods. Spherical particles were examined in Ref.37. It is well known that when illuminating these particles with laser radiation, scattered light components are not found in the plane normal to the vector of polarisation of the probing beam. If these components are present, their magnitude is used to evaluate the degree of deviation of the shape of the particles from spherical.

8.3 Measuring the flow and particle velocities

The plasma flow velocity is measured by the tracking particle method whose velocity is determined either by means of LDVMP[9,15] or by the time-of-flight method. Experience with the use of LDVMP for the diagnostics of heated gas flows, including plasma flows[9,15,39–41] shows that the use of conventional methods of photodisplacement of direct photo heterodyning for detecting the Doppler frequency shift in measurements in the plasma flows is associated with considerable difficulties due to a low signal:noise ratio caused by intense phonon radiation and also the disruption of mutual coherence and fluctuation of the phase of laser probing beams passing through the highly heterogeneous flow plasma. In addition, at small dimensions and high velocities of the light-scattering particles the energy of the LDVMP signal becomes so low that the signal can be analysed only using the methods of photon correlation spectroscopy. However, at velocities higher than 300 m/s and under the conditions of highly turbulent flows, the correlation methods are difficult to use because of difficulties in interpreting the correlation function and the need to use rapidly acting correlators (the upper limit of the analysed frequencies for the majority of correlators does not exceed 50 MHz).

Thus, the conventional methods of separating the Doppler frequency shift, based on photodisplacement or direct photodetection, are either unsuitable for examining turbulent two-phase plasma flows or are characterised by errors that are so large that the measurement results can be interpreted using the modelling method only.[10,41,42] These difficulties can be overcome using LDVMP with direct spectral analysis[42,46] which do not require the use of very complicated electronic devices for signal processing. In this case, the upper limit of the measured Doppler frequency shifts is not restricted, i.e. there is no upper limit of measured velocities. The spectral method can also be used to determine the direction of the velocity vector and is not sensitive to phase noise and the polarisation of scattered light. This is especially important in examining non-isothermal turbulent flows. LDVMP with direct spectral analysis can operate in a wide dynamic range of the intensity of light signals, including at a level of these signals where traditional LDVMP operate only with photon correlators. Examples of using the LDVMP with direct spectral analysis investigations of high-speed heated gas flows were described in Ref.45–49. In Ref.50, the LDVMP of this type was used to measure the particle velocity in a plasma jet.

When measuring the velocity of tracer particles (3 ± 1.5 µm in size) in the plasma flow good results are obtained using the time-of-flight method[10,13,50] when the intensity of the laser radiation scattered at such particles is no longer sufficient for reliable operation of conventional LDVMP. The main shortcoming of the time-of-flight anemometer, especially in measurements of turbulent flows, is the low rate of collection of the data limited by the small number of particles intersecting two measuring volumes (laser beam focused to 10–50 µm at a distance of 400–500 µm from each other). Experience with the use of high-speed plasma flows shows that the time-of-flight anemometer is preferred under these conditions to the conventional LDVMP. The method enables measurements of the velocity of up to 1000 m/s and is almost insensitive to gradients of the flow parameters and their pulsations.[10,50] The amplitude of the signal of the time-of-flight anemometer is 2–3 orders of magnitude higher than that of the LDVMP signal.

Objective comparison of the efficiencies of the time-of-flight anemometer and LDVMP with direct spectral analysis (in scanning or tracking regime) is not possible because there is almost no experience with the use of the latter in the diagnostics of plasma flows.

When using anemometers of various type based on the tracer particle method for measuring the velocity in plasma jets, arc discharges and other plasma objects characterised by high absolute values of temperature and velocity and also steep gradients of these parameters, there are large number of problems, with the main ones being: firstly, to measure the velocity, it is necessary to have particles that enter into heat- and mass-exchange processes with the plasma and influence its state: i.e., in this case, the method can no longer be regarded as a true contactless method.[39,40] Secondly, at relatively high temperatures the particles evaporate and the rate of evaporation increases with decreasing particle size. Therefore, on the one hand, the particle should be relatively large in order to avoid evaporation up to the moment when its measuring volume is reached and, on the other hand, it should be as small as possible in order to follow all pulsations of the flow velocity. In real situations where the diameter of the particle is $\delta \approx 2 \div 5$ µm, the ratio of the densities of the particle material in the surrounding plasma is $\rho/\rho_p = 10^3 \div 10^4$, with high gas velocities and high accelerations, the particle velocity can differ from that of the carrier phase ('sliding' effect).[10,39–41] To obtain the true profile of the average flow velocity it is necessary to introduce corresponding corrections based on modelling the laws of movement of the particles and results of measurement of their dimensions. Thirdly, the non-uniform distribution of the concentration of the particles in the flow results in additional errors of velocity measurements. It is therefore useful to examine the methods of measuring the velocity of the plasma flow based on the Doppler shift of emission spectral lines[51,52] and resonance Doppler anemometry.[53,54]

The measurement of the velocity of the dispersed phase particles in plasma flows does not differ greatly from the measurement of the velocity of tracer particles and, consequently, the previously examined methods of laser Doppler and time-of-flight anemometry can also be used in this case. When examining the two-phase plasma flows special attention is given to the problems associated with the processes of interphase heat and mass transport, the non-stationary interaction of the particles in the gas – the sliding effect of the particles and its dependence on various factors (material density, particle size and their concentration, etc.); interaction of the particles with each other leading to their coalescence or fragmentation, i.e. variation of the initial SDP; the effect of the dispersed phase and the structure of the carrier flow, etc. The use of the LDVMP for measuring the velocity of the particles in such flows requires overcoming not only technical but also system difficulties.[55–58] In previous sections we have stressed the promising nature of the LDVMP method for detailed measurements of the velocity and size distribution of the particles in the region of Mi scattering (i.e. at $\pi\delta/\lambda \gg 1$).[4,10,59] It is interesting to note that for initially homophase plasma flows in which a phase transition takes place measurements of the SDP by this method gives information on the form of the SDP also in the region of small particles which form an auto-modelling size distribution, for example, soot particles, quite rapidly from the moment of their formation.[60,61]

Thus, the currently available optical methods based on measuring the characteristics of the light scattered by the particles enable rapid advances to be made in solving the problem of measuring correlated size and velocity distributions of the particles. The situation with the measurement of particle temperature in the plasma flow is less satisfactory.

8.4 Particle temperature

The currently available methods can be used to measure the brightness surface temperature of particles with a relatively low accuracy. The error of measuring the true temperature of the particles is associated to a large extent with the inaccurate information on the emisivity of the particle material which also depends on the state of the particle surface and surface temperature.[62] The measurement error is reduced using the method of multicolour pyrometry.[63–66] In this case it is possible to carry out simultaneous measurements of the surface temperature, size and velocity of the particles.[32,67–69] The total error of the method of two–colour pyrometry is estimated to be ± 10–12%.[63,67,68,70] In variants of this method,[13,32,67–69] measurements were taken of the diameters of particles larger than 10 µm at a surface temperature higher than 1500 K (restrictions are caused only by the sensitivity of measuring apparatus) with the particle concentration in the flow not greater than 10^4 cm^{-3}.[32,67,68]

Further advances in the methods of measuring the particle surface temperature may be associated with using wide-aperture holographic gratings[13] to record the radiation of particles simultaneously in several areas of the spectrum. This will enable the methods of multicolour pyrometry to be used[62] and increase the overall sensitivity of apparatus and measurement accuracy.

Chapter 9

MEASURING THE PLASMA FLOW VELOCITY BY THE TRACER PARTICLE METHOD

In a number of studies, the plasma flow velocity was determined by measuring the reaction of a probe body F introduced into the flow[1-3]

$$F = C_D S \rho u^2 / 2, \qquad (9.1)$$

where ρ, u are the density and velocity of plasma, C_D is the drag coefficient that depends on the form and state of the body surface, S is the middle section of the probe body.

Sheets, discs and wires are used as probe bodies. The disadvantages of this method include the relatively large size and inertia of such sensors. The method was developed further by using free solid particles as probe bodies. Since such particles are thermally and electrically insulated, a number of sources of errors is eliminated. Particles accelerated in a centrifuge are thrown through the examined section of the plasma flows.[4] After passing through the flow the particles stick to the specially prepared surface. If the powder is sufficiently uniform, the particles reproduce the profile of the dynamic pressure of the plasma flow. Subsequently, the Able integral equation is used to determine the local flow velocity. In Refs.5 and 6, steel balls 1–5 mm in diameter were thrown through the plasma jet, and in Ref. 7 borosilicate balls 0.3–0.5 mm in diameter were used. Measuring the deflection of the ball after passing through the flow and using the Able equation, the authors calculated the local values of the deflecting force, and the equation (9.1) was used to calculate the profile of the plasma velocity with other parameters known.

The method of measuring the plasma velocity using tracer particles was proposed in Ref.8. It is based on measuring the acceleration of probed spherical particles introduced along the diameter of the cross section of an axisymmetric vertical plasma flow with the known properties

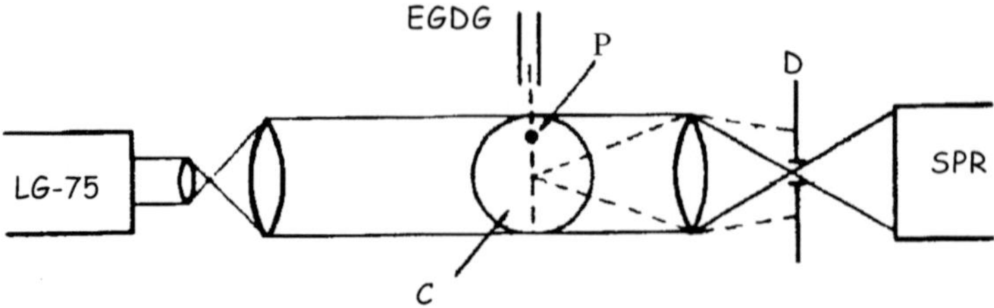

Fig.9.1 Diagram of photographic recording of particles moving in a free plasma jet. D is a screen with a point orifice, LG-75 is laser, C is the plasma jet (cross-section), SPR is the speed recording device, P is the particle.

$$m[a(x) - g] = \frac{\pi d^2}{8} C_D(x) \rho(x) [U(x) - V(x)],$$ (9.2)

where m, a, d is the mass, acceleration and the diameter of the sample particle; g is the free fall acceleration; v is the particle velocity in the direction of plasma movement; ρ, U is the density and axial velocity of the flow; C_D is the aerodynamic drag coefficient of the particle; x is the transverse co-ordinate (in relation to the movement of the flow). Two optical methods were developed to apply this method, for open jets and jets in channels, respectively. Figure 9.1 shows the diagram of photographic recording of the movement of particles in free plasma flows, Fig.9.2 – for plasma flows in the channels.

The particle was introduced into the flow using an electrogasdynamic gun (EGDG) (Fig.9.3) whose barrel was in the form of a medical needle with an internal diameter of $(0.1 \div 0.35) \cdot 10^{-3}$ m. A voltage of ~3 kV is applied to the gun electrodes. The pulsed charge (duration $\tau \sim 2 \cdot 10^{-6}$ s) between the electrodes expands the gas in the capillary tube

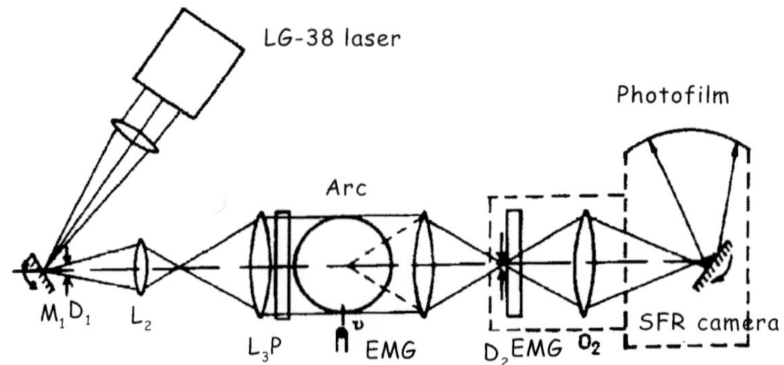

Fig.9.2 Diagrams of photographic recording of a particle moving in a flow in closed channel. D_1, D_2 are diaphragms, M_1, M_2 are mirrors, L_1, L_3 are lenses, O_1, O_2 are objectives, P is the picture frame, EMG is the electromagnetic gate.

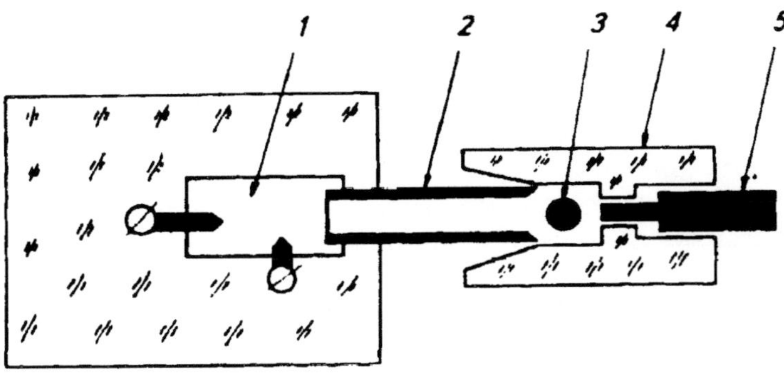

Fig.9.3 Schematic section of EDGD with a device for charging particles (4). 1) discharge chamber, 2) gun barrel, 3) spherical particle, 5) a pusher for charging particles into the barrel.

and pushes the particle from the barrel. The trajectory of the particle in the plasma flow is recorded frame by frame using an SFR-1 camera. The frames of the film record a shadow from the ball on the background of laser radiation. The pulsed discharges initiated synchronously with operation of the SFR chamber using standard systems of synchronisation and initiation of a high-voltage discharge. A single particle is shot each time through the flow and it is then trapped and used in repeated experiments. The velocity of rejection of the particle is selected from the condition that it is necessary to ensure that it does not melt during its stay in plasma. Inspection is carried out by examining the surface of the ball under a microscope after ejection and by weighing on an analytical balance. The particles are produced by pulsed electric melting aluminium wire in argon. Particles are carefully selected on the basis of surface roughness, spherical form and the diameter $d = (0.15–0.3) \cdot 10^{-3}$ m on a type IZA-5 comparator (IZA-7) with an error of $\pm 1 \cdot 10^{-6}$ m. The results of filming for each particle ejection are used to determine the trajectory of its movement $Z(x)$ and the horizontal velocity $W(x)$. Acceleration is determined as the second derivative of the particle trajectory in the direction of movement of the plasma flow:

$$a(x) = \frac{d^2[Z(x)]}{dt^2} = \frac{d^2[Z(x)W_0^2]}{dx^2}, \tag{9.3}$$

since $W(x) = W_0$ with an error of $\pm 1\%$, W_0 is the initial velocity of the particle.

Particle acceleration can be computed on the basis of approximating the experimental function $W_0^2 Z\,(x)$ by exponential polynomials, Chebyshef polynomials, modelling functions, etc. In the first case, we used the polynomial with degree N which describes the experimental function with an error

The unknown coefficients a_k are determined by the method of least squares from the system of algebraic equations of the N-th degree

$$\tau W^2 Z(x) \approx Y_N(x) = \sum_{k=0}^{N} a_k x^k.$$

where n is the number of experimental points. The number of terms of the series is determined from the minimum of the error of closure

$$\sum_{k=1}^{N} \left(\sum_{i=1}^{n} x_i^{j-1} x_i^{k-1} \right) a_k = \sum_{i=1}^{N} x_i^{j-1} W^2 Z(x), \quad j = \overline{1, N}$$

Acceleration in accordance with equation (9.3) is determined as follows

$$D_N = \frac{1}{n-N} \sum_{i=1}^{n} \left[\sum_{k=0}^{N} a_k x_i^k - W^2 Z(x_i) \right]^2.$$

A shortcoming of this method is the need to solve a system of algebraic equations; this increases the error at high N.

The use of Chebyshef polynomials greatly simplifies the calculation procedure. In this method, the experimental function $W^2 Z\,(x)$ is expanded into a series

$$a(x) \approx Y_N''(x) = \sum_{k=2}^{N} k(k-1) a_k x^{k-2}.$$

The polynomials $P_k\,(x)$ are computed from the known recurrent relationships

$$P_{k+1}(x) = (x + \beta_{k+1}) P_k(x) - \frac{H_k}{H_{k+1}} P_{k-1}(x),$$

$$P_0(x) = 1, \quad P_1(x) = x - \bar{x}, \quad H_k = \sum_{i=1}^{n} P_k^2(x_i),$$

$$\bar{x} = \frac{1}{n}\sum_{i=1}^{n} x_i, \quad \beta_{k+1} = \frac{1}{H_k}\sum_{i=1}^{n} x_i P_k^2(x_i).$$

Coefficients a_k are determined from the equation

$$a_k = \sum_{i=1}^{n} W^2 Z(x_i) P_k(x_i) / \sum_{i=1}^{n} P_k^2(x_i).$$

The number of the terms of the series is determined from the minimum of the error of closure

$$D_n = \frac{1}{n-N}\sum_{i=1}^{n}\left[Y_N(x_i)\right] - W^2 Z(x_i)^2.$$

Acceleration of the particle is determined in accordance with expression

$$a(x) \approx Y_N''(x) = \sum_{k=2}^{N} a_k P_k''(x).$$

For the polynomials $P_k''(x)$, we obtain the following recurrent relationships

$$P_{k+1}''(x) = 2 P_k'(x) + (X + \beta_{k+1}) P_k''(x) - \frac{H_k}{H_{k-1}}(x),$$

$$P_{k+1}'(x) = P_k(x) + (x + \beta_{k+1}) P_k'(x) - \frac{H_k}{H_{k-1}} P_{k-1}'(x),,$$

$$P_0'(x) = 0, \quad P_1' = 1, \quad P_1''(x) = 0, \quad P_2''(x) = 2.$$

In addition to the described methods, a method of determining second derivatives using modelling functions was developed. In this case, approximation is carried out in the form

$$W^2 Z(x) \approx Y(x) = \int\limits_0^{x} \int\limits_0^{x'} a(x)\,dx\,dx', \tag{9.4}$$

where $a\,(x)$ is the particle acceleration to be determined. The task is to select the distribution $a\,(x)$. Taking into account the axis symmetry of the plasma jet, we can use the three-parameter function

$$a(r) = a_0\left(1 - \frac{r}{\delta}\right)^N \left(1 + N\frac{r}{\delta}\right),$$

where δ is the radius of the plasma jet. The optimum set of the parameters a_0, δ, N and the position of the axis jet symmetry are determined from the minimum of the error of closure

$$D = \frac{1}{n}\sum_{i=1}^{n}\left[W^2 Z(x_i) - Y(x_i)\right]^2.$$

These methods were applied in a computer. To verify the processing methods, numerical calculations were carried out using a test function selected from apriori considerations in the form

$$a(r) = a_0\left(1 - \frac{r^2}{\delta^2}\right)\exp\left(-5\frac{r^2}{\delta^2}\right), \tag{9.5}$$

Random errors with a relative value of $\delta = 3$; 10% (Fig.9.4) were superimposed on the discrete values of $Y(x_i)$ computed from (9.4) with substitution from (9.5). It may be seen that the method of the modelling function gives a solution close to the accurate one even at $\delta = 10\%$ whereas the Chebyshef polynomials do not ensure sufficient accuracy at the peripheral parts of the flow, especially at high experiment errors. Statistical tests using the method of the modelling function at a random error of 3% show that the error of measuring acceleration in the near-axial zone of the jet does not exceed 5%, and in the peripheral section it is ~10%.

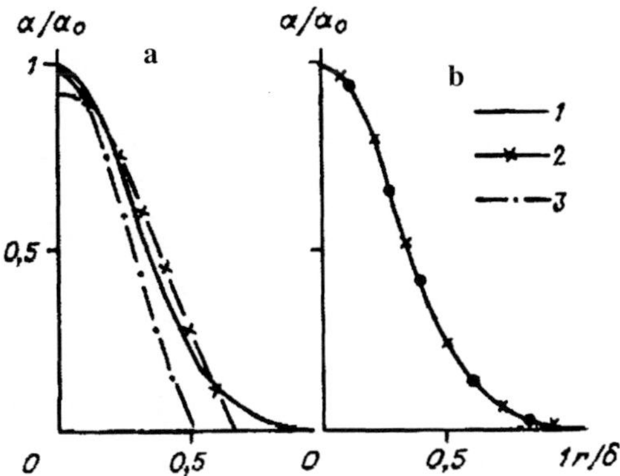

Fig.9.4 Restoration of the probe function (1) by Chebyshef polynomials (a) and modelling functions (b) at $\sigma = 3$ (2), 10% (3).

The resultant values of particle acceleration are then used to plot the plasma velocity profile, using equation (9.2), at other plasma parameters (temperature, density, viscosity) and the aerodynamic drag coefficient of the particle.

A special feature of this method is the direct determination of the local acceleration of the particle. Consequently, it is not necessary to use Abel's transformation and use the method for examining the non-axisymmetric flows. Advantages of the method include the possibility of determining the complete velocity profile in single channeling of the particle, a sufficiently high spatial resolution and small perturbations even in diagnostics of plasma jets with respect to a small transverse dimension. Examination of the plasma flows using tracing by spherical particles and the relationships (9.1) was carried out in Ref.6 for determining density, in Ref.5 to determine viscosity, in Ref.6 plasma velocity and in Ref.9 the aerodynamic drag coefficient of the particles. The error of the method is estimated from the equation

$$\frac{\delta u}{u} = \frac{1}{2}\left(\frac{\delta\rho}{\rho} + \frac{\delta a}{a} + \frac{\delta C_D}{C_D}\right).$$

The error of determining plasma density is determined by the error of temperature measurements and it can therefore be accepted that $\delta\rho/\rho \sim 5\%$. The error of measuring particle acceleration, determined by statistical tests, is ~5% at the axis and ~10% at the periphery of the plasma flow. The maximum area of estimating the head drag coefficient is close to 20–25%. Thus, the error of the velocity measurement method is 15–20%.

The aerodynamic drag coefficient. Using this method, measurements were taken of the coefficient C_D for a sphere in argon plasma. Experiments were carried out in the channel of a dc electric arc stabilised with cold walls with an internal diameter of 1.5 and 3 cm, arc current of 70–190 A, the flow rate of the plasma forming argon 0.2–2.75 g/s, and the diameter of probed particles 0.15–0.3 mm. The measured local average values are presented in Fig.9.5 in relation to the local Reynolds number calculated from the parameters of the non-perturbed incident flow. The maximum error of determining C_D and Re for random non-correlated measurement errors of initial values is 20–25%. The results of measurements of C_D at different values of arc current I, gas flow rate G and the diameter of the channel $2R$ and particles d were used to plot (using the method of least squares) the approximation dependence $C_D = f$ (Re) in a wide range of the values of the Reynolds number situated below the standard drag curve.

Figure 9.5 shows the standard drag curve for a sphere under the conditions of isothermal flow of a non-viscous incompressible gas (curve 1), the results of measurements in plasma[10–12] and also the expected dependence $C_D = 16.6\ \mathrm{Re}^{-0.75} + 0.2$ (curve 2) which approximates the experimental results obtained by the authors of this book and the data obtained in Ref.13 for Re = 0.4÷220. The same graph shows the calculated data from Ref. 14 where the authors analysed by numerical methods, the movement of aluminium particles with a diameter of (5–50)· 10^{-6} m in the laminar flow of argon plasma at the atmospheric pres-

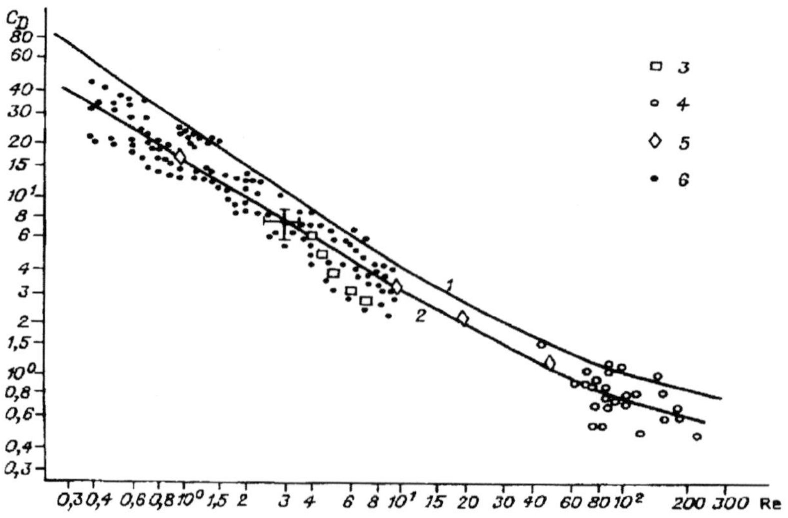

Fig.9.5 Dependence of the coefficient of aerodynamic resistance of the sphere on the Reynolds number in argon plasma. 1) standard drag curve, 2) approximating curve, 3) data from Ref.12, 4) data from Ref.13, 5) data from Ref.14, 6) this work.

sure taking into account the change of the plasma properties in the boundary layer and also showed a large decrease of the values of C_D in relation to the flow and particle temperature.

It is important to note that there are no experimental data for C_D for the plasma conditions in the range Re = 10–50. Taking into account the monotonic form of the dependence $C_D = f$ (Re), it may be assumed that in this range there will be no large deviations of the values of C_D from the values calculated from the equation approximating the experimental data for the Reynolds numbers Re = 0.4–220.

Measuring the velocity of the laminar plasma jet at the outlet from the plasma torch

Investigations were carried out in the laminar jet of argon plasma generated by a plasma torch[15] with the conditions: arc current $I = 80$ A, arc voltage $U = 80$ V, the flow rate of the plasma forming argon 2.5· 10^{-2} kg/s, the diameter of the outlet nozzle 3.3·10^{-3} m. The plasma jet was discharged into air. To prevent mixing of the jet with the surrounding air, it was necessary to apply co-axial blowing of the jet with an argon flow. The tracer particles were spherical and made of aluminium, copper and tungsten. The trajectory of the particles was recorded using the set up shown in Fig.9.1.

In calculating the plasma velocity using equation (9.2), the effect of gravity forces g and the axial component of the particle velocity $v \sim 10$ m/s was ignored (Fig.9.6). The error of determining the plasma velocity was obtained by varying the equation (9.2) taking into account the dependence $C_D = f$ (Re, M, γ) and equalled ~15% for the axis of the plasma jet and ~20% for the periphery of the latter.

The error of measuring acceleration is determined by the local scatter of the experimental data with respect to the extent of deflection of the particles, and by the method determining the second derivative. In addition, systematic errors are also important. In this case, they are associated with the variation of the dynamic pressure of the jet $\rho u^2/2$ within the range of the sphere size.

In movement across the plasma flow, the particle is subjected to the effect of the vertical dynamic pressure of the jet with a radial gradient. The presence of this gradient leads to lateral (in the direction of the gradient $\rho u^2/2$) displacement of the particle (the so-called profile effect[9]). The ratio of the transverse (horizontal) F_h and vertical F_v forces, acting on the particles, determined by the authors of this book within the limits of Newton's approximation[16] of the interaction of a spherical particle with the gas flow, is determined by the equation

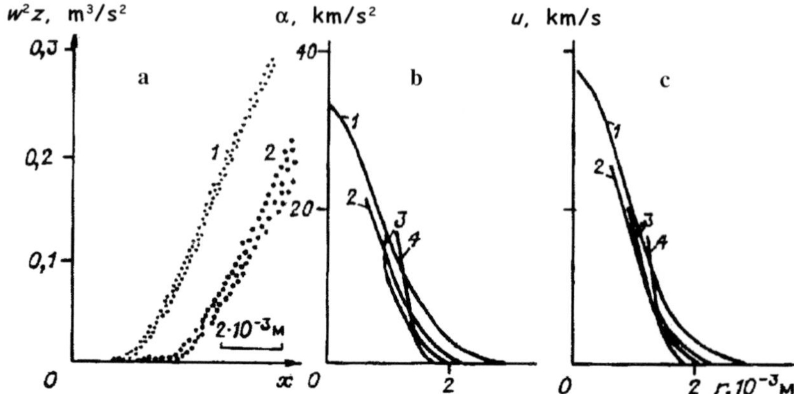

Fig.9.6 Experimental function (a), particle acceleration (b), and plasma velocity (c), obtained in shooting particles through the diameter and three chords of the cross-section y. $y = 0(1)$, 0.6×10^{-3} (2); 0.9×10^{-3} (3); 1.2×10^{-3}m (4).

$$\frac{F_h}{F_v} = 0.13\frac{\partial \ln \rho u^2}{\partial r}.$$

(9.6)

Under the given experiment conditions this ratio is on average equal to ~0.05. The influence of the profile effect increases in tracing along some chords of the cross section of the jet. The presence of the gradient of dynamic pressure causes ejection of the particle to the flow periphery. The particle is deflected from the initial direction and in turn affects the region of the flow where the dynamic pressure is 5–8% lower than the expected pressure (determined by the direction of introduction of the particles). To take into account the profile effect in tracing along the diameter it is sufficient to use the local velocity of the particle W (x) in plotting the experimental function W^2Z (x). Under the given experimental conditions, the error $\delta W/W$, determined by the profile effect, is, according to estimates, ~1.5%.

To determine more accurately the strength of the influence of the profile effect, particles were introduced in the same section of the jet $Z = 5\cdot10^{-3}$ m along the chords whose distance from the diameter was $(0.6, 0.9, 12)\cdot10^{-3}$ m. The data were processed by the method of modelling functions. Figure 9.6b, c shows the results of measurements of acceleration and the corresponding plasma velocity. It may be seen that the distributions obtained in introduction along the chords and the diameter differ. The systematically growing displacement of the distributions, obtained in introducing the particles along different chords, corresponds to increase of local values $\partial \ln \rho u^2/\partial r$.

In the flows with high radial gradients of the parameter the scatter in the direction of introduction of the particles in tracing along the chords increases the random error of the experimental data (Fig.9.6a). The influence of the profile effect in the experiments with the tracer particles will be strongest if processing is carried out using only the finely resultant deflection of the particle after intersection of the flow by it, and not its total trajectory.[4]

The undesirable consequences of the influence of the profile effect can also be observed in technological processes of plasma treatment of dispersed materials. In moving along an extended non-uniform flow, the particles are displaced under the effect of the radial ejection force to the periphery of the flow and stick to the walls of the reactor or the plasma torch channel. These consequences of the influence of the profile effect are evidently weakened by the superimposition of the displacing force acting in the opposite direction, for example, by means of annular blowing of an accompanying flow of cold gas into the plasma.

Measuring the velocity in plasma jets of a two-jet plasma torch

In the experiments, the anode and cathode jets at a distance $Z = 5$ mm from the outlet of the nozzle of the plasma torch were traced along the diameter with aluminium particles 150 µm in diameter. Plasma jets were discharged into the atmosphere. The operating regime of the two-jet plasma torch was characterised by an arc current of 105 A and the flow rate of the working gas (argon) of 0.075 g/s per head at an initial jet diameter 5 mm. These experiments were carried out using the measured spectroscopic radial temperature profile extrapolated at the periphery to the temperature of the environment. Local values of the velocity and

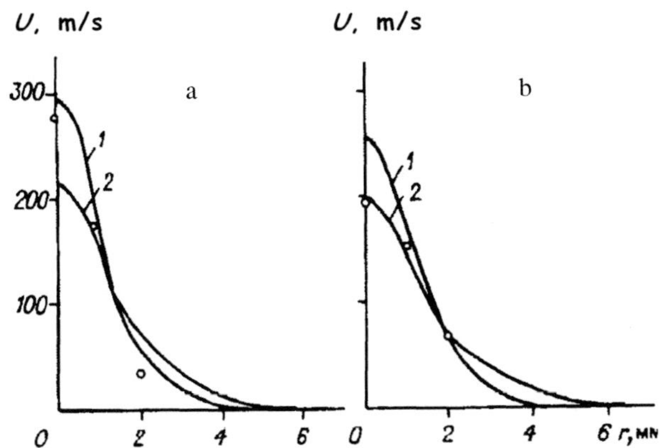

Fig.9.7 Profile of the plasma velocity in electrode jets of the two-jet plasma torch. a,b) cathode and anode jet, respectively; 1,2) calculations at $z = 2$ and 10 mm; points are the experimental data at $z = 5$ mm.

acceleration of the particles were computed from the method described previously. To evaluate the error of the results of measuring the velocity comparison was made with the calculated characteristics of the flow in individual plasma jets obtained on the basis of numerical solution of a system of MHD equations using the two-temperature model in the boundary layer approximation. The experimentally measured and calculated profiles of the plasma flow velocity are in good agreement (Fig.9.7).

Chapter 10

ELECTRIC PROBES IN NON-EQUILIBRIUM PLASMA

10.1 Introduction

The electric probe is a small metallic electrode placed in plasma and used to determine its characteristics. It usually measures the volt–ampere (probe) characteristics of the system, which includes a measuring probe, a reference electrode and a voltage source. The reference electrode can be represented either by one of the electrodes of the gas discharge system or by a specially introduced reference probe. An example of such a characteristic and typical potentials of the probe are shown in Fig.10.1.

If the system satisfies the relevant requirements, it can be used to determine the concentration of charged plasma particles, the energy

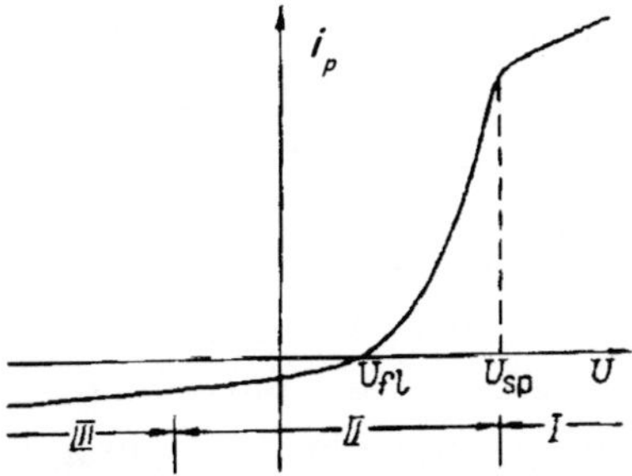

Fig.10.1 Volt–ampere characteristic of the probe i_p (U). U is the potential difference between the reference and measuring probes; U_{sp} is the plasma potential (space); U_f is the floating potential; $U_p = U - U_{sp}$ is the potential of the measuring probe in relation to plasma; I) $U_p \geq 0$ is the electronic saturation current; II) $U_p < 0$ is the electronic current on the probe; III) $U_p < 0$, $|U_p| > kT_e/e$ is ionic saturation current.

(velocity) distribution of the electrons, the plasma potential in the vicinity of the measuring probe, potential pulsations and the fluxes of charged particles. After appropriate modification, the method can also be used to examine chemical processes in plasma, such as plasma chemical polymerisation and etching. In contrast to the majority of other diagnostic methods, the probe method gives local values of the plasma characteristics.

Unfortunately, the use of probes for plasma diagnostics is faced with a number of difficulties which restrict the advantages of the method associated with its apparent simplicity. The probe and the plasma must satisfy a number of relatively stringent requirements, and only in this case the results of simple electrical measurements can be linked with the plasma parameters. These are also essential restrictions of the dimensions of the probes associated with plasma perturbation and fulfilment of the requirements of probe theory, and the need to take into account the variation of the plasma potentials, and a number of others. In addition, the probes are introduced into the plasma using various devices and it is therefore necessary to take into account their perturbing effect on the plasma. The use of a method must be justified in each specific case, otherwise both incorrect measurements and erroneous interpretation of these measurements may take place.

The studies of the theory of electric probes and apparatus used for the probe methods have progressed in the direction of expanding the number of plasma objects examined using probes. They are now used to examine plasma of dc discharges, radiowave and microwave plasma at pressures from 10^{-5} mm Hg to atmospheric pressure. The number of studies in this direction increases and this section cannot provide a complete pattern of the current state of investigations. We shall present only several reports on the probes associated with the problems of their practical application in non-equilibrium plasma and, in particular, to the problem of measuring the electron energy distribution function (EEDF) in the plasma. The EEDF is the most important characteristic of the electronic component of equilibrium plasma which determines the kinetics of all processes associated with the interaction of electrons with a heavy plasma component. Details of the probe method can be found in reviews and monographs[1-8] where extensive literature lists are also provided. In this book, the list is supplemented by the latest publications which, we believe, have made it possible to widen greatly the potential of the probe method. Special attention is given to the experimental aspects of probe diagnostics.

10.2 Probe measurements of the energy distribution of electrons

The theory of current on to a metallic body (probe), placed in plasma, which makes it possible to determine the electron temperature in the case of a Maxwell distribution of the velocity of charged particles, was developed by Langmuir and Mott–Smith.[9] At the arbitrary form of the energy distribution of the electrons $f(\varepsilon)$ $\left(f(\varepsilon) = n_e f_0(\varepsilon e), \int_0^\infty f_0(\varepsilon)\mathrm{d}\varepsilon = 1 \right)$ in the region of the negative potentials of the probe the electron current on the probe $i_e(U_p)$ is linked with $f(\varepsilon)$ by the relationship[1]

$$i_e\left(U_p\right) = \frac{2\pi n_e e}{m^2} \int_{eU_z}^{\infty} f_0(\varepsilon)\left(\varepsilon - eU_p\right)\mathrm{d}\varepsilon, \qquad (10.1)$$

where e, m is the charge and mass of the electron, ε is the electron energy, U_p is the potential of the probe in relation to plasma, n_e is the electron concentration. Equation (10.1) holds for isotropic EEDF and probes with a convex surface.

An important step in the development of probe diagnostics of plasma was the solution, by Druyvesteyn, of the problem of determination of the EEDF in the plasma on the basis of the second derivative of the electronic probe current i_e using the probe potential U_p $(U_p < 0)$[10]

$$n_e f_0(\varepsilon) = 2^{3/2} m^{1/2} e^{-3/2} S_p^{-1} U_p^{1/2} \frac{\mathrm{d}^2 i_e\left(U_p\right)}{\mathrm{d}U_p^2} = \mathrm{const}\, U_p^{1/2} \frac{\mathrm{d}^2 i_e\left(U_p\right)}{\mathrm{d}U_p^2}, \qquad (10.2)$$

where S_p is the probe area. This equation holds for isotropic EEDF and does not depend on the probe geometry if its surface is convex. It is known as the Druyvesteyn equation and represents the basis of the absolute majorities of studies concerned with the measurement of EEDF in non-equilibrium plasma.

Geometrical criteria of the theory of Langmuir probes. Theory[9] assumes that in measuring current the electrons move in the near-probe double electric layer without collisions in the potential field generated by an external source. The probes operating under such a regime are referred to as Langmuir probes.

The dimensions of the probe should be such that as to fulfil the assumption of the theory that enable the measured quantities to be linked with the plasma characteristics, and that the perturbations caused by the probe should be low. The requirement for a collisionless near-probe layer gives a relationship between the free path of the electrons λ and the Debye screening length λ_D. This also defines the lower boundary of the electron concentration in plasma at which probe measurements can be still be taken

$$\lambda n_e \gg 5 \cdot 10^5 T_e \left(N \langle \sigma(\varepsilon) \rangle \right)^2, \tag{10.3}$$

where T_e is the electron temperature, eV; N is the concentration of heavy particles, cm^{-3}; $\langle \sigma(\varepsilon) \rangle$ is the mean value of the collision section of the electrons with heavy particles.

Placing the probe in plasma leads to screening, by the probe, of some areas of the plasma in relation to others. The geometrical criterion of the smallness of perturbations of the plasma by the probe has the form

$$a_p \ll \lambda, \tag{10.4}$$

where a_p is the characteristic probe dimension. Thus, the condition $\lambda \gg a_p + \lambda_D$ should be fulfilled for Langmuir probes.

In additional to geometrical perturbations, two other causes of errors are associated with the probe dimensions: the sink of electrons to the probe, the effect of the finite resistance of plasma.

Sink of the electrons to the probe

The properties of plasma in the vicinity of the probe can change owing to the fact that the diffusion of electrons from the unperturbed plasma does not manage to compensate their losses associated with departure to the probe.[11,12] The VAC of the probe is distorted, and the extent of these distortions increases with the decrease of the difference between the potential of the probe and the plasma potential and with increasing sink parameter δ. The sink parameter depends on the ratio of the characteristic dimensions of the probe and the free path and also on probe geometry. For spherical and cylindrical probes the values of δ were determined in Ref.11 and 12, respectively

$$\delta_{\text{spher}} = \frac{3}{4} \left(\frac{r_p}{\lambda} \right)^2 / (1 + r_p / \lambda), \quad \delta_{\text{cylin}} = \frac{3}{4} \frac{r_p}{\lambda} \ln \frac{l_p}{2r_p}, \tag{10.5}$$

where r_p and l_p are the radius and the length of the probe. Since the free path λ depends on electron energy ε, then also $\delta = \delta(\varepsilon)$.

The sink of the electrons to the probe decreases the electron concentration calculated from the electronic saturation current, increases the electron temperature determined from the VAC, displaces the floating potential of the probe in relation to the plasma potential and distorts the second derivative of the probe current with respect to the probe potential. In particular, EEDF is not proportional to the second derivative of the probed current with respect to the probe potential, as at $\delta = 0$.

The effect of sink to the probe on the measurement results can be corrected by calculations. Thus, the distorted and true EEDFs at $\delta \ll 1$ are linked by the relationship[12]

$$f_m(\varepsilon) = f(\varepsilon)\big[1 - \theta(f(\varepsilon))\big],$$

$$\theta(f(\varepsilon)) = 2\int\limits_{\varepsilon}^{\infty} \frac{\delta f(\varepsilon')d\varepsilon'}{\varepsilon' f(\varepsilon)\big[1 + \delta(1 - \varepsilon/\varepsilon')\big]^3}, \tag{10.6}$$

and the concentration of the mean energies of the electrons by the relationship

$$n_e = (1 + 4\delta/3)n_{em}, \quad \bar{\varepsilon} = \bar{\varepsilon}_m(1 - \delta/2). \tag{10.7}$$

Analysis of the work of the probe at $\delta \gg 1$ show that also in this case the probe characteristic gives EEDF but it is proportional to the first (not second) derivative of electron current on to the probe with respect to the probe potential (see further sections).

The effect of end resistance in the probe circuit
When determining the probe characteristic (Fig.10.1), it was assumed that the entire voltage of the external source U is applied to the probe-plasma transition layer. Differences in the experimental conditions of this model may lead to an error in interpreting the probe measurements. This discrepancy appears if the probe circuit contains an element connected in series with the probe and the voltage drop on this element decreases the voltage drop in the near-probe layer during current passage (at a constant voltage of the external source).[7] This is indicated by examining equivalent circuits of the probe (Fig.10.2). Such elements are either the resistance of plasma between the measuring and refer-

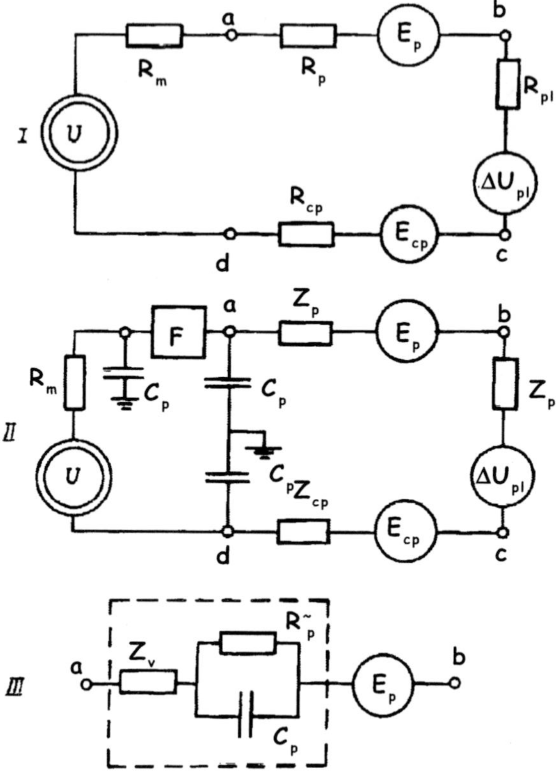

Fig.10.2 Equivalent diagram of the probe circuit for direct (*I*) and alternating (*II*) current and the equivalent circuit of the probe (*III*). R_p, R_{cp} and Z_p, Z_{cp} is the resistance and impedances of double electrical layers at the probe and the counter probe, respectively; E_p and E_{cp} is the emf of the layers at the probe and the counterprobe, to their floating potentials, determined by the plasma parameters in the areas in which they are situated; R_{pl} and Z_{pl} is the resistance and impedance of the plasma; ΔU_{pl} is the difference of the plasma potentials in the area of distribution of the probe and the counterprobe; C_p are the parasitic copacitances; R_m is the measuring resistance; U is the source of the probe voltage; Z_v is the impedance of the probe as the vibrators; C_p and R_p^N is the capacitance and resistance of the near-probe with respect to alternating current; Φ is the high-impendance element (filter).

ence probes R_{pl} or the resistance of conducting lines and resistances connected to the external part of the probe circuit. Consequently, the electron current to the probe in the circuit with a resistor is lower than without the resistor at the same U. The nature of the effect of this factor on the results of probe measurements is identical with the effect of the sink of the electrons to the probe.

Since the area of the counterprobe is usually considerably greater than the probe area (see later), neglecting the voltage drop in the counterprobe circuit, we can write the following equation for the probe circuit

$$i_e(U_p) = i_e^*\big[U(U_p)\big],$$

(10.8)

$$U = U_p + i_e(U_p)R_s + \Delta U,$$

(10.9)

where the asterisk denotes the values of quantities measured in the circuit with a resistor R_s connected in series; ΔU is the difference of the plasma potentials at the points of positioning the probe and the counterprobe.

Double differentiation of equation (10.8) with respect to U_p with an allowance made for (10.9) gives a relationship between the experimentally determined second derivative of the probe current with respect to the potential $(d i_e^2{}^*/dU^2)$ with the quantity determined by EEDF $(d i_e^2/dU_p^2)$:

$$\frac{d^2 i_e}{dU_p^2} = \frac{d^2 i_e^*}{dU^2}\left(1 - \frac{R_s}{R_p^{\sim *}}\right)^{-3},$$

(10.10)

where $R_p^{\sim *} = (d i_e^*/dU)^{-1}$ is the differential resistance of the probe–plasma layer calculated from the measured probe characteristic, $R_p^{\sim *} = R_p^{\sim *}(U)$. The right-hand part of equation (10.10) contains the quantities determined from the measured VAC of the probe.

The error in determining the second derivative of the probe current with respect to the probe potential decreases with decreasing ratio and $R_s/R_p^{\sim *}$. As a rule, the components of the external electrical circuit are selected in such a manner as to cause only small distortions in the probe measurements, and the main component of the R_s is the plasma resistance R_{pl}. The value of R_{pl} depends on both the plasma parameters and the geometry of the probe system. Calculations carried out for different structures of the probe show that at $r_p \ll \lambda$ the error of the probe measurement is proportional to the ratio r_p/λ, i.e. increasing r_p increases both the sink of electrons to the probe and the effect of end resistance of the probe circuit on the VAC of the probe.[7,14] Distortions of the VAC become greater with approach of the probe potential to the space potential because $R_p^{\sim *}$ is minimum at $U_p = 0$. In Ref.13, the authors examined the combined effect of electron sink and the resistance of the probe circuit on the VAC of the probe in a wide range of the sink parameter and R_s/R_p. The nature of distortions of the VAC of the probe is clear from Fig.10.3. In particular, the author of Ref.13 proposed a method of computing the temperature of plasma electrons (assuming Maxwell's energy distribution of the electrons) in which additional resistors are added specially to the probe circuit. Measurements are taken

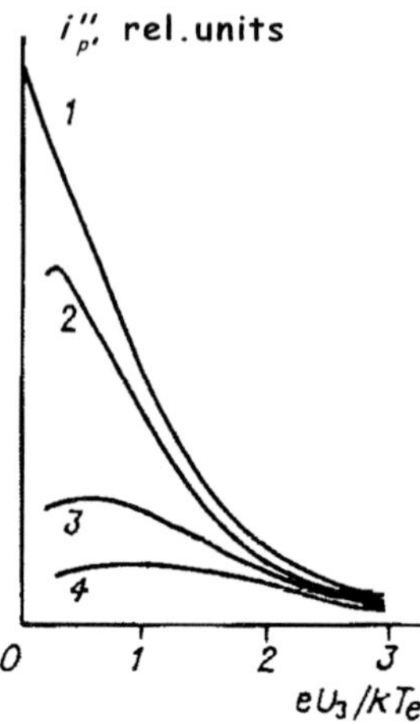

Fig.10.3 Qualitative pattern of the effect of electron sink on the probe of the resistance connected in the probe circuit R_p, on the second derivative of the electronic current to probe with respect to the probe potential (calculations)[13] ($\delta = 0.1$). 1) non-distorted derivative; $R_p i_e(0)/kT_e \times (1+\delta)^2 = 0(2);\ 0.5(3);\ 1.0(4)$.

of the potentials U_m of the maxima of the second derivative of the probe current with respect to the probe potential at two values of R_s and, consequently, T_e is described by the equation

$$T_e = -\frac{e}{k}\left(U'_m - U''_m\right)\left[R'_s / R''_s\right]^{-1}. \tag{10.11}$$

The effect of R_{pl} can be compensated by experiments if in plotting the VAC the potential is counted in relation to a special tracking probe situated in the vicinity of the measuring probe and subjected to the floating potential (the current of the tracking probe is equal to zero).[15]

Effect of the surface area of the counterprobe

In probe measurements in electrodeless discharges, such as radio- and microwave discharges, it is necessary to use two-probe circuits. To obtain a single-probe characteristic, the area of the reference probe (counterprobe) should greatly exceed the area of the measuring probe. This is indicated by the fact that the ionic saturation current of the counterprobe should be considerably higher than the saturation current of the probe. Estimates of the required ratio of the areas of the probe and the counterprobe from the equation from Ref.1 $\alpha = S_{cp}/S_p \gg (M_i/m)^{1/2}$ are too low owing to the fact that under the real conditions the dimensions of the counterprobe are greater than the free path of the charged particles and the random current to the counterprobe from the plasma is

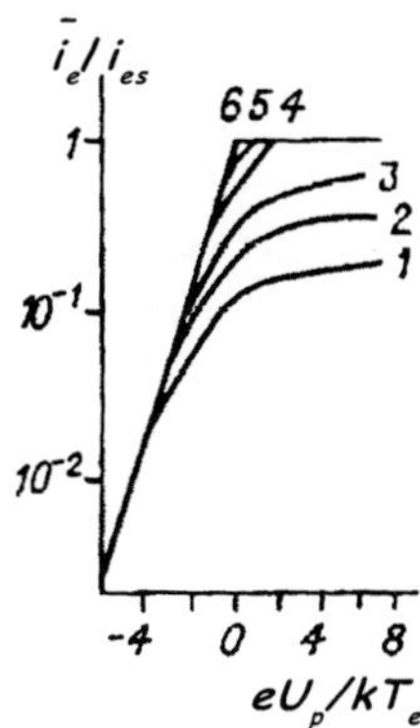

Fig.10.4 Effect of the ratio of the areas of the counterprobe and the probe on the VAC of the probe (calculated).[15] $R_p < \lambda_e, \lambda_i$; $\lambda_D/R_{cp} = 0.01$; $T_i/T_e = 0$; $\Sigma_\alpha = 100(1)$, 200 (2), 400 (3), 10^3 (4), 10^4 (5), ∞ (b); VAC normalised the electronic saturation current i_{es}.

lower than that calculated disregarding this fact (M_i is the mass of ions). Analysis of the problem carried out in Ref.16 for probes of different forms showed that the condition applied in most cases, $\alpha > 10^4$, should be replaced by a new condition $\Sigma_\alpha > 10^4$, where Σ_α is obtained by multiplying the value of α by the coefficient $C_\alpha \simeq 10^{-2}$, which takes into account the decrease of the random current to the probe and depends on its shape.

Thus, the new condition is considerably more rigid. The effect of Σ_α on the VAC of the probe and the nature of the distortion of the VAC due to the small area of the counterprobe are clear from Fig.10.4.

The influence of all these factors leads to 'stretching' of the VAC of the probe along the voltage axis, to smoothing of the VAC in the region of the plasma potential, and becomes weaker with decreasing ratio r_p/λ. The contribution of these factors to the total error of the probe measurement depend on the probe dimensions and the plasma characteristics. For example, in measurements in microwave plasma in atomic and molecular gases with a probe with $r_p = 5$ μm at pressures of 1–5 mm Hg, the errors associated with the sink of electrons to the probe and with the end resistance of the plasma were similar.

The dependence of the effect of all factors on the value of r_p makes it possible to correct the results of probe measurements if experiments with the probes of different thickness can be conducted. Extrapolating the results of measurements to a zero thickness probe, it is possible to reduce the error of measurements.[7]

Effect of alternating fields on VAC

A large procedural error may form in probe measurements in the plasma generated using electromagnetic fields (rf and microwave discharges) and also in the plasma with instabilities of different type leading to a periodic variation of the plasma potential with time. This error is associated with the fact that the probe–plasma layer (this layer yields the measured VAC of the probe) is a non-linear element and the variable signal on this layer which is rectified changes the measured VAC averaged out with respect to time.[1,7,17-20]

If a voltage of type

$$U_p = U_{p0} + u_0 \cos \omega t, \tag{10.12}$$

is applied to the near-probe layer, where U_{p0} and U_0 is the constant voltage and amplitude of the alternating voltage in the near-probe layer, then (assuming Maxwell's distribution of the electrons), the mean electronic current on the probe is expressed by the equation

$$i_e^* = i_{es} \exp\left(-\frac{eU_{p0}}{kT_e}\right) I_0\left(\frac{eu_0}{kT_e}\right), \tag{10.13}$$

where i_{es} is the electronic saturation current, $I_0(\ )$ is the modified Bessell's function of the zero order.

Equation (10.13) shows that identical values of the electronic current on the probe in the distorted characteristic (i_e^*) are obtained at higher negative values of U_{p0} than in the case of the non-distorted VAC ($u_0 = 0$), and the shift becomes greater with increasing U_0 (Fig.10.5). Correspondingly, the derivatives of the probe current with respect to the probe potential are also distorted (Fig.10.6 and 10.7). The nature of distortions is the same under the effect on the probe of hf and microwave signals, and the calculated results and experimental data are in qualitative agreement.

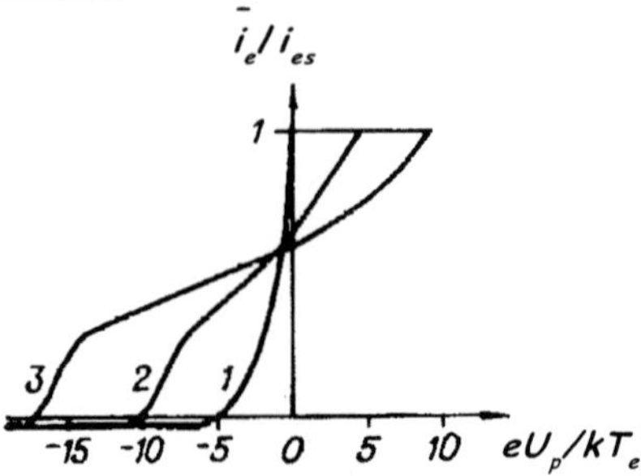

Fig.10.5 Effect of alternating voltage on the near-probe layer on the VAC of the probe (calculated assuming the Maxwell EEDF).[17] $eu_0/kT_e = 0(1); 5(2); 10(3)$.

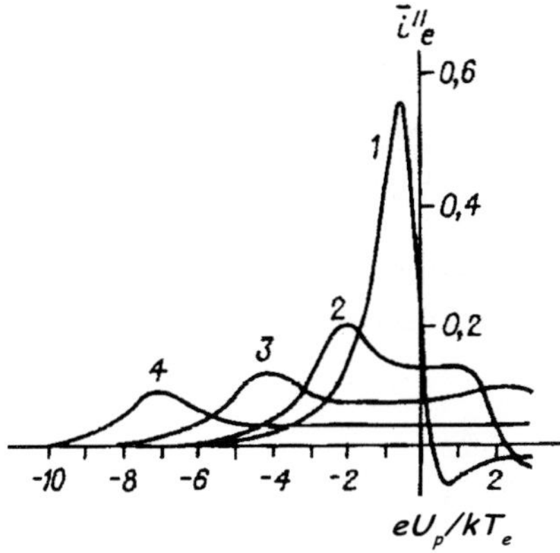

Fig.10.6 Effect of alternating voltage on the near-probe layer on the second derivative of probe current averaged with respect to period (calculated).[20] $eu_0/kT_e = 0(1)$, 2(2), 4(3), 7(4).

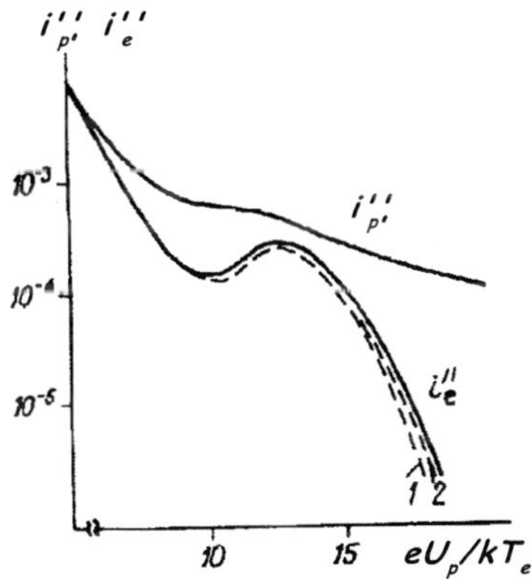

Fig.10.7 Effect of alternating field on the VAC of the probe (a) and the second derivative of the probe current with respect to probe potential (b) in microwave discharge, in N_2 under a pressure of 1.5 mm Hg (probe radius 5 μm).[7] 1) non-distorted, 2) distorted curves.

The methods of reducing the error of probe measurements, determined by the effect of alternating fields, are clear from the equivalent diagram of the probe circuit using alternating current (Fig.10.2). They are all associated with the decrease $u_0\left(U_p\right)= \Delta U^\sim Z_p\left(U_p\right)/ \sum_i Z_i$, where $\Delta U^\sim$ is the alternating voltage acting in the probe–reference electrode system; $Z_p\left(U_p\right)$ is the potential-dependent impedance of the probe; Z_i are the in-series connected elements of the equivalent diagram of the probe circuit (Fig.10.2). It is important to note that U_0 depends on U_p and, consequently, the VAC is not just simply displaced from the voltage axis but is also deformed.

One of the methods of decreasing U_0 is to position the probe and the counterprobe at equipotential (with respect to the alternating signal) points of the plasma. It should be noted that the counterprobe is larger and if it is placed in the vicinity of the probe it may alter the plasma parameters.

The second method is to increase the probe surface. This is accompanied by a reduction of its impedance

$$R_{\tilde{p}} = \frac{4kT_e}{1.5n_e \bar{v}_e S_p e^2} \exp^{-1}\left(-\frac{eU_p}{kT_e}\right),$$

(10.14)

$$C_p = \frac{2\pi l_p \varepsilon_0}{\ln\left(x/r_p\right)},$$

(10.15)

where $\bar{v}_e$ is the mean electron velocity, C_p is the capacitance of the near-probe layer at a floating probe potential, ε_0 is the permittivity of vacuum, l_p, r_p are the length and radius of the cylindrical probe, x is the thickness of the near-probe layer ($x \simeq (5\div7)\lambda_\mathrm{D}$).[7] It is obvious that, with other conditions being equal (all other components of the equivalent diagram of the probe circuit are constant), the alternating voltage acting on the layer decreases. It should be noted that when using this approach all errors examined in the previous sections become larger because to reduce them it would be necessary to reduce the probe dimensions.

The same effect can be achieved not by changing the impedance of the probe but by increasing the impedance of the external parts of the probe circuit. For this purpose, additional elements are added to the circuit (in the rf range it is filters with concentrated elements, in the microwave range – with distributed elements[7]).

There are active methods of reducing U_0 in which an alternating signal with the field frequency generating plasma is introduced into the probe circuit. The amplitude and phase of the signal are such that they compensate the alternating voltage on the probe generated by plasma (see, for example, Ref.20–22). For this purpose, the circuit shaping the required signal (and this signal is part of the voltage supplied to the discharge) should contain an attenuator and a phase-shifting device.

In the last and penultimate cases, it must be taken into account that the alternating voltage at the near-probe layer has the frequency of not only the external energy source but also multiple frequencies. This is associated with the fact that the discharge is a non-linear element in which frequency transformation takes place. There are cases in which the probe circuit must contain filters for both the first and second harmonics of the external voltage.

The criterion for evaluating the efficiency of the methods of decreasing u_0 is the sensitivity of the floating potential of the probe U_{fl} to the value of u_0. Equation (10.13) shows that the change of the probe potential (self-displacement of the probe) and, correspondingly, U_{fl} is linked with u_0 by the relationship

$$\Delta U_{fl} = -\frac{kT_e}{e} \ln I_0\left(\frac{eu_0}{kT_e}\right). \qquad (10.16)$$

Thus, reaching the minimum shift of the floating potential, it is possible to minimise the effect of u_0 on the results of probe measurements. In equation (10.16), the value U_{fl} is counted from the plasma potential (see Fig.10.1) which is not known prior to starting the probe measurements. In experiments, all potentials are usually determined in relation to the potentials of reference electrodes or electrodes of the gas discharge system. Therefore, the condition (10.16) can be reformulated for a specific type of discharge and a specific measurement system. For example, in Ref.22 a criterion in a capacitance rf discharge was the establishment of the maximum positive value of the contant potential of the probe in the absence of any current in relation to an earthed electrode.

Effect of ion current on EEDF

EEDF is determined from probe measurements using equations (10.1) and (10.2) if the dependence of electron current on the probe i_e on U_p at $U_p \leq 0$ is known. In the experiments, we measure not i_e but the total current to the probe $i_p = i_e + i_i$, where i_i is the ion current to the probe, and the contribution of the ion current to the measurement of

EEDF becomes significant at $\left|eU_p\right| \geq (3 \div 4)\bar{\varepsilon}$.[1,7,23]

The electronic component can be separated from i_p by extrapolating the ion saturation current to the region of lower probe potentials. However, this method has a number of significant shortcomings associated with the fact that the accurate dependence of the ion saturation current on the probe potential is not known and the extrapolation law is not available. This may cause a large uncontrollable error in the value of i_e.

In most cases, investigations were carried out using the derivatives of the probe potential and not current. Work in the region of high potential is possible due to the fact that the dependence of i_i on U_p is far

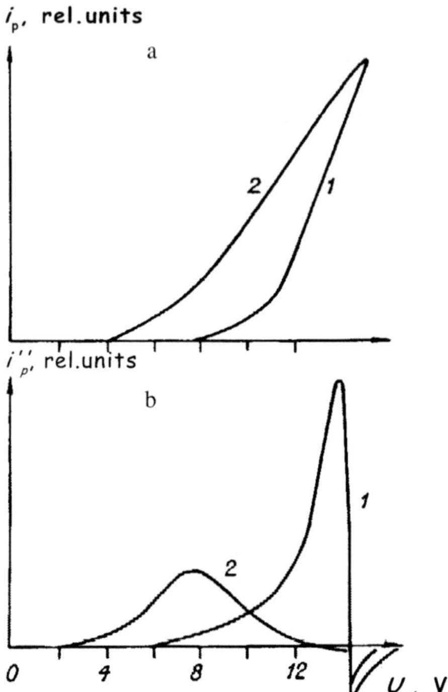

Fig.10.8 Restoration of the second derivative of electronic current on the probe with respect to the probe potential (i_e'') from the modelling second derivative of the probe current (i_p'') at non-monotonic EEDF. Numbers at the curves correspond to the numbers of iterations.[23]

weaker than the dependence i_e (U_p). Consequently, the effect of i_i on the curve of the second derivative of i_p with respect to U_p should be detected at values U_p higher than on the VAC of the probe.

The authors of Ref.23 described an iteration method of separating d^2i_e/dU_p^2 from the measured d^2i_p/dU_p^2 using the expressions for the ion current on the probe in limitation movement of the ions in the case of an arbitrary form of EEDF. In the first approximation and assuming that the measured curve $d^2i_p/dU_p^2 = d^2i_e/dU_p^2$, we determine the values that control the ion current, and also determine i_i and its second derivative. Subtracting this derivative from the measured curve, we carry out the first correction of the measurement results. The procedure is then repeated. Since the dependence of i_i on EEDF is weak, the procedure rapidly converges. An example of application of the method of a non-monotonic EEDF is shown in Fig.10.8.

Experimental methods of correcting the VAC of the probe for excluding i_i are available. The VAC can be produced using probes of

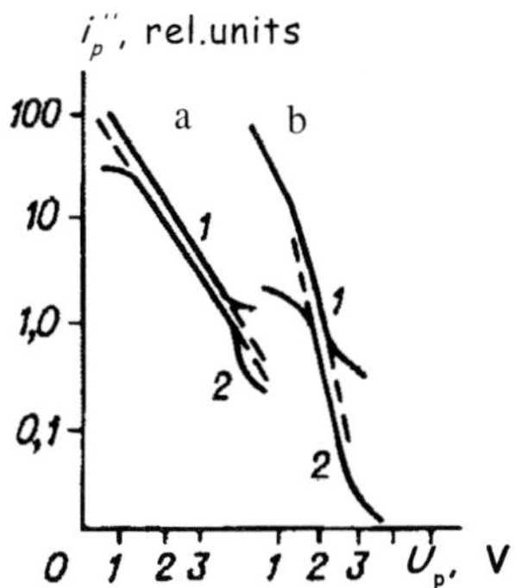

Fig.10.9 Application of two probes for widening the range of measurement of $i_p^{''}$ in the range of high energies in cesium vapours.[24] a) $2.8{\times}10^{-3}$ mm Hg, 200 mA; b) $1.8{\times}10^{-2}$ mm Hg, 100 mA; r_p = 0.025 (1), 0.25 mm (2).

different thickness and since the relative contribution of the second derivative of the ion current with respect to the probe potential decreases with an increase of its diameter (ion current of the flat probe does not depend on U_p),[2] measurements of $d^2 i_p/dU_p^2$ by the probes of different thickness make it possible to widen the range of measurement of EEDF towards higher electron energies (Fig.10.9). In this method, as in the previously described method of correcting the low-energy part of the EEDF using probes of different thickness, there are certain difficulties in combining the curves obtained by different probes along the voltage axis. This is caused by distortion of the VAC of the probe. The situation is made easier if the EEDF contains characteristic points, for example, inflections at specific energies.

The second method is based on measuring $d^2 i_p/dU_p^2$ to the probe potentials where this quantity completely determined by the ion current. The VAC of the probe is obtained at the same time and is approximated in the region of the ion current by some dependence on U_p (for example, $i_i \sim U_p^n$, where n is determined from the experimental curves; for example, in the case of a thick near-probe layer (where $r_p/(5{\div}7)\lambda_D \ll 1$), $n \simeq 0.5$).[7] The accuracy of approximation is verified by comparing the second derivative with respect to U_p of the resultant dependence for the ion current with the measured dependence. Subsequent subtraction of $d^2 i_i/dU_p^2$ from $d^2 i_p/dU_p^2$ makes it possible to separate $d^2 i_e/dU_p^2$. This method enables the EEDF to be measured in microwave plasma to electron energies of $\varepsilon \simeq 10\bar{\varepsilon}$ (Fig.10.10).

The ratio of the second derivatives of the ion and electron currents to the probe and, consequently, the electron energy to which the EEDF can be measured depends on the type of gas, the conditions in plasma (which determine the selection of ion current theory) and of the EEDF itself.

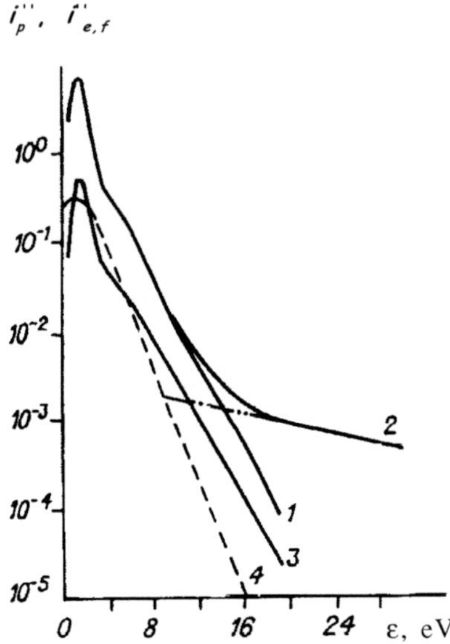

Fig.10.10 Restoration of the second derivative electronic current on the probe in respect to probe potential; 1) from the measured second derivative of probe current; 2) by subtracting the ionic components.[7] Microwave discharge plasma in CO_2 at a pressure of 1 mm Hg, $\overline{\varepsilon}$ = 1.9 eV, r_p = 5 µm; 3) EEDF, 4) Maxwell's distribution with the same mean energy.

The methods of determining EEDF from probe measurements

To determine the EEDF from probe measurements it is necessary to use the relationships (10.1) and (10.2). The Druyvesteyn method (equation (10.2)) is used most widely. In this method, the EEDF is linked with the second derivative of the electrode current on the probe with respect to the probe potential $d^2i_e/dU_p^2 = i_e^2$, and i_e^2 is determined using the following methods:

– double electrical differentiation of the VAC (modulation of VAC by special form signals, natural noise);

– double numerical differentiation of the VAC;

– double analogue differentiation;

– double graphical differentiation.

The **modulation methods** of differentiating the VAC of the probe are based on the non-linear properties of the near-probe layer and on the associated capacity to change the form, carry out displacement, and demodulation of the weak alternating signal $u(t)$ applied to it. The expansion of probe current $i_p[U_p + u(t)]$ into a Taylor series in these cases shows that different expansion harmonics contain the terms proportional to i_p''. The methods used for these cases are based on analysing various variable components of the probe current because they are less sen-

sitive to the instability of the discharge than the constant component.[1,4,7,25]
We shall present several examples of $u(t)$:

$$u_1(t) = a_1 \cos \omega t;$$

$$u_2(t) = a_2(1 + b \cos \omega_1 t)\cos \omega_2 t;$$

$$u_3(t) = a_3(b \cos \omega_1 t + c \cos \omega_2 t);$$

$$u_4(t) = a_4(1 + b\,\mathrm{sgn}(\cos \omega_1 t))\cos \omega_2 t; \qquad\qquad (10.17)$$

$$u_5(t) = a_5(1 + b\,\mathrm{sgn}(\cos \omega_1 t))\,\mathrm{sgn}\,(\cos \omega_2 t),$$

where $\mathrm{sgn}\,(x) = \{-1,\ x < 0; 0,\ x = 0;\ 1,\ x > 0\}$ is the signature function.

In modulating the signal of the type $u_1(t)$ the second derivative of the probe current is obtained using the second harmonic of frequency (the second harmonic method); for the signal of type $u_2(t)$ – the amplitude of the current with frequency ω_1 (demodulation method); for the signal of type $u_3(t)$ – the amplitude of the current of total or difference frequency $\omega_1 \pm \omega_2$ ('wobbling' method); at signals of type $u_4(t)$ and $u_5(t)$ – the amplitude of current with frequency ω_1 (modulation with a right-angled signal).

The error of the modulation method is linked with the fact that the amplitude of the measured harmonics contains not only i_p'' but also higher derivatives, for example, even derivatives of probe current for the second harmonics method. The contribution of higher derivatives becomes smaller with increasing amplitude of the modulating signal.

The second reason for the error is the distortion of the VAC of the probe by an alternating signal fed into the probe circuit. This has already been examined. The amplitude of the modulating signal should be small ($ea_i < \bar{\varepsilon}$).

The error of the measured values i_p'' is also caused by the end resistance of plasma elements of the probe circuit because this leads to a difference between $u(t)$ which depends on the probe potential, and to the corresponding voltage in the near-probe layer.

In Ref.26, comparative analysis was carried out on the basis of two parameters for the first out of four modulation signals described previously: the intensity of the signal S, proportional to i_p'' and the ratio D of the terms proportional to i_p'''' and i_p'' in the measured signal. In method 2, the maximum value of S is obtained at $b = 1$, and in method 3 at $b = c$. Comparison was carried out at the same amplitudes (from maximum to maximum) of the variable signal Δu. It was concluded that the the methods 1 and 4 have advantages. In the first of these methods, the maximum value is S, in the fourth the minimum value D. An amplifier makes it possible to compensate a small value of S, and the method 4 it is preferred.

The distorting effect of various methods of differentiation of the VAC (difference between the measured $i_p''^*$ and the true value) is taken into account by utilising the formalism of the apparatus functions (by analogy with optics) for each method.[26-28] The measured signal (for example, the voltage at the measuring resistance) is a convolution of the real secondary derivative i_p'' and the normalised apparatus function of the corresponding method A_i

$$i_p''^* = C\left(i_p'' \cdot A_i\right), \tag{10.18}$$

where C includes the values of measuring resistance, the gain factor, the amplitude of the modulating signal, etc. Examples of A_i for the signals of the type (10.17) are shown in Fig.10.11. The true i_p'' is restored as a result of solving the Fredholm equation of the first kind (10.18). Advantages of this approach are the possibility of describing by the same procedure all methods of obtaining i_p'' (see later), an increase of the signal:noise ratio by increasing the amplitude of the modulating signal and taking into account the resultant distortions of the measured results in the apparatus function, and also the possibility of taking into account various factors affecting the result of probe measurements (for example, reflection of electrons from the probe surface was taken into account in Ref.29).

Without using the previously described procedure for decreasing the distortions it is necessary to decrease $u(t)$. This impairs the signal:noise

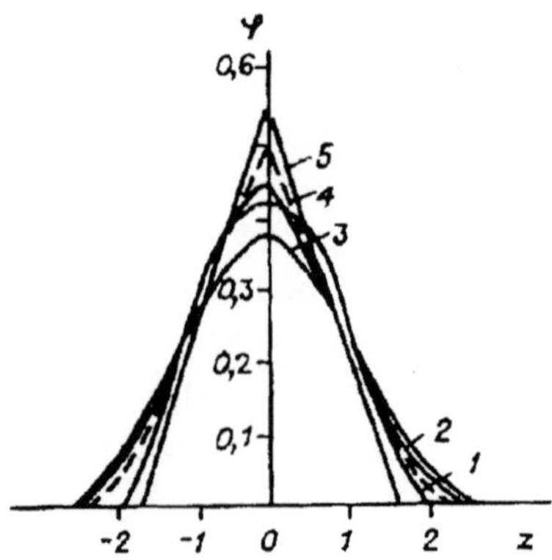

Fig.10.11 Normalised apparatus functions for modulated voltages of the type (10.17). The numbers of the curves correspond to the numbers of expressions in (10.17);[25] variable z links V_p and V_p'.

ratio. In Ref.30, it was proposed to determine EEDF by the Druyvesteyn method using the natural and minimum (for specific plasma) modulating signal – spontaneous fluctuations of the plasma potential (noise) $v(t)$. The probe–plasma system is regarded as an inertialess element carrying out non-linear transformation of the stationary random process $v(t) \rightarrow i_p[U_p + v(t)]$, and the Druyvesteyn equation is converted to the form giving the relationship of the EEDF with the statistical characteristic of the spontaneous functions of plasma measured in the experiments. The results obtained by this method are in satisfactory agreement with the data of double numerical differentiation of the VAC of the probe.

The **numerical methods** of differentiating the VAC are used with success for obtaining i_p'', and i_p is often approximated by polynomials. Comparison of the results of numerical and double electric differentiation of the VAC showed that the agreement between the two methods is obtained when using the polynomials with the factor $n \geq 11$. However, at $eU_p > (2.5 \div 3)\bar{\varepsilon}$ the error i_p'' and numerical differentiation remain large and non-monotonic features can also appear. This is associated with the error in measuring the VAC in the region of the ionic saturation current.[7]

Diagnostics of non-stationary plasma and plasma with a high electron concentration is often carried out using the pulse methods of obtaining the VAC and differentiation using analogue devices.[7] These measurements are based on applying a saw-like voltage to the probe and using in-series connected differentiation terms. This also results in substitution of differentiation with respect to U_p by differentiation with respect to time t:

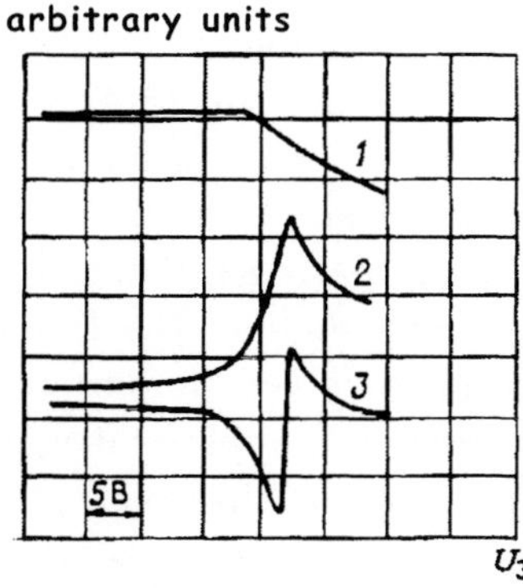

Fig.10.12 Oscillograms of VAC of the probe (1), i_p' (2), i_p''(3) in microwave discharge in argon obtained by the pulse method.[7]

$$\frac{\mathrm{d}^2 i_p}{\mathrm{d}t^2} = \frac{\mathrm{d}^2 i_p}{\mathrm{d}U_p}\left(\frac{\mathrm{d}U_p}{\mathrm{d}t}\right) + \frac{\mathrm{d}i_p}{\mathrm{d}U_p}\frac{\mathrm{d}^2 U_p}{\mathrm{d}t^2}. \tag{10.19}$$

Equation (10.19) shows that the error of this differentiation method is due to the linear form of the saw-like voltage used ($\mathrm{d}^2 U_p/\mathrm{d}t^2 \neq 0$). In addition, wide-band electronic circuits must be used to transfer the pulsed signal. Therefore, a shortcoming of the method is the small energy range of measurement of EEDF ($\varepsilon < 3\bar{\varepsilon}$) caused by the wide-band nature of the differentiating amplifiers and plasma noise. Examples of the VAC of the probe and of its first and second derivatives, produced by the pulsed methods, are shown in Fig.10.12.

The **graphical methods** of differentiation are used only very seldom because their error is very high.

The previously described formalism of the apparatus functions can be used to compare different methods of differentiation of the VAC. In Ref.31, this comparison was made for the modulation methods (equation (10.17) at $b = c = 1$), the double numerical differentiation (this will be denoted as method 6) and the double differentiation using differentiating amplifiers (method 7) at the same sensitivity of the methods (i.e. with selection of the corresponding values of A_i in equation (10.17)). According to the resolution power (the minimum energy interval in which two monoenergetic electron distributions are permitted), the introduced distortions and the quality of the restored EEDF (restoration of modelling EEDF from equation (10.8) using the regularising algorithm proposed by A.N. Tikhonov[32]), the examined method can be distributed in the following sequence with decreasing efficiency: 6, 5, 4, 2, 1, 3. The method 7 is not included in this sequence because it was the best method as regards the resolution power and the restoration of the EEDF and the worst method as regards the distortion caused by it.

When selecting the differentiation method, it is important to take into account other factors that are not linked or not taken into account by the apparatus functions. Some of them have already been mentioned, for example, difficulties in obtaining high-energy parts of the EEDF by the methods 6 and 7. In comparing the modulation methods, it is important to note that method 1 enables the first, second and higher derivative of the VAC (the latter is very important for examining the form of the EEDF) to be obtained by simple rearrangement of the frequency of the selective recording circuit. However, stringent requirements are imposed on the shape of the modulation signal (absence of higher harmonics). The methods 2–5 are less sensitive to the distortion of the signal form.

The EEDF is often determined using equation (10.1) by transferring from the Volterra equation (10.1) to the Fredholm equation of the first kind

$$i_e\left(U_p\right) = C\int_0^\infty K\left(\varepsilon, eU_p\right)f\left(\varepsilon\right)d\varepsilon,$$

(10.20)

where $K\left(\varepsilon, eU_p\right)$ is the kernel of the integral equation,

$$K\left(\varepsilon, eU_p\right) = \begin{cases} K\left(\varepsilon, eU_p, \quad \varepsilon > eU_p\right) \\ 0, \qquad\qquad e \le eU_p \end{cases}$$

(10.21)

As in the previous equation (10.18), the problem of determining the integrand function is incorrectly formulated. Regularisation methods are used to solve it.[32] The regularisation method was used for the first time for determining the EEDF using equation (10.20) in Ref.33 and it was developed further in Ref.23 and 34. The agreement between the restored and true EEDF depends on the method used to solve the incorrectly formulated problem. For example, the regularisation method based on the assumption of the smoothness of the initial solution gives the best agreement between the restored and true EEDF in the high-energy range.[34] The search for a solution on a large number of limited monotonic functions gives the best results in the low-energy range.[23]

The kernel of the integral equation can take into account different factors leading to the distortion of $i_p\left(U_p\right)$ and, at the same time, makes it possible to compensate them when restoring the EEDF. For example, taking into account the sink of electrons to the probe[23]

$$K\left(\varepsilon, eU_p\right) = \begin{cases} \dfrac{\varepsilon - eU_p}{1 + \delta(\varepsilon)\left(1 - eU_p/\varepsilon\right)}, & \varepsilon > eU_p \\ 0, & \varepsilon \le eU_p \end{cases}$$

(10.22)

where $\delta(\varepsilon)$ is the parameter of the sink and is determined by the equations of the type (10.5).

Plasma potential

To calculate EEDF and interpret probe characteristics, it is important to know the probe potential in relation to U_p, whereas the experiments yield the probe potential in relation to some reference electrode U and $U_p = U - U_{sp}$, U_{sp} is the plasma potential (see Fig.10.1). In accordance with the conventional assumptions regarding the VAC of the probe, the plasma potential is determined as the potential of the inflection point of the VAC in transition from $U_p < 0$ to $U_p > 0$.[2]

In the actual VAC of the probe there is no distinctive inflection point. This is associated with the effect of the variation of the plasma potential, the effect of a weak alternating signal for measuring i_p'' superimposed on the constant displacement of the probe, contamination of the probe surface, etc. Characteristic points on the derivative of the probe current with respect to the probe potential are used for determination. There are two approaches to determining U_{sp}: U_{sp} corresponds either to the probe potential at which i_p'' is maximum, or to the potential at which i_p'' passes through zero (i_p is maximum). This problem was analysed in, for example, Ref.1, 7, 13. Although views differ, in the majority of studies U_{sp} is determined by the condition $i_p'' (U_{sp}) = 0$.

Probe measurement of EEDF at medium pressures

These measurements are taken using conventional Langmuir probes, i.e. probes under the conditions where $\lambda >> r_p + \lambda_D$. This condition restricts the upper value of pressures at which EEDF can be calculated from the probe characteristics, at a pressure of 5–6 mm Hg and the EEDF is associated with electronic current to the probe or with its secondary derivative with respect to the probe potential using the equations (10.1) and (10.2).

However, it is possible to determine the EEDF from the probe measurements also at higher pressures[35,36] for which

$$\lambda_\varepsilon >> r_p + \lambda_D >> \lambda, \quad \lambda_p = \sqrt{4D_e\tau_{eff}}, \quad D_e = v\lambda/3,$$

$$\tau_{eff}^{-1} = v_e + (2m/M)v_{el} + (8B/T)v_r + (h\omega)/\varepsilon v_v + v^* = \kappa v_{eff}, \tag{10.23}$$

where λ_ε is the characteristic length of the electron energy loss, τ_{eff} is the effective relaxation time of the electron energy, D_e is the coefficient of electron diffusion, v_e, v_{el}, v_r, v_v, v^* are the characteristic frequencies of collisions of electrons with energy exchange between themselves, with energy transfer to translational, rotational, vibrational and electronic degrees of freedom of heavy particles, respectively; B is the rotational constant, $h\omega$ is the vibrational quantum, κ is the mean fraction of energy losses by the electron in a single collision, v_{eff} is the effective frequency of electron collisions. Since $\kappa \simeq 10^{-5} \div 10^{-2}$ then $\sqrt{\kappa}$ and this determines the expansion of the range towards higher pressures or, at the same pressure, the possibility of taking measurements using large-diameter probes.

The authors of Ref.35 derived a relationship of the EEDF and electronic current on the probe at $\lambda_\varepsilon \gg r_p + \lambda_D$ which gives in the limiting case $r_p + \lambda_D \ll \lambda$ relationship for the Langmuir probe, and at $r_p + \lambda_D \ll \lambda_\varepsilon$ the relationships of diffusion theory.[36] The equations are identical with those obtained in Ref.11, 13 but were derived on the basis of more general considerations. The equation for the current has the form of (10.20) with the kernel of the type (10.22).

The main difference of the probe theory at high pressures is the relationship of EEDF not with the second derivative of the electronic current with respect to the probe potential i_p'' but with the first derivative i_p''

$$n_e f_0(eU_p) = -\frac{3m\gamma_0}{8\pi e^2}\frac{\delta}{U_p^{1/2}}\frac{di_e}{dU_p}, \qquad (10.24)$$

where $4/3 \geq \gamma_0 \geq 0.71$ is a parameter which depends on r_p/λ ($\gamma_0 = 4/3$ at $r_p \ll \lambda$ and $\gamma_0 = 0.71$ at $r_p \gg \lambda$), $eU_p = \varepsilon$:

$$\delta(\varepsilon) = \frac{r_p C}{\gamma_0 \lambda(\varepsilon)}, \qquad (10.25)$$

where $C = 1$ for a spherical probe and $C = \ln (l/r_p)$ for a cylindrical probe (see equation (10.5)).

Analysis carried out in Ref.36 shows that EEDF can be computed from i_p'' at $\delta \leq 1$, and at $\delta \geq 10$ from i_p'.

At higher pressures or higher values or r_p, there is no relationship between the electron current to the probe and the non-perturbed EEDF.The values obtained in the probe measurements can be used to determine only the mean electron energy and the concentration of charged particles (for example, using the equations in Ref.6).

Measurement of EEDF with time resolution. Examination of non-stationary discharges faces the need to measure instantaneous EEDF, i.e. EEDF obtained during periods shorter than the characteristic period of variation of the plasma parameters.

For these purposes, it is possible to use pulsed methods of obtaining the VAC and its derivatives where saw-like or triangular voltage is supplied to the probe. This method has principal restrictions at the bottom on the duration of voltage pulse τ_v. This is due to the fact that the layer of the probe–plasma spatial charge manages to rearrange itself when the applied voltage changes. In the opposite case, the VAC is distorted by the transition processes.[7] The layer formation time is determined by the flight time τ_i of the slowest charge carriers (ions) to the layer with the characteristic dimension $x \simeq (5 \div 7)\lambda_D$

$$\tau_i \cong x / v_i \cong 10^{11}\left(M_i n_e\right)^{1/2}, \qquad (10.26)$$

which takes into account the acceleration of the ions in the prelayer; $v_i = (kT_e/M_i)^{1/2}$, M_i is the ion mass, kg; electron concentration, cm^{-3}. For example, for Ar ($M_i = 6.68 \cdot 10^{-26}$ kg) at $n_e > 10^{10}$ cm^{-3} $\tau_i < 5 \cdot 10^{-6}$ s. At these characteristic pulse times the results of the measurements are strongly influenced by the parasitic capacitance of the probe circuit (see, for example, Fig.10.2). Time τ_v is restricted at the top by the characteristic time of variation of the plasma characteristics T, i.e. $\tau_i < \tau_v < T$. The curves obtained in Fig.10.12 were obtained during a period of ~50 µs.

Measurements in plasma with periodically changing parameters can be taken using methods in which the probe receives pulses of variable amplitude under a negative potential so that the VAC can be recorded or the probe is connected to the measuring circuit for a short period of time. The pulses acting on the probe are synchronised by some method with the periodic process. The time resolution is determined by the duration of the pulses or switching time, with the minimum value of the latter, as previously, restricted to $\tau_v > \tau_i$. To obtain i_p'', we can use different methods. The authors of Ref.37 used the method of modulation of the signal of type (10.17) (second equation), and in Ref.38 measurements were taken by differentiation with analogue devices ($\tau_i = 4 \div 40$ µs).

Effect of the state of the probe surface and processes taking place on it on probe measurements

During probe measurements, various processes can take place on the probe surface: gas adsorption, deposition of conducting, semiconductor and dielectric films, removal of deposited coatings under the effect of fluxes of charged and excited particles from plasma, also due to heating of the probe, etc. The sources of contamination are either chemical compounds present in the gas medium, or vapours of organic compounds appearing in the discharge when using oil pumps for pumping, and also substances dissolved from the walls of the discharge chamber and from objects placed in the plasma. Consequently, the properties of the probe surface in plasma differ from those of the probe material and vary with measurement time. The effect of contaminants on the VAC of the probe is associated with the fact that the layers formed have an impedance which should be connected in series with a probe in the equivalent circuit in Fig.10.2, and also with the fact that the work function of the probe surface changes.[1,3,6,7]

In this case, the equation for the voltage of the near-probe layer in relation to plasma U_p differs from that mentioned previously

$$U_p = \left(U - i_p R_l\right) - U_{sp} - \left(\varphi_p - \varphi_{c.p}\right), \tag{10.27}$$

where R_l is the resistance of the layer on the probe surface, φ_p, $\varphi_{c.p}$ is the work function of the surfaces of the probe and the counterprobe taking the layers on them into account. All the considerations expressed previously regarding the effect of a resistance connected in the probe circuit apply to the resistance formed by the surface layers.

The variation of the properties of the probe surface shifts the VAC along the voltage axis, causes deformation of the VAC (as a result of a change in the properties of coatings during the measurement time) and smoothes the inflection point of the VAC at U_{sp}, and results in a hysteresis.

The sensitivity of the probe to contamination depends on the probe material. Typical examples of the effect of contaminants on the VAC are shown in Figs.10.13 and 10.14. The characteristic shown in Fig.10.14

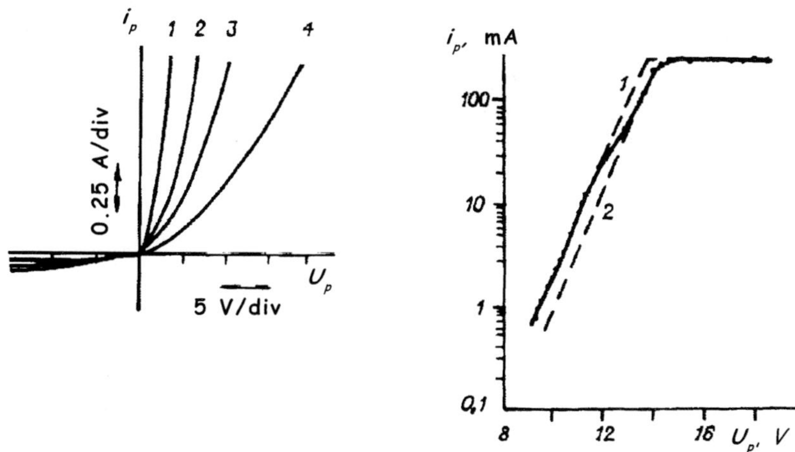

Fig.10.13 Effect of contaminants on the VAC of the probe in measurements in dc glow discharge in argon at a pressure of 0.167 mm Hg. Time from the moment of cleaning the probe: 0 (1), 3 (2), 7 (3), 20 min (4).[39]

Fig.10.14 Variation of the VAC of a tungsten probe when changing the properties of its surface.[14]

contains a region of a rapid variation of the slope. One of the possible explanations is the variation of the work function of the probe surface on approaching U_{sp} as a result of its heating and, consequently, the variation of U_p (equation (10.27)). This is also confirmed by the fact that with a rapid increase of U_c the VAC is described by curve 1 and when U_p rapidly decreases – by curve 2. It is evident that the derivatives of the distorted probe current should be non-monotonic in such cases.

In Ref.41, non-monotonic i_p'' was observed in investigations with a heated probe in nitrogen plasma with cesium vapours at low probe temperatures. The authors attributed this to the formation of cesium spots on the tungsten surface of the probe and by representing the VAC by a superimposition of probe characteristics displaced by $\Delta\varphi$ due to different work functions of clean areas and areas coated with cesium. In Ref.20, it is reported that U_{sp} can be shifted as a result of the formation of oxide films on the surface of the tungsten probe and the aluminium counterprobe and the associated variation of the work function.

The nature and strength of the effect of contaminants on the results of probe measurements must be analysed in every case. For example, in Ref.2 it was concluded that the formation of a film of amorphous hydrogenised silicon $a = Si:H$ at temperatures higher than $200°$ C has no effect on the results of probe measurements.

One of the important problems of probe diagnostics is the problem of criteria of contamination of the probe because on the basis of the form of the VAC and its derivatives it is difficult to draw any conclusions regarding the presence of contamination. Thus, hysteresis may not form if the rate of contamination is relatively high.

The low-frequency impedance of the probe is an objective parameter enabling the presence of contaminants to be evaluated and inspect their size.[43] Cleaning of the probe increases the active and decreases the capacitance component of probe impedance.

In discharge system where the probe surface can be contaminated, advantages are often offered by pulsed measurements. They are less sensitive to contamination due to a higher capacitance component of the impedance; in addition, the probe surface does not manage to change during short periods of time. The inflection at U_{sp} on the VAC produced by the pulsed method is more distinctive than on the static characteristics.

In probe measurements in plasma with spattering of the metal or with the formation of conducting coatings, the VAC may be distorted as a result of an increase of the probe area in depositing conducting films on the elements of probe insulators that are in contact with the probe. This can be avoided by moving the area of contact of the probe with the insulator away from the plasma region.

The main methods of cleaning the probes is ion bombardment at high negative potentials of the probe and heating using an external source. Usually the probe is cleaned for a long period of time and is then connected to the measuring system for a short period of time.

The appearance of coatings on the probe and the variation of its surface is regarded as a phenomenon that complicates the measurements. However, the sensitivity of the probe to the surface condition makes it an efficient tool for examining the processes of plasma chemical polymerisation and etching (see Ref.7 and also Chapter 13 of this book).

Another reason for the errors should be mentioned. It is associated with the emission of electrons from the probe under the effect of ions at high negative probe potentials.[3,6] This distorts the potential distribution in the near-probe layer and increases the apparent ion current and may cause an error in the ion concentration determined from the ion current. Since it is very complicated to consider and detect this phenomenon, it is necessary to use materials with a low secondary emission coefficient and work with low negative values of U_p.

10.3 Determination of plasma parameters from probe measurements

Mean electron energy $\bar{\varepsilon}$ is the first moment of the EEDF and is described by the relationship

$$\bar{\varepsilon} = \int_0^\infty \varepsilon f(\varepsilon)\,\mathrm{d}\varepsilon \,/\, \int_0^\infty f(\varepsilon)\,\mathrm{d}\varepsilon, \quad \bar{\varepsilon} = \int_0^\infty \varepsilon f_0(\varepsilon)\,\mathrm{d}\varepsilon, \tag{10.28}$$

where $f(\varepsilon) = n_e f_0(\varepsilon)$ is linked with the probe current by the relationships (10.12) or (10.24) (dimension $f(\varepsilon)$ – cm$^{-3}\cdot$eV^{-1}). To calculate $\bar{\varepsilon}$ with an error lower than 5%, it is sufficient to measure EEDF to energies lower than $(3 \div 4)\bar{\varepsilon}$.

Assuming Maxwell's EEDF, we can calculate the electron temperature T_e

$$\frac{kT_e}{e} = -\left(\frac{d(\ln i_e)}{dU_p}\right)^{-1}. \tag{10.29}$$

Since the real EEDF differs from Maxwell's EEDF, T_e calculated from equation (10.21) can be both too high and too low in comparison with

$T_{eff} = 2\bar{\varepsilon}/3$, and this difference depends on the probe potential at which T_e.[7,23] When T_e is determined at a floating potential $T_e > T_{eff}$. It was shown in Ref. 27 that the best agreement between T_e and T_{eff} is obtained when using the VAC in the vicinity of the plasma potential U_{sp}. It should be taken into account that this region of probe potentials is characterised by the maximum possible distortions of the VAC as a result of the sink of the electrons to the probe, the finite resistance of the plasma and elements of the probe circuit.

Electron concentration n_e is computed by integrating equations (10.2) and (10.24) with respect to energy

$$n_e = \int_0^\infty f(\varepsilon)d\varepsilon, \tag{10.30}$$

For the Langmuir probe (equation 10.2)

$$n_e = \text{const}\int_0^\infty U_p^{1/2} i_p'' dU_p. \tag{10.31}$$

If $f(\varepsilon)$ is not determined in the absolute measure, the equations (10.30) and (10.31) give the relative value of n_e.

The electron concentration can be determined from the random current on the probe and the plasma potential

$$n_e = 4i_e\left(U_{sp}\right)/e\bar{v}S_p, \tag{10.32}$$

where $\bar{v} = 5.9\cdot 10^5(\bar{\varepsilon})^{1/2}$ is the mean electron velocity in m/s; $\bar{\varepsilon}$ in eV, all other quantities are determined in the SI system.

Ion concentration n_i is determined from the VAC in the region of the ion saturation current. The problem of the relationship of the plasma parameters with the ion current on the probe is one of the most complicated in the probe diagnostics and is not examined here. Several approximate solutions of this problem have been described in Ref.1-6.

Equations of the following type

$$i_i = S_p e n_i\left(\frac{kT_e}{2\pi M_i}\right)^{1/2}\left(\frac{eU_p}{kT_e}\right)^n, \tag{10.33}$$

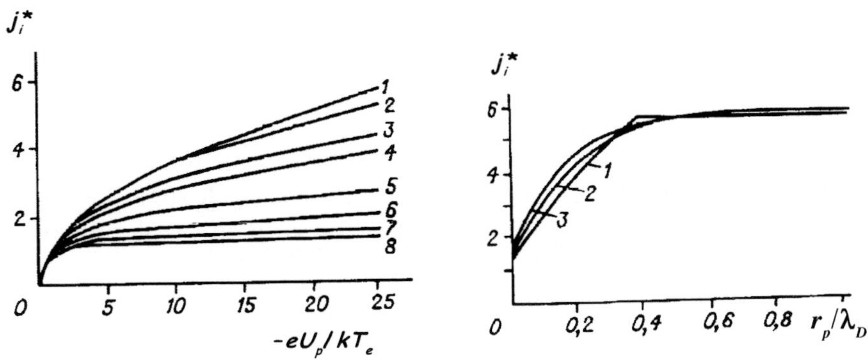

Fig.10.15 Dependence of the normalised ion current on the cylindrical probe j_i^* on probe potential. r_p/λ_D = 0 (1), 3 (2), 4 (3), 5 (4), 10 (5), 20 (6), 50 (7), 100 (8); T_i/T_e = 0.[44,45]

Fig.10.16 Dependence of the normalised ion current on the cylindrical probe on r_p/λ_D. eU_p/kT_e = –25; T_i / T_e = 0 (1), 0.5 (2), 1.0 (3).[44,45]

are used quite often. Here n is determined by experiments, M_i is the ion mass. For a thin probe and a collisionless layer $n = 0.5$.[3,4,7] Equation (10.33) takes into account ion acceleration in the preliminary layer, i.e. the fact that the ions arrive at the layer boundary with the probe at velocities corresponding to T_e and, consequently, the information on ion temperature T_i recorded in this manner in the ion current is lost.

It is preferred to use numerical solutions of the equations describing the current on the probe from the surrounding plasma. The most detailed examination of this problem has been carried out by Laframboise.[44,45] In the case of a collision-less layer in plasma, consisting of neutral particles, positive ions and electrons with Maxwell distributions, the ion current on the probe is determined by the equation

$$i_i = en_i S_p \left(\frac{kT_e}{2\pi M_i} \right)^{1/2} j_i^*,$$

$$(10.34)$$

where $j_i^* = j_i^* (r_p/\lambda, r_p/\lambda_D, T_i/T_e)$ is the normalised current density on the probe which also takes into account the dependence of ion current to the probe on the probe potential. The values of j_i^*, calculated for $T_i/T_e = 0$ for a cylindrical probe, are shown in Fig.10.15 and the effect of T_i is evident from Fig.10.16.

To calculate n_i from the VAC of the probe in the region of the ion saturation current, independent measurements must be taken to determine the ion composition of plasma and ion temperature T_i. In addition, the expression for the ion current includes T_e and U_p which can be determined if the VAC is measured in the region of not only the ion current but also electron current up to the plasma potential (U_p is measured in relation to U_{sp}).

In quasi-neutral plasma in the absence of negative ions $n_i = n_e$, and the ion parts of the VAC can be used to determine n_e. It should be noted that since the value λ_D, required for calculations, includes the unknown value n_e, an iteration procedure must be used to determine this value. Iterations can be avoided if we use the method proposed in Ref.46. In this procedure, the known value of ion current i_i at the selected probe potential (sufficiently high to eliminate the effect of electron current) and T_e are used to calculate $j_i^* (r_p/\lambda_D)^2$ which in contrast to j_i^* depends only on i_i and T_e

$$j_i^* \left(r_p / \lambda_D \right)^2 = \frac{i_i r_p}{l_p \varepsilon_0} \left(M_i / 2\pi e \right)^{1/2} \left(kT_e / e \right)^{-3/2}, \tag{10.35}$$

where ε_0 is the base of natural logarithms. From the graph showing the dependence of j_i^* on $j_i^* (r_p/\lambda_D)^2$ at the corresponding probe potential we determine j_i^* and then from equation (10.36) the value of n_i

$$n_i = i_i \left[j_i^* e r_p l_p \left(2\pi k T_e / M_i \right)^{1/2} \right]^{-1}. \tag{10.36}$$

Many attempts have been made to verify the ion current theories, including Laframboise theory. They were based on comparing the values of n_i, determined from ion current, with n_e measured by independent methods (microwave or from electronic saturation current). Some of the results from this comparison are presented in Ref.6. Good agreement was obtained using the Laframboise theory. However, there are also data indicating differences in the values of n_i and n_e, with n_i being higher than n_e obtained from the VAC of the plasma potential. For example, in Ref.47 the values of n_i were systematically three times higher than n_e. This may prove to be important in determining the concentration of charged particles in plasma containing negative ions because their concentration is assumed to be equal to $n_i - n_e$.

The results of calculations of n_i from the ion current may be influenced by the difference between the EEDF and the Maxwell EEDF that is usually used in calculating i_i. The authors of Ref.23 examined the effect of the EEDF on the Debye screening length and i_i on an example of a wide range of modelling functions. It is shown that although

λ_D depends on the EEDF, the role of high-energy part of the EEDF is small. An equation was derived determining the dimensionless multiplier in the equation of type (10.33) for the current taking into account the form of the EEDF. Calculation of these multipliers from the EEDF measured in hf discharge in inert gases shows that they differ by up to 30% from the values obtained assuming the Maxwell EEDF.

Collisions of the ions together and with neutral particles change the VAC of the probe in the ion current region.[3,6] For example, the former leads to an increase of ion current, the latter to its decrease, with the exception of the case of a single collision in the layer where the current may increase as a result of disruption of orbital motion.

It should again be noted that the apparent ion current can be increased by electron emission from the probe under the effect of different factors and this may become significant at high negative probe potentials in ion bombardment of the probe surface.

In electrodeless discharges, there are considerable difficulties in positioning the counterprobe with a large surface to obtain a single-probe characteristic. Double probes are used extensively under these conditions.[48] They consist of a system of two probes of the same area placed in plasma at a distance at which there is no mutual screening. If the plasma properties in the areas where these probes are positioned do not differ, the volt–ampere characteristic of such a system has the form shown in Fig.10.17. The abscissa gives the voltage between the probes $U = U_{p1} - U_{p2} + \Delta U$, where ΔU equals the difference of the plasma potentials in the areas in which the probes are positioned and the difference of the contact potentials of the probes. At $\Delta U \neq 0$, the characteristic is displaced along the voltage axis by ΔU. The maximum current in this system is determined by the ion saturation current on the probe under high negative potentials. The second probe is also at a negative potential which is such that the electron current to this probe

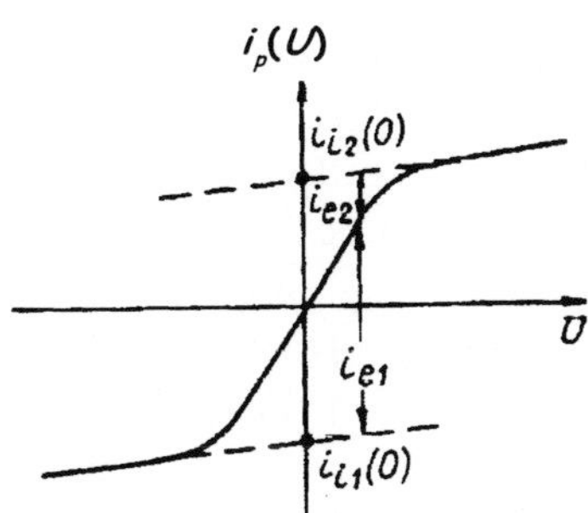

Fig.10.17 VAC of a symmetric double probe.

is equal to ion current of the first probe. When measuring the polarity U, the probes change their roles.

The double probe has a number of advantages in comparison with a single probe. Since the probes are usually quite close, in measurements in hf plasma the alternating voltage acting between them is greatly reduced thus reducing the degree of distortion of the VAC. In addition, the effect of alternating voltage on the ion part of the VAC is considerably smaller than that on the electron part. It must be considered that parasitic capacitances (Fig.10.2) lead to the appearance of high-frequency currents and distort the VAC.

In addition, the collisions in the near-probe layer have a smaller effect on the double probe in comparison with a single probe so that the double probe can be used to determine the plasma parameters under the conditions in which the single probe gives inaccurate results.[6]

To obtain the plasma characteristics from the VAC of the double probe we can use the assumption of the Maxwell's distribution of electrons in plasma. There are several methods of processing the VAC (see, for example, Refs.4 and 5). One of them is similar to the procedure used in the method of the semi-logarithmic graph for single probes

$$\frac{e}{kT_e} = \frac{d}{dU}\ln\left(\frac{i_{i1}+i_{i2}}{|i_{e1}|}-1\right),$$

(10.37)

where i_{i1}, i_{i2}, i_{e1} are the ion electron currents on the probe. The principle of determination of these currents is clear from Fig.10.17 (linear approximation of ion current was used).

The ion concentration is calculated by the previously described methods of processing the ion parts of the VAC.

10.4 Selected problems of probe diagnostics

It is not possible to examine here all types of application of probes for investigating the plasma. We shall therefore mention only some of them and quote the literature where further information can be found, and we shall examine in greater detail only the applications which have not been described sufficiently in the review of literature. Many plasma devices are based on plasma in a magnetic field. The magnetic field greatly complicates the probe diagnostics. The velocity of the particles along and across the field differs and there are a large number of problems associated with plasma anisotropy (the measurement results depend on the probe of orientation). In the presence of a magnetic field the electron current at the plasma potential decreases, the transition from the region $U_p < 0$ to the region $U_p > 0$ is smoothed out, the section of the VAC at $U_p < 0$ is distorted (in this section the current on the probe is still determined by the electrons). The effect of the magnetic field on the ion part of the VAC is considerably weaker than that on the electron part. At present, there is no complete theory of probes in the magnetic field, and some of the attempts made to develop such a theory have been described in Ref.1, 3, 4, 6, 13.

Specific problems appear in probe diagnostics of plasma flows under the conditions of a collisionless layer, the transition regime and the continuum regime.[2-6]

Probe measurements in plasma containing negative ions

Regardless of the fact that the non-equilibrium plasma of electronic active gases is used widely and there are large number of investigations in which the results of probe measurements have been published, this region of application of probes is one of the least examined. The presence of negative ions leads to a number of effects.

1. The difference between electron and ion saturation currents and between the floating potential and plasma potential decreases.[1,6,49-52] Consequently, the effect of the layer of current on the VAC of the probe is observed at almost all probe potentials, and in calculating the parameters of the electronic component it is necessary to deduct the ion current. The role of this factor becomes important with increasing n^-/n_e, where n^- is the concentration of negative ions. A decrease of the difference of the ion electron currents of saturation simplifies the requirements imposed on the ratio of the areas of the probe and the counterprobe examined in (10.2).

2. In determining n_e from the electronic saturation current (equation (10.32)) it is important to take into account that $i_p (U_{sp})$ can be distorted by the current of negative ions i^-. The ratio of these currents is determined by the equation

$$i^- / i_e = \left(n^- / n_e\right)\left(T_i^- m / T_e M_i^-\right),\qquad(10.38)$$

where T_i^- and M_i^- is the temperature and mass of the negative ions. At $m/M_i^- < 10^{-4}$, $T_i^-/T_e < 10^{-1}$ the effect of ions with $i_p(U_{sp})$ can be ignored if $n^-/n_e < 10^{-3}$.

3. The ion part of the VAC of the probe in the regime of the collisionless layer at the electron temperature considerably higher than the ion temperature, and at $n^-/n_e < 2$ does not change and ion 'collection' takes place in the same manner as in the plasma without negative ions. When determining n_i from the ion saturation current, we can use the

well-known equations in which the positive ions travelled to the boundary of the near-probe layer with the velocity of ion sound ($(kT_i/M_i)^{1/2}$). At $n^-/n_e > 2$ the structure of the electric field in the layer is distorted and the velocity of the positive ions at the boundary of the layer is $(kT_i/M_i)^{1/2}$.[51]

4. The second derivative of the probe current with respect to the probe potential i_p'' at probe potentials close to U_{sp} can contain information on the energy distribution of the negative ions[1]

$$U_p^{1/2} i_p''(U_p) \cong \mathrm{const}\left[(e/m)^{1/2} n_e f_0^e(eU_p) + (e/M_i^-)^{1/2} n^- f_0^-(eU_p)\right], \quad (10.39)$$

where f_0^e and f_0^- are the energy distributions of the ions and electrons, respectively. Examples of the distribution of the negatively charged particles (obtained from the second derivative of the distributions) in the stratified discharge of direct current and in oxygen and in a plasma-beam discharge in SF_0 are shown in Fig.10.18 (i_p'' was determined by modulation methods using signals of the type (10.17)) (the fourth equation) (O_2) and (10.17) (the first equation) (SF_6)). Peaks in the energy range ~0.1 eV are attributed to negative ions. Integration of the resultant distributions with an allowance made for m and M_i^- gives absolute concentrations n^- and n_e.[50,52]

5. The presence of negative ions has no influence on the position of the potential of the inflection point of the VAC in transition from the region with $U_p < 0$ to $U_p > 0$ and the inflection at U_{sp} is more marked than without the negative ions.[49]

It should be noted that the presence of the peak of negative ions on

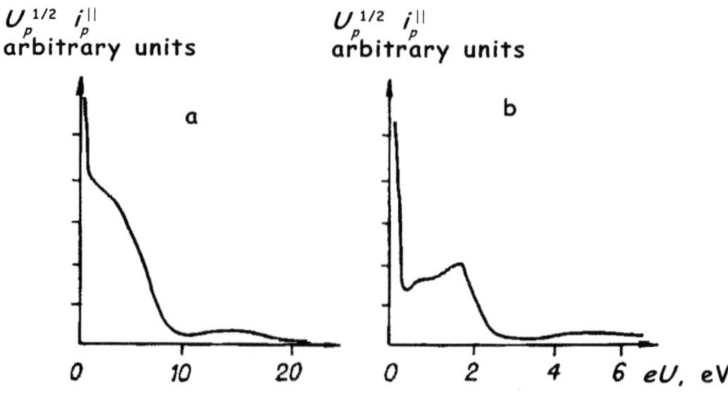

Fig.10.18 Energy distributions of negatively charged particles in the plasma of a stratified discharge in oxygen at $n^-/n_e \simeq 20(a)$[51] and in plasma-beam discharge in SF_6 at $n^-/n_e = 150$ (b).[50]

i''_p was not observed in most investigations (see, for example, for plasma of the same gases but in rf discharge[20,53]). The authors of Ref.49 described their method of probe investigation of the plasma of electronic active gases with the energy distribution of electrons slightly differing from Maxwell's distribution.

Probe measurements in anisotropic plasma

We previously assumed that the distribution of the electrons in the non-perturbed plasma region is isotropic. Under the conditions of real gas discharge plasma there are gradients of the strength of the electric field, the beams of electrons in the near-electrode regions of dc discharges and rf discharges, the plasma can be sustained in magnetic fields, the plasma often moves in relation to the probe, etc. All this leads to plasma anisotropy and, consequently, the previously described methods of probe diagnostics cannot be used. The results of probe measurements must depend on the probe orientation.[1]

There are several approaches to solving the problem of probe measurements in anistropic plasma.

In axially-symmetric plasma, the electron velocity distribution function (EVDF) in the spherical co-ordinate system with the polar axis coinciding with the axis of symmetry of the plasma, can be represented by an expansion with respect to Legendre polynomials

$$f(v, \theta) = \sum_{j=0}^{\infty} f_j(v) P_j(\cos \theta), \qquad (10.40)$$

where v, θ is the velocity modulus and the polar angle. Coefficient f_0 determines the velocity modulus distribution of electrons, f_1 is linked with the convective velocity of the electrons. The representation of $f(v, \theta)$ only through f_0 and f_1 is known as the Lorentz approximation. Equation (10.40) shows that determination of the EVDF is reduced to determining the coefficients in the expansion $f(v, \theta)$.

In Ref.54, this problem was examined with special reference to various types of probes. It was shown that for axial symmetry of plasma, the determination of the finite number of the expansion coefficients $f(v, \theta)$ is possible when taking probe measurements for fixed probe orientations. Measurements taken using flat two-sided and cylindrical probes enable even expansion coefficients to be determined. Flat one-sided and flat double probes can be used to determine even and odd expansion coefficients. The spherical probe is not sensitive to plasma anisotropy and gives f_0.

The possibilities of this approach were illustrated in Ref.55 on an example of investigating a plasma-beam low-voltage arc in He using cylindrical probes, with some of the probes oriented along the axis of symmetry of the discharge gap and others normal to this axis.

At $U_p < 0$ and with all the requirements of probe diagnostics fulfilled, with the exception of isotropic distribution, the density of electronic current on the probe is determined by the equation

$$j(eU_p) = \frac{2\pi n_e}{m^2} \int\limits_0^{2\pi} d\varphi \int\limits_{eU_p}^{\infty} \varepsilon d\varepsilon \int\limits_0^{\arccos\left(\sqrt{eU_p/\varepsilon}\right)} f(\varepsilon, \theta_1, \varphi)\sin\theta_1\cos\theta_1 d\theta_1. \tag{10.41}$$

where θ_1, φ are the polar and azimuthal angles in the spherical co-ordinate system whose axis is directed along the external normal to the probe surface; $f(\varepsilon, \theta_1, \varphi)$ is the distribution function normalised with respect to 1. After integrating equation (10.41) and taking into account (10.40), with double differentiation with respect to U_p, we have

$$i''_{e\parallel}(eU_p) = \frac{2\pi e^2 n_e S_p}{m^2} \sum_{j=0}^{\infty} F_{2J}(eU_p)P_{2j}(0),, \tag{10.42}$$

$$i''_{e\perp}(eU_p) = \frac{2e^2 n_e S_p}{m^2} \sum_{j=0}^{\pi} P_{1j}(\cos\theta)d\theta, \tag{10.43}$$

$$F_{2j}(eU_p) = ef_{2j}(eU_p) - \int\limits_{eU_p}^{\infty} f_{2j}(\varepsilon)\frac{\partial}{\partial U_p} P_{2j}\left(\sqrt{eU_p}\right)d\varepsilon, \tag{10.44}$$

where the signs $\parallel$ and $\perp$ indicate the axial and perpendicular position of the probe.

Equations (10.42)–(10.44) show that i''_e in anisotropic plasma depends on probe orientation. In addition, the signs of the terms of the series in the equations (10.42)–(10.44) change and, consequently, i''_e may assume negative values. These facts are used as a basis for experimental verification of the anisotropy of EVDF. It should also be noted that the current on the cylindrical probe does not depend on odd harmonics, i.e. it is independent of convective current. Examples of i''_e for two probes at different points of the arc are shown in Fig.10.19. Anisotropy of plasma of the cathode and isotropic transformation of the EVDF at a distance from it are visible. It was assumed in Ref.55 that $f_2 \gg f_4$, and f_0 and f_2 were also determined.

This method was developed further in Ref.56–58. The characteris-

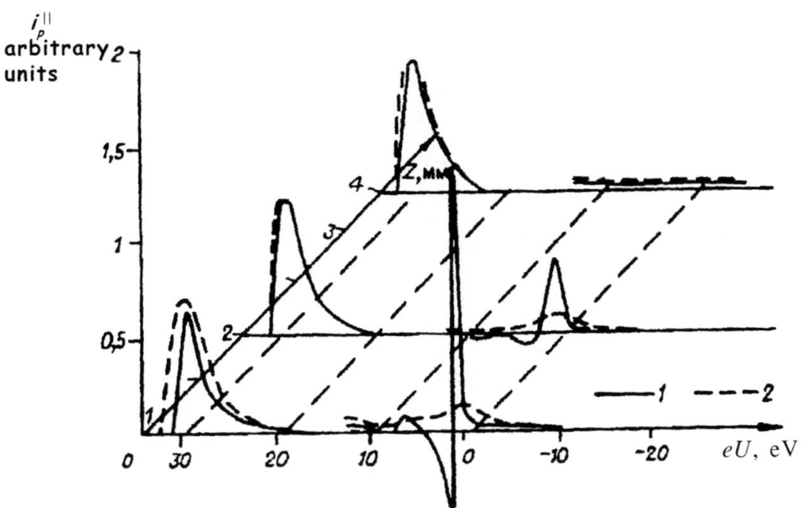

Fig.10.19 Second derivatives of probe current i''_p at different distances z from the cathode in a low-voltage arc in helium at a pressure of 1 mm Hg and a discharge current of 0.4 A. Spacing of the electrodes 6 mm, $U = 0$ is the cathode potential, the scale of the high-energy part is magnified 10 times; 1) perpendicular probe; 2) parallel to the discharge axis.[55]

tics of the cylindrical and flat probes in anisotropic plasma were also studied in Ref.59.

The authors of Ref.60 and 61described the method of measuring the directional part of electron distribution f_1 (v), if the distribution is represented in the Lorentz approximation

$$f(\vec{v}) = f_0(v) + f_1(v)\cos\theta. \tag{10.45}$$

Measurements should be taken using two flat probes oriented in the opposite directions, or using a double flat probe. Measuring the current difference of two probes at $\theta = 0$, we can determine f_1 (v)

$$f_1\left(\sqrt{2eU_p/m}\right) = \frac{m^2}{4\pi n_e e^3 S_p} U_p^{1/2} \frac{\mathrm{d}}{\mathrm{d}U_p}\left[U_p^{-1/2}\frac{\mathrm{d}\left(\Delta i_p\right)}{\mathrm{d}U_p}\right], \tag{10.46}$$

or, after differentiation

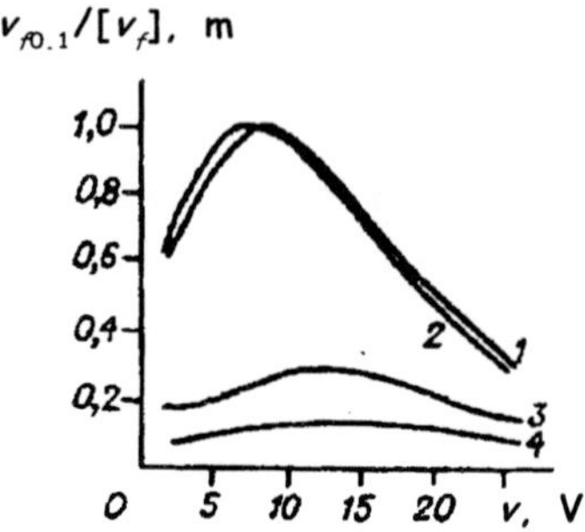

Fig.10.20 Functions $vf_0(v)/[vf(v)]_m$ (1,2) and $vf_1(v)/[vf(v)]_m$ (3,4) in discharge in helium at a pressure of 0.11 mm Hg.[61] $[vf(v)]_m$ is the maximum value of the Maxwell function at a current of 30 mA; 1,3) 40 mA; 2,4) 30 mA.

$$f_1\left(\sqrt{2eU_p/m}\right) = \frac{m^2}{8\pi e^3 S_p}\left[2\frac{d^2(\Delta i_p)}{dU_p} - \frac{d(\Delta i_p)}{dU_p}\right]. \tag{10.47}$$

The equations (10.46) and (10.47) were obtained assuming the identical plasma potentials at the areas where the probes were situated, i.e. $\Delta U_{sp} = 0$. If $\Delta U_{sp} \neq 0$, the expressions for f_1 are greatly complicated and an addition appears. The value of this addition depends on U_p and is maximum at $U_p \sim \Delta U_{sp}$. The equations which enable corrections for $\Delta U_{sp} \neq 0$ to be made are presented in Ref.61 and they were analysed and determined more accurately in Ref.13. The application of the double flat probe is illustrated in Fig.10.20.

If the EVDF in a system with cylindrical symmetry is represented in the form $f(v) = f(v_z, v_\perp)$, where v_z and $v_\perp$ are the velocities of the electron in the direction of the z axis and the perpendicular plane, and the longitudinal $f_1(v_z)$ and transverse $f_2(v_\perp)$ EVDF, expressed by the equations

$$f_1(v_z) = 2\pi\int_0^\infty f(v_z, v_\perp)v_\perp dv_\perp,$$

$$f_2(v_\perp) = \int_{-\infty}^\infty f(v_z, v_\perp)dv_z, \tag{10.48}$$

are introduced, the EVDF can be determined using the method proposed in Ref.13 and 62. Measurements with a flat one-sided probe, oriented in the direction normal to the z axis, and the cylindrical probe oriented along the z axis can be used to determine $f_1(v_z)$ and $f_2(v_\perp)$

$$f_1(v_z) = \frac{m \, \mathrm{d}i_p}{n_e e^2 S_p \mathrm{d}U_p},$$

(10.49)

$$f_2(\varepsilon_\perp) = \frac{m^{3/2}}{2^{3/2} n_e e^2 S_p} \int_{\varepsilon_\perp}^{\infty} \left(eU_p - \varepsilon_\perp\right)^{-1/2} \frac{\mathrm{d}^2 i_p}{\mathrm{d}U_p^2},$$

(10.50)

where $\varepsilon_\perp = mv^2/2$.

The VACs of the flat and cylindrical probes make it possible to calculate the 'longitudinal' and 'transverse' temperatures using equation (10.28). Orienting the normal to the surface of the flat probe along and across the z axis, the drift electron velocity can be determined from the current difference at the space potential

$$v_{dr} = \Delta i_p / n_e S_p.$$

(10.51)

In Ref.1 it was concluded that as a result of screening the plasma with a probe, the value v_{dr}, computed from (10.51) may be too low.

Chapter 11

ELECTRIC PROBES IN CONTINUUM REGIME

11.1 Introduction

Electric probes as a means of diagnostics of low temperature plasma have a number of advantages, the main ones being simple experimental procedures and the possibility of determining local parameters. Unfortunately, there is no general theory of electric probes suitable for interpreting the results of probe measurements over a wide range of the conditions.

The parameters characterising the operating regime of the probe (i.e. the regime of passage of current in the near-probe region) are numerous. Usually, the regimes are subdivided in relation to the ratio between the free path of the ions and electrons under the conditions of elastic collisions λ_i, λ_e and the linear dimension of the near-probe region of perturbation. To simplify considerations, it will be assumed that the thickness of the near-probe layer of the volume charge is not greater (as regards the order of magnitude) than the characteristic damage of the probe a (this assumption holds if the Debye radius is comparable with the probe size or is smaller, and the probe potential is not too high). Consequently, the probe size a can be regarded as the linear scale of the near-probe region of perturbation.

When λ_i, $\lambda_e \gg a$, the collisions in the near-probe region are not significant. These operating conditions of the probe are referred to as collisionless or molecular. The theory of these regimes was developed in the twenties[1] and described in the well-known studies[2-8] (we should also mention Ref.9 which contains, in particular, an extensive bibliography; please also refer to Chapter 10 of this book). It should be noted that, in this case, the volt-ampere characteristic of the probe contains also information on the non-perturbed function of electron distribution, in addition to information on the parameters such as, for example, the electron concentration in non-perturbed plasma or plasma potential.

At λ_i, $\lambda_e \lesssim a$, important kinetic scales are the length of the energy relaxation of the electrons as a result of collisions with neutral particles $\lambda_u = \lambda_e/\sqrt{\delta}$ and the length of Maxwellisation as a result of interelectron collisions $\lambda_m = \lambda_e(v_{en}/v_{ee})^{1/2}$ where δ is a parameter characterising energy transfer from the electrons to neutral particles (for elastic collisions, this parameter is equal to the value of the double ratio of the electron mass to the mass of the neutral particle, for non-elastic collisions it is equal to the same value multiplied by the coefficient of inelastic losses), v_{en}, v_{ee} are the local values of the frequency of elastic collisions of electrons with neutrals and the frequency of interelectron collisions (it is assumed that the plasma is weakly ionised and $v_{en} \gg v_{ee}$). It should be noted that usually $\delta \ll 1$ and, consequently, $\lambda_u \gg \lambda_e$; in the examined case of weakly ionised plasma we also have $\lambda_m \gg \lambda_e$.

In the case with λ_i, $\lambda_e \lesssim a \ll \lambda_u \lambda_m$, collisions in the near-region are significant but the change of electron energy as a result of collisions is small. The theory of electron current on the probes under these regimes was developed in Ref.10 and 11. It should be noted that in this case the VAC retains information on the energy distribution of electrons on non-perturbed plasma.

In the opposite limiting case where at least one of the lengths λ_u, λ_m is considerably smaller than the probe size, λ_i, $\lambda_e \ll (\lambda_u^{-1} + \lambda_m^{-1})^{-1} \ll a$, collisions in the near-probe region are the more significant process. Evidently, there is no relationship between the VAC of the probe and the type of non-perturbed electron distribution function in this case; on the other hand, the non-perturbed distribution function under these conditions should differ only slightly from the Maxwell's function. To describe the distribution of both ions and electrons in a large part of the near-probe region (outside the layer with a thickness of the order of $(\lambda_u^{-1} + \lambda_m^{-1})^{-1}$ adjacent to the surface of the probe) we can use the hydrodynamic approximation in this case. These regimes of operation of the probe are referred to as the continuum regimes or hydrodynamic regimes. Hydrodynamic equations were used for analysis of the characteristics of electric probes in Ref.12; in Ref.13 and 14, the hydrodynamic equations were used in the entire near-probe region of perturbation without dividing this region in advance into a quasi-neutral region and a volume charge layer. Reviews of the hydrodynamic theory of electric probes were published in Ref.6, 8, 15-17.

In this chapter, we examine probes in the hydrodynamic regimes; its aim is to add to the reviews mentioned previously and describe the current state of probe diagnostics procedures.

11.2 A system of equations and boundary conditions of the hydrodynamic theory of electric probes

We shall examine stationary moving or still slightly ionised plasma containing neutral particles (generally speaking, of several different types), ions (in a general case also of several types) and electrons. The plasma contains a conducting solid (electric probe). The effect of charged particles on the flow field of neutral components and plasma as a whole will be ignored. It will be assumed that the degree of dissociation and ionisation of the main neutral components of the plasma (i.e. the components whose contribution to collisions with the charged particles is considerable) are low and the molar fractions x_q ($x_q = n_q/n$, where n_q is the concentration of the q-th component, n is the total concentration of plasma particles) changes only slightly in the vicinity of the probe. It will be assumed that the frequencies of elastic collisions of the charged particles with the neutral particles greatly exceed the frequencies of collisions of charged particles.

The system of equations which determine the distributions of the concentrations of charged particles n_m, the densities of their fluxes J_m and electrostatic potential φ, includes the equations of conservation of the charged particles and Poisson's equation (see, for example, Ref.6 and 8):

$$nv\nabla\frac{n_m}{n} + \nabla J_m = w_m, \quad m = i, e, \tag{11.1}$$

$$\Delta\varphi = -4\pi e \sum_{m=i,e} z_m n_m, \tag{11.2}$$

and also the hydrodynamic equations of transfer which can be written in the form (see, for example, Ref.18)

$$J_m = -\mu_m\left(\nabla\frac{P_m}{e} + z_m n_m \nabla\varphi\right), \quad m = i, e \tag{11.3}$$

Here v is the mean mass velocity of the plasma, w_m is the rate of variation of the concentration of the m-th component and the result of volume reactions, μ_m, z_m, P_m is the mobility, charge number and the partial pressure of the m-th component, index i assumes the values corresponding to different types of ions, the index e is given to the electrons. The quantities n, v as well as the temperature of neutral particles T, can be determined by solving the corresponding gas-dynamic problem of flow around the probe and by neglecting the effect of charged particles. It is assumed that they are given functions of the coordinates. The first term in the left-hand part of the equations of conservation of the charged particles (11.1) describes convective transfer, the second term – diffusion and drift transfer, the term in the right-hand part - the formation or disappearance of the charged particles of the given type as a result of volume reactions. Poisson's equation (11.2) links the volume charge of plasma with the induced electric field. First term in the brackets in the right-hand part of the hydrodynamic transfer equation (11.3) describes the concentration diffusion, the second term - the drift in the electric field. It should be noted that when writing the last equations the terms describing thermal diffusion were omitted together with some terms taking into account baric diffusion; usually, these effects are small and they can be taken into account only in special cases.[6,19-21]

The hydrodynamic equations of transport (11.3) for the ions ($m = i$) are applicable at $\lambda_i \ll L$, where L is the local macroscopic scale, i.e. the linear dimension of the examined hydrodynamic zone. In writing the diffusion term we can accept that $P_i = n_i kT$, where k is the Boltzman's constant.

In the case of a weak field ($eE\lambda_i \ll kT$, where E is the strength of the electric field) these equations can be regarded as the partial case of the Stefan–Maxwell equations for multicomponent diffusion[22-24] (see also Ref.19 and 25). If only one component is the main in the group of neutral plasma components, then the quantity $\mu_i n$ is the function of only temperature and, to a first approximation, is equal to $3e/(16m_{iq}\Omega_{iq}^{(1,1)})$, where $\Omega_{iq}^{(1,1)}$ is the collision integral,[23,26] m_{iq} is the reduced mass; to determine this quantity we can use either the experimental data or the results of calculating the collision integral (these calculations can be carried out for the known transport scattering cross section or the ion–neutral particle interaction potential). If the number of the main neutral component of the mixture exceeds unity, quantity μ_{in} also depends on the molar fractions x_q; to a first approximation, these dependences are described by Blanc's law[19,27] and the mobility is expressed through mobility in pure gases

$$\mu_i = \left(\sum_q \frac{1}{\mu_{iq}} \right)^{-1} , \quad \mu_{iq} = \frac{3e}{16n_q m_{iq} \Omega_{iq}^{(1,1)}}.$$

In the case of a strong field ($eE\lambda_i \gtrsim kT$), the diffusion term of the transfer equations has the order λ_i/L in relation to the drift term and under the conditions of applicability of the hydrodynamic approximation is low. It should be noted that this circumstance determines in the majority of examined problems the use of the Einstein equation in writing the diffusion term of the equation (11.3) at the temperature of the neutral particles. The validity of this equation in strong fields is violated. The quantity $\mu_i n$ in the case of the strong field is the function of also E/n, and Blanc's law, generally speaking, is no longer valid. To determine this quantity, it is necessary to determine the experimental data of the results of solving the kinetic equation.

An extensive review of the data on the mobility of ions in pure gases was published in Ref.28–30.

The hydrodynamic transfer equation (11.3) for the electrons ($m = e$) is valid if at least one of the following two conditions is fulfilled:

$$\lambda_u \ll L, \quad \lambda_m \ll L. \tag{11.4}$$

The first of the conditions is discussed in Ref.31, the second follows from Ref.32–34. The mobility of electrons at the known electron-neutral particle transport collision sections and the known electron distribution function or, more accurately, its main isotropic part f^0 is described by the Lorentz equation for slightly ionised plasma.[18,31,35,36]

If the first condition (11.4) is fulfilled, the function f^0 (and, consequently, the quantity $\mu_e n$) is given by the local values of the parameters T, E/n, x_q, x_e; when writing the diffusion term of the transfer equation we can accept $P_e = n_e kT$. When $E \ll kT (e\lambda_u) f^0$ is the Maxwell function with the temperature of the neutral particles and the dependence of $\mu_e n$ on E/n, x_e disappears. When $E \gtrsim kT (e\lambda_u)$ at $v_{ee}/v_{en} \lesssim \delta f^0$ is not, generally speaking, Maxwell's function;[18,31,35,36] at $v_{ee}/v_{em} \gg \delta$, f^0 is the Maxwell function with temperature T_e determined by the local balance of Joule heat and energy exchange with the neutral particles. It is important to stress that when $E \gtrsim kT (e\lambda_u)$ the diffusion term of the electron transport equation has the order λ_u/L in relation to the drift term so that the Einstein equation can be used in the majority of examined problems at the temperature of neutral particles.

If only the second condition (11.4) is fulfilled (this is possible if $v_{ee}/v_{em} \gg \delta$), then f^0 is the Maxwell function with the temperature T_e. To determine this temperature, the system (11.1)–(11.3) should be supplemented by a differential (containing spatial derivatives) equation of electron energy.[6,18] When writing the diffusion term of the transport equation, it should be accepted that $P_e = n_e kT_e$.

A large number of data on the mobility and transport sections of electron scattering was published in Ref.37–41.

The boundary conditions with the concentration of the charged particles and the potential on a probe surface (which is assumed to be ideally absorbing and non-emitting) and away from the probe have the form:[6]

On the probe surface

$$n_m = 0, \quad m = i, e; \quad \varphi = \varphi_w, \tag{11.5}$$

away from the probe

$$n_m = n_{m\infty}, \quad m = i, e; \quad \varphi = 0, \tag{11.6}$$

where φ_w, $n_{m\infty}$ is the potential of the probe surface in relation to plasma and the concentration of the charged particles in the non-perturbed plasma (given values).

We shall examine briefly the justification of the first boundary conditions (11.5), initially for the ions. Evidently, the most consistent conclusion is the one based on the asymptotic analysis of the Boltzmann equation.[42] In this approach, the entire region of the plasma adjacent to the surface of the probe is subdivided into the Knudsen layer, i.e. the wall region of the plasma with the thickness of the order λ_i, and the adjacent hydrodynamic region with the characteristic linear size $L^g \gg \lambda_i$. In each of these zones, we construct its own asymptotic expansion of the distribution function with respect to the small parameter λ_i/L^g. Combining the expansions gives, in particular, the condition that must be satisfied by the first term of the asymptotic expansion which holds in the hydrodynamic region. This condition is also the required macroscopic boundary condition.

For the case in which the field in the near-wall region is weak, the first condition (11.5) was justified in Ref.42 and can be formulated as follows. Let it be that n_i^k, n_i^g are the characteristic values of the ion concentration in the Knudsen layer and the hydrodynamic region. The flux of the ions to the surface in the Knudsen layer can be evaluated (taking into account the anisotropy of the distribution function) with regard to the order of magnitude as $n_i^k C_i$, and in the hydrodynamic region as $D_i n_i^g / L^g$, where C_i, D_i are the characteristic thermal velocity and the diffusion coefficient of the ions. Since the flux across the Knudsen layer is maintained (with the accuracy to the ratio of the frequencies of ionisation, recombination, adhesion, etc. to the frequency of elastic collisions), these estimates are compared and we obtain $n_i^k/n_i^g \sim \lambda_i/L^g \ll 1$. To enable combining, the first term of the asymptotic expansion of the distribution function in the hydrodynamic region can convert to zero on the surface of the probe. This leads to equation (11.5).

In the case of a strong field, the diffusion term of the equations of ion transfer in the hydrodynamic region is, as shown previously, small in comparison with the drift term and can be omitted. The order of the degenerate system of the hydrodynamic equations, compiled in this manner and, consequently, the number of boundary conditions required to solve it decrease. Evidently, in this formulation of the problem for the ions drifting to the probe, the boundary condition on their surface is superfluous; for the ions drifting from the probe the first condition (11.5) applies. On the other hand, the solution obtained within the framework of this formulation, coincides, with the accuracy λ_i/L^g, with the solution described by the initial formulation that takes into account the diffusion term and uses the first condition (11.5). An exception is only the distribution of a concentration of the ions drifting to the probe in the near-wall region with the thickness of the order of λ_i; if the concentration in the framework of the degenerate problem is constant to a first approximation, then in the framework of the problem with diffusion it decreases to zero (in fact, it should be noted that both the first and second solutions are physically incorrect in the given region in the general case and to determine the true distribution of the concentration it is necessary to find the first term of the asymptotic expansion of the distribution function in the Knudsen layer[43]). In order to provide standard and accurate description (this is important, for example, when carrying out numerical calculations), equation (11.3) with the diffusion term and the first boundary condition (11.5) can also be used in the presence of a strong field; however, it should be remembered that the solution obtained in this approach is not reliable in the region of diffusion-induced decrease of the concentration of the ions drifting to the surface.

For the electrons, the structure of the near-wall non-hydrodynamic region is more complicated;[32-34] in addition to the Knudsen layer with a thickness of the order of λ_e in which the distribution function is isotropic, there is another kinetic layer in which the distribution function is isotropic to a first approximation but is not local and is not of the Maxwell type. To determine this function, it is necessary to solve the equation with spatial derivatives.[18,31,35] The thickness of this kinetic layer is considerably greater than λ_e and is determined by the smaller of the scales λ_u, $\lambda_m^{kin} = \lambda_e(v_{en}/v_{ee}^{kin})^{1/2}$, where v_{ee}^{kin} is the frequency of electron collisions evaluated on the basis of the characteristic concentration of electrons in the kinetic layer n_e^{kin}.

If $\delta v_{en} > v_{ee}^{kin}$, the thickness of the kinetic layer is of the order of λ_u. In the adjacent hydrodynamic region, the first condition (11.4) must be fulfilled. When $E \ll kT\ (e\lambda_u)$, estimating the fluxes of electrons in the kinetic layer and the hydrodynamic region as $n_e^{kin} C_e \sqrt{\delta}$ and $D_e n_e^g/ L^g$ (n_e^g, C_e, D_e is the scale of the electron concentration in the hydrodynamic region, thermal velocity and diffusion coefficient of the electrons) and equating these estimates (with the accuracy to the ratio of the ionisation frequencies, and so on, to δv_{en}), we obtain $n_e^{kin}/n_e^g \sim \lambda_u/ L^g \ll 1$, and the boundary condition (11.5) is valid. As in the case of the strong field for the ions, this condition can also be used at $E \gtrsim kT/(e\lambda_u)$.

If $\delta v_{en} \ll v_{ee}^{kn}$, the thickness of the kinetic layer is of the order of λ_m^{kin}. Only the second condition (11.4) is fulfilled in the adjacent hydrodynamic region because the quantity λ_u in this case is itself the macroscopic scale – at the distance of this order the transport of electron energy by thermal conductivity is comparable with the energy exchange with the neutral particles. Estimating the fluxes in kinetic layer and the hydrodynamic region as $n_e^{kin} C_e (v_{ee}^{kin}/v_{en})^{1/2}$ and $D_e n_e^g/L^g$, we obtain $n_e^{kin}/n_e^g \sim \lambda_m^{kin}/L^g \ll 1$, and (11.5) is again valid.

Thus, the boundary condition (11.5) is also substantiated for the electrons. It should be noted that when $\delta v_{en} \ll v_{ee}^{kin}$, to close the hydrodynamic formulation of the problem it is also necessary to formulate the boundary condition on the probe surface for the electron energy equation. Taking into account that this equation has a singularity[6] on the probe surface, we find the asymptotics of its solution close to the surface. The asymptotics includes two terms, one of which is restricted and the other one algebraically increasing so that the restriction of T_e on the probe surface should be used as the required boundary condition. Writing the differential equation of the first order (which is satisfied by the first of the terms mentioned previously) for the surface, we obtain the equivalent form of the accepted condition:

$$\left(\frac{K_{he}}{kh_e D_e} + \frac{5}{2}\right) k \frac{\partial T_e}{\partial y} = \frac{5}{2} \frac{ew_e}{\mu_e} \left(\frac{\partial n_e}{\partial y}\right)^{-1} - eE_y,$$

where K_{he} is the coefficient of electronic heat conductivity, and the y axis is directed along the normal from the probe surface; this equation can be regarded as a generalisation of the conditions derived by several other methods in Ref.44 and 45.

In the general case, the conditions of applicability of the solution, determined within the framework of the described hydrodynamic formulation of the problem, are as follows. The characteristic size of the probe a must satisfy one of the conditions (11.4). When the resultant solution describes hydrodynamic zones with different linear scales, each scale must satisfy one of the conditions (11.4).

11.3 Volt–ampere characteristics of probes under hydrodynamic regimes

We shall analyse the form of probe VACs, predicted by hydrodynamic theory, on a frequently encountered example of a probe in a plasma flow where the gas-dynamic regime of flow of plasma around the prob is the regime of a viscous boundary layer (i.e. $Re = v_\infty a/v_\infty \gg 1$, where Re is the Reynolds number, v is the coefficient of kinematic viscosity of plasma, the indices ∞ and w here and below are given to the values of the corresponding quantities in non-perturbed plasma and on the probe surface, respectively), and the Debye radius h, estimated from the parameters of non-perturbed plasma, is considerably smaller than the scale of the thickness of the viscous boundary layer in the vicinity of the probe surface $\Delta = a/\sqrt{Re}$. It will be assumed that the flow-around is chemically non-equilibrium, i.e. the duration of reactions in which the charged particles of each type form and disappear is comparable with the hydrodynamic (time of flight) time a/v_∞. It should also be noted that the first condition (11.4) is fulfilled in each of the hydrodynamic zones. It will be assumed that the Schmidt numbers of the ions v/D_i under normal conditions have the order of unity.

The asymptotic solutions of the problem of this type were examined in Ref.46–48. We shall present several asymptotic estimates resulting from these solutions.

The entire near-probe region can be divided into the region of non-viscous flow, the quasi-neutral region of the near-probe viscous boundary layer and the near-wall layer of volume charge (Debye layer, or DL) (Fig.1.11).

In the quasi-neutral part of the boundary layer and in the DL the electric field and fluxes of the charged particles are directed to a first approximation along the normal to the probe surface (i.e. along the y axis). In the region of non-viscous flow the distribution of electric field and fluxes are three-dimensional in the general case.

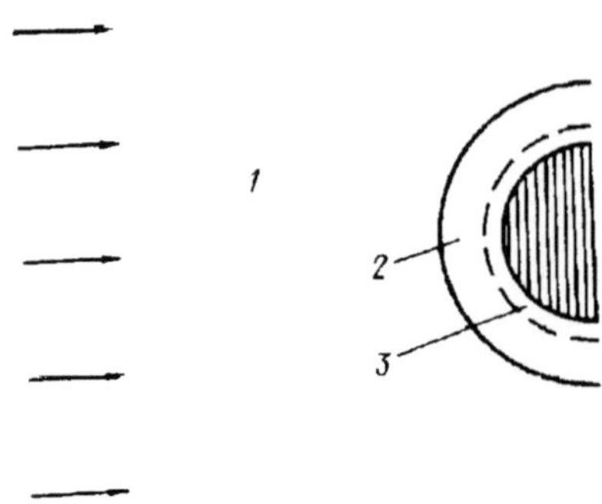

Fig.11.1 Asymptotic structure of the near-probe region. 1) region of non-viscous flow; 2) quasi-neutral part of the near-probe viscous boundary layer; 3) DL (Debye layer).

In the region of non-viscous flow and the quasi-neutral part of the bounadry $n_m \sim n_{m\infty}$. The scale of the flux density in all three zones is the same and is determined by the order of the diffusion term of the equations (11.3) in the quasi-neutral part of the boundary layer: $J_{my} \sim D_m n_{m\infty}/\Delta$. The order of magnitude of the electric field in the region of non-viscous flow and in the quasi-neutral region of the boundary layer is equal to $kT/(e\Delta)$. In the region of non-viscous flow, the convective transport of the charged particles is considerably greater than the drift transport which, in turn, is far more extensive than the diffusion transport. The convective, drift and diffusion transports in the quasi-neutral part of the boundary layer are comparable.

When estimating the orders of magnitude of the quantities in the DL, we shall into account the results of asymptotic analysis which has been carried out many times in the literature for collision layers of the volume charge, starting with Ref.13 and 14; in particular, the authors of Ref.49 and 50 obtained a solution by the method of combined asymptotic expansions with respect to the small parameter (when mentioning Ref.49 it should be noted[50,51] that the more exact definitions made in Ref.50 and cited in Ref.6 regarding the conditions of applicability of the solution [14] and the presence of two (not one as in Ref. 13) transition regions between the main part of the DL and the quasineutral region have not been substantiated). The convective transport of the charged particles in the DL does not play any significant role. When the DL is homogeneous (i.e. has no internal structure)[14,50] the diffusion and drift transport in the DL are comparable and have the order $D_m n_{m\infty}/\Delta$, which gives the equations for the orders of magnitude of n_m, E_y in the layer through the scale of its thickness y_D: $n_m \sim n_{m\infty} y_D/\Delta$, $E_y \sim kT/(ey)$. Substituting these estimates into the Poisson equation, we determine $y_D \sim (h^2\Delta)^{1/3} \ll \Delta$.

When the thickness of the DL is considerably greater than $(h^2\Delta)^{1/3}$, but its order of magnitude is not higher than Δ, the layer is heterogeneous[13,49] and includes the main part of and the transition region and the zone of the diffusion layer of the concentration of the particles drifting to the probe separating it from the quasi-neutral region and the probe surface. The estimates made previously, i.e. $n_m \sim n_{m\infty} (h/\Delta)^{2/3}$, $E_y \sim kT/[e(h^2\Delta)^{1/3}]$, hold in the transition region, and the scale of the thickness of this region is $(h^2\Delta)^{1/3}$ and considerably smaller than y_D. In the main part of the layer the concentrations of the particles repulsed by the probe field are low. The controlling transport mechanism for the attracting particles is a drift mechanism and the concentration of the particles is $n_{m\infty} h (y_D\Delta)^{-1/2}$; $E_y \sim (y_D/\Delta)^{1/2} kT/(eh)$. The thickness of the main part is, because of the small thickness of the adjacent regions, close to the thickness of the DL in the whole y_D. In the zone of the diffusion decrease, the concentrations of particles are the same as regards the orders of magnitude as in the main part, the drift transport of the attracting particles is compared with the diffusion transport, the strength of the electric field is constant, the scale of the thickness of this zone is equal to $h(\Delta/y_D)^{1/2}$ and considerably smaller than y_D as well as the scale of the thickness of the transition region. Evidently, with decreasing parameter y_D and with this parameter becoming comparable with $(h^2\Delta)^{1/3}$, the structure of the layer degenerates and the estimates transfer to corresponding estimates for the case of a homogeneous DL.

These considerations show that the condition of fulfilling the first inequality of (11.4) in all hydrodynamic zones is equivalent to the condition $\lambda_u \ll h(\Delta/y_D)^{1/2}$ which means that in all zones f^0 is the Maxwell function with temperature T.

In the case of a thin DL ($y_D \ll \Delta$), the intensity of generation of the particles in the DL is low due to the assumption that the characteristic duration of the volume reactions is comparable with the time of flight. The fluxes of the charged particles on the probe and, consequently, the probe current are determined by the fluxes from the quasi-neutral region to the external boundary of the DL. We shall examine the problems that describe the distribution of the concentration of the charged particles in the region of non-viscous flow and the quasi-neutral part of the viscous boundary layer. In the region of the non-viscous flow, the distributions of the concentration of the charged particles are described by the problem

$$nv\nabla \frac{n_m}{n} = w_m, \tag{11.7}$$

and away from the probe

$$n_m = n_{m\infty}. \tag{11.8}$$

It is important to stress that the electric field does not influence in the first approximation the distribution of the concentration of the charged particles in the region of non-viscous flow.

The system of equations described in the distribution of the concentration of the charged particles, the density of their fluxes and also the strength of the electric field in the quasi-neutral part of the viscous boundary layer have the form

$$nv\nabla \frac{n_m}{n} + \frac{\partial J_{my}}{\partial y} = w_m; \quad \sum_{m=i,e} z_m n_m = 0, \tag{11.9}$$

$$J_{my} = \mu_m \left(-\frac{kT}{e} n \frac{\partial}{\partial y} \frac{n_m}{n} + z_m n_m E_y \right). \tag{11.10}$$

We shall formulate the boundary conditions for the given system of equations. The concentration of the charged particles at the outer boundary of the viscous boundary layer tends to the values resulting from the solution of the problem (11.7), (11.8) on the probe surface. The boundary conditions for the concentration of the charged particles on the probe surface should be determined from combination with the asymptotic expansion of the concentrations which holds in the transition region (or in the DL as a whole, if it is homogeneous). As indicated by the above considerations, expansion starts with the terms of the order of $n_{m\infty} (h/\Delta)^{2/3} \ll n_{m\infty}$, and the quasi-neutral concentrations on the probe surface can assumed to be equal to zero. The insufficient boundary condition for the system of equations (11.9) and (11.10) can be written in the form[52]

$$y \to 0: \quad E_y = \frac{K-q}{K+1} \frac{kT_w}{e} \frac{1}{y} + \ldots,$$

where $K = \left(\sum_{m=i_-,e} J_{my}/\mu_m \right)_w \left(\sum_{m=i_+} J_{my}/\mu_m \right)_w^{-1}$ (the first and second sums

are taken with respect to the components with the negative or positive charge, respectively), and all ions are assumed to be having the same charge. Evidently, the quantity K determined in this manner can assume only non-negative values.

It is important to stress that in the problem formulated here which specifies the distribution of the concentration of the charged particles, the densities of their fluxes and the strength of the electric field in the quasi-neutral part of the viscous boundary layer, the effect of the probe potential is manifested only through parameter K. At $K \to 0$ $(K \to \infty)$, i.e. when suppressing the flux of the particles with the negative (positive) charge to the probe, this effect disappears and, consequently, the probe current tends to some constant value which will be denoted I_+ (I_-).

Thus, when the DL is thin, the probe current is restricted by the limiting values I_+, I_-. If the probe current is assumed to be positive in cases in which it is directed from the probe into plasma, then $I_+ < 0$,

$I_- > 0$. Evidently, $|I_+| \sim eD_i n_{e\infty} a^2/\Delta$, $I_- \sim eD_e \cdot n_{e\infty} a^2/\Delta$. It should be noted that $|I_+|/I_- \sim \beta << 1$, where β is a parameter whose order of magnitude is equal to the ratio of the mobility of the ions to the mobility of the electrons (this parameter usually does not exceed 10^{-2}).

Initial considerations enable the VAC of the probe to be analysed. In the current range $I_+ < I < I_-$ the DL is homogeneous and the main contribution to the total difference of the probe–plasma potential comes from the region of non-viscous flow. Since the distribution of the concentration of charged particles in this region does not depend on the probe potential, and the current transport is of the drift type, the VAC of the probe is linear in this range.

In the range $I > I_-$ we have $y_D \sim \Delta$, and the contributions to the probe–plasma potentials come from the region of non-viscous flow in the DL. As regards the order of magnitude, the contributions are equal to respectively $(kT/e)\,a/\Delta$ and $(kT/e)\,\Delta/h$. The ratio between them depends on the parameter ah/Δ^2 (it should be noted that the last parameter, as regards the order of magnitude, is equal to Re h/a or, which is the same, Mh/λ_i, where M is the Mach number of the incident flow). In the case with $ah/\Delta^2 >> 1$, the controlling contribution is the contribution of the region of non-viscous flow, and the VAC remains linear. When $ah/\Delta^2 >> 1$ the contribution of the DL is dominant, and the characteristic value of $dI/d\varphi_w$, which gives the slope of the VAC in relation to the potential axis, is $\sigma_\infty a^2 h/\Delta^2$ (σ is the conductivity of plasma), which is considerably lower than the characteristic value of this quantity in the range $I_+ < I < I_-$, equal to $\sigma_\infty a$. In other words, when the probe current reaches the value I_- the slope of the VAC rapidly decreases. Naturally, this variation of the slope is referred to as saturation and the value of the current I_- is the saturation current. When $ah/\Delta^2 \sim 1$ the contributions of the region of non-viscous flow and the DL to the total potential difference between the probe and the plasma in the range of currents $I > I_-$ are comparable and the VAC has an intermediate shape.

In the range $I < I_+$, we have $y_D \sim \Delta$, and the contributions of the region of non-viscous flow on the DL with respect to the order of magnitude are $\beta\,(kT/e)\,a/\Delta$, $(kT/e)\cdot\Delta/h$. Depending on the order of magnitude of $\beta ah/\Delta^2$ the VAC is linear, although there is saturation at $I = I_+$ or the VAC has the intermediate shape.

It should be noted that as a result of the non-uniformity of the problem the densities of the limiting currents in different areas of the probe surface differ and are obtained at different values of the probe potential. Because of the limited space, this effect is not discussed here, see Ref.48.

These estimates show that the floating potential with the error of the order kT/e coincides with the plasma potential. More accurate analysis gives the following equation for the value of the floating potential of the probe in relation to plasma (to a first approximation, i.e. with the error to terms of the order of kT/e):

$$\varphi_f = \frac{kT_w}{e} \ln\left[\left(\frac{h}{\Delta} \right)^{2/3} \beta \right]. \tag{11.11}$$

Thus, the characteristic special features of the VAC in the examined case are: the presence of a linear section; the difference of the saturation criteria in the regions of positive and negative potentials of the probe, with the saturation criterion at positive potentials, i.e. on the electronic part of the VAC, which has the form $h \ll \Delta^2/a_1$, being considerably more 'rigid' than the saturation criterion at negative potentials ($h \ll \Delta^2/(\beta a)$, Δ); the fact that the floating potential of the probe is close to the plasma potential. It should be stressed that, when discussing the linear section, we are concerned with the relatively long part of the VAC and not, for example, the small section of the VAC in the vicinity of the inflection point. Saturation here refers not to the absence of the dependence of the probe current on its potential but to the change of the curvature of the VAC which is caused by a rapid increase of the resistance of the near-probe region under the conditions in which the capacity of diffusion as regards supply of charged particles to the probe surface is exhausted and a further increase of current can take place only as a result of increasing the thickness of the near-probe layer of the volume charge.

Previously, it was assumed that the condition $\lambda_u \ll h\,(\Delta/y_0)^{1/2}$ can evidently be 'weakened': it is sufficient if $\lambda_u \ll \Delta$. It is evident that under the last condition, the hydrodynamic approximation remains valid in the region of non-viscous flow, the quasi-neutral part of the viscous boundary layer and (if $y_D \sim \Delta$) and the main part of the DL, i.e. in these regions which determine the main elements of the VAC. Evidently, if the given condition is fulfilled, the function f^0 in the region of non-viscous flow and the quasi-neutral part of the viscous boundary layer is a Maxwell function with temperature T.

It has been assumed that the electron concentration in non-perturbed plasma is comparable with the concentration of positive ions and electrons remain the main carriers of the negative charge. All these considerations, with exception of equation (11.11), remain also valid in the case of the plasma of the electronegative gas when the contribution of electrons to the transfer of the negative charge is not controlling. In this case, parameter β should be set equal to unity, and the currents I_+ and I_- have the same order of magnitude as the saturation criteria in the region of positive and negative potentials (evidently, the presence of the ions should decrease the asymmetry of the VAC; for example, if there are no electrons, and the mobility of the negative and positive ions are identical, then the function $I(\varphi_w)$ is odd, i.e. there is no asymmetry in the VAC).

Previously, we examined the case in which the plasma flows around the probe under the regime of the chemically reacting non-equilibrium viscous boundary layer. It may easily be seen, on the other hand, that the proposed theory is also applicable for the case in which the flow-around takes place under the regime of the viscous boundary layer but is chemically 'frozen' (evidently, in this case, the source terms in equations (11.7) and (11.9) should be omitted).

The theory can also be applied to the case of stationary or moving plasma with rapid reactions where the reaction times are short in comparison with both the time of flight a/v_∞ and the diffusion time of the ions (or, which is the same as regards the order of magnitude, the ambipolar diffusion term) in the near-probe a^2/D_i. In this case, analogues of the region of non-viscous flow and the near-probe viscous boundary layer are the region of ionisation equilibrium and the near-probe layer of non-equilibrium ionisation. The convective terms in the equations (11.7) and (11.9) should be omitted. Instead of the parameter Δ, the estimates will include a linear scale characterising the reactions, for example, the recombination length d calculated from the parameters of non-perturbed plasma (evidently, in this case, this scale is considerably smaller than the probe diameter and, if the Reynolds flow-around number is high, the thickness of the near-probe viscous boundary layer). It is only important that the Debye radius remains small in comparison with the previously mentioned scale, and the equilibrium concentration of the charged particles in the vicinity of the probe is not too small in comparison with that in the non-perturbed plasma.

In particular, the theory described previously is applicable to the case of a spherical probe in stationary plasma containing neutral particles, ions of the same kind and the electrons, with volume ionisation and recombination under the condition that $h \ll d \ll a$ (in this case a is the probe radius). Figures 11.2 and 11.3[53] show the corresponding electron and ion parts of the VAC, obtained by numerical solution of the initial problem (11.1)–(11.3), (11.5), (11.6) (in this case, the problem is, because of spherical symmetry, unidimensional and its numerical solution is not difficult). In the calculations it was assumed that the ionisation of the neutral particles is carried out by an external ioniser or

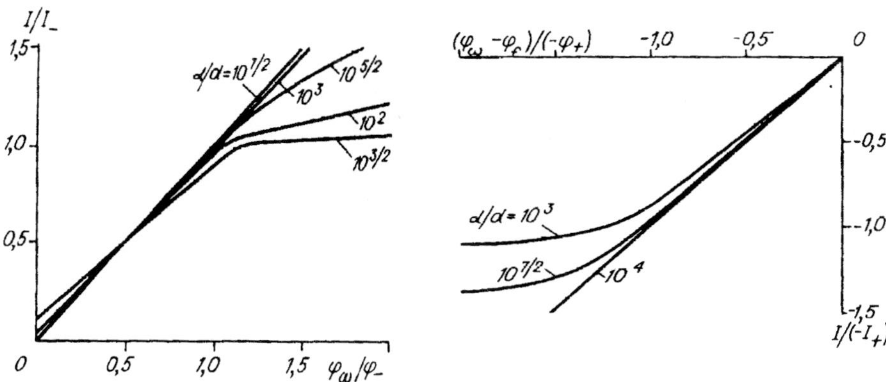

Fig.11.2 Electron branch of the VAC of a spherical probe in stationary plasma with fast reactions (numerical calculations).
Fig.11.3 Ion branch of the VAC of a spherical probe in stationary plasma with fast reactions (numerical calculations).

in collisions with neutral molecules; recombination – in collisions with the neutral molecules or dissociative recombination; the plasma temperature, the rate of volume ionisation, the coefficient of recombination and the mobility of the ions and electrons are constant in the near-probe region; $h/a = 10^{-5}$, $\beta = 0.01$. Both VAC branches are normalised with respect to the values of the saturation currents computed from the corresponding asymptotic equation. The normalised multipliers for the probe potential $\varphi_{\pm} = I_{\pm}\,(4\pi a\sigma_{\infty})$ are selected taking into account the asymptotic results in such a manner that the angle of inclination of the linear section, predicted by the asymptotic theory, is 45°. Each curve in Fig.11.3 is displaced along the potential axis by the value computed from the asymptotic equation (11.11). Taking this normalisation into account, it may be seen that the numerical calculations are in complete agreement with the asymptotic results. The normalised values of the floating potential for the probe of the given case, obtained by the numerical solution, are given below.[53] It may be seen that they differ only slightly from the values determined from equation (11.11):

a/d	10	$10^{3/2}$	10^2	$10^{5/2}$	10^3	$10^{7/2}$	10^4
$e\varphi/kT$ (numerical solution)	-10.8	-10.0	-9.20	-8.43	-7.66	-6.86	-6.03
(11.11)	-10.7	-9.98	-9.21	-8.44	-7.68	-6.91	-6.14

It is evident that the relationship between the Debye radius, the scales Δ and d and the size of the probe at which the previously described qualitative analysis is valid, can be written in the form $h \ll l \ll a$, where l is the quantity which coincides with the smaller of the scales Δ, d. In the case of stationary or slowly moving plasma with slow or frozen reactions in the near-probe region, when the order of magnitude of the Reynolds number does not exceed unity, and the duration of the reactions is not small in comparison with the diffusion time of the ions in the near-probe region, the right-hand inequality is violated. In this case, the VAC does not contain the distinctive linear section. The saturation conditions in the region of positive and negative potential coincide and have the form $h \ll a$. The VAC is situated at a large distance from the origin of the co-ordinates; for example, for the probe in stationary plasma with frozen reactions and constant temperature, we have $I(0) = (I_+ + I_-)/2$ ($I(0)$ is the value of current on the probe at the potential equal to the plasma potential).

Finally, in the case in which the concentration of charged particles in the plasma is not high and the Debye radius is comparable with the smallest of the scales Δ, d, a, there is no justification for expecting the existence of a distinctive region of saturation at positive or negative probe potentials. The conditions of existence of the linear section of the VAC has the form $l \ll a$ in this case.

These considerations show that the effect of movement of the plasma on the VAC of the probe under the examined conditions is characterised by parameter Re. Since, as already shown, the Schmidt number of the ions under the normal conditions has the order of unity, this number is naturally interpreted in the probe theory as the ratio of the velocity of the incident flow to the diffusion rate of the ions in the near-probe region. On the other hand, the order of magnitude of this parameter is equal to the ratio of the Mach number of the incident plasma flow to the Knudsen number λ_i/a. Since the Knudsen number under the examined conditions is small, the influence of the convective effects under the hydrodynamic conditions becomes more significant at considerably lower values of the velocity of the incident flow than under collisionless conditions. For example, for a probe with a characteristic size of 1 cm in the plasma of slightly ionised air at 3000 K and at the atmospheric pressure, the movement of plasma must be taken into account already from the velocities of the order of units of centimetres per second (under these conditions, the Reynolds number becomes equal to unity at $v_\infty \approx 8$ cm/s).

The influence of the convective effects on the VAC of the probes under the hydrodynamic conditions is, as indicated previously, of the dual type. On the one hand, the movement of plasma results in a general increase of the current on the probe caused by an increase of the diffusion rate of the charged particles to the probe surface resulting from a decrease of the thickness of the region in which this diffusion is localised. On the other hand, at relatively fast movements (Re≫1) the form of the VAC changes: a linear section appears, and so on. Evidently, this change is caused by the transformation of the structure of the near-probe region of perturbation: the quasi-neutral region becomes of the two-scale type, i.e. includes two regions with greatly differing scales (the region of non-viscous flow and the quasi-neutral region of the viscous boundary layer). The volume reactions in the near-probe region have the same dual effect.

11.4 Diagnostic methods

As shown by the results of qualitative analysis described in the previous section, the physics of the passage of current in the near-probe region of high-pressure slightly ionised plasma is relatively complicated. It is not surprising that there are no universal solutions of the problem (11.1)–(11.3), (11.5), (11.6) that would enable the diagnostic methods to be developed for the entire range of hydrodynamic conditions.

The problem of numerical solution has been examined with special reference to unidimensional problems (see Ref.8 and the studies quoted in there, and also Ref.20, 53–63 and Chapter 10 of this book). At present, effective numerical methods have been developed for these problems and enable the influence of different physical effects, such as thermal diffusion,[20] high-intensity volume ionisation and recombination,[53,58] the emission of electrons from the probe surface,[8,59,63] the presence in the plasma of negative ions in addition to neutrals, positive ions, electrons[8,61,62] to be considered.

There are only a few studies concerned with the accurate numerical solution of the problems of the hydrodynamic theory of electrical probes in non-unidimensional formulations. In addition to the studies quoted in Ref.6 and 8, we should mention the studies in Ref.64 and 65 which present the results of two-dimensional numerical calculations for the case in which the concentration of charged particles in the plasma is not high and the Debye radius is comparable with the probe size.

Evidently, in this situation, the methods of accurate numerical solution can be used efficiently as the main means of analysis of the VAC in the undimensional problems, for example, the problems of spherical, double cylindrical (in the form of two coaxial cylinders) and double flat probes in stationary plasma. It should be noted that when constructing the numerical algorithms, it should be taken into account that at low values of the Debye radius the Poisson's equation is not suitable for determining the potential[58,66] because the only term of the equation which contains the potential is small in the main part of the plasma volume. In non-unidimensional problems, including the problems of probes in moving plasma, it is more efficient to use the approximate approaches which enable specific regions of the VAC to be calculated. On the other hand, the degree of accuracy and the region of applicability of these approaches can be efficiently evaluated using the numerical solutions of simulation unidimensional-problems.

Qualitative analysis, described in the previous paragraph, leads to three relatively simple approaches within which we can calculate approximately either the linear section of the VAC or the sections of currents of positive or negative particles $I < I_+$, $I > I_-$ (in this case, we take into account the voltage drop only in the main part of the DL) or the saturation current of the positive or negative particles I_+, I_-. Figure 11.4 shows schematically the VAC of the probe (solid line) and the linear section (line 1), the sections of the currents of positive or negative particles (lines 2, 3, respectively) and the values of the saturation currents, calculated using these approaches. These approaches will now be discussed in greater detail.

The first approach. To calculate the linear section of the VAC, it is initially necessary to find the distribution of the concentration of the charged particles in the region of non-viscous flow (or, in the case of plasma with fast reactions, regions of ionisation equilibriurm) described by the problem (11.7), (11.8). Subsequently, in this region we determined the distribution of the conductivity of plasma and solve the problem for the distribution of the potential, neglecting the voltage drop in the near-probe viscous boundary layer (the non-equilibrium ionisation layer)

$$\nabla \cdot \left(\sigma \nabla \varphi \right) = 0. \tag{11.12}$$

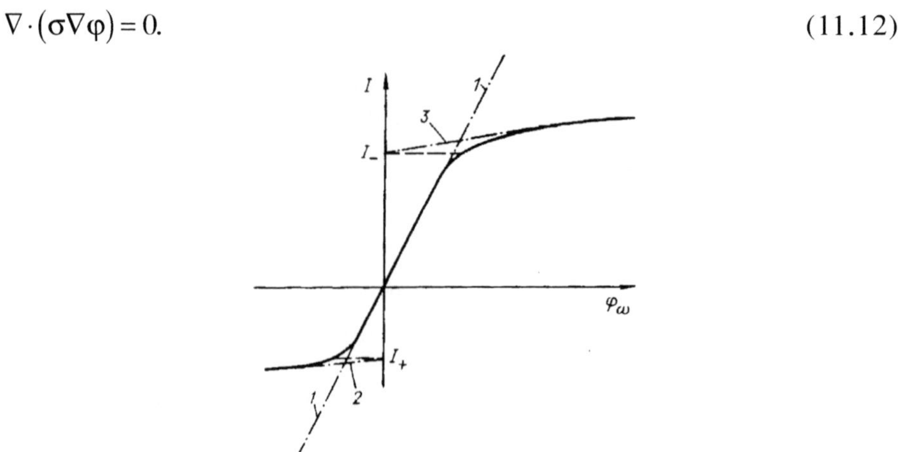

Fig.11.4 VAC of the probe and its elements, calculated using approximate approaches.

On the surface of the probe

$$\varphi = \varphi_w, \tag{11.13a}$$

and away from the probe

$$\varphi = 0. \tag{11.13b}$$

In a number of cases, changes of the pressure and temperature of plasma in the examined region are small. For example, for stationary or slowly moving plasma, this situation is realised in the case of a non-cooled probe. If the Reynolds flow-around number is high, such as in the case in which the Mach number of the flow is moderate (for example, for a diatomic gas at $M \leq 0.6$ the changes of pressure and temperature in the region of non-viscous flow do not exceed approximately 28 and 7%, respectively). For trans- and supersonic flows, this situation occurs in particular in the case of a thin probe, for example, in the form of a disc or flat sheet. If the flow is hypersonic, $M \gg 1$, and the probe is 'blunted', the extent of withdrawal of the head shock wave, formed in front of the probe, is often not large and the controlling contribution to the resistance of the region of non-viscous flow comes from the non-perturbed region in front of the shock wave.

In these cases, the changes of the concentration of the charged particles and, consequently, plasma conductivity in the examined region are not large. Cases can also occur in which the variations of the pressure or temperature in the given region are large but they do not cause any change of conductivity. In particular, in cases in which the flow-around the probe is chemically 'frozen', the molar fractions of the charged particles in the region of non-viscous flow are constant; if the value $\mu_e n$ in the examined temperature range is constant (this approximation is satisfactory for, for example, the plasma of combustion products), the conductivity of plasma in the region of non-viscous flow is constant regardless of the dependence on M.

Since the conductivity of plasma in the given region is constant, the proposed approach used in these cases is greatly simplified: equation (11.12) becomes the Laplace equation, and the expression describing the linear section of the VAC has the form

$$\frac{I}{\varphi_w} = 4\pi C \sigma_\infty, \tag{11.14}$$

where C is the electrical capacitance of the probe. This quantity depends only on the form of the probe and its value is well known for the probes of the simplest configurations. In particular, for a spherical probe with radius a we have $C = a$; for a probe in the form of a thin cylinder with height H and radius R ($R \ll H$) we have $C = H/[2 \ln (H/R)]$; for a disc-shaped probe with radius a it is $C = 2a/\pi$; the equations for the probe in the form of an ellipsoid of revolution are also known (see, for example, Ref.67, 68); for a wall probe in the form of a circle with radius a, positioned flush on the non-conducting plane, $C = a/\pi$; for a hemispherical probe with radius a positioned on a non-conducting plane, $C = a/2$, and so on. It should be noted that using the results published in Ref.69, the equation (11.14) is generalised to the case of a probe in a magnetic field.

As indicated by the results obtained in the previous paragraph, the discussed approach can be used at $h \lesssim l \ll a$; $\lambda_u \ll l$. It should also be noted that when λ_u is comparable with the scale l and the last inequality is violated, function f^0 in the examined region will, generally speaking, differ from the Maxwell function and will depend on the local value E/n and, consequently, on the probe potential. Therefore, in the given case the analysed section of the VAC is no longer linear. On the other hand, we can attempt to calculate this section on the basis of the combined solution of the problems (11.7), (11.8) and (11.12), (11.13).

The second approach. In the majority of studies carried out within the framework of the second and third approach, it is assumed that the plasma contains charged particles of two types – electrons and positive ions of the same type, and we recalculate the region of the ion current of the VAC or ion saturation current. The condition of applicability of these approaches is written in the form of inequalities h, $\lambda_u \ll l, a$; $\beta ha \ll l^2$.

To calculate the region of the section of ion current, it is necessary to find initially the distribution of the concentration of the charged particle in the quasi-neutral region and determine the value of the ion flux on the outer boundary of the DL as the function of the thickness of the DL. This is followed by solving the problem in the DL, and the diffusion transport of the charged particles is ignored in comparison with the drift transport and we determine the voltage drop in the DL which is regarded as identical with the total difference of the probe-plasma potentials. It should be noted that the function $I (\varphi_w)$, obtained within the framework of this approach, which describes the region of ion current of the VAC, assumes the value I_+ if $\varphi_w = 0$ (Fig.11.4).

When solving the problems in the DL in the case in which the inequality $\lambda_i \ll h$ is not fulfilled, it may be necessary to take into account the dependence of the mobility of the ions on the local value of E/n. The electron concentration in the DL is considerably lower than the ion concentration. However, in certain situations it is essential to take into account the presence of the electrons in the DL (in particular, as a result of the high energy of the electrons an important role is played by ionisation by the electron impact, for example, Ref.70); it should therefore be noted that the dependence of function f^0 in the DL on the local value of E/n becomes significant at $\lambda_u \gtrsim h$.

This approach will be illustrated on an example of a spherical probe (discussed in the previous paragraph) in still plasma with volume ionisation and the recombination and constant properties in cases in which h, $\lambda_u \ll d \ll a$, and within the framework of the model of the kinetics described previously when examining Figs. 11.2 and 11.3. In the region of ionisation equilibrium, the concentration of the ions and the electrons is constant and equal to that in the non-perturbed plasma: $n_i = n_e = n_{i\infty} = n_{e\infty}$ (here and when examining the model of the plasma with the charged particles of two types – positive ions and electrons – the ion or electron concentration in non-perturbed plasma will be denoted by $n_{c\infty}$: $n_{c\infty} = n_{i\infty} = n_{e\infty}$). In the quasi-neutral part of the near-probe layer of non-equilibrium ionisation the distribution of the concentration of charged particles n_c ($n_c = n_i = n_e$) is determined by the problem

$$D_a \frac{d^2 n_s}{dy} = k_r n_c^2 - S, \tag{11.15}$$

$$y = y_D: \ n_c = 0; \ y \to \infty: \ n_c \to n_{c\infty}, \tag{11.16}$$

where $D_a = 2D_i D_e/(D_i + D_e) \approx 2D_i$ is the coefficient of ambipolar diffusion, S is the volume ionisation rate, k_r is the recombination coefficient (evidently, $k_r n_{c\infty}^2 = S$). In accordance with the assumption $d \ll a$ the curvature of the probed surface is not taken into account here.

Equation (11.15) will be multiplied by dn_c/dy and integrated. Taking into account the second boundary condition (11.16) gives

$$\frac{D_a}{2}\left(\frac{dn_c}{dy}\right)^2 = \frac{k_r}{3} n_c^3 - S n_c + \frac{2}{3} S n_{c\infty}. \tag{11.17}$$

Dividing the equation (11.13) for the ions and electrons by D_i, D_e, respectively, and combining the results, in the quasi-neutral approximation we obtain

$$\frac{J_i}{D_i}+\frac{J_e}{D_e}=-\frac{2}{T}\nabla(n_cT).$$

(11.18)

Consequently, for the examined model, using (11.17) and taking into account the first boundary condition (11.16), the determined relationship between the fluxes of the ions and electrons on the outer boundary of the DL:

$$\frac{J_i(y_D)}{D_i}+\frac{J_e(y_D)}{D_e}=-2\frac{n_{c\infty}}{d},\qquad d=\left(\frac{3D_a}{4k_r n_{c\infty}}\right)^{1/2}.$$

(11.19)

The problem in the DL has the form (ignoring the effect of the curvature – this is justified when the absolute value of the probe potential is not too high and $y_D \ll a$):

$$\frac{dJ_i}{dy}=S,\quad \frac{dJ_e}{dy}=S,\quad \frac{dE}{dy}=4\pi e(n_i-n_e),$$

(11.20)

$$J_i=\mu_i n_i E,\quad J_e=-\mu_e n_e E,$$

(11.21)

$$y=0:\ \ J_e=0;\quad y=y_D:\ \ E=0.$$

(11.22)

When writing the equations of conservation of the charged particles, recombination was ignored because the concentration of these particles in the DL was low. The first boundary conditions indicates that the probe is under the high negative potential and the electron flux on its surface is suppressed, the second condition corresponds to neglecting the electric field at the outer boundary of the DL in comparison with the values inside the layer. The equation (11.19) is the insufficient boundary condition.

If the dependence of the mobility of ions and electrons on the strength of the electric field is ignored, the problem (11.19)–(11.22) is easily solved. In particular, the solution shows that $n_e/n_i \sim \beta \ll 1$, and, consequently, the presence of the electrons in the DL can be ignored with the error of the order of β. In other words, the second term in the left-hand part of equation (11.19) and the second term in the brackets in the right-hand part of the third equation (11.20) are insignificant and can be omitted. Consequently, the problem (11.19)–(11.22) is split up: the functions n_i, J_i, E are determined independently of the electron distribution, and this is followed by determining n_e, J_e; the solution obtained using this procedure coincides with the error β with the solution of the complete problem (11.19)–(11.22).

The equations for the dependence I (y_D) and also the equation for the dependence φ_w (y_D), obtained by integrating the solution for the electric field in the range from 0 to y_D, have the form

$$I = -4\pi e a^2 \left(S y_D + 2 D_i\right)\frac{n_{c\infty}}{d}, \tag{11.23}$$

$$\varphi_w = -\frac{16\sqrt{6}\pi}{9} e n_{c\infty}\, hd\Phi\left(\frac{3 y_D}{8d}\right) \tag{11.24}$$

with the function $\Phi = \Phi\,(u)$ is given by the equation

$$\Phi = \Phi(u) = 2(2u+1)\left(u^2 + u\right)^{1/2} - \ln\left[2\left(u^2 + u\right)^{1/2} + 2u + 1\right].$$

These equations represent the parametric description of the section of ion current of the VAC. The terms in the brackets in the right-hand of (11.23) are naturally interpreted as the contributions to the probe current of the generation of charged particles in the DL and of the ion diffusion from the quasi-neutral region. Evidently, the absence of the dependence of the flux from the quasi-neutral region on the probe potential is associated with neglecting the effect of the curvature and by the fact that the plasma properties in the near-probe region are constant within the framework of the examined model. The influence of the curvature effects under the conditions in which the generation of charged particles in the DL is 'frozen' was analysed in Ref.71–73 (the authors of Ref.72 also took into account the dependence of the mobility of ions on the local strength of the electric field). Under the conditions in which the generation in the DL is the controlling process, the influence of the curvature effects was examined in Ref.74. The influence of the variability of the plasma properties has been discussed in Ref.75 and 76.

The authors of Ref.77 and 78 calculated the regions of the ion current on the VAC of wall probes on sheets and also conical probes, in plasma flows with frozen ionisation and recombination at Re>> 1. In addition to these studies where the authors examined the case where

$y_D \gg \Delta$ (the so-called sheath–convection regime); the appropriate studies have been reviewed in Ref.15 (see also Ref.79–83). A distinguishing feature of the last case is the simple calculation of the ion flux on to the outer boundary of the DL: in this case, the flux is equal to $n_{i\infty} v_{\perp}$ ($v_{\perp}$ is the projection of the mean mass velocity on the normal to the upper boundary of the DL). To explain this special feature, we shall examine briefly the buffer zone dividing the region of the non-viscous flow and the main part of the DL. It will be assumed that $h \ll \Delta \ll y_D \lesssim a$, and we shall take into account the results published in Ref.84. At $hy_D/\Delta^2 \ll 1$, the zone is identical with that in the case where $y_D \sim \Delta$ and includes the quasi-neutral diffusion region in which the convective, drift and diffusion transports of the charged particles are comparable, and the transition region in which the non-quasi-neutrality is significant and the intensity of convective transport is low. Taking into account that $v_{\perp} \sim v_{\infty} y_D/a$, we shall determine the scale of the thickness of the regions: Δ^2/y_D, $(h^2\Delta^2/y_D)^{1/3}$. At $hy_D/\Delta^2 \sim 1$, the scales coincide: the buffer zone is homogeneous and all the previously mentioned effects are simultaneously significant in this zone. At $hy_D/\Delta^2 \gg 1$, the buffer zone again becomes inhomogeneous and includes the layer of decrease of the electron concentration in which the electrons are repulsed by the probe fields, the concentration of the ions remains constant, and the convective-drift layer in which the ion concentration decreases from $n_{i\infty}$ to the values corresponding to the main part of the DL (it is assumed that $\beta y_D h/\Delta^2 \ll 1$). The scales of the thickness of the layers are equal to h, $h^2 y_D/\Delta^2$.

Using these estimates, it can be easily be shown that the flux of the ions across the buffer zone is considerably stronger than their convective transport in the longitudinal direction. Therefore, the density of the ion flux across the buffer zone changes only slightly and is determined by the convective and drift fluxes at distances from the zone considerably greater than its thickness. Finding the electric field from the condition of suppression of the electron flux, we obtain that the density of the ion fluxes $n_{i\infty} v (1 + \mu_i/\mu_e)$, and since μ_i/μ_e is low we obtain the expression described previously.

The voltage drop in the DL at $\Delta \lesssim y_D \lesssim a$ has the order $(kT/e) y_D^2/(h\Delta)$ (for comparison, we shall show that at $(h^2\Delta)^{1/3} \lesssim y_D \lesssim \Delta$ the drop is of the order of $(kT/e) y_D^{3/2}(h\Delta^{1/2})$). To neglect the voltage drop in the region of non-viscous flow, it must be that $\beta ah/(\Delta y_D) \ll 1$. If $y_D \ll a$, the DL is treated as locally uniform; if $y_D \sim a$, non-unidimensional effects must be taken into account.

The third approach. The saturation current is determined by solving the problem of the distribution of the concentration of charged particles in the quasi-neutral region. It is important to stress that the specific features, associated with the presence of the electric field in the plasma, are not very important in this problem and the formulation of this problem is similar to the formulation of the well-known problem of the flow of a mixture of neutral gases around the solids. To solve these problems, the gas dynamics and heat and mass exchange theory offer sufficiently developed methods, both analytical and numerical.

Initially, we examine the model of plasma with two types of charged particles (electrons and positive ions). Assuming that the electron flux on the probe surface (or, more accurately, on the outer boundary of the DL) is equal to zero, the equation (11.18) gives the relationship between the density of ion saturation current and the quasi-neutral concentration of the charged particles

$$j_+ = -2e \left(D_i \frac{\partial n_c}{\partial y} \right)_w. \tag{11.25}$$

From equation (11.18) we exclude the density of the electron flux using the equation $J_e = J_i - j/e$ (j is the density of electric current). Expressing J_i from the resultant equation and substituting this result into (11.1) for the ions, we obtain the equation for n_c

$$n v \nabla \frac{n_c}{n} - \nabla \left[\frac{D_a}{T} \nabla (n_c T) \right] + \frac{j}{e} \nabla \frac{D_i}{D_i + D_e} = w_c, \tag{11.26}$$

where w_c is the rate of variation of the concentration of the ions or electrons as a result of volume ionisation or recombination.

The corresponding boundary conditions have the form:
on the probe surface

$$n_c = 0, \tag{11.27a}$$

and away from the probe

$$n_c = n_{c\infty}. \tag{11.27b}$$

It may be assumed that under the ion saturation regime the density of the electric current in the entire near-probe region of perturbation is equal to the right-hand part of (11.25) (as regards the order of magnitude). Since the third term in the left hand part (11.26) has the order β in relation to the second part, the third term can be ignored. Consequently, the problem (11.26) and (11.27) becomes closed: solving this problem and using equation (11.25), we determine the density of saturation current. It is interesting to note that in the case in which the coefficients of the diffusion of ions and electrons depend on plasma temperature in the same manner, the distribution of the concentration of charged particles in the quasi-neutral region is independent of the probe potential not only in the saturation regime but also in the entire potential range in which the DL remains thin (the third term in the left-hand part of the (11.26) is actually equal to zero).

When volume ionisation and recombination are 'frozen', equation (11.26) (here and in further references to this equation it is assumed that the third term in its left part is omitted) is linear and the saturation current is proportional to the concentration of ions or electrons in non-perturbed plasma $n_{c\infty}$. Analysis of the results of a large number of experiments shows that the effect of changes in the surface temperature of the probe of the ion saturation current on the probe in still plasma or plasma moving at a moderate Mach number is quite weak. This effect has been observed many times in experiments. It is caused by the fact that the changes of saturation current, caused by the change of the field of gas-dynamic parameters (primarily gas density) and by changes of the diffusion coefficient of ions, compensate each other to a first approximation. In this situation the ion saturation current is calculated to a first approximation ignoring the changes of the transport properties of plasma in the near-probe region, and the overall expression for the current on the probe of the given geometry is written in the form

$$I_+ = -\mathrm{Sh}\, en_{c\infty}D_{i\infty}a. \tag{11.28}$$

The dimensionless coefficient Sh (Sherwood number) introduced here, is a function of Schmidt ambipolar number $\mathrm{Sc} = v_\infty/D_{a\infty}$ and of the diffusion Peclet number $\mathrm{Pe} = \mathrm{Re}\,\mathrm{Sc} = v_\infty a/D_{a\infty}$. The coefficient is determined using either the experimental results or by solving the approximate hydrodynamic problem. In particular, at $\mathrm{Pe} \ll 1$, if the characteristic size of the probe is represented by its electrostatic capacitance, the equality[68] is fulfilled with the accuracy to terms of the order $\mathrm{Pe}^2 \ln (\mathrm{Pe}^{-1})$

$$\mathrm{Sh} = 4\pi\, (2 + \mathrm{Pe}). \tag{11.29}$$

It is interesting to note that this equation makes it possible to determine the concentration of charged particles in non-perturbed plasma using the measured dependence of the ion saturation current on the speed of the probe in relation to plasma without using information on the values of the diffusion coefficient under the given conditions:

$$\frac{d\left(-I_+\right)}{dv_\infty}\bigg|_{v_\infty=0} = 2\pi e C^2 n_{c\infty}.$$

At Pe >> 1 when plasma flows around the probe in the regime of the viscous boundary layer, Sh $\approx F\sqrt{Pe}$, and equation (11.28) can be written in the form

$$I_+ = -\frac{F}{\sqrt{2}}en_{c\infty}\left(v_\infty D_{i\infty}a^3\right)^{1/2}, \tag{11.30}$$

where $F = F$ (Sc) is the dimensionless coefficient which depends on Schmidt's number. The studies concerned with applying equation (11.30) to probes of different geometry are reviewed in Ref.17. In particular, the distribution of the density of ion saturation current along the 'windward' surfaces of a sphere and a thin cylinder can be calculated using the equations[85]

$$j_+(\alpha) = j_+(0)\cos^2\frac{\alpha}{2}\left(1 - \frac{2}{3}\sin^2\frac{\alpha}{2}\right)^{-1/2},$$

$$j_+(0) = -\frac{1.3}{Sc^{1/6}}en_{c\infty}\left(\frac{v_\infty D_{i\infty}}{a}\right)^{1/2}, \tag{11.31}$$

$$i_+(\alpha) = j_+(0)\cos\frac{\alpha}{2}, \quad j_+(0) = -\frac{1.1}{Sc^{1/6}}en_{c\infty}\left(\frac{v_\infty D_{i\infty}}{R}\right)^{1/2}, \tag{11.32}$$

where α is the polar angle counted from the direction opposite to the vector of the speed of the non-perturbed incident flow (i.e. from the front critical point).

These equations can be used to calculate the integral ion saturation current on spherical and cylindrical probes whose collecting surfaces are represented by any segments of the 'windward' surface. In particular, when the collecting surface is the entire 'windward' surface (i.e. the front hemisphere or front half cylinder respectively), the integral saturation currents are described by the equation

$$I_+ = -\frac{6.7}{Sc^{1/6}} en_{c\infty}\left(v_\infty D_{i\infty} a^3\right)^{1/2}, \quad I_+ = -\frac{3.2}{Sc^{1/6}} en_{c\infty}\left(v_\infty D_{i\infty} R\right)^{1/2} H. \quad (11.33)$$

When the entire surface of the sphere or cylinder is collecting, theoretical determination of the contribution of the 'leeward' surface is associated with difficulties due to taking into account the separation of the boundary layer and formation of dead zones. In this situation this contribution can be evaluated on the basis of an analogy between mass and heat exchange and using the experimental data for the Nusselt number. Using the semi-empirical equations for the Nusselt number, presented for the sphere and cylinder in Ref.86, 87, the integral points on the total surface of the probe are described by the relationships similar to (11.33) where the first factors in the right hand part are replaced by the factors 6.8 and $(3.3/Sc^{0.13})$. Using semi-empirical equations published in Ref.88 gives equations that are also identical (11.33) but the numerical coefficients 6.7 and 3.2 are replaced by 8.3 and 3.3. Taking into account that the above equations are similar to (11.33), it can be concluded that the contribution of the rear hemisphere or rear half-cylinder to the integral saturation current is not large. It should be noted that this conclusion is confirmed by the results of measuring the distribution of the density of ion saturation current around the circumference of the cylindrical probe described in Ref.89, 90, see also Ref.91. It should be noted that the contribution of the ends to the integral saturation current on the cylinder is also insignificant under these conditions.

The distribution of the density of ion saturation current along the surfaces of a flat sheet or cone is calculated using equations[6]

$$j_+(x) = -\frac{0.47}{Sc^{1/6}} en_{c\infty}\left(\frac{v_\infty D_{i\infty}}{x}\right), \quad j_+(x) = -\frac{0.81}{Sc^{1/6}} en_{c\infty}\left(\frac{v_{x\delta} D_{i\infty}}{x}\right)^{1/2}, \quad (11.34)$$

where x is the distance from the edge of the sheet or tip of the cone, $v_{x\delta}$ is the speed at the external boundary of the boundary.

The density of saturation current at the critical point of a 'blunted' solid is determined from the equation[6]

$$j_+(0) = -\frac{2^\varepsilon \cdot 0.66}{Sc^{1/6}} en_{c\infty}\left[\frac{dv_{x\delta}}{dx}(0)D_{i\infty}\right]^{1/2}, \quad (11.35)$$

where $dv_{x\delta}/dx$ (0) is the value at the critical point of the derivative of the speed at the outer boundary of the boundary layer, $\varepsilon = 0$ or

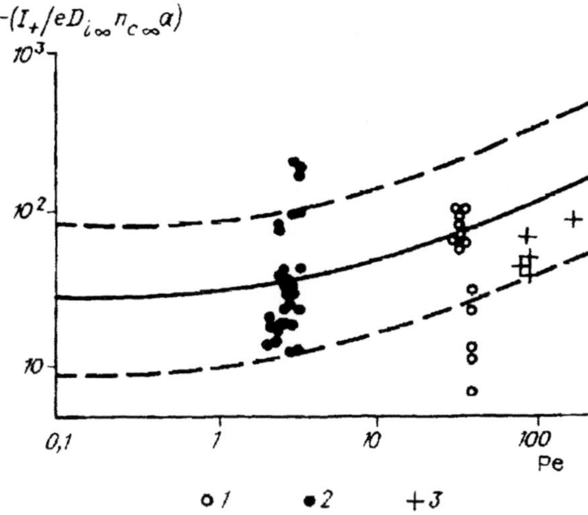

Fig.11.5 Normalised ionic saturation current on the spherical probe in a subsonic flow. solid line - calculations; points - experiments in Ref.122(1), 151(2), 153(3).

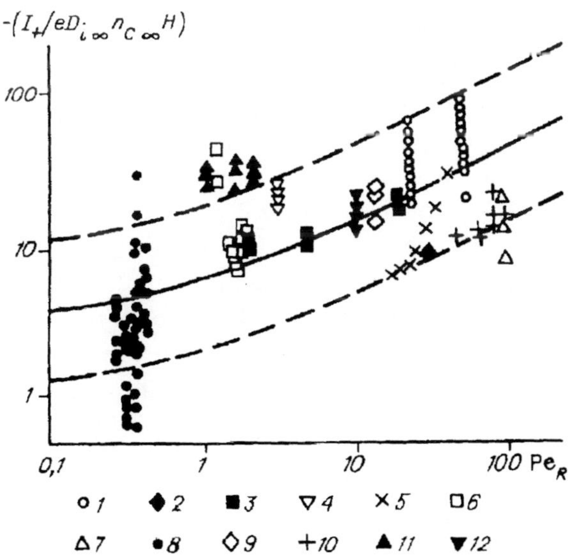

Fig.11.6 Normalised ionic saturation current, a cylindrical probe in a subsonic field . solid line – calculation; points – experiments. Ref.52 (1), Ref.90 (2), Ref.113 (3), Ref.118 (4), Ref.148 (5), Ref.149 (6), Ref.150 (7), Ref.151 (8), Ref.152 (9), Ref.153 (10), Ref.154 (11), Ref.155 (12)

$\varepsilon = 0.5$ respectively, in cases in which the flow-around the probe is two-dimensional or axisymmetric. It should be noted that the value of $dv_{x\partial}/dx$ (0) for spherical and cylindrical probes is equal to $3v_{x\infty}/(2a)$ and $2v_\infty/R$, respectively, and equatioan (11.35) for these probes is similar to the second equation of (11.31), (11.32) (the numerical coefficients

1.3 and 1.1 are replaced by the coefficients 1.2 and 0.94).

To calculate the saturation current at Pe ~ 1 it is necessary, strictly speaking, to find (numerical) solution of the elliptical problem (11.26), (11.27). On the other hand, the approximate analytical expression for Sherwood's number for this case is obtained[92] by algebraic interpolation starting with the asymptotics (11.29), (11.30):

$$\mathrm{Sh} = 8\pi + \frac{4\pi F \mathrm{Pe}}{F + 4\pi\sqrt{\mathrm{Pe}}}. \tag{11.36}$$

Note that in the case of a spherical probe (for which in accordance with the first equation (11.33) it can be assumed that $F = 9.5\mathrm{Sc}^{-1/6}$) this equation agrees with the accuracy to 20% with a similar equation derived on the basis of the correlation equation for Nusselt's number published in Ref.86.

In the case of a probe in the form of a cylinder the condition of validity of equation (11.29) has the form $v_\infty H/D_{a\infty} = \mathrm{Pe}_h \ll 1$ (if the probe is characterised by several (not 1) greatly differing dimensions, the criterion of validity of equation (11.29) is the smallness of the Peclet number estimated from the largest of these dimensions). On the other hand, the validity of the second equation (11.33) (which is the variant of equation (11.30) for a cylindrical probe) is limited by the range of high values of the Peclet number estimated from the radius $v_\infty R/D_{a\infty} = \mathrm{Pe}_R \gg 1$. In this situation it is difficult to hope for a rational accuracy of the interpolation (11.36). Therefore, for approximate calculations of the integral ion saturation current on the thin cylindrical probe whose entire surface is collecting, the following procedure is recommended at Pe_R of the order of unity.[99] The second equation (11.33)

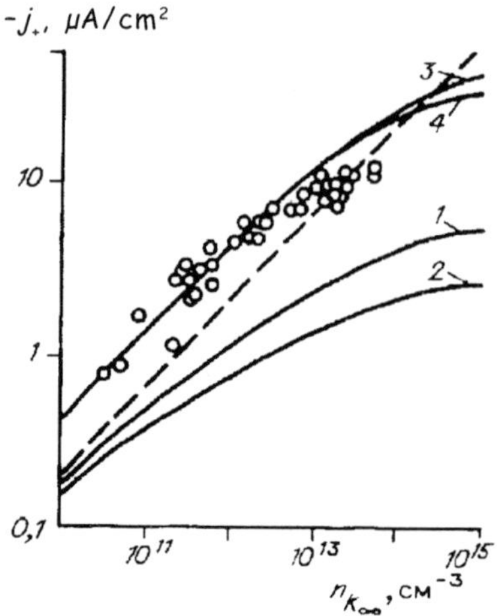

Ref.11.7 Density of saturation current at a critical point of a cylindrical probe in the flow of plasma of combustion products with an addition of potassium in relation to the concentration of potassium atoms in non-perturbed plasma. Lines - calculations: $T_w = 800(1,3)$, 400 K$(2,4)$; points - experiments.

is used, in accordance with the data in Ref.87, at $Pe_R > 20$. For the range $0.5 < Pe_R < 20$, taking into account the correlation relationship for the Nusselt number[87] the following equation holds

$$I_+ = -\frac{6.3}{Sc^{0.03}} Pe_R^{0.4} en_{c\infty} D_{i\infty} H. \qquad (11.37)$$

Finally, for the range $R/H \ll Pe_R \leq 0.5$

$$I_+ = -\frac{13 en_{c\infty} D_{i\infty} H}{\ln\left(Pe_R^{-1} + 4.2\right) + 0.8}. \qquad (11.38)$$

This formula is in agreement with the two-term asymptotic expansion of the Sherwood number with respect to parameter $1/\ln (Pe_R^{-1})$ and agrees at the point $Pe_R = 0.5$ with equation (11.37).

It should be stressed that in the equations (11.29), (11.31)-(11.35), (11.37), (11.38) the dependence on the ambipolar Schmidt number Sc is either not present or is very weak. It should be assumed that it also remains weak in the general case, i.e. the Sherwood number in (11.28) depends on approximately only on Peclet number. In other words, the quantity $I_+/(en_{c\infty} aD_{i\infty})$ for the probe of the given geometry is a universal function of the Peclet number. These functions for the spherical and cylindrical probes calculated using equations (11.36) and the second equation (11.33) and equations (11.37), (11.38) are presented by solid lines in Fig.11.5 and 11.6.[93]

The validity of the first equation (11.34) is not limited by the assumption on the moderate values of the Mach number of the incident flow. If the index ∞ in the second equation (11.34) and in equation (11.35) is replaced by the index δ (given to the values of the corresponding functions at the external boundary of the boundary layer), the validity of the equations is also not limited by this assumption; of course, in the case trans- or supersonic flows these equations link the saturation current with the parameters not in the perturbed plasma but at the outer boundary of the boundary layer. The dependence of $n_{c\infty}$ of the ion saturation current of the spherical probe in the flow with high M and moderate Re, when plasma flows around the probe in the regime of the viscous impact layer, were obtained in Ref.94, 95; the authors of Ref.96 presented the results of numerical calculations of the distribution of the concentration of charged particles and the potential at the axis of symmetry of the viscous impact layer.

We have examined the situation in which the recombination length is considerably greater than the smaller of the scales a, Δ and the volume ionisation and recombination in the near-probe region are 'frozen'. If this equality is not fulfilled, the ionisation recombination must be taken into account. Equation (11.26) becomes non-linear (due to the non-linear dependence of the recombination rate on the concentration of charged particles) and the current is no longer proportional to the concentration of ions or electrons in non-perturbed plasma $n_{c\infty}$.

At the limit of the high rate of the reactions ($d << a$; Δ) the equation (11.26) is solved analytically.[97,98] However, the solution links the saturation current not with $n_{c\infty}$ but with some value of the equilibrium concentration in the perturbed region so that the saturation current can be used to determine $n_{c\infty}$ only under conditions in which the equilibrium concentration in the vicinity of the probe is constant (as an example of the corresponding diagnostic equation we shall mention the equation (11.23) at $y_D = 0$).

Calculations of the saturation current on the probes with plasma flowing around them under the regime of the viscous boundary layer with non-equilibrium volume ionisation and recombination ($d \sim \Delta$) were calculated in Ref.52, 99–103. In all these studies, with the exception of Ref.101, the authors used numerical methods; in Ref.101, calculations were carried out using the approximately linear (with respect to the concentration of charged particles) equation for the rate of recombination. As an example, Fig.11.7[52] shows the results of calculations of the density of ion saturation current at a critical point of a cooled cylindrical probe in the flow of the plasma of combustion products with a potassium addition: the broken line was calculated using the second equation (11.32), lines 1 and 2 - numerical calculations taking into account ionisation and recombination in the volume of the boundary layer (these calculations, the only type of ions present in the plasma were the ions of atoms of the addition; in particular, the quantity $n_{c\infty}$ in the second equation (11.32) for each value of the concentration of free potassium atoms in non-perturbed plasma $n_{k\infty}$ was computed using Saha's equation). Evidently, taking into account volume reactions reduces the rate of increase of saturation current with increasing concentration of the charged particles in non-perturbed plasma $n_{c\infty}$ (this concentration is proportional to $\sqrt{n_{k\infty}}$). The reasons for this effect is clear: in the case of a cooled probe in plasma with thermal ionisation the ionisation rate in the vicinity of the probe is low and the current is influenced to some extent by recombination. Therefore, with increasing $n_{c\infty}$ the recombination time decreases and the current increases slower than $n_{c\infty}$. The recombination time increases with increasing temperature and this effect weakens. This also explains the appearance of the dependence on the surface temperature of the probe T_w.

It should be noted that for a non-cooled probe an increase of $n_{c\infty}$ decreases the spatial scale of the measurement of the concentration of charged particles at a probe surface proportional to the recombination length. Consequently, the current increases more rapidly than $n_{c\infty}$ (for example, within the framework of the model corresponding to the equation (11.23) the saturation current is proportional to $n^{3/2}_{c\infty}$).

We have discussed the effect of volume reactions on the value of the ion saturation current under the conditions in which the distribution of plasma temperature in the vicinity of the probe is monotonic. The equation for the saturation current on to a wall probe on a sharp cone under the conditions in which the local Mach number is high and the temperature distribution in the boundary layer contains a maximum to which ionisation takes place was presented in Ref.6 and derived more accurately in Ref.104.

Until now, we examine the model of plasma with two types of charged particles. In cases in which there are large numbers of charged particles of more than two types, to determine the saturation current of the particles with a positive (negative) charge it is necessary to solve numerically a system of equations (11.1), (11.3) with an additional condition of quasineutrality, with the first boundary conditions (11.5), (11.6) and the boundary condition

$$y \to 0: \quad E_y = -\frac{kT_w}{e}\frac{1}{y}+... \left(E_y = \frac{kT_w}{e}\frac{1}{y}+... \right).$$

As an example illustrating the effect of the presence of negative ions on the level of the saturation current of the positive ions, the lines 3 and 4 in Fig.11.7[52] show the results of numerical calculations in which a number of negative ions, especially HCO_3^- ions, were considered in addition to the K^+ ions and electrons. It can be seen that the effect of the negative ions under these conditions is relatively strong, in particular: the range of concentration of the addition atoms in which the saturation current is proportional to $n_{c\infty}$ is considerably wider; saturation current generally increases; the dependence of saturation current on the surface temperature of the probe is greatly weakened. In Ref.5, on the basis of the analysis of the calculation results it was established that an important role in the mechanism of this effect is played by both the presence of the negative ions in the non-perturbed incident flow and by the 'sticking' of electrons to the neutral molecules in the volume of the boundary layer. It is interesting to know that the difference of the dependence j_+ $(n_{k+\infty})$, determined within the framework of this model, from a similar dependence calculated using the second equation (11.32) (with $n_{c\infty}$) replaced by $n_{k+\infty}$) is relatively small – around 3040% at $n_{k\infty} \lesssim 10^{14}$ cm^{-3}.

The authors of Ref.105 calculated numerically the ion saturation current onto a flat probe in a still reacting plasma within the framework of a model which takes into account the presence in plasma of electrons and two types of positive ions.

Previously, it was assumed that the degree of ionisation of plasma is relatively small and the frequencies of elastic collisions of charged particles with the neutral particles greatly exceed the frequency of collisions of charged particles together. However, it may be expected that the validity of a number of the previously examined diagnostic results is not limited by this assumption. In particular, taking into account ion-electron collisions, the authors of Ref.48 substantiated the first of the previously examined approaches, including (11.14). Similarly, the equation (11.18)[104] considering the ion–electron collisions is also valid (the terms of the equations of ion and electron transfer, taking into account ion–electron collisions, are shortened when deriving (11.18). Consequently, the method of calculating the ion saturation currents was in the framework of the model of plasma with two types of charged particles, based on the equations (11.25)–(11.27) is valid, together with subsequently derived diagnostic equations. It should be noted that, in some cases, the equations for the ion saturation current, obtained within framework of the previously examined model in which the function of the energy distribution of the electrons in the quasi-neutral region – Maxwell's function with the temperature of heavy particles, are approximately accurate also in the framework of the model with the 'frozen' electron temperature.[6]

11.5 Special features of experimental procedure

The procedure of the probe experiment has been sufficiently developed with respect to both measuring systems and probe design.[2-5,7,8,16] In particular, there are systems in which a computer is used for controlling the experiment, and collecting, processing and storing the results (see, for example, Ref.106, 108).

We shall examine briefly some of the problems often arising in carrying out probe experiments in high-pressure plasma. In many situations strong heat flows travel to the probe from plasma and the probe surface has a tendency to heat up rapidly during the experiment. This heating often causes a number of undesirable consequences. Firstly, it results in intensive thermal electronic emission from the probe surface which influences the ion part of the VAC (the effect of electroemission on the electronic part of the VAC is usually less marked). In interpreting the measurements it is usually difficult to take this emission into account because the work function under the conditions typical of real technical devices is not known. Secondly, insulation is less effect and this results in both an increase of the effective collecting surface of the probe and in the appearance of strong leakage currents. Finally, at relatively high probe temperature the probe fails. Therefore, if experiments are not pulsed, then in these situations it is necessary to use either cooled probes or reduce the time during which the probe stays in plasma (for example, the time-of-flight probe can be used). It should be noted that since the density of thermoemission current depends quite strongly on the surface temperature of the probe, the strong sensitivity of the ion part of the experimental VAC to the change of the surface temperature under the conditions in which thermal ionisation in the vicinity of the probe surface is insignificant, indicates that emission exerts an effect.

It is important to examine the problem of a reciprocal electrode through which the probe current is closed. Within the framework of the theory of a single probe examined in the previous sections, the distance between the probe and the return electrode is considerably greater than the size of the probe, and the probe-plasma potential difference greatly exceeds the plasma-return electrode potential difference. Under these conditions, the probe potential should be deducted from the return electrode. Evidently, the second condition can be fulfilled more efficiently if the surface area of the return electrode and the temperature of the surface is increased. It should be noted that the return electrode is represented by an electric arc, either already present in the given equipment (for example, the working arc of the plasma torch) or a specially produced arc. If diagnostics is carried out by analysing the ion part of the VAC, it is then quite easy to fulfil the second condition. In cases in which the first of the approach is discussed in the previous section is used for diagnostic, and it is difficult to fulfil the second condition, an additional (reference) electrode under a floating potential can be introduced into the plasma and the probe potential can then be deducted from the potential of this electrode. The distance from the probe to the reference electrode should be greater than the probe size.

Under the conditions typical of real technical devices the probe surface contains various contaminants. Deposition of non-conducting materials decreases the effective collecting surface of the probe and that of conducting materials increases this area and also causes leakage currents to appear. The presence of contaminants is also reflected in a change of the VAC in relation to the holding time of the probe in the plasma. It is also rational to examine the dependence of the VAC of the collecting surface of the probe. If the probe operates in the electronic part of the VAC and the saturation criterion of the VAC in the region of the negative potentials is satisfied, then the presence of strong leakage currents is indicated by the absence of any change of the slope of the VAC in transition to the ion part. It should be noted that in some cases the effect of contamination can be minimised by ion bombardment of the probe surface.[6]

Organisation of probe measurements in high-speed and high-energy flows of chemically active plasma in large-scale technical devices, such as MHD generators of electric energy,[109-112] has its specific features.

11.6 Comparison of theoretical and experimental results

Both the qualitative and quantitative comparison of the results of the hydrodynamic theory of electric probes with the experimental data has been carried out quite often in the literature (see, in particular, reviews in Ref.6, 8, 15, 16. We shall discuss briefly comparison of the theoretical results examined in section 11.4 with the experimental results.

The diagnostic method, based on equation (11.14) was optimised in Ref.48, 53, 113, 114). The parameters of plasma in which these experiments were carried out are given in the table on p.295 of this book. The shape of the VAC recorded in experiments, is in complete agreement with the shape predicted for these conditions by theory in section 11.3 (in particular, the floating potential proved to be close to the plasma potential, is also confirmed by the measurements[115-118]). The slope of the linear section of the experimental VAC in each case is consistent with the values estimated from equation (11.14).

There are a relatively large number of studies in which sections of the current of positive ions were measured and compared with the experiments, see, in particular, Ref.15, 74, 78-83, 116, 119-125.

The number of studies in which the saturation current of positive ions was measured was even larger. When analysing these measurements it is natural to use, as the basic model, the model of plasma with two types of charged particles and 'frozen' volume ionisation and recombination. As shown in section 11.4, in this model, the normalised ion saturation current on a probe of a given configuration in a subsonic flow can be regarded approximately as a universal function of the Peclet number whose form does not depend on the type of gas, parameters of non-perturbed plasma, etc. Therefore, within the framework of this model it is natural to examine, from the same position, all experimental data available in the literature on the saturation currents of positive ions on the probes of the most frequently used configurations – spherical and cylindrical. The results of this interpretation[93] are presented in

Combustion products(CP)	$n_{e\infty}$, cm^{-3}	T_∞, 10^{3}K	v_∞, m/s	Source
Air + K	10^{13}–10^{14}	2.5	300	[48]
CP C$_3$H$_8$ in air + Na, K	10^{10}–10^{12}	2.2	6	[52]
CP C$_3$H$_5$OH in air + K	$3 \times 10^{12} - 6 \times 10^3$	2.8	400	[53]
Ar	10^{17}	13	230	[90]
Air	$10^{13} - 10^{15}$	5 – 7	100	[113]
N$_2$ + O$_2$	$8 \times 10^7 - 2 \times 10^9$	0.35	0	[114]
Ar	2×10^{17}	15	130	[118]
CP C$_3$H$_{18}$ in air + Na, Cs	$3 \times 10^{10} - 10^{13}$	2.3	5	[122]
Ar	$5 \times 10^{15} - 2 \times 10^{17}$	9 – 15	80	[148]
CP C$_3$H$_8$ in air + Li, Na, K, Rb, Cs	$10^{10} - 2 \times 10^{12}$	2.2	4	[149]
Ar + K	$(1 \div 4)10^{13}$	2.3	25	[150]
CP CO in O$_2$ + N$_2$, Ar + Na, K, Cs	$9 \times 10^9 - 2 \times 10^{12}$	2.4	2	[151]
CP H$_2$ in air + Cs	$5 \times 10^{10} - 8 \times 10^{11}$	2.5	40	[152]
CP C$_3$H$_8$ in air + K	$7 \times 10^{10} - 4 \times 10^{11}$	2.1	8 – 20	[153]
CP C$_2$H$_2$ in air + Na	$(2 \div 4) \, 10^{10}$	2.4	4	[154]
CP CH$_4$ in O$_2$ + N$_2$ + K	$2 \times 10^{11} - 10^{12}$	2.2	10	[155]

Comment. 1. Experiments (with the exception of those in Ref.114) were carried out in thermal plasma at atmospheric pressure, the experiments in Ref.114 were carried out in the active zone of a nuclear reactor at a pressure from 0.5 to 2 atm. 2. CP – combustion products.

Fig.11.5, 11.6. The sources of experimental data in the plasma parameters at which all these experiments are carried out are given in the Table. The concentration of charged particles n_c used to normalise the saturation current, were either taken from the sited or calculated using Saha's equation from the normal values of the plasma temperature and the concentration of ionising atoms. The values of the densities of saturation currents are the critical points of spherical and cylindrical probes, presented in Ref.153, 52, were converted to the saturation current to the front hemisphere and the front half cylinder using the first equations (11.31) and (11.32).

It can be seen that the simple analytical equations in section 11.4 describe a large part of the experimental data with the accuracy to the factor 3 (the boundaries of the corresponding corridor are shown by dashed lines in Fig.11.5, 11.6). It should be stressed that all experimental points distributed above this corridor were obtained in the experiments with gas torches at low concentrations of charged particles (to $4 \cdot 10^{10}$ cm^{-3}) were, as shown by estimates, it is not possible to expect ionisation equilibrium in the plasma incident on the probe and the evaluation of the concentration of the charged particles using Saha's equation, used in normalising the saturation current, may prove to be too low, as is usually the case under these conditions. On the other hand, the points situated below this corridor were obtained at higher concentration values where the ratio of the recombination length to the radius of the probe or (at Pe > 1) to the radius divided by Pe are less than unity, and the saturation current is strongly influenced by the recombination in the volume of the boundary layer.

As an example of interpreting the measurements of saturation current using a more complicated model we can mention Fig.11.7.[52] It must be stressed that in the probe surface temperature range 400-800 K which also includes the experimental data shown on this Figure,[52] the ion currents are independent (within the experiment range) of the surface temperature and material. Evidently, the results of numerical calculations within the framework of the model which takes into account the effect of negative ions are consistent with the experimental data.

It should be stressed that the presence of negative ions as a probable reason for the special features of the ion saturation currents in the plasma of combustion products was mentioned in Ref.126, whereas HCO_3^- ions were linked with the effect in Ref.52 after publishing thermodynamic data.[127] Although examination of the role of negative ions in the plasma of combustion products has a very long history, sufficiently reliable experimental data on the concentration of charged particles in the temperature range in the order of 2200 K and lower which would enable an unambiguous confirmation or rejection of this conclusion, have appeared only recently – see Ref.128 in which a specially developed high-sensitivity laser interferometer was used to measure the electron concentration in the plasma of combustion products with an addition of potassium in the temperature range 2000–2500 K. At temperatures lower than approximately 2300 K this concentration was (in the case of a stoichiometric mixture) considerably lower than that calculated using Saha's equation from the measured values of the concentration of the free potassium atoms and temperature. Since the deviation from equilibrium is, according to estimates, small under these conditions, the result evidently shows the presence of a large number of negative ions, as confirmed by the conclusions made in Ref.126 and 52. Thus, electric probes in high-pressure plasma can be used not only for controlling the operation and experimental equipment but also for examining the processes in plasma.

11.7 Conclusion

Of the three diagnostic methods discussed previously and based on the linear section of the VAC, the section of the saturation current of positive ions, respectively, the simplest method is often the first one. Indeed, equation (11.14) can be used to calculate directly the value of σ_∞ from the slope of the linear section of the VAC measured in the experiments, and it is not necessary to know accurate values of the transport and kinetic plasma coefficients. Of course, if it is required to determine $n_{e\infty}$ after determining σ_∞, it is necessary to know μ_e. This method is especially efficient in inspecting the conductivity of plasma directly during the experiments.[48,53]

Within the framework of the plasma model which takes into account the types of charged particles (positive ions and electrons), the second and third method can also often be used in the form of explicit equations which link the expression for the section of the ion current or the value of the ion saturation current with $n_{c\infty}$. In particular, such equations are available for a number of situations in which ionisation recombination in the near-probe region are frozen. It should be noted that the values of the diffusion coefficients of the ions included in these equations are often known or can be estimated with the acceptable accuracy (with the error not exceeding 10%).[28] Within the framework of the model of the multicomponent chemically active plasma the calculation method is greatly complicated and requires taking into account the kinetics of reactions in the near-probe region. Information of the kinetics is often insufficient so that the possibilities of using this method for determining the concentration of charged particles in non-perturbed plasma are limited. On the other hand, the method can be used to examine the discussed kinetics.

Numerical methods are an efficient means of calculating and analysing the probe VAC for unidimensional problems.

It should be noted that in developing a method of probe diagnostics of high-pressure plasma for new conditions the formulation experiment should be as complex as possible: it is rational to carry out experiments with probes of different configuration, using different diagnostic methods, at different surface temperatures and holding time of the probe in the plasma, etc.; it is also efficient to combine the probe method with other diagnostic methods. On the one hand, this increases the reliability of interpreting the results and, on the other hand, enables additional information on the plasma parameters and properties to be obtained.

In conclusion, it is convenient to mention some studies concerned with the problems of probe diagnostics of high-pressure plasma which are outside the framework of this book (see also books in Ref.6, 8): in Refs.57, 60, 63-65, 129–133 the authors examined non-stationary probe measurements in Ref.134 measurements were taken in plasma with an applied electric field, in Ref.135–139 measurements were taken in a magnetic field, in Ref. 16, 140 in turbulent plasma, in Ref.145, 146 in a plasma with a higher degree of ionisation, and in Ref. 145, 146 measurements were taken in flames. As regards the problem of double probes, it is necessary to mention Ref.147.

Chapter 12

PROBE METHODS OF DIAGNOSTICS OF CHEMICALLY REACTING DENSE PLASMA

The probe method of plasma diagnostics is a simple and efficient means of obtaining information on the local plasma parameters. A large amount of information has been accumulated on the operation of electric probes in greatly differing situations. This information has been systematised in Ref.1, 4 and 12. The probe theory in the case of collisionless plasma where the mean free path of the electrons and ions in the plasma is considerably longer than the characteristic size of the probe has been sufficiently developed and is a reliable means of obtaining information on the properties of such plasma. The situation is complicated in transition to collisional plasma where the mean free path of charged particles is of the same order or smaller than the characteristic probe size.

In this case, the electric probe starts to disturb significantly the examined plasma in its vicinity and greatly changes the concentration of charged particles in this region in comparison with non-perturbed quasineutral plasma. This perturbation can no longer be regarded as small and, in addition, it generates the probe current measured in the experiments which carries information on the plasma parameters. The task of theory in this case is to learn how to calculate, using the given parameters of the plasma and electric probe, electric layers around the probe, electron and ion currents, the total current on the probe, and also develop approaches to solving an inverse problem, *i.e.* determine the plasma parameters from the probe current.

It should be noted that in the case of dense plasma attention is usually given to situations with frozen or non-frozen chemical reactions, in the presence or absence of negative ions or ions of different types, and moving or still plasma.[1]

Each specific case of dense plasma requires separate accurate theoretical development because the relationship of various plasma parameters with probe currents is non-linear. We shall pay special attention to the case of an electric probe placed in dense stationary plasma formed as a result of chemical ionisation processes in molecular gas mixtures.

12.1 Formulation of the problem and solution method

We shall examine low-temperature ($T = 2000 \div 3500$ K) dense still plasma (with the pressure equal to approximately atmospheric pressure) in which charged particles form as a result of reactions of chemical (associative) ionisation and they annihilate during dissociative recombination processes. These processes are the most efficient processes of formation and annihilation of charged particles in low-temperature plasma without an external ionisation source.[3,15] A cylindrical probe with radius R_p and length L is placed in plasma. The probe is under potential φ_p in relation to the earthed walls of the chamber in which the examined plasma is generated.

We shall now list assumptions which must be made when formulating the problem.

1. It is assumed that the probe problem has cylindrical symmetry. The asymptotic boundary conditions at $r^* \to \infty$ are replaced by conditions at some finite point $r^* = r_m^*$ ($r_m^* \gg 1$). The boundary conditions at $r^* = r_m^*$ correspond to the assumption according to which the plasma sufficiently far from the probe is not perturbed and there are no external electric fields and currents.

2. The effect of the cold probe surface on the thermophysical properties of plasma is ignored. The solution of the problem of the cooled probe in stationary plasma, obtained in Ref.2, taking into account the change of the transport properties, shows that in the region of negative values of the probe potential in plasma the effect of temperature perturbation is associated mainly with the thermodiffusion effect and can decrease the ion current on the probe.

3. The probe surface is assumed to be ideally catalytic, i.e the reactions of recombination of the charged particles on the surface occur at an infinite rate and surface ionisation or ionisation by an electron impact at the boundary of the Knudsen layer in the vicinity of the surface are regarded as insignificantly small. The numerical solution of the problem shows that defining the quantity n_i^* (l, t^*) $= n_{iw}^*(t^*) \neq 0$ on the surface has only a slight effect on the current at $\varphi_p^* < 0$. As shown by calculations, the probe is not heated during short examination times of the order of 100–300 µs and there is no thermoelectronic emission. The intensity of autoelectronic emission is low because of the relatively low strength of electric fields on the probe surface.

4. It is assumed that the equilibrium distribution of the velocity of all particles is established in the plasma and the temperatures of the

electrons, ions and neutral plasma or neutral particles are equal.

5. Plasma is regarded as still, and the convective transport of charged particles during the measurement period is ignored.

The equations that describe, within the framework of the electrodynamics of solid media, the distribution of the electric potential φ and the concentration of singly-charged positive ions n_i and free electrons n_e in plasma around a cylindrical probe, will be written in the dimensionless form to facilitate numerical solution of the problem[27]

$$\frac{\partial n_s^*}{\partial t^*} + \frac{\delta_s^*}{r^*}\frac{\partial(r^* j_s^*)}{\partial r^*} = W^*(t^*) - \alpha^* n_i^* n_e^*, \tag{12.1}$$

$$\frac{\varepsilon^*}{r^*}\frac{\partial}{\partial r^*}\left(r^*\frac{\partial\varphi^*}{\partial r^*}\right) = n_e^* - n_i^*, \tag{12.2}$$

$$f_s^* = -\frac{\partial n_s^*}{\partial r^*} - \text{sign}\,(e_s)n_s^*\frac{\partial\varphi^*}{\partial r^*}, \quad s=i,e,$$

$$j^* = j_i^* - \frac{\delta_c^*}{\delta_i^*}j_e^* - \frac{\varepsilon^*}{\delta_i^*}\frac{\partial^2\varphi^*}{\partial r^*\partial t^*}. \tag{12.3}$$

The dimensionless variables, denoted by the asterix, are introduced in the following form

$$r^* = r\,/\,R_p, \quad t^* = t(W_0\alpha_0)^{0.5}, \quad W^* = W\,/\,W_0, \quad \alpha^* = \alpha\,/\,\alpha_0,$$

$$\varphi^* = \frac{e\varphi}{kT}, \quad n_s^* = n_s\left(\frac{\alpha_0}{W_0}\right)^{0.5},$$

$$j_s^* = j_s\,/\,j_{sd}, \quad j_{sd} = \frac{kTb_s\,W_0^{0.5}}{\alpha^{0.5}\,R_p}, \quad j^* = j\,/\,j_{id}, \quad s=i,e, \tag{12.4}$$

here W is the effective ionisation rate that changes with time in non-stationary plasma; α is the effective recombination coefficient; W_0, α_0 are the characteristic (normalisation) values of these quantities; R_p is the probe area; T is the equilibrium plasma temperature; k is the Boltzmann constant; b_s is the coefficient of mobility of the ions and electrons; j is the density of total current recorded by the probe; j_s is the density of currents of charged components; e_s is the charge of the particles of the s-th kind. In this definition of the dimensionless parameters, three dimensionless coefficients are introduced into the system of equations (12.1)–(12.3) and

$$\varepsilon^* = \frac{\alpha_0^{0.5} kT}{4\pi e^2 W_0^{0.5} R_p^2}, \quad \delta_s^* = \frac{kTb_s}{eR_p^2 (W_0\alpha_0)^{0.5}}, \quad s=i,e. \qquad (12.5)$$

Parameters δ_s^* in the continuity equation of the components (12.1) have the meaning of the ratio of the characteristic time of development of the chemical process ($\tau_{ch} = (W_0\alpha_0)^{-0.5}$) to the characteristic diffusion time of the charged particles ($\tau_{sd} = R_p^2/D_s$, where $D_s = kTb_s/e$ is the diffusion coefficient of component s).

Parameter ε^* in Poisson's equation (12.2) represents the ratio of the square of the characteristic Debye length R_d to the square of the probe radius ($R_d^2 = kT/4\, e^2 N$, $N = W_0^{0.5}/\alpha_0^{0.5}$). It should be noted that in the examined non-stationary problem the true length of Debye's screening, determined from the instantaneous value of the concentration of charged particles non-perturbed by the probe, changes over a very wide range, because the concentration of the ions and electrons in the gas at the start of the ionisation process is low. The introduced dimensionless variables (12.4) are suitable for the numerical solution of the problem:

Initial conditions

$$t^* = 0, \quad n_s^*\left(r^*, 0\right) = 0, \quad s=i, e; \qquad (12.6)$$

boundary conditions

$$r^* = 1, \quad n_s^*\left(1, t^*\right) = 0, \quad \varphi^*\left(1, t^*\right) = \varphi_p^* = \text{const},$$
$$r^* = r_m^*, \quad \varphi^*\left(r_m^*, t^*\right) = 0, \quad \partial n_s^* / \partial r^* = 0. \qquad (12.7)$$

From the solution of the problem (12.1)–(12.7) it is necessary to find the time dependence of the total current on the probe $I_p = Sj_{id}j^*$ at the given values of the quantities W^* (t^*) and α^*. The geometrical dimensions and the probe potential are assumed to be given.

The problem was solved numerically on the basis of the method developed in Ref.2. The difference system implicit with respect to time was useful. The boundary problem for the system for the resultant difference equations was solved by the method of matrix run with iterations and variable with respect to the spatial co-ordinate with the integration step. This was carried out using a special method of defining the previous iteration based on analysis of solution at points with the highest gradients of the parameters. The accuracy of the solution was inspected using the integral of the system of equations (12.1)–(12.3) $r^*j^* = f^*j^*$ (t^*), where f^* (t^*) is some function that depends only on time and does not depend on the spatial co-ordinate.

The programme makes it possible to find the accurate solution of the problem for arbitrary functions $W^*(t^*)$ and α^*. If $W^* = $ const, $\alpha^* = $ const is considered, the solution at relatively high t^* reaches the stationary level.

As a typical example, we examine the results of solving the problem for a methane–oxygen mixture

$$0.5\% \ CH_4 + 2\% \ O_2 + 97.5\% \ Ar. \tag{12.8}$$

We examine the conditions behind a reflected shock wave at temperatures of 2000–3000 K and the atmospheric pressure.

Under these conditions, the main ionisation process is the reaction[11]

$$CH + O \rightarrow CHO^+ + e \tag{12.9}$$

The appearance of the primary ion CHO^+ causes a whole series of subsequent ion–molecular reactions, especially overcharging reactions. Several types of positive ions form in the mixture: CHO^+, H_3O, NO^+ and, possibly, some others.[3] The quantities n_i^* and $W^*(t^*)$ in the equations (12.1)–(12.3) correspond to the total concentration and the total rate of formation of ions of all types. It is assumed that the properties of the transport of ions of different types are similar.[5–7]

For the examined conditions, it can be assumed that the values of the function $W^*(t^*)$ are determined by the rate of the process of associative ionisation (12.9). The corresponding values of the rate of formation of CHO^+ positive ions or free electrons at $T = 2750$ K and the atmospheric pressure are presented in Fig.12.1 (curve 1). They were obtained as a result of solving a direct kinetic problem describing the process of chemical ionisation in this system.[3] The initial approximation for the function $W^{(0)}$ (t) in the probe problem was the rate of ionisation from the direct kinetic problem $W(t)$.

The low degree of ionisation under the conditions examined here enables us to ignore the interaction between charged particles when calculating the transport properties.

We examined the results of the calculations for the studied meth-

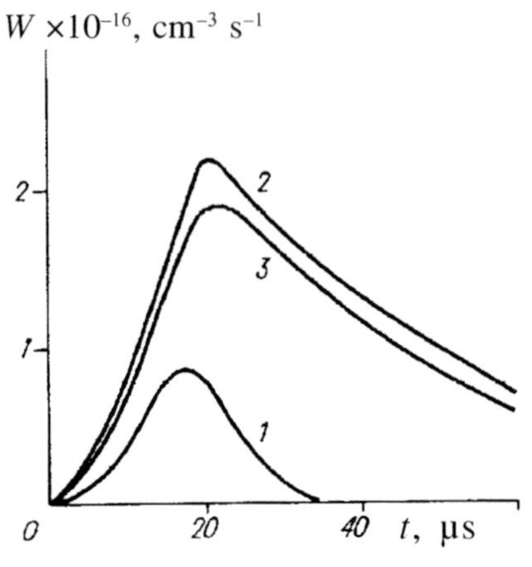

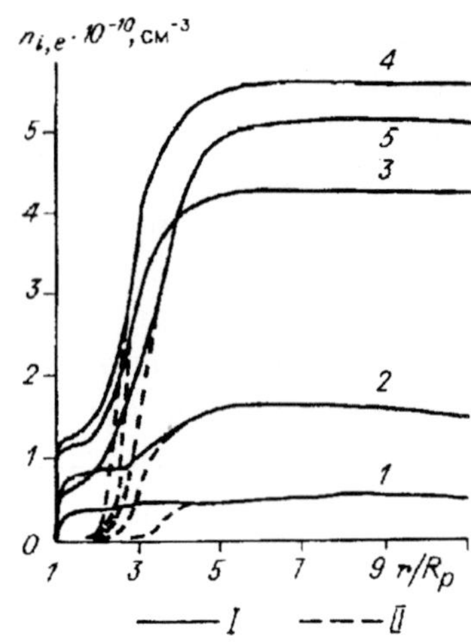

Fig.12.1 Dependence of the rate of formation of charged particles on time for a mixture of 0.5% CH_4 + 2% O_2 with argon at 2750 K and atmospheric pressure behind a reflective shockwave. 1) $W^{(0)}(t)$ from the kinetic calculation of ionisation processes,[3] 2) $W^{0l}_{(t)}$ from dependence (12,11), 3) $W^{(2)}_{(t)}$ is the corrected velocity.

Fig.12.2 (right) Space–time distribution of the concentration of positive ions (I) and free electron (II) around the probe. $t = 10$ (1), 14 (2), 20 (3), 23 (4), 39 μs (5).

ane-oxygen mixture at $T = 2750$ K and atmospheric pressure. The following data were used in the calculations: $b_i = 33.6$ cm^2/(V·s);[5,6] $b_e = 1.07 \cdot 10^4$ cm^2/(V·s);[6,8] $\alpha^* = 1$, $\alpha_0 = 6 \cdot 10^{-7}$ cm^3/s, $W_0 = 10^{16}$ cm^{-3}·s^{-1}; $\varphi_p = -9$ V, $R_p = 0.015$ cm, $S = 1.51$ cm^2.

In calculating the coefficient of mobility of the ions b_i it was assumed that the main contribution to the interaction of the ion with the neutral particle at $T = 2000 \div 3000$ K is provided by the polarisation interaction, and the section of elastic collisions was estimated using the procedure described in Ref.5. The coefficient of mobility of the electrons b_e was calculated assuming that the elastic collisions of the electrons with the argon atoms are dominant. The decrease of the section in the temperature range 2000–3000 K was considered.[8] As shown by the calculations, at the negative probe potentials the variation of quantity b_e within the limits of several orders of magnitude influences only slightly the intensity of total current on the probe.

12.2 Numerical solution results

Figure 12.2 shows the values of the concentration of positive ions and electrons in the vicinity of the probe at different moments of time. From the qualitative point, the distribution resembles the situation in stationary plasma.[1] Figure 12.2 can be used to examine the dynamics of the change of the layer of the free charge around the probe. The region of perturbation of the plasma by the probe equalling $(5 \div 6)\ R_p$ is clearly visible.

The distribution of the strength of the electric field is shown in Fig.12.3. The maximum strength of the field at the probe surface for the given parameters is of the order of 10^3 V/cm. The electric field changes to the maximum extent inside the charged layer, although it does penetrate into the region of quasi-neutral plasma.

Figure 12.4 shows the distribution of the electric potential φ for different moments of time, and the calculated total current to the probe is shown in Fig.12.5 (curve 3).

12.3 Experiments

To compare the calculated and experimental values of probe currents experiments were carried in a shock pipe designed in Ref.3. At the same time, the non-stationary concentration of the electrons was recorded using a microwave interferometer with a two-lead line as a probe system[3] and the current on the probe was determined using the methods described in Ref.13. This was achieved by combining the two-lead line with the electric probe: the conductor is made of stainless steel and parts of the waveguides to which they were connected were electrically insulated from the remainder of the microwave circuit using thin teflon gaskets which have no effect on the parameters of the microwave wave. Both wires received the same electric potential (constant with time) in relation to the earthed walls of the shock pipe. Consequently, the electric probe consisted of two wires 0.3 mm in diameter

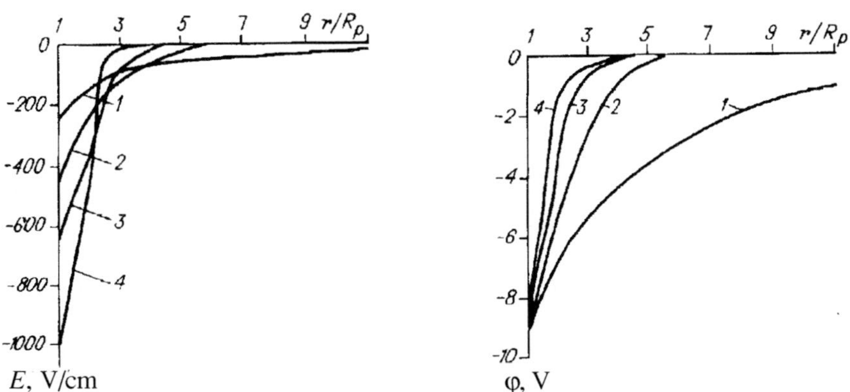

Fig.12.3 Distribution of the strength of electric field in the vicinity of the probe surface. $t = 1.44$ (1), 5.76 (2), 10 (3), 20 µs (4).

Fig.12.4 (right) Distribution of the electric potential in the vicinity of the probe surface. $t = 1.22$ (1), 5.76 (2), 10 (3), 20 µs (4).

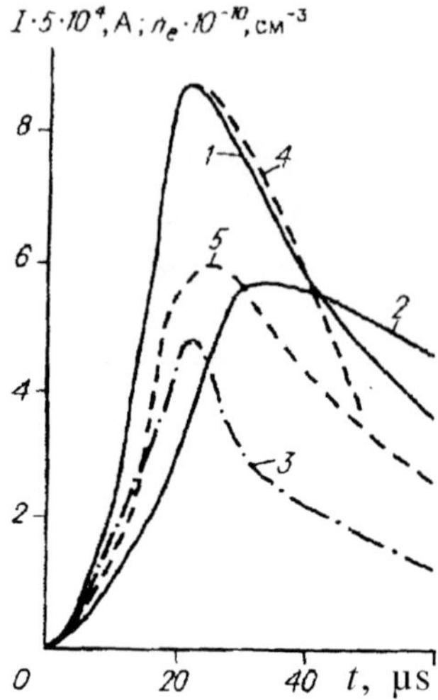

Fig.12.5 Experimental and calculated dependences of probe current and electron concentration for a mixture 0.5% CH_4+2% O_2 with argon at $T = 2750$ K and the atmospheric pressure behind the reflected shockwave. Experiments: 1) probe current $I(t)$, 2) electron concentration $n_e(t)$; calculations: 3) $I_p^{(0)}(t)$ using the initial approximation of the function $W^0(t)$, 4) $I_p^{(2)}(t)$ using the second approximation of the function $W^{(2)}(t)$, 5) concentration of electrons $n_e(t)$ with functions $W^{(2)}(t)$.

and 80 mm long separated by a distance of 4.5 mm. The wires were stretched in the direction normal to the direction of the speed of propagation of the shockwave at a distance of 10 mm from the end of the shock pipe.

The results of one of the typical experiments are presented in Fig.12.5. It can be seen that from the moment of arrival of a reflected shockwave to the probe current (curve 1) increases initially with time and then, passing through the maximum ($I_{max} = 1.76 \cdot 10^{-4}$ A at $t = 21$ µs) starts to decrease. The electron concentration (curve 2) reaches the maximum value slightly later, at $t = 30$ µs, and the decrease of the electron concentration due to the recombination process is slower than the decrease of probe current as a result of a rapid decrease of the ionisation rate.

12.4 Discussion of the results

Figure 12.5 shows that the time to establishment of the maxima of the calculated and experimentally measured currents coincides but the calculated values of the total current $I_p(t)$ in the region of the maximum, obtained using the initial approximation of the function $W^{(0)}(t)$ is approximately half the experimental values (curves 1, 3).

To understand the effect of $W(t)$ and α on the probe current, we carried out calculations with different values of these parameters. The results show that the current depends mainly on $W(t)$, and the variation of α over a relatively wide range has only a slight effect on current. The maximum current with respect to time almost completely coincides with the maxima of the ionisation rate. These special features of the effect of $W(t)$ and on the current enables the ionisation rate to be corrected with respect to the experimental current using the following procedure. From the numerical solution of the problem with the initial approximation $W^{(0)}(t)$ we can plot the dependence of the calculated current $I_p(t)$ on $W(t)$ (Fig.12.6). To consider a wider range of the variation of $W(t)$, in calculations, the function $W^{(0)}(t)$ was multiplied by the constant coefficient 20. It can be seen that there is a peculiar hysterisis loop in the dependence of $I_p(t)$ on $W(t)$, if the values of current prior to and after the maximum are plotted. This indicates that the current is also affected by the diffusion and mobility processes which change the structure of the layer of the volume charge with time. However, to a first approximation, this effect can be ignored and the dependence of current on the ionisation rate can be regarded as an unambiguous function. If we examine the situation in which the ionisation rate changes from experiment to experiment while other plasma parameters (the coefficient of diffusion and mobility, temperature, pressure, etc.) remain almost constant, the dependence $I_p(W)$ can be approximated by the relationship

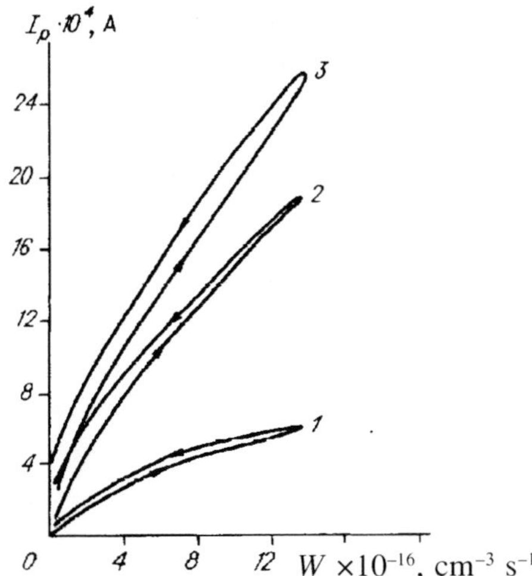

Fig.12.6 Dependence of calculated current on probe $I_p(t)$ on the actual value of the rate of formation of ions $W(t)$ for different values of probe potential. $\varphi_p = -9$ (1), -48 (2), -72V (3); the arrow on the curves indicates the direction of the passage of time.

$$I_p(t) = 1.72 \cdot 10^{-17} S \left[\varphi_p\right]^{0.6} \left[W(t)\right]^{0.75}, \qquad (12.10)$$

where I_p is in A, φ_p is the probe potential and V;W (t) is in $cm^{-3} \cdot s^{-1}$; S is the surface area in cm^2. The dependence of current on the negative potential is determined by processing the results of calculations carried out for several different values of the probe potential.

The relationship (12.10), which is valid at all times, with the exception of the very late stages of the processes of disappearance of charged particles, represents an approximate volt–ampere characteristic of the cylindrical probe for non-stationary plasma with chemical reactions. Because of the dependence of the function W (t) on time, the non-stationary volt–ampere characteristic (12.10) is a surface in space (I_p, φ_p, t). The dependence of total current on the probe on plasma temperature and the transport coefficients will be determined in examining stationary plasma with chemical reactions where dynamic equilibrium is established between the process of formation and annihilation of the charged particles.

Calculations show that in the equation (12.10) it is necessary to correct the value of the constant numerical coefficient in order to obtain the best agreement with the calculated values of current. This depends on the gas mixture examined and the specific experiment conditions. It is also possible to obtain agreement between the calculated values of current and the values calculated from equation (12.10) in the range 20%.

Converting the resultant dependence and using the experimentally measured probe current I (t) instead of I_p (t), we can determine the dependence of the ionisation rate in the system in the first approximation

$$W^{(1)}(t) = 2.25 \cdot 10^{22} S^{-4/3} \left[\varphi_p\right]^{-0.8} \left[I(t)\right]^{4/3}. \qquad (12.11)$$

The values of $W^{(1)}$ (t), obtained from the equation (12.11) for the experimental conditions shown in Fig.12.5, are presented in Fig.12.1 (curve 2). The calculations show that a small correction of the first approximation $W^{(1)}$ (t) in the region of the maximum current is sufficient to obtain agreement between I_p (t) and I (t). Curve 3 in Fig.12.1 has the corrected profile $W^{(2)}$ (t). For this function there is a good agreement between I_p (t) and I (t) (curves 1 and 4 in Fig.12.5) for almost all times, with the inspection of the late stages of the process.

For the non-perturbed region of plasma away from the probe, the equations of continuity for the probes and electrons (12.1) give

$$\partial n_{i,e} / \partial t = W(t) - \alpha n_i n_e, \quad n_i = n_e. \tag{12.12}$$

Substituting here the values of $W(t) = W[I(t)]$ we can select the value of the effective recombination coefficient α which is such that the profile of the electron concentration $n_{ep}(t)$, determined by solving this equation, is in satisfactory agreement with the experimental profile from microwave measurements (curve 2 in Fig.12.5) within the experiment error range ($\pm 20\%$). The profile obtained in this manner is shown in the same graph (curve 5). However, the recombination coefficient is several times higher than the literature values for the dissociative recombination coefficient.[14–16]

The intensity of the chemical ionisation processes depends strongly on temperature.[3] We carried out a series of experiments at different temperatures behind a reflected shockwave with the same mixture and using the same procedure with simultaneous measurement of the probe current and the concentration of free electrons at a constant probe potential, $\varphi_p = -9$ V. Using the relationship (12.11), the experimental values of the probe currents were used to determine the ionisation rate values $W^{(1)}(t)$ shown in Fig.12.7. Dashed curves correspond to the corrected values of $W^{(2)}(t)$ at which there is good agreement between $I_p(t)$ and $I(t)$. Solution of the equation (12.12) with the functions $W^{(2)}(t)$ and selection of the corresponding values of α to obtain good agreement between the calculated and experimental electron concentrations (Fig.12.8) leads to the following dependence of the effective recombination coefficient α temperature:

$$\alpha = 4 \cdot 10^{-2} T^{-1} - 9.8 \cdot 10^{-6},$$

where α is in $\text{cm}^{-3} \cdot \text{s}^{-1}$.

As previously, the value of α is too high. When the temperature increases from 2250 to 2900 K, the values of α change from $8 \cdot 10^{-6}$ to $4 \cdot 10^{-6}$ $\text{cm}^3 \cdot \text{s}^{-1}$.

This method can be used to determine the values of the effective ionisation rate $W(t)$ and the effective recombination coefficient from the probe current and electron concentration measured in the experiments. The values of α are higher than the literature data, and if the literature data for α and the ionisation rate at which the currents are in good agreement are considered, there is a difference between the calculated and measured electron concentrations. The calculated concentration is approximately twice as high as the concentration meas-

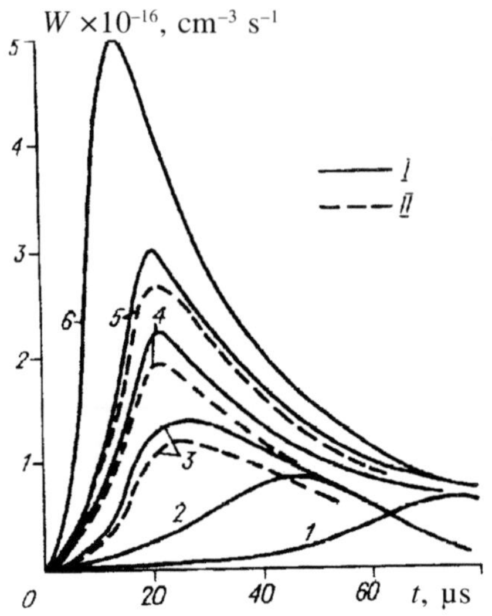
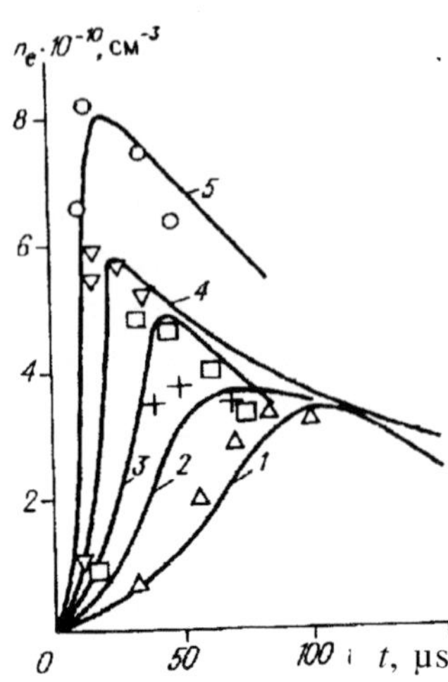

Fig.12.7 Dependences of the rate of formation of charged particles $W(t)$ on time for a mixture of 0.5% CH_4+2% O_2 with argon at different temperatures and atmospheric pressure behind the reflected shockwave. I) $W^{(1)}(t)$, obtained from (12.11); II) $W^{(2)}(t)$ is the corrected velocity; T = 2250 (1), 2440 (2), 2600 (3), 2750 (4), 2800 (5), 2900 K (6).

Fig.12.8 (right) Dependence of the concentration of free electrons on time for a mixture of 0.5% CH_4+2% O_2 with argon at different temperatures and atmospheric pressure. Lines – experiments for temperatures of: 2250 (1), 2400 (2), 2600 (3), 2600 (4), 2800 K (5); points – calculated electron concentrations.

ured in experiments in the region of the maximum. Combining these curves by decreasing the ionisation rate leads to calculated current values which are several times lower than the experimental data.[29]

Possible reasons for the differences in these values were examined theoretically: the presence of a large number of negative ions in plasma, the difference between the electron temperature and the temperature of heavy particles, the presence of convective transport of the charged particles by the gas flow, the chemical reactions of ionisation under probe surface in the diffusion layer around the surface, possibly with electronically excited particles taking part.[35]

We believe that the most likely reason for this difference are the ionisation reactions of neutral particles (possibly electronically excited) on the probe surface or in its vicinity. Further analysis is required and the model of the work of the probe in non-stationary chemically reacting plasma must be improved.

However, on the basis of the results it is already possible to propose a semi-empirical method of determining the electron concentration and the effective ionisation rate from results of probe measurements. The correcting factor must be introduced into the $W^{(1)}(t)$ calculated from equation (12.11). This factor can be evaluated by direct comparison of the chemical ionisation rates obtained in experiments simultaneously by probe and microwave measurements. Such a comparison for a methane–oxygen mixture (12.8) and the same mixture with helium in the temperature range 2150–2900 K at the atmospheric pressure behind the reflected shockwave leads to the relationship $W_{micro} = \kappa W^{(1)}(t)$, where $\kappa = 0.13$. The quantities W_{micro} and $W^{(1)}$ are found for the same moment of time corresponding to the maximum of probe current: the first – from the first time derivative of the region of the time dependence of the free electron concentration in the region of the maximum derivative, the second – from equation (12.11). Using the constant correction factor $\kappa = 0.13$ for the entire time distribution of the probe current, we obtain an approximate profile of the effective ionisation rate. The resultant solution of equation (12.12) with this profile is the calculated profile of the electron concentration n_e (t) which is in good agreement with the experimental profile $n_e^{micro}(t)$ for the literature values of the recombination coefficient.

As already mentioned, for constant ionisation rates and recombination coefficient a dynamic equilibrium is established in plasma after some time between the rates of formation and annihilation of the charged particles. Their concentrations become stationary and the probe current also ceases to depend on time.

In this case, the solution of the problem (12.1)–(12.5) is determined completely by the dimensionless parameters δ_i^*, δ_e^*, ε^* and the dimensionless probe potential (φ_p^*). Calculations were carried out in which these parameters were measured in a relatively wide range and the results were used to determine the dependence of total current on the probe on these parameters. The dependence can be described approximately as an exponential function of these parameters

$$I = 8 \cdot 10^{-10} e W_0 R_p S \left(\delta_i^*\right)^{0.66} \left(\varepsilon^*\right)^{0.12} \left|\varphi_p^*\right|^{0.4}, \tag{12.13}$$

where I is in A; S in cm^2; R_p is cm; W_0 in $cm^{-3} \cdot s^{-1}$.

As indicated by (12.13), the total current at the negative probe potentials is almost independent of parameter δ_e^* which is not present in (12.13). The resultant relationship is an analytical approximation of the general self-modelling solution of the problem of the probe in chemical reacting equilibrium plasma and is highly universal for any case of ionisation processes, with the exception of electron impact ionisation. Since the ionisation rate becomes a complex self-consistent (with other parameters of the problem) function of the distance from the probe surface, this case requires special examination.

If the expressions for the dimension parameters δ^*_i, ε^*, $|\varphi^*_p|$ are substituted into equation (12.13), we obtain the dependence of total current on the main parameters of the examined plasma

$$I = 3.367 \cdot 10^{-19} (R_p)^{0.44} L |\varphi_p|^{0.4} b_i^{0.66} W^{0.61} \alpha^{-0.27} T^{0.38}, \tag{12.14}$$

where I is in A; b_i in cm²/(W s); W in cm^{-3}·s^{-1}; φ_p in W; T in K; α in cm³ s^{-1}; R_p and L are in cm.

If it is taken into account that the equation (12.12) in the case of stationary plasma gives a simpler relationship between the ionisation rate, the recombination coefficient in the concentration of heavy particles in the plasma non-perturbed by the probe, the total current on the probe can be expressed by the equilibrium concentration of charged particles

$$I = 3.367 \cdot 10^{-19} (R_p)^{0.44} L |\varphi_p|^{0.4} b_i^{0.66} n_{i,e}^{1.22} \alpha^{0.34} T^{0.38}, \tag{12.15}$$

where n_{ie} is in cm^{-3}, other parameters as in (12.14).

If we use an independent method to determine the electron concentration in the equilibrium plasma (with a microwave interferometer or some other method), the measurements of the probe current at the known plasma temperature enable us to determine the coefficient of mobility of the ions b_i from equation (12.15).

12.5 Electric probes in special conditions

In the literature dealing with the probe methods of plasma diagnostics it has become traditional to examine the group of the following problems in the section under this name: the electric probe in the transition regime, the probe in multicomponent plasma with different types of positive ions of different masses, with positive and the negative ions, the effect of electronic emission on the probe characteristics, the probe under the conditions with the strong effect of convection in the molecular regime and in the continuum regime, the probe in a magnetic field.

In the transition regime, the mean free path of the charged particles coincides, as regards the order of magnitude, with the characteristic size of the probe. In the general formulation, the problem of the probe in this case includes a complete system of Boltzmann equations and Maxwell equations. This problem has not as yet been solved. Usually, experimental data are processed using different interpolation equations. A comprehensive list of the studies concerned with the probes in the transition regimes can be found in the reviews in Ref.1, 4 and 12.

Investigations of the behaviour of the probe in multicomponent plasma were carried out because of a number of circumstances. Firstly, if the multicomponent nature of the plasma is ignored, the physico-chemical processes, used to model the real properties of the plasma during the experiment, can be greatly distorted. Secondly, with the exception of probe methods, at present there are no reliable means of determining the concentration of negative ions under real plasma conditions. The attempts to use mass spectrometric methods for this purpose encountered principal difficulties: very rapid 'sticking' of the free electrons to the electronegative particles in the zone of the cooled sampler of the device.

In most cases, the multicomponent plasma forms in electric discharge and chemical lasers, in plasma chemical reactors in different etching processes with halides taking part, in upper layers of the Earth's atmosphere, in the products of combustion of chemical fuel, in the wake of a shockwave formed during movement of solids in the atmosphere at a high velocity. The method of producing chemically active media using the energy of free electrons is used on an increasing scale. The energy of the external electric field through the free electrons can be used for selective excitation of different quantum states of the atoms of molecules thus stimulating atomic–molecular transformations. By varying the plasma parameters, especially the concentration and electron temperature, the chemical process can be directed towards the formation of useful products. The transition from the electron to the negative ion is capable of changing the course of recombination and will lead to a loss of energy which had been previously used to form the free electrons. It is therefore important to examine the problem of ensuring sufficiently active measurements of the concentration of electrons and negative ions in the plasma where a large number of reactions of both the formation and annihilation of the electrons and different ions take place.[15,23,24]

The effect of the multicomponent nature of the plasma on the probe characteristics in the molecular regime has been examined in Ref.18, 19,12. The appearance of negative ions leads to changes in the structure of the layer of the volume charge in the vicinity of the probe, the distribution of the electric field changes and the ion and negative probe potentials increases by 30–50%.[12]

In the continuum regime for plasma with frozen chemical reactions, the problem of the probe in the mutlicomponent plasma has been examined in Ref.1, 12, 17. The main result of these investigations was the conclusion according to which the increase of the 'sticking' electrons slightly widens the zone perturbed by the probe, the current density of positive ions slightly increases, and the electronic current decreases in accordance with the decrease of the concentration of free electrons.

We shall examine in more detail the method of determining the concentration of negative ions in a chemically reacting high-pressure low-temperature plasma, in the continuum regime, on the basis of the volt-ampere characteristic of the electric probe. We shall examine the conditions under which the main processes of formation of the primary charged particles, i.e positive ions and free electrons, are the processes of chemical associative ionisation $A + B \rightarrow AB^+ + e$. The negative ions then form as a result of sticking of the electrons in $C + e + M \rightarrow C^- + M$ ternary collisions. Reversed reactions represent the processes of rapid dissociative electron-ion recombination and separation of the electrons from the negative ions. Within the framework of the examined model we can also investigate the thermal ionisation and dissipative sticking of the electrons. At high concentrations of the negative ions it is necessary to consider the ion–ion recombination under the actual plasma conditions, depending on its composition, temperature and parameters, situations can arise in which the parameter $\lambda = n^0_- / n^0_e$, characterising the ratio of the equilibrium concentrations of the negative ions and the electrons, changes in a very wide range – from zero when there are almost no negative ions, to very high values where almost all free electrons formed transform to negative ions.

In this system of reactions the rates of variation of the concentration of charged particles in plasma ω_s can be described using five effective parameters:

$$\omega_e = W + \gamma n_- - \alpha n_+ n_e - \kappa n_e,$$
$$\omega_{i+} = W - \beta n_+ n_- - \alpha n_+ n_e,$$
$$\omega_{i-} = \kappa n_e - \gamma n_- - \beta n_+ n_-.$$

(12.16)

Here W is the effective rate of the formation of the electrons and positive ions in the associative ionisation processes; α is the electron-ion recombination coefficient, κ is the effective frequency of sticking of the electrons to the electronegative components, γ is the effective frequency of separation of the electrons, β is the ion–ion recombination coefficient, n_e, n_{i+}, n_{i-} is the concentration of the electrons, positive and negative ions. The values W, γ, κ depend on the concentration n_s of the negative particles.

$$W = k_1 n_a n_b, \quad \kappa = k_2 n_c n_M, \quad \gamma = k_3 n_M.$$

(12.17)

Here k_1, k_2, k_3 are the constants of the rates of associative ionisation, three-particle sticking and separation of the electrons. It is assumed that in (12.17) the concentrations of the neutral particles are constant in the vicinity of the probe. The effective parameters W and κ can also describe thermal ionisation in the collisions of neutral particles and dissociative sticking of the electrons $CD + e \rightarrow C^- + D$. k_1 is the constant of the rate of thermal ionisation, and parameter κ is written in the form $\kappa = k^*_2 n_{CD}$, where k^*_2 is the constant of the rate of dissociative sticking.

The equations describing the distribution of the densities of charged particles and the electron potential φ^* in the vicinity of the cylindrical probe in the approximation of the electrodynamics of the continuum in the dimensionless variable[27] have the form

$$\frac{\partial n_s^*}{\partial t^*} + \frac{\delta_s^*}{r^*}\frac{\partial}{\partial r^*}(r^* j_s^*) = \omega_s, \quad s = e, i+, i-,$$

$$\frac{\varepsilon^*}{r^*}\frac{\partial}{\partial r^*}\left(r^*\frac{\partial \varphi^*}{\partial r^*}\right) = n_e^* + n_{i+}^* - n_{i-}^*,$$

$$j_s^* = -\frac{\partial n_s^*}{\partial r^*} - \text{sign}(e_s)n_s^*\frac{\partial \varphi^*}{\partial r^*}, \tag{12.18}$$

$$j^* = j_{i+}^* - \frac{\delta_{i-}^*}{\delta_{i+}^*}j_{i-}^* - \frac{\delta_c^*}{\delta_{i+}^*}j_e^*.$$

Here j_s^*, j^* are the densities of the currents of the components and the total current. When writing the equations of continuity of the components the convective transport of current to the probe is ignored. The left-hand parts of the equation (12.18) include for dimensionless parameters.

$$\varepsilon^* = (R_d / R_p)^2, \quad \delta_s^* = \tau / \tau_{ds}, \quad s = e, i+, i-,$$

$$R_d^2 = kT / (4\pi e^2 n_0), \quad \tau_{ds} = R_p^2 / D_s. \tag{12.19}$$

Here R_d is the characteristic Debye length, τ_{ds} is the characteristic diffusion time of component s. The right-hand parts of the equations of continuity of the components in the variables (12.4) have the form

$$\omega_e^* = 1 + \gamma\tau n_{i-}^* - n_e^*(n_{i+}^* + \kappa\tau),$$

$$\omega_{i+}^* = 1 - n_{i+}^*\left(n_e^* + \beta n_{i-}^* / \alpha\right), \qquad (12.20)$$

$$\omega_{i-}^* = \tau\kappa n_e^* - n_{i-}^*(\gamma\tau + \beta n_{i+}^* / \alpha).$$

The boundary conditions for the equation (12.18):

$$r^* = 1, \quad n_s^* = 0, \quad \varphi_p^* = \varphi^*, \quad s = e, i+, i-,$$

$$r^* = r_m^*, \quad \varphi^* = 0, \quad \partial n_s^* / \partial r^* = 0, \quad r_m^* \gg 1. \qquad (12.21)$$

The initial conditions for the equation (12.18):

$$t^* = 0, \quad n_s^*(r^*, 0) = 0, \quad s = e, i+, i-. \qquad (12.22)$$

The problem (12.18)–(12.22) was solved numerically on the basis of the method proposed in Ref.2. The main assumptions, made in formulating the problem and also the special features of solving the boundary

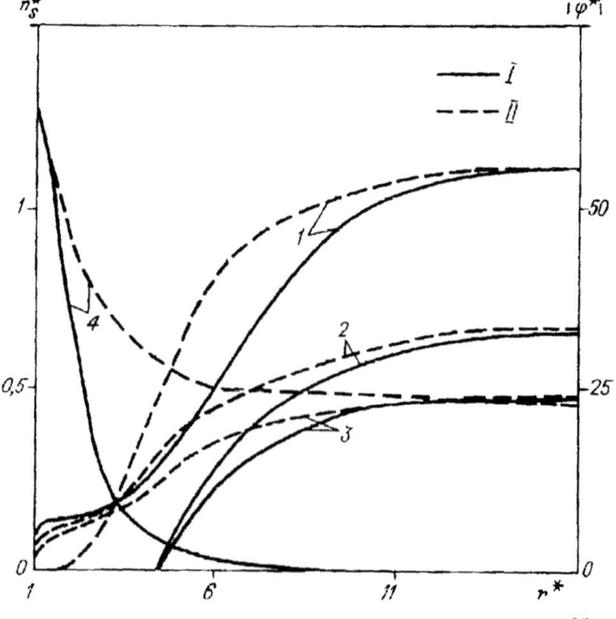

Fig.12.9 Distribution of the concentration of charged particles and electric potential for $\varphi_p = -9$ (I), $+ 9$ V (II). 1) n_{i+}^*, 2) n_e^*, 3) n_{i-}^*, 4) $|\varphi^*|$.

problem for the system of the resultant different equations, were discussed previously in this book and described in Ref.2 and 22.

The distribution of the concentrations of the charged particles and the electric potential in the vicinity of the probe for one of the variance of calculations are presented in Fig.12.9. Calculations were carried out for the following variants of the parameters: $\varepsilon^* = 2.7 \cdot 10^{-2}$, $\delta_{i+} = \delta_{i-} = 6.7$, $\delta_e = 6.7 \cdot 10^3$, where $W = 10^{13}$ cm$^{-3} \cdot$s^{-1}, $\alpha = 10^{-7}$ cm$^3 \cdot$s^{-1}, $\beta = 5 \cdot 10^{-8}$ cm$^3 \cdot$s^{-1}, $\gamma = \kappa = 1.5 \times 10^3$ s^{-1}, the ratio of the equilibrium concentration of the negative ions and electrons $\lambda = 0.73$. These data simulate the situation in air with an addition of CCl_4 vapours at a temperature of $T = 2200$ K and the atmospheric pressure.

For these values of the parameters, the size of the region of electric perturbation of the plasma by the probe under the positive potential extends over a large distance, over more than 100 R_p. In the case of the negative potential of the probe, the main decrease of the electric potential takes place on the length $\approx 5 R_p$. Under the examined conditions, the dimensions of the layer of the volume charge equal $(2-4)R_p$, and the size of the entire region where the concentration of the charged particles greatly differs from the non-perturbed values is $10R_p$. It should be noted that the analytical and numerical solutions of the probe problem, available in the literature, were obtained mainly for the limiting cases where the size of the layer of the volume charge is either small in comparison with the radius of the probe or, on the other hand, is considerably greater.

The problem (12.18)–(12.22) contains five parameters characterising the ionisation processes, and three transport coefficients – the effective coefficients of mobility b_{i+}, b_{i-}, b_e. For the conditions examined below, $b_e >> b_{i-}$, b_i and the conditions b_{i+}, b_{i-} are of the same order of magnitude. It is also assumed that b_{i+}, $b_{i-} = b$.

The numerical solution of the problem (12.18)–(12.22) enables the distribution of the quantities n_e, n_{i+}, n_{i-}, φ in the region perturbed by the plasma to be determined at the given probe potential and calculate the current on the probe for each set of the given parameters, i.e solve a direct problem. However, our main task is to solve the inverse problem, i.e to develop a method of determining the plasma parameters on the basis of experimental data for the current on the probe. We shall examine the possibilities of fulfilling this task, as previously, by constructing analytical approximation dependences of the probe current on the main plasma parameters.

Under the chemical equilibrium conditions for each specific initial gas mixture the definition of temperature and pressure determines the values of the kinetic parameters W, α, β, γ and κ, and also the values of the coefficients of mobility of the charged particles. The kinetic parameters are linked by the chemical equilibrium conditions which include the equilibrium concentrations of the neutral components formed as a result of a large number of chemical transformations. In the experiments, the concentrations of the neutral components, taking part in the formation of charged particles in accordance with (12.17), can vary over a relatively wide range by varying the initial composition of the mixture and the experimental condition as regards temperature and pressure. Analysis of the literature data makes it possible to indicate the ranges of the values of the kinetic parameters and the transport coefficients which in principle can occur under the experiment conditions.

The values of the effective coefficients, describing the annihilation of the charged particles (the coefficient of electron–ion recombination α and ion–ion recombination β) when the temperature varies from 1000 to 3500 K are in the range[15,24,25]

$$10^{-8} \leq \alpha \leq 10^{-6},$$
$$5 \cdot 10^{-9} \leq \beta \leq 5 \cdot 10^{-7}, \qquad (12.23)$$

where α and β are in $cm^3 \cdot s^{-1}$.

The parameter W, which describes the processes of chemical and thermal ionisation, has the values in the range[15,26]

$$1.5 \cdot 10^{1} \leq \kappa \leq 1.5 \cdot 10^{8},$$
$$1.5 \leq \gamma \leq 1.5 \cdot 10^{4}, \qquad (12.24)$$

where W is in $cm^{-3} \cdot s^{-1}$.

The parameters κ and γ which characterise the rates of 'sticking' and separation of the electrons depend on the concentration of the electronically active particles present in the plasma. To examine the effect of the negative ions on the volt–ampere characteristic of the probe, we examine the following ranges of the values of these parameters

$$1.5 \cdot 10^{1} \leq \kappa \leq 1.5 \cdot 10^{8},$$
$$1.5 \leq \gamma \leq 1.5 \cdot 10^{4}, \qquad (12.25)$$

where κ and γ is in s^{-1}.

The corresponding change of the parameter λ is:

$$2 \cdot 10^{-3} \leq \lambda \leq 2 \cdot 10^{4},$$

$$\lambda = n_{i_-}^{0} / n_{e}^{0}. \tag{12.26}$$

The values of the coefficients of mobility of the ions and electrons for the gas mixtures at the pressure of the order of the atmospheric pressure are in the range[7]

$$1 \leq b \leq 100,$$

$$10^{3} \leq b_{c} \leq 10^{5}, \tag{12.27}$$

where b and b_{e} are in $cm^{2}/(V\ s)$.

At the negative probe potentials where $|\varphi^{*}_{p}| >> 1$, the change of the coefficients of mobility of the electrons has almost no effect on the probe current.

These parameters correspond to a relatively wide spectrum of possible experiment conditions and, consequently, the equations derived below can be used for the diagnostics of a large number of partially ionised media which are of interest for practice.

We shall examine the solution of the problem (12.18)–(12.22) at negative probe potentials. A large number of numerical calculations were carried out for different combinations of the parameters including the problem, which correspond to possible experiment conditions. We shall use the results obtained in kinetic modelling the processes of chemical ionisation in complex molecular systems, such as mixtures of oxygen with hydrocarbons methane and acetylene,[3] air with different additions,[27], *etc.*

Analysis shows that the results of numerical solution of the problem of a probe with the accuracy of 20% in relation to accurate numerical calculations can be approximated by the correlation equation:

$$I_{-} = 5.5 \cdot 10^{-19} C \cdot R_{p}^{0.44} L W^{0.68} b^{0.6} T^{0.34} \left| \varphi_{p} \right|^{0.55},$$

$$C = \alpha^{-0.07} \kappa^{0.018} \gamma^{0.004} \beta^{-0.02} b_{e}^{0.02}. \tag{12.28}$$

where I_{-} is the total current on the probe at the negative potential, A; R_{p}, L is the radius and length of the probe; cm; T is the plasma temperature, K; φ_{p} is the probe potential, V.[20]

Under the examined conditions, the layer of the volume charge in the vicinity of the probe is quite long. Chemical ionisation reactions take place inside this layer; positive ions and electrons form from neutral components in these reactions. The electric field displaces the electrons from the layer thus preventing the reactions of electron–ion recombination and electron sticking. These circumstances result in a relatively strong dependence of I_- on the ionisation rate and a very weak dependence on the parameters α, β, γ, κ, which control the formation of negative ions.

At the negative probe potential in the plasma with no negative ions, the probe current is considerably stronger than the current of the same negative potential.[1] However, if the plasma does not contain free electrons, and the transport properties of the positive and negative ions are similar, the volt–ampere characteristic of the probe is almost completely symmetric. Therefore, it may be expected that the maximum sensitivity of the probe current to the presence of the negative ions must be in the region of the positive probe potentials.

Analysis of the results of numerical calculations in this region shows that the dependence of the total current I_+ at $\varphi_p^* \gg 1$ on the kinetic parameters W, α, β, and temperature T is almost the same as in equation (12.28). However, the relationship of I_+ with the mobility coefficients b, b_e and with the effective rate of electron sticking is more complicated. The dependence on these parameters changes with the change of the ratio of equilibrium values of the concentration of negative ions and electrons λ. Therefore, the parameter $\lambda = n_{i-}^0 / n_{e-}^0$ will be used as the controlling parameter. In our case of equilibrium plasma, λ is related with the kinetic parameters, included in (12.16), by algebraic relationships.

The results showing that there is a small difference in the dependences of the currents I_+, I_- on W, α, β, T indicate that it is rational to use the dimensionless quantity $\Psi = I_+(|\varphi_p|)/I_-(-|\varphi_p|)$ which is almost independent of these parameters. On the basis of numerical calculations for W we can write the following equation

$$\Psi = f_1(\lambda)\eta^{f_2(\lambda)}, \quad \eta = b_e / b. \tag{12.29}$$

Functions f_1, f_2 are shown in Fig.12.10. It can be seen that higher sensitivity of the ratio of the currents to the concentration of the negative ions is at $\lambda \geq 1$. In the region where $\lambda \gg 1$, the functions $f_1 \to 1$, $f \to 0$ so that the ratio of the currents $\Psi \to 1$, and the volt–ampere characteristic of the probe is close to symmetric. When $\lambda \ll 1$, the ratio of the currents becomes

$$\Psi = b_e / b. \tag{12.30}$$

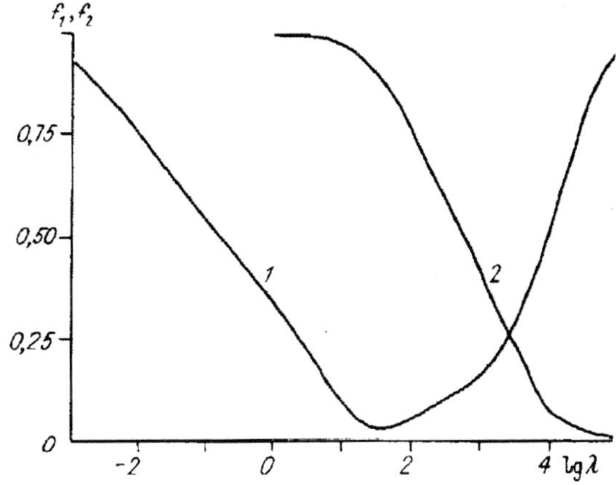

Fig.12.10 Dependences $f_1 = f_1(\log \lambda)$ (1), $f_2 = f_2(\log \lambda)$ (2).

Consequently, in reacting plasma with a small number of negative ions the ratio of the currents measured at the positive and the same negative potential $(e|\varphi_p|/kT \gg 1)$, can be used to determine the ratio of the mobility coefficient b_e/b.

Equation (12.30) was derived assuming that at $\varphi_p > 0$ the probe induces slight electric perturbations in the plasma, i.e there is a counter probe of a relatively large area at some distance from the probe. At the known value of b_e/b the relationship (12.30) makes it possible to inspect the accuracy of probe measurements at the positive probe potential.

We shall now return to equation (12.29). It gives the dependence of the ratio of the probe currents at the same absolute probe potential in relation to the plasma on two parameters: the ratio of the equi-

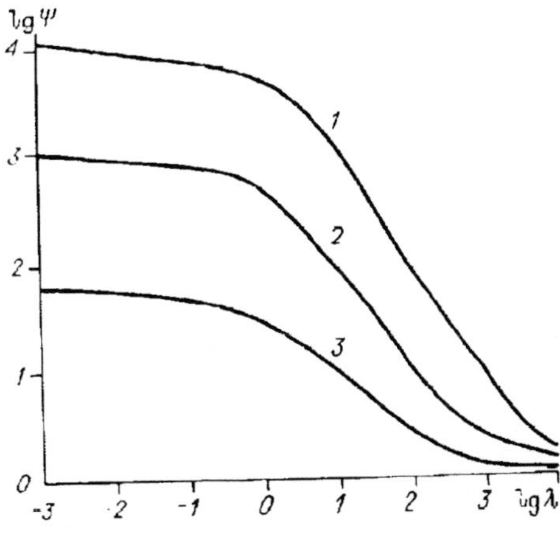

Fig.12.11 Sections of the surface log Ψ (log λ, log η) by planes η = const. η = 10^4 (1), 10^3 (2), $\eta = 10^2$ (3).

librium concentrations of the negative ions and electrons λ and the ratio of the coefficients of mobility of electrons and ions η. Figures 12.11 and 12.12 shows the cross section of the surface $\log \Psi$ ($\log \lambda$, $\log \eta$) by planes $\lambda = $ const, $\eta = $ const. Using the functions f_1, f_2, shown in Fig.12.10, it is easy to construct a graph of the function $\log \Psi$ ($\log \lambda$) also for the intermediate values η. We can propose the following method of processing the probe measurements in the plasma with negative ions.

We assume that the ratio of the mobility coefficients b_e/b is known. Using the dependence of $\log \Psi$ or $\log \lambda$ as shown in Fig.12.11, from the ratio of the currents $\Psi_1 = I_+/I_-$ measured in the experiments we determine the value $\log \lambda$ which is the abscissa of the point of intersection of the $\log \Psi = \log \Psi_1$ with a curve $\log \Psi$ ($\log \lambda$) which corresponds to the value $\eta = \eta_1$. The determined value of λ gives the ratio of the concentrations of the negative ions and electrons in the examined plasma.

Dashed lines in Fig.12.12 show the values of α, corresponding to the cases in which the mobility coefficients are determined by the collision sections of the charged particles with the He and Ar atoms. It can be seen that depending on the type of gas (diluent) $\log (\Psi)$ changes relatively appreciably.

When adding a relatively small amount of an electronic active addition to the plasma, the effective ionisation rate W and the recombination coefficient α remain almost unchanged. Equation (12.28) shows that the probe current for the negative potentials is almost constant, although the equilibrium value of the concentration of positive ions slowly

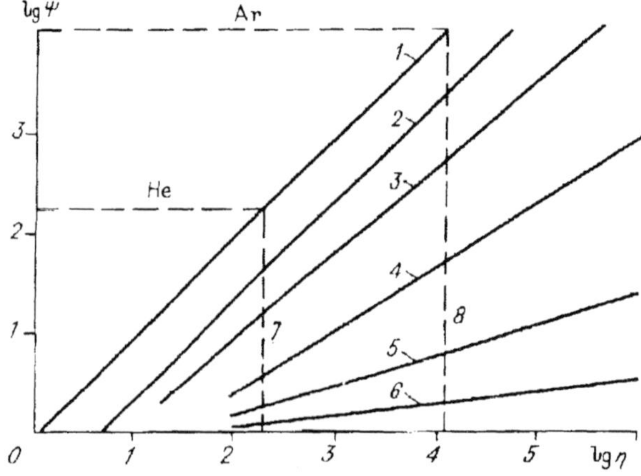

Fig.12.12 Sections of the surface $\log \Psi$ ($\log \lambda$, $\log \eta$) by the planes $\lambda = $ const., $\lambda = 10^{-27}$ (1), 1 (2), 10 (3), 10^2 (4), 10^3 (5), 10^4 (6), 7) η for helium, 8) η for argon.

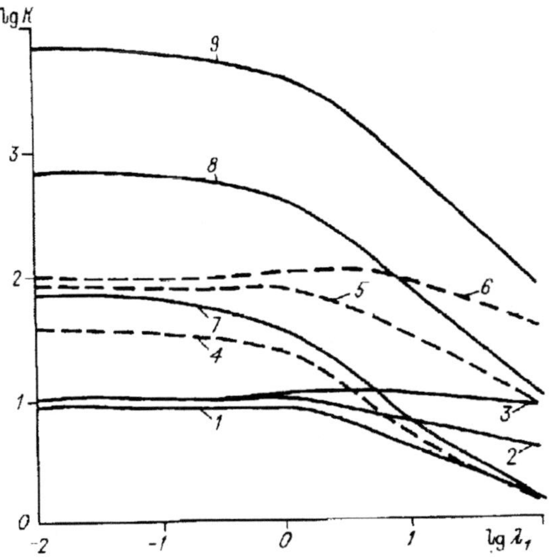

Fig.12.13 Dependence of log K on log λ_1. 1) $\eta = 10^2\,(\theta =10)$, 2) $\eta = 10^3(\theta = 10)$, 3) $\eta = 10^4\,(\theta =10)$, 4) $\eta = 10^2(\theta = 10^2)$, 5) $\eta = 10^3(\theta =10)$, 6) $\eta =10^4(\theta = 10^2)$, 7) $\eta = 10^4\,(\theta =10^2)$, 8) $\eta = 10^4\,(\theta =10^3)$, 9) $\eta = 10^4(\theta =10^4)$.

increases as a result of the shift of the chemical equilibrium in the formation of negative ions. Calculations carried out using the examined reaction system show that the variation of the equilibrium concentration of the positive ions does not exceed 50% of the entire range (12.25) of the values of κ.

It is evident that the formation of negative ions leads to a change in the electron concentration. We introduce parameters $\theta = n_{e1}/n_{e2}$, $K = \Psi_1/\Psi_2$, where n_{ej} is the equilibrium electron concentration, Ψ_j is the ratio of the probe currents at the same absolute value of the probe potential in the initial plasma ($j = 1$) and in the plasma with the electron negative addition ($j = 2$). Since at $\varphi_p < 0$ the current remains almost unchanged, parameter K is equal to the ratio of the currents of the positive potential

$$K = I_1(+\varphi_p)/I_2(+\varphi_p).\qquad(12.31)$$

Figure 12.13 shows the dependences of log K on log λ_1 for different values of the parameters θ and η, where λ_1 is the ratio of equilibrium concentration of the negative ions and electrons in the initial plasma.

Figure 12.13 shows that the dependence of log K on the parameter $\eta = b_e/b$ becomes stronger when the magnitude of the change of the electron concentration increases, i.e. with increasing parameter θ. At

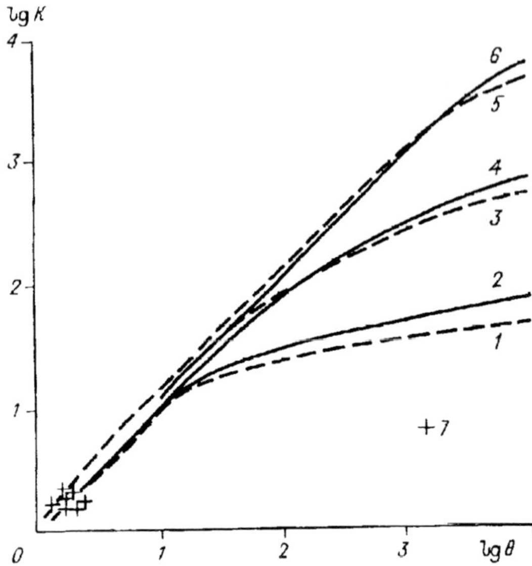

Fig.12.14 Dependence of log K on log θ. 1) $\eta = 10^2$ (λ_1), 2) $\eta = 10^2$ $(\lambda_1=10^{-2})$, 3) $\eta = 10^3$ $(\lambda_1=1)$, 4) $\eta = 10^3$ $(\lambda_1 = 10^{-2})$, 5) $\eta = 10^4$ $(\lambda_1 = 1)$, 6) $\eta = 10^4$ $(\lambda_1 = 10^{-2})$, 7) experiments.

$\theta = 10$ the dependence of log K on η up to the value log $\lambda_1 = 0$ is very weak and it can be assumed that there is a linear relationship between K and θ, as clearly indicated by Fig.12.14. The values of log K change only slightly with the variation of λ_1 in the range log $\lambda_1 <$ 0.5. This is due to a very significant contribution of the electrons to the current at a negative probe potential. Consequently, in the region where $\lambda_1 < 0.3$, the probe measurement at the positive potential can be used to determine the relative variation of the equilibrium concentration of the electrons after adding electron-absorbing components to the plasma. When log λ_1 increases the dependence of K on λ_1 shows a non-linearity which is determined by the increase of the contribution of the negative ions to the current formation process. At higher values of λ_1, i.e high initial concentration of the negative ions in the plasma, the contribution of the negative ions to the current becomes controlling and the value of K tends to unity. As indicated by Fig.12.13, at log $\lambda_1 \geq 1$ the dependence of log K on θ and η is relatively complicated and to determine θ from the value of log K, it is necessary to know the value η. At log $\lambda_1 \geq 2.5$ log K tends to zero and the sensitivity of K to the parameters θ and η rapidly decreases. Therefore, at very high initial concentrations of negative ions in plasma it is not possible to determine the change of the concentration of electrons using probe measurements at a constant probe potential.

Figure 12.14 shows the dependence of log K on log θ for different values of the parameters η and log λ_1. The value of K is directly proportional to the parameter θ up to the values $\theta = 10$ and only slightly

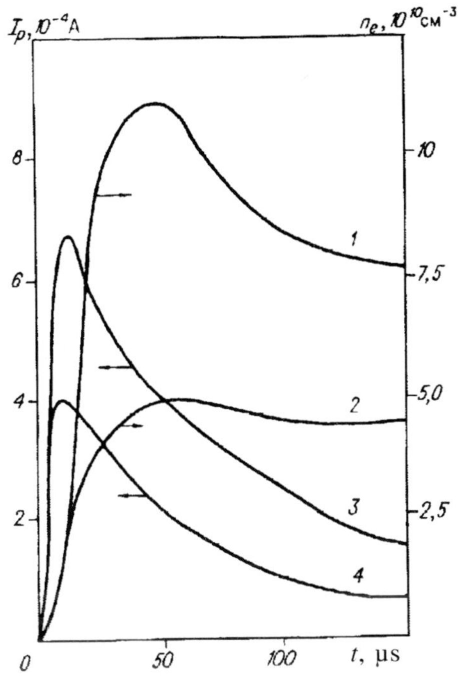

Fig.12.15 Experiment to determine dependence of the concentration of electrons n_{ep} and probe current on time. 1,2) $n_{ep}(t)$ for mixtures I and II, respectively; 3,4) (t) mixutres I and II.

sensitive to the change of η. Consequently, probe measurements at $\theta < 10$ can be used to determine the variation of the electron concentration when the electronegative component is added, even if there is no information on parameter η.

Experimental verification of the calculated dependences was carried out on a shock pipe in low-temperature plasma formed behind the front of the reflected shock wave. Simultaneously, probe currents on the concentration of free electrons were measured in the same cross section of the shock pipe using a microwave interferometer with a high spatial resolution. Experimental equipment and procedure were described in Ref.22. An acetylene–oxygen mixture 0.27% C_2H_2 + 1.4% O_2 + 98.3% Ar were used. In this mixture, the process of chemical ionisation, ensuring a relatively high electron concentration (of the order of 10^{10}–10^{11} cm^{-3}) occurred at the relatively low temperatures of the order of 2000–2500 K and pressures of the order of the atmospheric pressure. The electromagnetic addition was carbon tetrachloride CCl_4.

Figure 12.15 shows the dependences of the electron concentration and probe currents, measured with the microwave interferometer, at φ_p = 9 V, on time in the mixture without (I) and with the addition of CCl_4 (II). Temperature T = 2500 K, pressure − atmospheric.

The results of processing a series of experiments at temperature of 2200–2600 K and the atmospheric pressure using the method described previously in the mixtures I and II are presented in Fig.12.14. The maximum deviation of the experimental points from the calculated dependence is around 30%.[21]

12.6 Diagnostics of dense chemically reacting still plasma with electric probes with a modulated potential

A single measurement in plasma diagnostics using an electric probe with a constant potential can provide information on the amplitude characteristics of the probe signal associated with the plasma parameters. In modulation of the probe potential by specific pulses, in addition to the amplitude characteristics of the signal we obtain the phase characteristics, especially the phase shift between the current signal and the probe potential. The function of dependences of the probe method widen in this case.

The method of pulsed probe characteristics is used widely for examining different non-stationary plasma formations. At a high formation content, at high operating speed and simple calibration, this method is characterised by the appearance of surges of probe current at pulsed probe excitation. These surges greatly exceed the stationary value. The effort of experimentators is directed mainly to developing measuring systems in which information on the current to the probe during the transition process was excluded in order to eliminate the effect of non-stationary current surges. For example, in Ref.28 each consecutive change of the pulsed voltage on the probe was carried out only after completing the transient process caused by a previous change. However, this resulted in a loss of valuable information on the plasma parameters and required relatively complicated measuring systems. We shall now examine how to utilise the 'interfering' surge of probe current in pulsed probing for determining the plasma parameters.[34]

We shall examine the case of dense still plasma in which chemical reactions of ionisation and recombination take place. The law of variation of the probe potential is shown in Fig.12.16. The dynamic properties of the probe–plasma system will be examined in the stationary section of the probe current when the total current to the probe becomes constant and the concentration of charged particles in the plasma region not perturbed by the probe is also constant under the conditions of establishment of equilibrium of the ionisation and recombination rates.

The equations determining the distribution of electric potential and the concentration of singly-charged positive ions and electrons in the vicinity of the cylindrical probe can be written, on the basis of the electrodynamics of solids, in the dimensionless form similar to the equa-

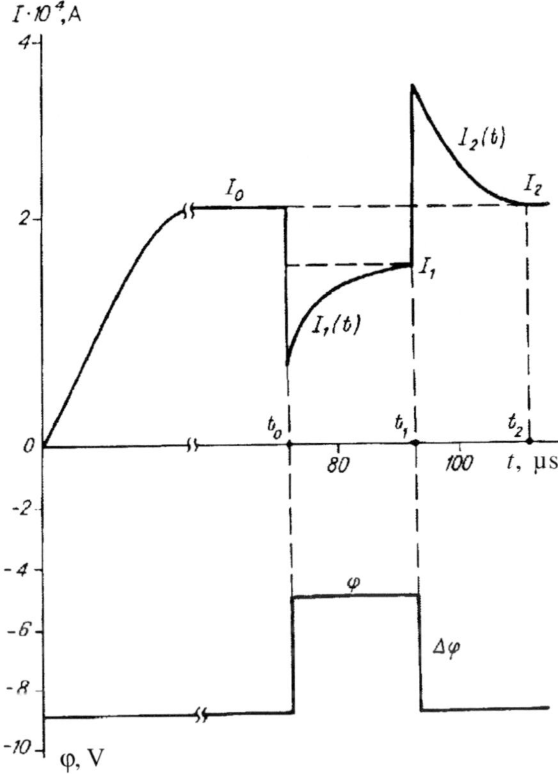

Fig.12.16 Calculated dependence of the probe current on time for the pulsed variation of probe potential. $W = 10^{16}$ cm^{-3} s^{-1}, $n_{1,e} = 4.25 \times 10^{11}$ cm^{-3}.

tions (12.1)–(12.5). The dimensionless variables are introduced as in the case of equation (12.4). The initial conditions remain in the form (12.6). The boundary conditions will be written in the form:

$$r^* = 1, \quad n_s^*(1, t^*) = 0, \quad s = i, e,$$

$$\varphi^*(1, t^*) = \varphi_p^* = \text{const}, \quad t^* < t_0^*,$$

$$\varphi^*(1, t^*) = \varphi_p^* + \Delta\varphi^*, \quad t_0^* \leq t^* < t_1^*,$$

$$\varphi(1, t^*) = \varphi_p^*, \quad t^* \geq t_1^*,$$

$$r^* = r_m^*, \quad \varphi^*(r_m^*, t^*) = 0, \quad \partial n_s^* / \partial r^* = 0, \quad s = i, e. \tag{12.32}$$

The moments of time t^*_0 and t^*_1 correspond to the time of supplying a pulse to the probe which changes the probe potential by the value $\pm\Delta\varphi^*$. Assumptions made in formulating the problem, and details of this numerical solution are identical with those discussed in the problem (12.1)–(12.7).

We shall examine the results of the numerical solution of the problem for plasma formed behind a reflected shock wave in air at $T = 3200$ K and the atmospheric pressure. The main ionisation process under these conditions is the reaction

$$N + O \rightarrow NO^+ + e^-. \tag{12.33}$$

The primary ion NO^+ is dominant.[30] Therefore, when calculating the coefficient of mobility of the ions we use the sections of elastic collisions of the NO^+ ion with N_2, O_2 molecules calculated taking into account the polarisation interaction of the particles by the method described in Ref.5. The coefficient of mobility of the electrons is determined using the data on the sections of elastic collisions of the electrons with nitrogen and oxygen molecules published in Ref.32. In the numerical solution of the problem of the probe, the value of b_i was varied to explain the nature of the dependence of the current on the probe. The following data were used in the calculations: $b_i = 38.6$ cm²/(V·s); $b_e = 2\cdot10^4$ cm²/(V·s); $\alpha^* = 1$; $\alpha_0 = 5\cdot10^{-8}$ cm³/s;[31] $W_0 = 10^{16}$ cm⁻³·s⁻¹; $R_p = 0.015$ cm; $S = 1.51$ cm²; $\varphi^0_p = -9$ V; $\varphi_p = -5$ V. The variants with $\varphi_p = -11$; -12.8 V were also considered.

Figure 12.16 shows the results of numerical calculations of the time dependence of probe current for a stepped change of the negative potential of the probe in relation to the earthed walls of the shock pipe.

For quantitative analysis of the calculation results of transient currents it was convenient to transfer to a new dimensionless variable:

$$\Psi_1 = \frac{I_1^c - I_1(t)}{eW_0R_pS}, \quad \Psi_2 = \frac{I_2^2 - I_2(t)}{eW_0R_pS}, \quad \xi = \frac{\tau_{ch}}{t - t_{imp}} = \frac{1}{\Delta t(W_0\alpha_0)^{1/2}}, \tag{12.34}$$

where $I_1(t)$ and $I_2(t)$ are the instantaneous values of current; t_{imp} are the moments of variation of the potential (t_0 or t_1, respectively); $\Delta t = t - t_{imp}$.

Figure 12.17 shows, in the form of dependences $\Psi_1(\xi)$ and $\Psi_2(\xi)$, the results of calculations of the transient currents $I_1(t)$ and $I_2(t)$ on the probe for different values of the equilibrium concentrations of the charged particles of the plasma under the stationary regime conditions. The concentration values varied by changing the ionisation rate at a constant recombination coefficient.

The numerical solution of the problem (12.1)–(12.32) enables us to

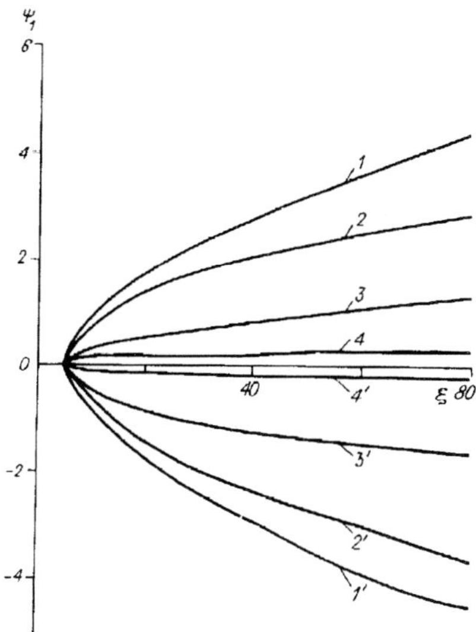

Fig.12.17 Dependence of Ψ_1 and Ψ_2 on ξ.

determine the dependence of the current of the probe on the parameters which control the state of the plasma. For this purpose, we carried out calculations in which the values of W, α, δ_i, $|\Delta\varphi|$ were varied at different values of the initial probe potential. In all variants the value $|\Delta\varphi|$ was taken as lower than $|\Delta\varphi^0_p|$.

The non-stationary current in the transient sections $t_0 \leq t < t_1$ and $t_1 \leq t < t_2$, excluding a short period of time of the order of 0.5 μs where bias currents play a significant role, depends mainly on the concentration of the charged particles, non-perturbed by the probe in the quasi-neutral region of the plasma and on the mobility coefficient of the ions. Separating the dependences on the varied parameters, we obtain the approximation relationships

$$\Psi_1 = 0{,}025\delta_i^{0.5} n^* \left|\Delta\varphi^*\right|^{0{,}35} F(\xi), \tag{12.35}$$

$$\Psi_2 = 0.0375\delta_i^{0.5} n^* \left|\Delta\varphi^*\right|^{0.35} f(\xi), \tag{12.36}$$

The functions $F(\xi)$ and $f(\xi)$ are presented in Fig.12.18. They depend only slightly on the parameters mentioned previously, and can be regarded as constant for a set of dependences 1–4 and 1′–4′, shown in Fig.12.17 and corresponding to a wide range of the variation of the parameters n^* and δ_i.

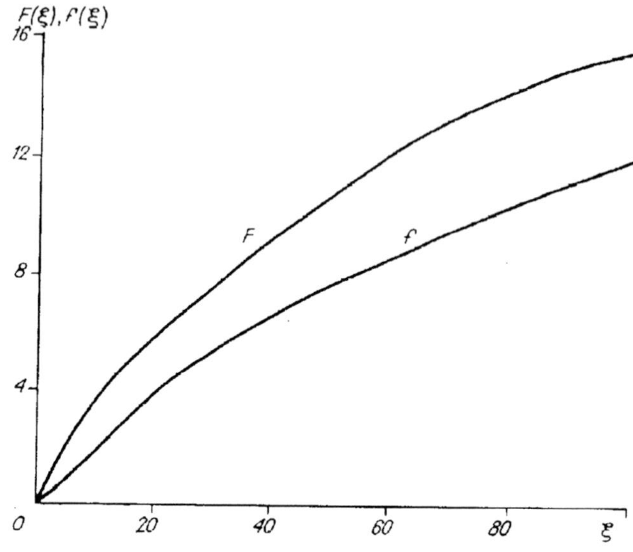

Fig.12.18 Dependence of F and f on ξ.

The relationships (12.35) and (12.36) include the main plasma parameters: W, α, b_i. These relationships can be used as diagnostic equations for determining any plasma parameters both directly and together with relationships for stationary probe currents. In this case, we already have two independent equations and can determine two unknown plasma parameters.

The function $F(\xi)$ and $f(\xi)$ can be approximated by analytical relationships with the accuracy sufficient for measurements in practice. Consequently, we obtained the following dimensionless dependences on the plasma parameters for transition currents:

$$\Delta I_1 = 1.11 \cdot 10^{-20} R_p L W^{0.75} \alpha^{-0.25} b_i^{0.5} T^{0.15} |\Delta\varphi|^{0.35} \left(\frac{1}{(W\alpha)^{0.25}(t - t_1)^{0.5}} - 1 \right),$$

$$(12.37)$$

$$\Delta I_2 = 1.39 \cdot 10^{-20} R_p L W^{0.75} \alpha^{-0.25} b_i^{0.5} T^{0.15} |\Delta\varphi|^{0.35} \left(\frac{1}{(W\alpha)^{0.25}(t - t_2)^{0.5}} - 2 \right).$$

$$(12.38)$$

Here R_p, L are the radius and length of the cylindrical probe, cm; b_i is the coefficient of mobility of the ions, cm^2/(V s); T is the equilibrium plasma temperature, K; $\Delta\varphi$ is the amplitude of variation of the probe potential, V; W is the ionisation rate, cm$^3\cdot$s^{-1}; α is the recombination factor, $cm^{-3} \cdot s^{-1}$. The method of determining ΔI_i and $t - t_i$ is shown in Fig.12.19.

The plasma temperature and the coefficient of mobility of the ions will be regarded as known. Consequently, from the equations (12.37) and (12.38) we can derive equations for determining the ionisation rate and the effective recombination coefficient

$$W = \frac{\Delta I_i^0}{(i) A_i (\Delta t_i^0)^{0.5}}, \quad \alpha = \frac{A_i}{(i)^3 \Delta I_i^0 (\Delta t_i^0)^{1.5}}, \tag{12.39}$$

$A_i = B_i R_p L b_i^{0.5} T^{0.15} |\Delta\varphi|^{0.35}$, $B_1 = 1.11 \cdot 10^{-20}$, $B_2 = 2.78 \cdot 10^{-20}$, $i = 1.2$.

The method of determining the values ΔI_i^0 and Δt_i^0 is shown in Fig.12.19. It can be seen that in the co-ordinates $|I_i - I_i(t)|$ $1/(t - t_i)^{0.5}$ the time dependences of the transient currents are straightened out. The points of intersection of these straight lines with the ordinates and abscissa give the values of ΔI_i^0 and Δt_i^0, respectively. Index 1 relates to the case in which the pulse potential reduces the absolute value of the probe potential, and the index 2 corresponds to the opposite case. These cases are completely equivalent when determining W and α and either of them can be used.

Experimental verification of the resultant relationships was carried

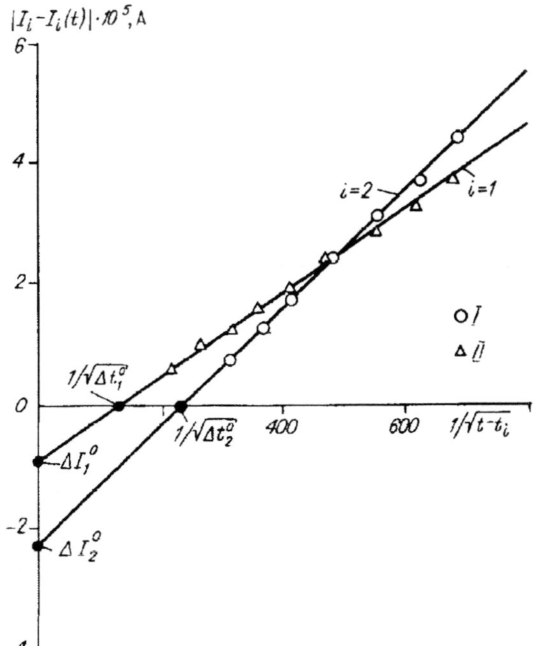

Fig.12.19 Dependence of transition currents on time for determining ΔI_i^0 and Δt_i^0. I, II relate to the currents in Fig.12.16.

out on a shock pipe in a series of experiments with air in the temperature range $T = 2700 \div 3200$ K and at the atmospheric pressure behind the reflected shock wave. The gas parameters were calculated from the measured velocity of the incident shock wave taking into account the chemical reactions of dissociation and ionisation using the method proposed in Ref.33. the following conclusions can be drawn as a result of comparing the calculated and experimental results.

1. The concentrations of charged particles, obtained on the basis of the transient currents with the pulsed change of the probe potential, are in satisfactory agreement with the concentrations determined from the results of probe measurements at a constant probe potential in the section of the stationary concentrations of charged particles in the plasma.

2. Determination of the recombination factor of the charged particles using the two methods gives values close to the literature data.

3. The pulsed probe can be used to determine the coefficient of mobility of the positive ions if the recombination factor of the charged particle is available.

4. The concentration of charged particles, obtained from the probe measurements, is 1.5–2 times higher than the concentration of electrons measured using the microwave interferometer.

We shall examine the method of diagnostics of chemically reacting plasma by using an electric probe working in the regime of potential modulation by high-frequency sinusoidal voltage.

The methods available at present are based on using approximate analytical solutions of the problem of probe impedance.[36–39] This problem is complicated by the need to calculate the propagation of weak perturbations caused by the potential modulation, through the highly heterogeneous state of the medium in the vicinity of the probe. Therefore, the procedure of determining the analytical solution is accompanied by a large number of simplifying assumptions whose validity for specific experimental conditions cannot always be checked. The analytical solution derived as a result which link the impudence characteristics with the plasma parameters are often complicated and cumbersome and cannot in many cases be resolved in the explicit form in relation to the required plasma parameters. It is possible to solve only a direct problem – obtain agreement between the theoretical and experimental data on impedance by substituting, into analytical relationships, appropriately selected values of the parameters of the medium examined in the experiment.[39]

The solution of the inverse problem, i.e determination of the plasma parameters from the impedance measured in the experiments, requires further considerable simplifications which can greatly restrict the range of applicability of the final relationships.

We shall examine plasma consisting of one type of positive ions and free electrons. The rate of formation of the charged particles ω is described in this model by two effective parameters $\omega = W - \alpha\, n_i n_e$, where W is the effective rate of chemical ionisation, α is the coefficient of electron–ion recombination. The values of these parameters give the equilibrium quasineutral density of the charged particles.

The system of equations which determines the distribution of the densities of charged particles and electric potential φ in the approximation of the solid medium, is formulated in the region of the cylindrical probe in the same manner as the equations (12.2)–(12.5).

The boundary conditions are written in the form

$$r^* = 1, \quad n_i^*(1, t^*) = n_e^*(1, t^*) = 0,$$

$$\varphi^*(1, t^*) = \varphi_p^* + \Delta\varphi^* \sin(2\pi f^* t^*),$$

$$\varphi_p^*, \quad \Delta\varphi^* = \mathrm{const}, \quad \Delta\varphi^* > 0, \tag{12.40}$$

$$r^* = r_m^*(r_m^* \gg 1), \quad \partial n_s^*/\partial r^* = 0, \quad s = i, e, \quad \varphi^*(r_m^*, t^*) = 0.$$

Here $\Delta\varphi^*$ is the amplitude of the variable component of probe potentials; φ_p^* is the constant component of the potential; $f^* = f\tau_{\mathrm{ch}}$, f is the modulation frequency of the potential. We shall analyse the case of negative potentials $\varphi_p^* < 0$, the absence of convective transport of charged particles, and thermal diffusion. Inertia terms in the relationships for the current density of charged components are not investigated. The latter circumstance restricts the modulation frequency of the probe potential: $f \ll \nu_i$, where ν_i is the frequency of collisions of ions with the neutral particles in plasma. The ionisation rate and the recombination factor are constant with time.

The problem was solved numerically using the difference method implicit with respect to time. The solution of the non-linear system of different equations in each time layer was determined using iterations in relation to the modulation frequency of the potential. Usually, the ratio of the time step to the period of potential change was not greater than $5 \cdot 10^{-3}$. It was thus possible to determine the phase shift of current on the probe with the error not greater $\pm 2°$.

The numerical solution gives the distribution of the quantities n_i^*, n_e^*, φ^* in the vicinity of the probe at every moment of time. These distributions were used to calculate the value j^* and the time dependence of the total dimension current on the probe $I = 2\pi R_p L j_{id} J^*$. The calculations show that under conditions in which $|\Delta\varphi^*/\varphi^*_p| < 0.2$, the function $j^*(t^*)$ can be approximated with sufficient accuracy using their relationship

$$j^*(t^*) = j^*_0 + \Delta j^* \sin(2\pi f^* t^* + \Psi), \quad \Delta j^* > 0, \tag{12.41}$$

where j^*_0 is the stationary value of the dimensionless density of the current on the probe at a constant probe potential φ^*_p, Ψ is the phase shift between the current and the potential; Δj^* is the amplitude of the variable component of current density.

We shall examine the results of the numerical solution of the problem obtained in the values of the parameters: $b_i = 38.6$ cm²/(V s); $b_e = 2\cdot10^4$ cm²(V s); $\alpha = 1.5\cdot10^{-8}$ cm³ s⁻¹; $W = 10^{16}$ cm⁻³ s⁻¹; $n_{i,e} = 4.47\cdot10^{11}$ cm⁻³; $R_p = 1.5\cdot10^{-2}$ cm; $\varphi = -9$ V; $\Delta\varphi = 1$ V; $f = 10^5$ Hz. These parameters correspond to air plasma at $T = 3000$ K and atmospheric pressure. It should be noted that for these conditions the simplifying conditions at which the authors of Ref.37–39 obtained the approximate solutions of the impedance problem are not fulfilled.

In accordance with equation (12.3), the total current density recorded by the high-frequency probe consists of the conduction current (the first two terms in the right-hand part of (12.3)) and the bias current. In the numerical solution it is possible to find separately each component and analyse its frequency dependence. With increasing modulation frequency the phase difference between current and potential tends to 90° because the layer of the volume charge resulting in the appearance of an effective capacitance in the probe circuit does not manage to react to the potential change. The amplitude of the oscillations of bias current increases and that of the conduction current decreases.

The aim of solving the impedance problem is to obtain relationships for determining the plasma parameters from the current amplitudes (measured at different frequencies) on the probe ΔI and the phase shift Ψ. The equations (12.1)–(12.4), (12.40) show that the values of Ψ and Δj^* (amplitude Δj^* is proportional to the measured amplitude ΔI) are determined completely by the values of the dimensionless parameters δ^*_i, δ^*_e, ε^*, φ^*_p, $\Delta\varphi^*$, f^*.

The numerical solutions enables us to find Δj^* and Ψ for each set of the values for these parameters. The values of φ^*_p, $\Delta\varphi^*$ at the known plasma temperature are given, other quantities depend on the plasma properties and the modulation frequency of the potential f. It is possible to estimate the range of variation of these parameters by examining the permissible range of the values of W, α, b_i, b_e for different gas mixtures at the atmospheric pressure and a temperature of 2000–3500 K. For the negative probe potential the variation of δ_e in the range $5\cdot10^2 - 1\cdot10^4$ has only a very slight effect on the values of j_0^*, Δj^*, and therefore, calculation of δ_e from the data on the value of Δj^* is incorrect. The range of variation of other parameters will be restricted on the basis of the estimates by the following inequalities:

$$0.1 \leq W / W_0 \leq 50; \quad 1 \leq \alpha / \alpha_0 \leq 100; \quad 0.1 \leq b_i / b_0 \leq 10; \quad 2 \leq T / T_0 \leq 4;$$

$$W_0 = 10^{16} \text{ cm}^{-3} \cdot \text{s}^{-1}; \quad \alpha_0 = 5\cdot10^{-8} \text{ cm}^3 \cdot \text{s}^{-1};$$

$$b_0 = 40 \text{ cm}^2 / (W \cdot \text{s}); \quad T_0 = 10^3 \text{ K}.$$

(12.42)

These inequalities determine the relatively wide range of the values of the effective parameters W, α, b_i and enable the following relationship to be used for a large number of gas mixture compositions which are interesting for practice.

The given values of φ_p, $\Delta\varphi$, f were varied in the following ranges: -45 V $\leq \varphi_p \leq -1$ V; 0.5 V $\leq \Delta\varphi \leq 5$ V; 10^5 Hz $\leq f \leq 10^7$ Hz. In calculations, the value of $\Delta\varphi$ for every value of φ_p was selected in such a manner as to fulfil the relationship $|\Delta\varphi/\varphi_p| < 0.2$. The dimensionless quantities ε^*, δ_i^* varied in the range $10^{-4} < \varepsilon < 2\cdot10^{-2}$; $0.1 < \delta_i < 30$.

Analysis of the results of numerical calculations, the amplitude of the variable component of the current density on the probe and the phase shift between the current and the potential can be written in the form

$$\Delta j^* = \left[(\Delta j_p^*)^2 + (\Delta j_c^*)^2\right]^{1/2},$$

$$\Psi = \arctan (\Delta j_c^* / \Delta j_p^*),$$

(12.43)

where Δj_c^*, Δj_p^* are the amplitudes of the density of bias current and a variable component of the conduction current density, respectively. In the examined range of the variation of the parameters, the results of the numerical solution can be approximated by the following relationships

$$\Delta j_p^* = 0.967(\delta_i^*)^{-0.2}(\varepsilon^*)^{-0.06}\left|\varphi_p^*\right|^{0.9}(f^*)^{-0.1},$$

$$\Delta j_c^* = 8.12(\delta_i^*)^{-1.04}(\varepsilon^*)^{0.74}\left|\varphi_p^*\right|^{-0.4}\left|\Delta\varphi^*\right|(f^*).$$

(12.44)

The resultant equations (12.43) and (12.44) can be used to determine the effective rate of ionisation W and the coefficient of mobility of the ions b_i on the basis of the results of measuring the amplitude of the variable component of current on the probe and the phase shift between the current and the potential.

We shall write the relationships (12.44) in a dimensionless form

$$\Delta I_p = 2\pi R j_{id} \Delta j_p^* = A_p f^{-0.1},$$

$$\Delta I_c = 2\pi R_p L j_{id} \Delta j_c^* = A_c f,$$

$$A_p = 8.74 \cdot 10^4 R_p^{0.52} L \Delta\varphi^{0.9} b_i^{0.8} W^{0.68} T^{0.44} \left|\varphi_p\right|^{-0.6} \alpha^{-0.38},$$

$$A_c = 2.615 \cdot 10^{-6} R_p L \Delta\varphi b_i^{-0.04} W^{0.15} T^{0.1} \left|\varphi_p\right|^{-0.4} \alpha^{-0.11}.$$

$$(12.45)$$

Here ΔI_p, ΔI_c are the amplitudes of the total conduction current and of the bias current on the probe.

Solving the equations (12.45) in relation to the quantities b_i and W, we obtain

$$b_i = C_b \Delta I_p \Delta I_c^{-4.62},$$

$$W = C_W \Delta I_p^{0.27} R_p^{2.24} L^{3.6} \Delta\varphi^{3.7} f^{4.72} \left|\varphi_p\right|^{-1.24} \alpha^{-0.12},$$

$$C_W = 9.5 \cdot 10^{28} R_p^{-3.4} L^{-5.7} f^{-5.4} \left|\varphi_p\right|^{2.33} \alpha^{0.7} T^{-0.66} \Delta\varphi^{-5.67}.$$

$$(12.46)$$

The values of ΔI_p, ΔI_c are determined from the amplitude of the variable component of total current on the probe (measured at a given frequency) ΔI and the phase shift between the current and the potential Ψ using the relationships

$$\Delta I_p = \Delta I \cos \Psi; \quad \Delta I_c = \Delta I \sin \Psi. \qquad (12.47)$$

When the measurement of the phase shift at frequency f is difficult, the values of ΔI_p, ΔI_c can be obtained from the values of ΔI_1 and ΔI_2 – the amplitudes of the total current measured at two different frequencies f_1 and f_2. Consequently, to determine A_i and A_c we obtain

$$A_p = D(\Delta I_1^2 \cdot f_2^2 - \Delta I_2^2 \cdot f_1^2)^{1/2}, \quad f_1 < f_2,$$

$$A_c = D(\Delta I_2^2 \cdot f_1^{-0.2} - \Delta I_1^2 \cdot f_2^{-0.2})^{1/2},$$

$$D = (f_2^2 \cdot f_1^{-0.2} - f_1^2 \cdot f_2^{-0.2})^{-1/2}. \tag{12.48}$$

The values of ΔI_p and ΔI_c are calculated from the resultant values of A_p and A_c for the corresponding frequency. The equations (12.45) - (12.48) can be used to estimate the error in determining the values of b_i, W at a known error of measuring the current amplitudes at each of the frequencies. For example, if the error of measuring the current is 10%, the error of determining b_i and W is approximately 100% of the frequency of $f = 10^6$ Hz.[40]

Thus, equations (12.46) at a known temperature and the recombination factor can be used to calculate the coefficient of mobility of the ions and the effective rate of ionisation in equilibrium plasma. For this purpose, it is necessary to measure, at the selected frequency, the amplitude of the variable component of the current on the probe and the phase shift between the current and the probe potential or measure the current amplitude at two different frequencies. The unperturbed equilibrium concentration of the electrons is then determined from the relationship $n_0 = (W/\alpha)^{1/2}$ assuming the mechanism of chemical ionisation and dissociative recombination.

Chapter 13

ELECTRIC AND THERMAL PROBES IN THE PRESENCE OF CHEMICAL REACTIONS IN NON-EQUILIBRIUM PLASMA

13.1 Electric probe

Recently, non-equilibrium electric discharges at reduced pressures have been used on an increasing scale in the technology of depositing and removing (etching) thin layers on the surface (from the surface) of solids.[1-3] Layers can be dielectric, semiconducting or conducting, and have optical, diffusion and other properties interesting from the applied viewpoint.

It was therefore necessary to carry out diagnostics of the electronic and ionic components of the plasma under the conditions when dielectric or other coatings grow on the probe surface or when the dimensions of the probe decrease as a result of etching its surface layers. Similar changes can also occur in structural elements of the probe device (current conducting lines, insulation, etc.). Consequently, deviations occur from the normal operating regime of the probe, i.e disruption of some important postulates of the theory of the probe method. In order to resolve these contradictions, it was proposed to use the electric probe also for the diagnostics of the process of growth of the dielectric polymer layers[4,5] and then for diagnostics of the kinetics of growth and removal (etching) of layers of different nature.[6-10,25-27] At present, the electric probe in the film-forming or etching plasma is used for the diagnostics of electronic and ionic components of the plasma and also for examining the kinetics of the layers on the surface of the probe/substrate.

In the first case, theoretical fundamentals of the method, developed for plasma diagnostics without chemical reactions, remain unchanged. However, during measurements it is necessary to take measures to clean the surface to remove dielectric deposits, and take into account the

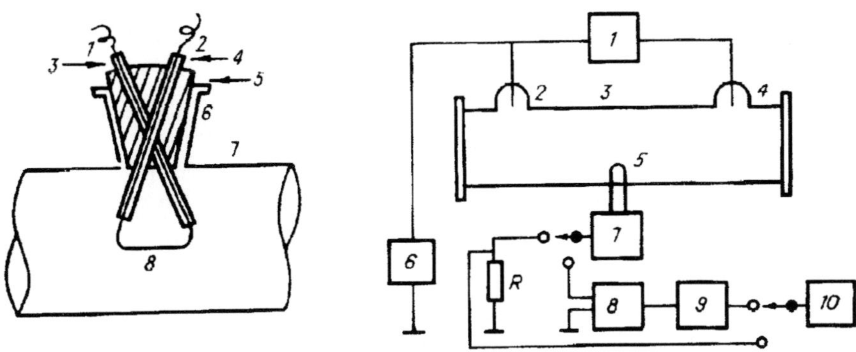

Fig.13.1 The probe-wire device. 1,2) connected conductors; 3,4) insulating capillaries; 5) conical plug; 6) conical ground joint; 7) discharge tube; 8) working part of the probe.
Fig.13.2 (right) Block diagram of the probe circuit for measuring U_p, r_{pf} and VAC of the probe.[8,10] 1) power source for the discharge; 2,4) electrodes; 3) discharge tube; 5) probe; 6) bias voltage source of the probe; 7) unit for setting and measuring probe temperature; 8) current stabiliser; 9) unit for matching the stabiliser with the automatic recording device and setting I_p; 10) automatic recording device.

change of the dimensions of the probe as a result of deposition on its surface of conducting or semiconducting layers or as a result of etching the probe material.

In discharges in hydrocarbons, the probe can be efficiently cleaned to remove the polymer film grown on it during rapid heating of the probe to a temperature of $T_p \simeq 900 \div 1000$ K by a current from a special current source. The probe suitable for this operation has two lead-outs (Fig.13.1). The diagram of the electric circuit of the probe, used in dc glow discharge,[8,10] is shown in Fig.13.2. After rapid heating, the volt-ampere characteristic or its derivatives with respect to voltage are measured in the pulsed regime during the period in which the probe surface remains clean. Measurements with a heated probe can also be taken. The temperature of the probe is regulated by the filament current and should be selected on the level at which no deposits form on the probe surface. For discharging hydrocarbons, the temperature should be around 800–900 K.

The discharges in the vapours of fluorine- and chlorine-containing compounds are used to etch metallic and semiconductor materials.[3] Under these conditions, it is possible to etch metals, including noble ones, which are often used to produce electric probes. The etching rate of noble metals is usually $r_{etch} \lesssim 0.1$ nm/s, and the radius of the probe-wire $r_p \simeq 10$ μm.

The variation of the dimensions of the probe made of gold or platinum remains unchanged over an operating period of 1 h. This estimate requires verification by experiments.

Thus, the occurrence of the processes of deposition of the layers on the probe or etching of the probe material complicates the application of the probe method of diagnostics of charged particles in practice. After all, we have mentioned here only some partial cases of the processes of etching and depositing the layers. Their number rapidly increases.[1,3,9]

The kinetics of growth and etching of the layers on the surface of the solids under the conditions of electric discharges at reduced pressure has been examined using the methods of optical interferometry, microscopic, weighing methods, etc. Each method has its advantages and disadvantages.[11] The electric probe method[8,10] is suitable for examining the level of charged plasma particles in heterogeneous stages of the process.[8–10,25] It can also be used to examine some electrophysical properties of the dielectric layer during their growth.[6,8,10,12,25–27]

We shall examine the principle of this method on an example of a dc discharge in hydrocarbons, where the polymer film with the dielectric properties grows on the surface of the probe/substrate, and the probe has the form of a thin wire with a diameter of $2r_p \simeq 10 \div 20$ µm and a length of $l_p \simeq 1$ cm (Fig.13.1). In this case, the method has the following characteristics mentioned in Ref. 7,8,10,12–15,25.

1) high sensitivity, the capacity to record changes of film thickness of several nanometers;

2) reliable recording of the first monolayers of the film on the probe-substrate;

3) deposition of a film at the given densities of electron fluxes W_e and ion fluxes W_i of the plasma on its surface, where the range of specified values of W_e and W_i is more than 10 orders of magnitude;

4) localisation of measurements determined by the probe diameter;

5) negligible (at the given probe dimensions) thermal and geometrical perturbations of the plasma in the vicinity of the probe;

6) definition of the temperature of the probe/substrate T_p in a wide range $T_{melt} \gtrsim T_p \gtrsim T_{0p}$, where T_m is the melting point of the probe material, T_{0p} is the probe temperature in the plasma with preheating current switched off (T_{0p} can differ from the gas temperature of plasma);[8,10]

7) it is convenient to combine in the same structural device of the electric probe for investigating the kinetics of film growth, the Langmuir–Mottsmith probe for diagnostics of charged plasma particles and a microcalorimeter for examining the thermal effects on the surface of the probe/substrate (in the latter case, the probe temperature is measured on the basis of its electrical resistance using, for example, the device shown in Fig.13.3);

8) modification of the method for high-frequency discharge,[9,25–27] for diagnostics, by this method, of the etching processes,[9] for using, as the probe, flat substrates[9,25–27] non-conducting substrates, etc.[8]

Figure 13.4 shows a set of the VAC of the probe/wire in a dc glow discharge in a mixture of neon with methane at different moments of growth of the polymer film (PF) on its surface. It is assumed that the thickness and all other properties of the PF are the same over the entire working surface of the probe (the probe–plasma contact surface), that the plasma is homogeneous in the vicinity of probe and that the probe is placed in the equipotential plane of the electric field in plasma. We shall make a section through the set of the VAC by the straight line I_p = const, where I_p is the probe current. The same current also flows through the PF, thus $I_p = I_{PF}$. The transition from one VAC to another at I_p = const is possible at an appropriate change of the voltage U of the external source (see Fig.13.2) $|\Delta U| = I_p R_{PF}$, where R_{PF} is electric resistance of the PF at a given moment of time. In other words, with the growth of the PF the densities of the currents W_e and W_i of the charged particle from the plasma on the external (facing the plasma) surface of the PF remain unchanged

$$I_p = W_e + W_i = f\,(\varphi_{pl} - \varphi_{PF}) = \text{const},\qquad(13.1)$$

is the difference of the potentials of this surface φ_{PF} and the plasma φ_{pl} is constant. It is evident that

$$(\varphi_{PF} - \varphi_{pl}) = \text{const at } |\Delta U| = I_p R_{PF}.\qquad(13.2)$$

Of course, all other the quantities characterising the probe, the plasma and the film are constant with time. The distribution function of the plasma electrons with respect to energy is arbitrary, but not necessarily

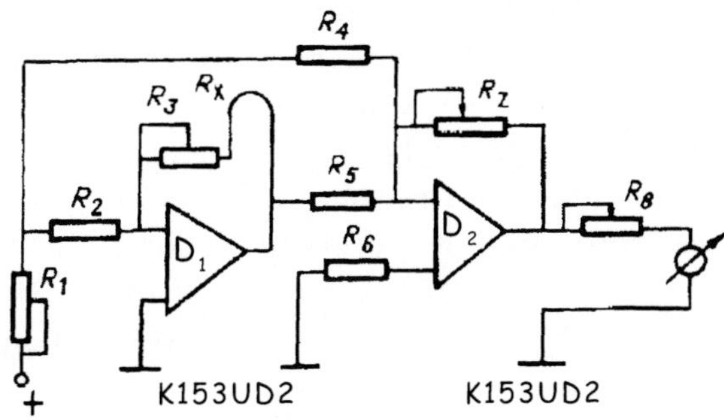

Fig.13.3 Principal electrical diagram of a device for measuring the temperature of the probe–wire. R_x=30÷100 ohm is the probe resistance, R_1=10 kohm, R_2=40 ohm, R_3=100 ohm, R_4=R_5=R_6=12 kohm, R_7=1 Mohm, R_8=10 kohm; D_1,D_2 are amplifiers.

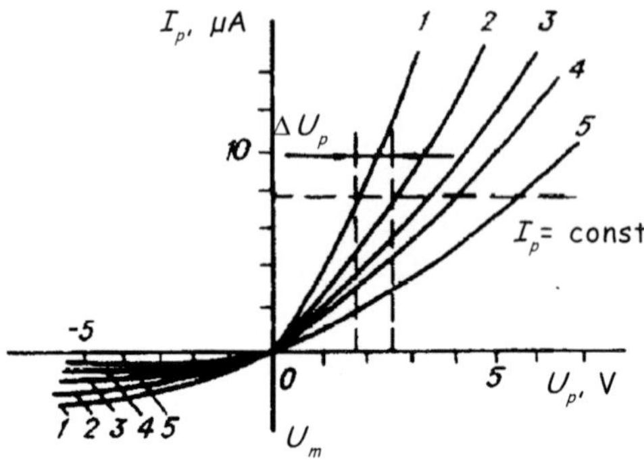

Fig.13.4 Time dependence of the VAC of the probe with growing dielectric film. Growth time of the film $t_5>t_4>t_2>t_1$.

Maxwellian. The condition required for applying the theory of the Langmuir–Mottsmith probe and, in particular, the ratio of the Debye length λ_D and the mean length λ_e of the free path of the electrons in the plasma $\lambda_e \gg \lambda_D$, may not be fulfilled. The film thickness d_{PF}, satisfy the condition

$$d_{PF} \ll R_p \tag{13.3}$$

The size of the working surface of the probe-wire, coated with the film, does not depend on the film thickness (with the accuracy of fulfilling the condition (13.3)). If the conditions (13.1) and (13.3) are simultaneously fulfilled and the plasma parameters are constant, the condition (13.2) is also fulfilled. For a flat probe with any thickness of the PF the area of the outer surface of the film is constant with time. We do not examine the growth of the first monolayers of the film when the properties of the probe such as the work function of the electron changes. This disrupts the condition (13.2). Evidently, the large changes of the useful signal ΔU, observed in the initial period of the growth of the PF[7,26] are caused mainly by this effect.

The VAC of polymer films, measured during its growth[8,10,12,25–27] proved to be non-linear and non-symmetric in relation to the origin of the co-ordinates (Fig.13.5). The value of R_{PF}, included in equation (13.2), does not have the meaning of the 'ohmic' resistance. In Ref.7, 8, 10, 13, 25, 26, it was shown that at the given discharge conditions of the current probe of I_p the value ΔU is directly proportional to the film thickness of the probe d_{PF} at $d_{PF} \lesssim 100$ nm. Consequently, it can be assumed that

$$R_{PF} = \rho d_{PF} \text{ at } I_p = \text{const}, \tag{13.4}$$

where ρ has the meaning of 'specific' resistance of the film. Under these conditions, the value ρ does not depend on the thickness and, consequently, the mechanism of conductivity through the film remains unchanged.

The problem of the mechanism of conductivity through the thin ($d_{PF} \lesssim 100$ nm) polymer films grown in glow discharge remains unchanged. Until now, all investigations into this problem were carried out after extracting the films from the reactor and depositing the upper electrodes on them by some method. Experimental data shown in Fig.13.5 are unique in that they were obtained during film growth where one electrode of the measuring cell was the probe/wire, the second electrode with the plasma around the probe. The method of producing the VAC during film growth will be examined below.

The mechanism of passage of the electric current through dielectric polymer films up to 250 nm thick, produced by plasma chemical polymerisation of organic substances, was investigated in Ref.16,17, 18,19 and 20. A detailed analysis of the experimental and theoretical data on the passage of current through the thin directive films was published by Simmons.[21] It was shown here that the electrical characteristics of the electrode–thin dielectric film–electrode cell are determined not only by the special features of the dielectric material of the film. They also depend on the nature of the electrode–dielectric contact.

At a low strength of the field in the film, the passage of current can be governed by the Ohm law but the current intensity in this case is very low.[21] A current of high intensity appears at a field strength of the film of $E_{PF} = 10^4 \div 10^5$ V·cm^{-1}. The VAC shown in Fig.13.5 was measured at $E_{PF} \gtrsim 10^5$ cm^{-1}. In such high fields, the current

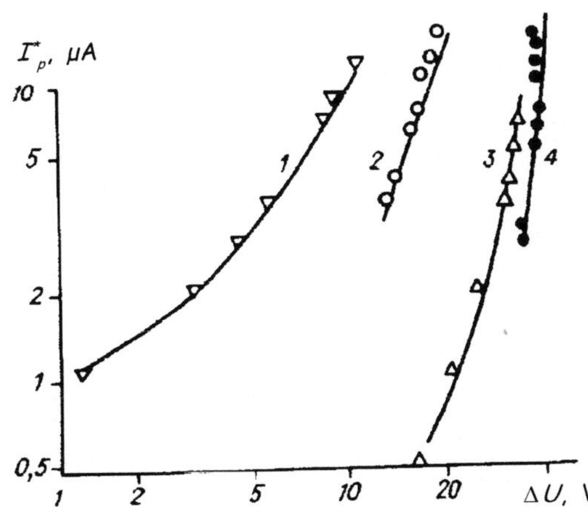

Fig.13.5 VAC of polymer films produced in Xe+1.7% vol. CH$_4$ 1 min after start of growth. Current I_p at which the polymer film grows: 1) 10 µA, 2,4) 0, 3) +10 µA; "measuring current" I^*: 2) $I^*_p = |I_e|-|I_j|>0$, 4) $I^*_p = |I_e|-|I_j|<0$; $P = 240$ Pa, $I_p = 50$ mA.

through a thin polymer film is transported by the electrons (according to the view of a large number of investigators) and 'the bottleneck' of the transport process is the generation of current carriers. This generation takes place as a result of the emission of Schottky electrons from the electrode into the film and/or the melt (ionisation) of acceptor-donor centres in the film in a strong homogeneous electric field by the Poole–Frenkel mechanism.[20,21] The transport mechanism in which the current is restricted by the spatial charge (CRTS) does not correspond to the experimental data, as shown in Ref.20 and 21.

For the conductivity mechanism by the CRTS, the dependence of current through the film on the voltage of the cell U_{PF} and the film thickness d_{PF} is described by the equation[20]

$$I_{PF} \sim U_{PF}^{l+1} / d_{PF}^{2l+1}, \quad l \gtrsim 1,$$
(13.5)

or

$$\ln I_{PF} \sim (l + 1) \ln U_{PF} \text{ at } d_{PF} = \text{const.}$$

When restricting the current by the generation of current carriers by the Schottky and Poole-Frenkel mechanisms

$$\ln I_{PF} = (\beta / kT)U_{PF}^{1/2}d_{PF}^{-1/2} + C,$$
(13.6)

where k is the Boltzmann constant; T is the cell temperature; C is the constant; $\beta = (e^3/\alpha\pi\varepsilon\varepsilon_0)^{1/2}$, $\alpha = 4$ for the Schottky mechanism and $\alpha = 1$ for the Poole–Frenkel mechanism; e is electron charge; ε is the dielectric constant of the film; ε_0 is the permittivity of the vacuum. Thus, $\beta_{PF} = 2\beta_{SCH}$.

The majority of experimental data are in agreement with the dependence $\ln I_{PF} \sim U^{1/2}$, i.e they confirm that the current is restricted by the generation of current carriers by the Schottky and/or Poole-Frenkel mechanism.[20] For example, in Ref.22, for films obtained in plasma chemical polymerisation of the styrol better agreement was obtained regarding the value of β with the hypothesis on the conductivity by the Poole–Frenkel mechanism, and, in addition to this, the VAC of the film did not depend on polarity of the electrodes. This is typical of the Schottky emission. The electrical conductivity of films of styrol was examined in greater detail in Ref.23. The VAC measured in Ref.10, 12, 25 and 26 (Fig.13.5) was non-symmetric in relation to the variation of the current direction. However, the specific feature of the 'measuring cell' where plasma can be used as one of the electrodes, does not make it possible to make an unambiguous conclusion on the occurrence of the Schottky conductivity mechanism in this case.

Equation (13.6) shows that a constant current I_{PF}, the voltage on the film U_{PF} and its thickness d_{PF} are proportional in relation to each other: $U_{PF} = B d_{PF}$, where B is a constant typical of the given cell. Consequently, the value ρ, included in equation (13.4) is related with the value V by the ratio $\rho = B/I_{PF}$. It is evident that $B = \overline{E}_{PF}$ is the mean strength of the field in the film with respect to its thickness. It is not known in advance how the value of B changes with the variation of the conditions of plasma chemical polymerisation: pressure, chemical composition of the flow rate of the plasma forming gas, specific energy contribution to the plasma, the density of the flows of charged particles of the plasma on the surface of the growing film, etc. In the experiments with discharge the mixtures of xenon on neon with methane[10,13] and argon with perfluorobutane[25,27] the authors compared the data on the growth rate of the film, obtained by the probe and optical methods. They show that under the operating conditions[10,13] a variation of the pressure of the mixture from 0.3 to 4 torr and a discharge current of 20 to 75 mA and a film thickness of $d_{PF} = 1 \div 100$ nm the value of B remain constant with an error not greater than 15–25%. For other discharge conditions, the constancy of the value B must be verified by experiments. If $B = $ const is fulfilled, the probe method can be used to examine the kinetics of growth of the film and some electrophysical properties of the latter.

For the conditions of discharge in the mixture of xenon or neon with methane the electrical resistance of the film growing on the probe changes by only 1–2%, and the variation of the temperature of the probe–substrate by $\Delta T_p = 50$ K and $T_p \cong 350$ K. Consequently, the probe method can be used to examine the dependence of the kinetics of film growth on substrate temperature.[8,10,15] Since the thermal inertia of the probe/wire is not high – around 0.1s, on the basis of the variation of ΔU_{PF} at the moment of variation of the probe temperature we can estimate the activation energy of the electrical conductivity of the film during its growth. In Ref.6,10,12–15 it was of the order of $E_a \sim 0.1$ eV.

As already mentioned, the main advantage of the probe method is that it enables the film to be grown on the probe at different densities of the electron and plasma ion fluxes on its surface and, at the same time, reveals the role of the charged particles in heterogeneous stages of plasma chemical polymerisation. The following procedure is used: the film is grown at a given discharge conditions at different values of I_p, i.e at different points of the VAC of the probe. However,

$I_p = I_{PF}$, i.e a displacement of the probe along the VAC indicates simultaneous displacement on the VAC of the film. Since the VAC of the film is non-linear, direct comparison of the growth rates at different values of I_p on the basis of the rate of variation ΔU_{PF} is incorrect. This problem is solved by two methods.

The film is grown at unchanged current I_p during a relatively long period of time. The probe/wire is then extracted from the reactor and its thickness is measured using an optical microscope. The procedure is repeated for a number of other values of I_p. From the VAC of the heated probe we determine the correspondence of I_p to the densities of the electron fluxes W_e and ion fluxes W_i so that we can determine the dependence of the growth rate of the film r_{PF} on W_e and W_i. This method can also be used to examine the kinetics of growth of the conducting and semiconducting films.

It is also suitable for dielectric substrates, for example, in the form of glass filaments.[8] In the latter case, measurements are simple only at $I_p = 0$, i.e at the point of 'the floating' potential, when $W_e = W_i$. Replacing the glass filament by the capillary with an inserted wire-heater enabled the dependence of the growth rate on substrate temperature to be examined. The use of the thin capillary with the diameter smaller than the mean length of free path of the molecules enables extensive thermal disruption of the plasma to be avoided. This method can be used to examine efficiently the distribution of the growth rate over the volume of the reactor.

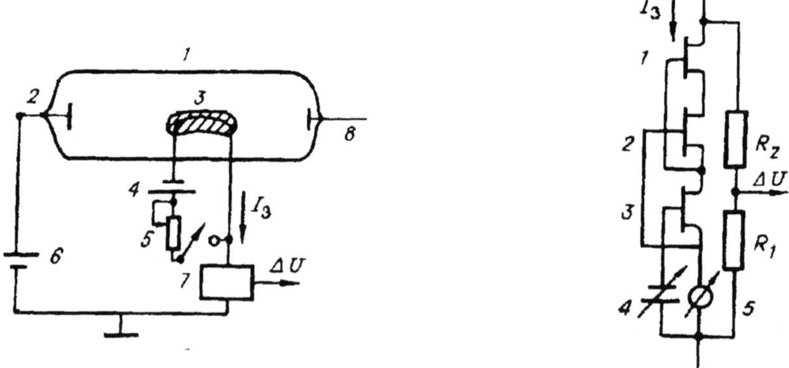

Fig.13.6 Electrical block diagram of the probe circuit. 1) reactor; 2,8) electrode; 3) the probe coated with the dielectric film; 4) heating battery of the probe; 5) resistance for regulating the heating current; 6) source of the bias voltage of the probe; 7) stabilising of probe current I_p.

Fig.13.7 (right) Principal electric diagram of current stabiliser. 1-3) transistors with an isolated gate; 4) regulated voltage source; 5) microammeter, R_1, R_2 is the matching voltage divider.

The second method is more complicated but provides a large amount of information. A current stabiliser (Fig.13.6, 13.7) is introduced into the probe circuit. The decrease of the voltage at the stabiliser is identical with the amount of increase of the voltage drop on the film. To compare the growth rates of the films at different I_p (different densities of the flows W_e and W_i), it is necessary to measure the rate of variation of their resistances at the same point of the VAC of the film, i.e. at the same current I_p^*. (It is assumed that the 'specific resistance' at a given I_p^* is the same for the films grown at different I_p. This assumption must be verified by experiments.) The following procedure can be used. Two kinetic curves $\Delta U_{PF}(t)$ for currents I_p and I_p^* are recorded in an automatic recording device (Fig.13.8). The value of current I_p is selected in such a manner as to ensure that the surface of the growing film receives the required electron W_e and ion W_i fluxes. The current is selected using the VAC of the 'pure' probe measured in advance (see above). The regime of switching over the current stabiliser from I_p to I_p^* is specified in such a manner as to ensure that the film grows at current I_p for 90–95% of the entire growth period. Switching to the 'measuring current' I_p^* is carried out over short periods of time, equalling 5–10% of the total duration of the growth process. The procedure is repeated for another value of I_p, and I_p^* remains unchanged. Consequently, this gave the dependences of the growth rate on the density of the electron ion flux in the discharge in mixtures of neon or xenon with methane[6,8,10,12] and argon with perfluorobutane.[25-27] If different values of the 'measuring current' I_p^* are applied quite rapidly for short periods of time whilst retaining unchanged film growth conditions, i.e

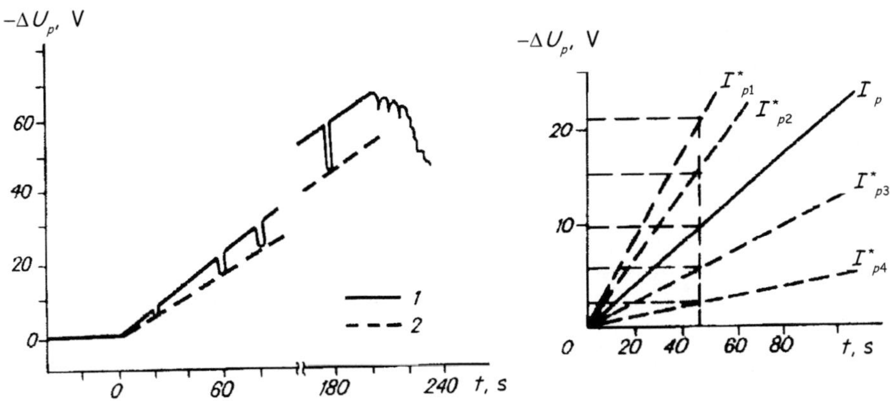

Fig.13.8 Kinetic voltage curves on the film. $U_{PF} = -U_p$; 1) U_{PF} for I_p; 2) U_{PF}^* for I_p^*.
Fig.13.9 (right) Construction of the VAC characteristic of growing film on the basis of kinetic curves obtained for different measuring currents $I_p^*(k)$ (polymer film is grown at probe current I_p).

I_p, then the resultant data can be used to measure the VAC of the film (see Fig.13.5) during its growth at different moments of time (Fig.13.9). The absolute values of the growth rates of the film are determined by calibrating the probe method by the first method (see above) or by some other independent method. To facilitate measurements of both parts of the VAC of the film we can use the 'bridge' circuit of the current stabiliser.

Cleaning the probe to remove the polymer film is carried out by rapid heating the probe with a current from a special source (see Fig.13.2). To set and measure the temperature of the probe-wire produced in the form of a loop (see Fig.13.2) we can use a device whose main electrical circuit is shown in Fig.13.3. The circuit of the device was proposed in Ref.24. A cartridge with resistances can be used as a variable resistor R_3. The values of R_2 and R_3 for the probe-wire made of platinum wire 10–20 μm in diameter and approximately 1 cm long are around 20–50 Ω. Regulating R_1 we set a current through the probe I_{heat} and its temperature T_p. The cartridge of the resistances R_3 is used to calibrate the output device in the values of the resistance of the probe at each current I_{heat}. The transition from the values of the probe resistance to its temperature is evident:

$$R_3 (T_p) = R_3 (T_0) (1 + \alpha \Delta T_p) \tag{13.7}$$

where α is the temperature coefficient of resistance of the wire.

Measurements of the dependence of the growth rate of the film on T_p at given I_p and the discharge regime (Fig.13.10) enable the effective activation energy of heterogeneous stages of the film growth to be determined.

At a film thickness of the order of $d_{PF} \simeq 100 \div 200$ nm the authors of Ref.7 observed an electrical breakdown of the film at a given

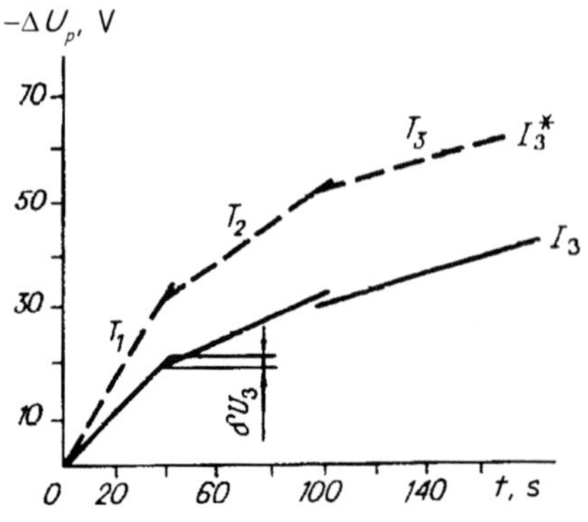

Fig.13.10 Measurement of the dependence of the rate of growth on dielectric film on substrate temperatures of the given I_p. $T_3 > T_2 > T_1$ is the temperatures of the probe - substrate, ΔU_p is the voltage jump on the film with the variation of temperature.

I_p. Evidently, the conductivity mechanism changed here, other avalanche-like processes developed and caused heating and destruction of the film (Fig.13.8).

Until now, no experiments have been carried out to examine the problem of the process of polarisation of the dielectric film during its growth and the effect of polarisation on the error of measurements of the growth rate by the method described previously. When switching from current I_p^* to I_p the ratio (13.2) is disrupted if an electric polarisation field forms in the film. This also relates to the case in which the probe method is used to examine the etching kinetics of dielectric films. Evidently, to examine the film polarisation it is convenient to use the probe method based on alternating current.[8] It can also be used to examine the kinetics of growth (etching) the film and its VAC.

The method is based on the following procedure. A source of sinusoidal voltage of low amplitude $u_0 \sin \omega t$, where $eu_0 << \bar{\varepsilon}$, $\bar{\varepsilon}$ is introduced into the probe circuit (see Fig.13.2) in series with the dc source; here $\bar{\varepsilon}$ is the mean energy of plasma electrons expressed in electron-volts; e is the electron charge. A condenser is connected in parallel to the current stabiliser. The value of the capacitance and the frequency ω of the weak signal are selected such that the stabiliser 'does not smooth' the current components formed in the probe circuit with a frequency equal to or multiplicable by ω. The amplitude of the first harmonics of the variable component of the current is proportional to dI_p/du_p. Here u_p is the potential of the probe-substrate at a point of the VAC of the probe coated with a film corresponding to constant current I_p. The value of the current is set by the current stabiliser. The amplitude of the component of the current with a frequency of 2ω is proportional to d^2I_p/du_p^2.[8] It is evident that the resistances of the probe (i.e dual electric layer in the plasma at its vicinity) of the film are connected in series. If at the initial moment of time the probe is clean, the variation of the variable signals with a frequency ω, 2ω, *etc.* with time is proportional to the variation of the first, second, and other derivatives of the VAC of the film with respect to time. Recording of these signals makes it possible to examine the kinetics of film growth and fine effects of the variation of its VAC during growth (etching).

The amplitude and procedure for measurements using alternating current were described in the section concerned with the probe measurements of the function of the energy distribution of the electrons and also in Ref.8.

Frequency ω can easily be measured over a wide range which in principle enables polarisation effects of the film to be detected. The main difficulty in measurements using alternating current is associated with a sensitivity to the 'noise' of the plasma and instability of the discharge.

13.2 Thermal probe

In the conditions of non-equilibrium electric discharges at reduced pressure, the thermal probe method is used to determine the gas temperature of the plasma and thermal effects of the recombination and deactivation reactions on the surface.[8] Two design variants of the thermal probe are used – a thermocouple coated or not coated with a dielectric screen, and thin wire with two lead-outs. In the first case, the temperature of the thermal probe is determined from the measured value of the emf of the thermocouple, in the second case it is determined from the change of the electric resistance of the wire. The first variant is far simpler as regards practical application. The second variant has a considerably higher information content as a result of the possibility of regulating its temperature by means of an external current source during the process of measuring the heat flows on the probe. It is also convenient that the same device can be used as a microcalorimeter, the electric Langmuir–Mottsmith probe and the probe for investigating the kinetics of growth (etching) of the layers. Both variants use in the explicit form models of the mechanisms of the processes of heating plasma and energy exchange between the probe, plasma and the reactor. This is the most important feature of the microcalorimeter (thermal probe) method when used for problems of non-equilibrium plasma diagnostics. We shall examine it on an example of the thermal probe in the form of heated wire immersed into the plasma of dc glow discharge, under the deposition regime of a polymer film.

The thermal balance of the thermal probe-wire is described by equation

$$Q_{h.t} + Q_e + Q_{rad} + Q_{he} + Q_{Ip} + Q_{eq} + Q_{ch} + Q_{PF} + Q_c + Q_{dea} = 0$$

$$(13.8)$$

Here $Q_{h.t}$ is the heat flow through the side surface of the thermal probe as a result the change of the kinetic energy of the neutral plasma particles in collision with the surface of the probe, Q_e is the heat flow through the ends of the probe-wire on the holders, Q_{rad} is the heat exchange of the probe with the environment through radiation, Q_{he} is the heating of the probe-wire by current by the external source, including in the operation of measuring the wire temperature from its electrical resistance (see Figs.13.2, 13.3), Q_{eq} is the heating of the thermal probe as a result of passage of probe current I_p through it when the thermal probe like the Langmuir–Mottsmith electric probe is not

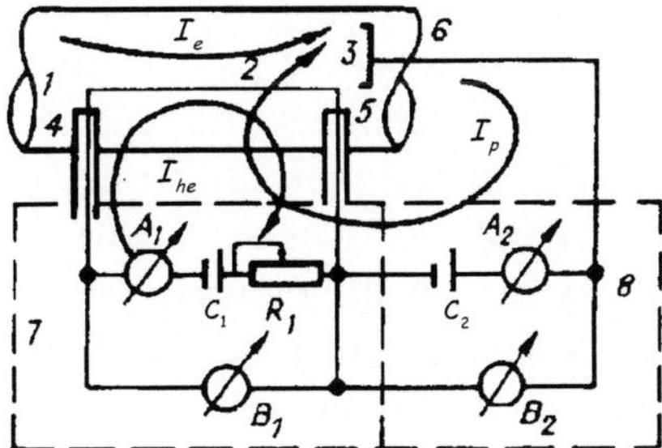

Fig.13.11 Simplified principal electrical diagram of the microcalorimeter. 1,3) leadouts of the microcalorimeter; 2) microcalorimeter – wire; 4,5) holders; 6) electrode–anode of the discharge tube; 7) device for setting and measuring heating current I_{he} of microcalorimeter measuring its temperature; 8) circuit for setting probe current I_p and measuring the VAC of the probe–mircocalorimeter; A_1, A_2 are the current measurement device, B_1, B_2 are the voltage measurement devices, C_1 and C_2 are power sources, R_1 is the regulated resistor, I_e is the equalisation current.

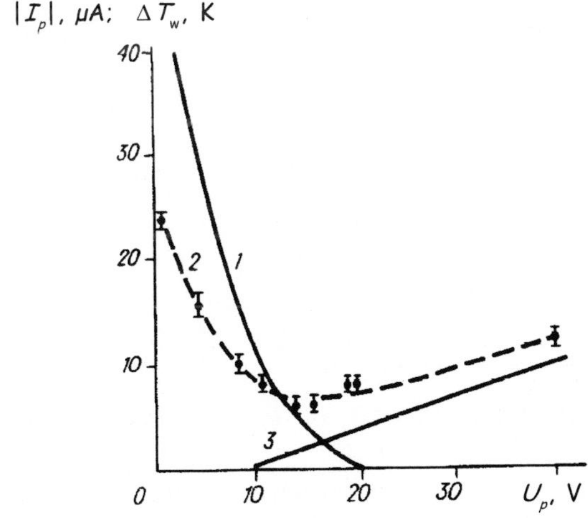

Fig.13.12 Dependence of probe temperature and fluxes on the probe of ions I_j and electrons I_e on probe potential U_p in direct current discharge in Ar + 10% vol H_2. Pressure $P = 0.3$ torr, discharge current $I_D = 25$ mA, diameter of the discharge tube 5 cm; 1) I_e, 2) $T_{pc} = T_p - T_w$, T_w is the wall temperature of the tube; 3) I_j (measurements by V.I. Timakin).

at the 'floating' potential (Figs.13.11, 13.12), Q_{eq} is the heating of the thermal probe by the 'equalisation' currents caused by differences in the potentials of the plasma along the probe, Q_{ch} is the heat flow through the side surface of the thermal probe as a result of chemical reactions on the surface (recombination of radicals, hydrogen atoms, *etc.*), Q_{PF} is the heat generation in the polymer film growing on the surface of the thermal probe, in passage of equalisation current I_{eq} through it and the probe current I_p, Q_c is the heat generation in recombination of the charged particles and accommodation of the kinetic energy on the surface of the thermal probe, Q_{dea} is the heat generation in deactivation of the electronically and vibrationally excited plasma particles on the surface of the thermal probe. It will be assumed that $d_p \ll \lambda$, where d_p is the diameter of the thermal probe, λ is the mean path of the neutral particles in the plasma. This simplifies calculations of the fluxes of the particles from the plasma on the probe surface.

The equation (13.8) can be used in practice only if a number of simplifying assumptions are made.

The term $Q_{h.t}$ cannot be measured directly in the plasma or calculated with sufficient accuracy. In the gas of the same chemical composition and pressure, $Q_{h.t}$ can be easily measured as the function $Q_{h.t} = f(T_p - T_g)$, where T_g is the gas temperature, T_p is the temperature of the probe/wire, and $T_p > T_g$. The quantity T_p can be specified and measured using a device shown in Fig.13.3. It can assumed that at small differences of the temperatures T_p and T_g the modulus $Q_{h.t}$ does not depend on the sign of the difference $T_p - T_g$ and in transition from the gas to the plasma with the same gas temperature and the chemical composition $Q_{h.t}$ does not change. It is assumed that the characteristics of the probe surface in both cases are the same.

The total value $(Q_e + Q_{rad}) = \varphi(T_p, T_w)$ is measured by heating the probe-wire with current I_{he} at the pressure in the reactor without discharge at which the term $Q_{h.t}$ can be ignored. According to the experimental data published in Ref.10 it is sufficient to set the pressure below $(2 \div 3)10^{-2}$ torr. The temperature of the reactor walls T_w should be set in the range of its values under the experimental discharge conditions. It should be noted that a given temperature T_w the temperature of the walls and the contacting leads under the discharge condition is slightly higher than T_w. The dimensions of the thermal probe should be selected such as to ensure that the end losses are minimum (large length and small probe diameter).

The measured quantities $Q_{he} = I^2_{he}R_p$, where R_p is the resistance of the probe-wire at a given current I_{he}, represent part of the procedure of measuring the probe temperature from the probe resistance using a device whose diagram is shown in Fig.13.3. It was assumed that the resistance of the conducting leads is far less than R_p.

The upper limit of quantity Q_{I_p} can be easily estimated assuming that the probe current I_p flows along the entire length of the thermal probe. Usually, the value of Q_{I_p} can be ignored if the probe is not covered by the dielectric film. This also applies to Q_{eq} because the equalisation current is of the same order of magnitude as I_p.

If the thermal probe is covered with a dielectric film, the situation does not change. For example, for a probe with a length of 1 cm and a diameter of 10 µm in a discharge in the mixtures of argon with methane at a pressure of 1 torr $Q_{h.t} \simeq 10^{-3} \div 10^{-2}$ W. For the probe current $I_p \simeq 10^{-5}$ A and the thickness of the dielectric film $d_{PF} \approx 100$ nm the voltage drop in the film reaches $V_{fi} \approx 100$ V, *i.e.* $Q_{PF} \simeq 10^{-3}$ W is comparable with $Q_{h.t}$. According to the estimate, the extent of heating the film with the equalisation current can also be significantly high.

The term Q_{ch} describes heat generation in the probe during chemical reactions on its surface. It can be reliably measured in discharges and mixtures of inert gases with hydrogen. Here it is interpreted as the thermal effect of recombination of the hydrogen atoms under probe surface: $Q_{ch} = \gamma_{rec} W$ (H) ε_{rec}, where γ_{rec} is the thermal coefficient of recombination; W (H) is the flux of hydrogen atoms onto the probe surface; γ_{rec} is the binding energy of the atoms in the H_2 molecules. The values of ε_{rec} are available for some materials, for example platinum.[29] If $d_p \ll \lambda$ (see previously), then $W(H) = 1/4[H]\bar{v}_t$, where [H] is the atom concentration in the plasma; $\bar{v}_t$ is the mean thermal velocity of the atoms. Evidently, interpretation of Q_{ch} does not cause any difficulties in other cases as well.

Q_p can be evaluated by experiments by measuring the temperature of the probe-wire T_p at different points of the volt–ampere characteristic of the probe, i.e at different values of the probe current I_p (see Fig.13.11). The minimum T_p is observed in the vicinity of the 'floating' potential of the probe where the electronic and ionic components I_p are equal as regards their modulus. The thermal effects of the ions and electrons can be separated only away from the 'floating' potential. The mechanism of heating the probe as a result of the kinetic and potential energy of the charged particle remains insufficiently examined because only approximate polarisation from remote parts of the VAC to the 'floating' potential of the contributions of the ions and electrons is possible. For reliable measurements of other terms (usually $Q_{h.t}$ and/or Q_{ch}) it is important to verify by experiment that Q_p is small.

When interpreting the meaning of the term Q_{da} it is important to remember that the following condition is fulfilled in the plasma of non-equilibrium electric discharges at reduced pressure:

$$T_g \approx T_{rot} \lesssim T_{vib} < T_{el}$$

where T_g, T_{rot}, T_{vib}, T_{el} is the gas (translational), rotational, vibrational and electronic 'temperatures' of the plasma particles, respectively (the concept of the temperature of the internal degrees of freedom of the neutral particles in the non-equilibrium plasma is not always correct,[1,8,10,13] as indicated by placing the term in inverted commas). It is usually assumed that $T_g \cong T_{rot}$. In discharges in nitrogen, hydrogen and mixtures of nitrogen with CO_2, T_{vib} was considerably higher than T_g.[30] Under these conditions, the heating of the thermal probe in the deactivation process of vibrationally-excited molecules on its surface can be considerable.[31] In non-equilibrium plasma of hydrocarbons, as a result of rapid vibrational–translational relaxation, according to the estimates, $T_{vib} \cong T_g$ at a pressure of $P \gtrsim 0.1$ torr and a specific energy contribution of $W \lesssim 1$ W·cm^{-3}.[10] The calorific effect of the deactivation of electronically excited particles should be taken into account in, for example, discharges with a low energy contribution $W \lesssim 10^{-2}$ W·cm^{-3} in inert gases with small ($\lesssim 1$ mol.%) additions of molecular gases. The concentrations of the metastable excited atoms in such a plasma reach $A_M \simeq 10^{10} \div 10^{11}$ cm^{-3}.[10,32]

Figure 13.12 illustrates the meaning of the terms Q_p, Q_{eq}, Q_{he}, Q_{PF}.

Thus, the relationship between the quantities included in (13.8) depends on the external parameters of the discharge: pressure, chemical composition of the plasma forming gas, specific energy contribution, the presence of the dielectric coating on the surface of the thermal probe, etc. Preliminary evaluation of the thermal effects can be carried out on the basis of the order of magnitude or even more approximately. Therefore, the use of the method is associated with carrying out a number of procedure experiments: measuring $Q_{s.t}$ in the gas without discharge, $Q_e + Q_{rad}$ in vacuum, Q_e in plasma at different points of the VAC of the thermal probe in the electric probe regime. It may be useful to compare the measurements in the given reactor in the discharge in the inert gas (where there are no heterogeneous chemical reactions on the probe surface) and in the 'working' gas, *etc.*

Another useful the method is used mainly for examining the thermal effects of chemical reactions on the surface under the condition that the gas temperature of the plasma must be obtained by other experimental or calculation methods. This comment applies to both types of the thermal probe: thermocouple and the heated wire. The use of the same design of the device enables three diagnostic methods to be used (see previously) widens the possibility of each method under the conditions of investigating the extremely complicated object – non-equilibrium plasma – in the presence of chemical reactions.

The method can be used efficiently for measuring the gas temperature of the plasma T_g in cases in which in equation (13.8) we can ignore the terms Q_{da}, Q_p, Q_{PF}, Q_{ch}, Q_{eq}, Q_{lp}. Most difficulties are associated with the preliminary evaluation of the value of the term Q_{ch}. Assuming that the thermal probe is thin

$$Q_{ch} / Q_{ht} \cong \left(\frac{1}{4} \bar{v}_t [R] \gamma_{rec} \varepsilon_{rec} \right) / \left(\frac{1}{4} \bar{v}_t [N] \alpha k (T_g - T_p) \right), \tag{13.9}$$

where α and γ_{rec} have the thermal effects of accommodation and recombination on the surface, respectively; $\bar{v}_t$ is the thermal velocity of the heavy particles of the plasma, ε_{rec} is the binding energy (recombination energy) or the radicals, k is the Boltzmann constant, [N] are the concentration of radicals and the total concentration of the heavy plasma particles. For example, for a thermal probe made of platinum in a discharge in a mixture of argon with hydrogen at a pressure $P \simeq 1$ torr we can accept that: $\alpha \simeq 1$; $\gamma_{rec} \simeq 10^{-2}$; $\varepsilon_{rec} \simeq 4$ eV; $T_g - T_p \simeq 10 \div 100$ K; $[H] \simeq 10^{14} - 10^{15}$ cm^{-3}; $[N] \simeq 10^{16}$ cm^{-3}. Substituting into (13.9) gives $Q_{ch}/Q_{h.t} \simeq 1 \cdot 10^{-2} \div 1$, i.e. Q_{ch} and $Q_{h.t}$ are the same order of magnitude. If the platinum thermal probe-wire is coated with a dielectric film (for example, by polymerisation in a discharge), then ε_{rec} should greatly decrease.[33] Consequently, $Q_c/Q_{h.t} << 1$ so that $Q_{h.t.}$ and T_g can be measured.

In high-frequency discharge the thermal probe is heated by the electromagnetic field. Heating can be easily evaluated by experiments by switching the hf field at low pressure in the reactor when there is no gas discharge. Special features of using the electric probe in rf and microwave discharges have been examined in Chapter 10.

Chapter 14

MEASUREMENTS OF THE ENTHALPY OF HIGH-TEMPERATURE GAS FLOWS

14.1 Introduction

Enthalpy is one of the most important characteristics of high-temperature media, especially moving ones, in the presence of ionisation. Enthalpy data are especially important in applying various technologies.

Determination of enthalpy using different devices started at the end of the 50s when work was carried out in many countries to develop high-temperature processes and equipment. The largest number of developments of the measuring devices was reported in the 60s and 70s.[1-60] Measurement methods were proposed and substantiated, appropriate sensors were developed and subjected to successful tests in measuring the enthalpy of different high-temperature media: inert, hot, plasma. At present, the methods and devices for measuring the enthalpy are usually developed along the path of improving and automating the measurements.

In the group of the methods of measuring enthalpy there are usually three most important groups:[8,21,37] 1) sound flow, 2) direct determination of the enthalpy by taking gas samples, 3) heat flow – determination of the enthalpy from the heat exchange of the solid with the gas flow. The first of these methods, which used special thermal probes, was not used widely because of complicated design of the probes and limited possibilities in comparison with the other two methods[8] which are used extensively. The main type of device are enthalpy probes (or sensors) based on gas samples. The enthalpy probe is a device introduced into the gas flow in order to measure the flow enthalpy. The enthalpy sensor is an enthalpy probe in a complex with a sensitive element ensuring reception of an electric signal.

No detailed analysis of the design of probes and methods of measurement have been carried out from the moment of development of the first devices for enthalpy measurements. Some studies in this direction are of limited nature and reflect only partial aspects of measuring the enthalpy of high-temperature media. It should be noted that the enthalpy of the low-temperature flows is measured by conventional methods, for example, using thermocouples, thermal resistors and other heat-sensitive elements.

14.2 Main methods of measuring the enthalpy of high-temperature gas

The currently used methods of measuring the enthalpy of high-temperature gas flows are based on the second and third method.

The second method has been used widely; it makes it possible to obtain high accuracy of measurements and develop various devices for enthalpy measurements.

Determination of enthalpy by calorimetric measurements. The current available methods of measuring the enthalpy by calorimetric measurements of the gas sample in the application range relate to two different groups   stationary and non-stationary. Historically, the stationary method was the first one to be developed it was then used on a basis for developing various devices – enthalpy probes, sensors, etc. The founder of the method is seen to be Grey[2,3,6,8] who founded its own company in the USA (Grey-rad) producing enthalpy sensors included in recording devices. Subsequently, another USA (Calprobe) started to produce sensors on the basis of Grey's licence.

It should be noted that the devices for determining the characteristics of the flow by taking gas samples were constructed a long time ago. For example, Fig.14.1 shows a probe[1] for taking gas samples and examining their chemical composition. This probe is very similar to enthalpy probes. Grey[8] also reported that prototypes of enthalpy probes existed prior to 1962 when a method of measuring the enthalpy using probes was substantiated for the first time.

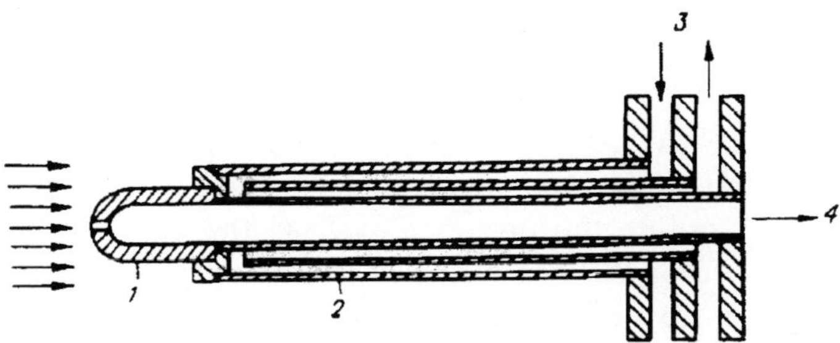

Fig.14.1 Microprobe for examining flames. 1) attachment, 2) calorimeter, 3) inlet and outlet of water, 4) discharge of gas.

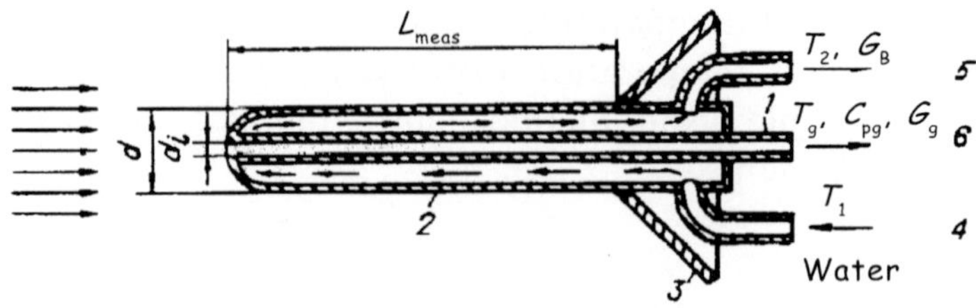

Fig.14.2 Stationary enthalpy probe. 1) measuring, 2) outer tube, 3) screen, 4,5) inlet and outlet of water, 5) discharge of gas.

Stationary method. The device using the stationary method of enthalpy measurement is shown in Fig.14.2. The main elements are the internal tube 1 for sucking in (taking sample) of the gas and the outer tube 2 for protecting the sensor against failure in the high-temperature flow and for ensuring that the shape of the sensor is such that it minimises the flow perturbation. The tail part of the sensor usually contains tubes for introducing and discharging the cooling liquid, sucking away the gas and the screen 3 for protecting auxiliary devices against the effect of the flow. Enthalpy determination is based on two measurements. The first measurement makes it possible to find the heat flow Q_1 on the probe without sucking the gas through the internal tube of the probe, i.e. the amount of heat received by the probe per unit time only as a result of an exchange of the flow with the outer surface. The result of the second measurement – heat flow Q_2 on the probe with sucking away the gas with flow rate G_g when the probe is in contact with the high-temperature flow on both the outer and inner surfaces. Taking into account the results of these measurements, the enthalpy of the flow is calculated from

$$h_0 = \frac{Q_2 - Q_1}{G_g} + C_{pg}T_g,$$
(14.1)

where $Q_2 = G_1 C_{pl}\Delta T_2$; $Q_1 = G_{11}G_{pl}\Delta T_1$; $\Delta T_2 = T_2 - T_0$, $\Delta T_1 = T_1 - T_0$; G_1, C_{pl}, T is the flow rate, specific heat capacity and the temperature of the liquid cooling the probe; C_{pg}, T_g is the specific heat capacity and temperature of the gas probe at the exit from the probe.

Usually, the flow rate of the cooling liquid is maintained constant, i.e. $G_{11} = G_{12} = G_1$, and

$$h_0 = \frac{C_{pl}G_l}{G_g}(T_2 - T_1) + C_{pg}T_g, \qquad (14.2)$$

It should be noted that h_0 is the total enthalpy including the enthalpy of the frozen flow and its kinetic energy

$$h_0 = h + v^2/2, \qquad (14.3)$$

where h, v are the enthalpy and speed of the flow.

Ignoring the kinetic energy of the flow increases the enthalpy of the frozen flow.

The speed of the incident flow can be determined using the same sensor. During measurement of the heat flow on the probe without sucking away the gas, a device for measuring the pressure is connected to the measuring tube, i.e. the enthalpy probe operates as a pressure sensor p:

$$p = p_0 + C_d\rho v^2/2, \qquad (14.4)$$

where p_0 is the static pressure in the region of the inlet orifice of the probe. The value p_0 is measured by other sensors; for jet flows with subsonic speed at the atmospheric pressure p_0 close to atmospheric pressure and, consequently,

$$p = C_d\rho v^2/2. \qquad (14.5)$$

Coefficient C_d which takes into account 'sticking of the gas' at low Reynolds numbers, as determined from the dependence $C_d = f$ (Re).[37] In a number of cases Re > 10 $C_d \simeq 1$, and with satisfactory accuracy $\rho = \rho v^2/2$, i.e. $v^2 = 2p/\rho$, where $\rho \simeq A/h$;[37] A is a constant. Thus, h is determined from the transcendental equation

$$h_0 = h + 2p/\rho, \quad \rho = f(h), \qquad (14.6)$$

where h_0 and p are measured by the same enthalpy probe.

The problems of design of the probes, sensitivity and error of measurements will be examined in detail below and recommendations will be made for using other probes.

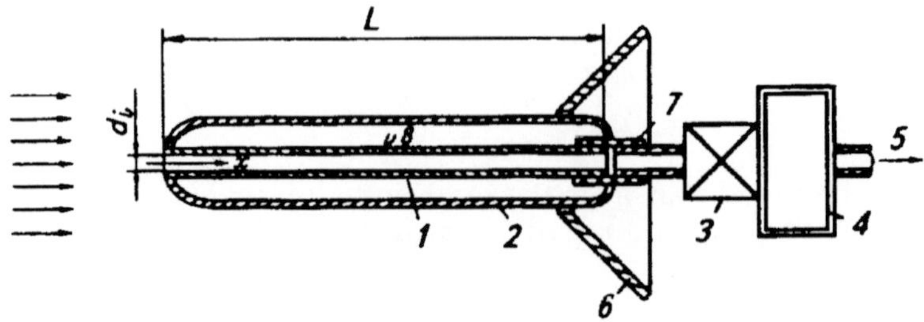

Fig.14.3 Nonstationary enthalpy probe. 1) specimen measuring tube, 2) shielding shell-screen, 3) high-speed valve, 4) evacuated cavity, 5) connection of vacuum pump, 6) shielding screen, 7) thermal insulation, 8) thermocouple.

The non-stationary method. The design of the device illustrating the non stationary method of determining gas enthalpy[19] is shown in Fig.14.3. The main element here is the standard measuring tube 1 with the wall thickness δ and the base length L (the section in which its temperature is measured using, for example, a thermocouple). On the outside there is a shielding shell whose shape defines, as in the stationary method, the nature of heat exchange of the probe with the flow and ensures the minimum heat flow. At the outlet of the probe there is a high-speed valve, an evacuated container and a system for sucking away the gas.

The measuring tube is used in such a manner that the entire energy of the gas flow, sucked away by the pump, is absorbed in the tube. In this case, the temperature of the tube continuously increases when taking gas samples. It is important to prevent the penetration of the heat flow through the outer tube, *i.e.* the dimensions of the air or some other space between the tubes 1 and 2 should be such as to ensure that the heat flow to the tube 1 during the measurement arrives only from the gas sample. The equation, describing the variation of the enthalpy of the gas, passing through the measuring tube, has the form

$$h - h_{out} + \frac{v^2}{2} = \frac{C_{ph}\rho_h V_h}{G_g} \frac{\partial}{\partial t} \frac{1}{L} \int_0^L T_h(x)\,dx \pm \frac{Q_{loss}}{G_g},\qquad (14.7)$$

where h_{out} is the enthalpy of the gas at the outlet of the probe, C_p, ρ, V, T is the heat capacity, density, volume, gas temperature at the inlet into the probe, index t relates to the measuring tube, g to the gas. Q_{loss}/G_g characterises the error of determining the enthalpy as a result of losses or additional supply of heat. The value h_{out} is in the majority of cases close to zero because the probe is designed in such a manner that the entire energy of the sucked away gas is absorbed by the measuring tube. The derivative determines the variation of the mean temperature of the tube, i.e.

$$\frac{\partial}{\partial t}\frac{1}{L}\int_0^L T(x)dx = \frac{\partial T_m}{\partial t}.$$
(14.8)

Consequently

$$h - h_{out} = -\frac{v^2}{2} = \frac{C_{ph}\rho_h V_h}{G_g}\frac{dT_m}{dt} \pm \frac{Q_{loss}}{G_g}.$$
(14.9)

Producing an enthalpy probe using a specific procedure it is easy to ensure that $h_{out} \approx 0$ and the speed v can be measured with the same probe in the regime without sucking away the gases; the measuring tube is connected to a high-speed pressure recording device, for example, a condenser microphone (at low pressures) or to a pressure gauge. The value Q_{loss}/G_g with other conditions being equal should be insignificantly small. Under these conditions

$$h + \frac{v^2}{2} = \frac{C_{ph}\rho_h V_h}{G_g}\frac{dT_m}{dt}$$
(14.10)

The flow rate of the gas to the measuring tube is calculated measuring the rate of variation of pressure in constant volume 4 (see Fig.14.3) from the equation

$$G_g = A dp / dt,$$
(14.11)

where dp/dt is the rate of variation of the pressure in the filled volume, and $A = $ const for the given volume and the given gas. Thus, we obtain

$$h + \frac{v^2}{2} = \frac{C_{ph}\rho_h V_h}{A}\frac{dT_m / dt}{dp / dt} = B\frac{dT_m}{dp}$$
(14.12)

The value dT_m/dp is determined by experiments; if the flow speed is known, the enthalpy of the gas in the examined flow is calculated.

This approach was used as a basis for developing a large number of devices. Their design, special features of application and application areas will be described below in a separate section.

Determination of enthalpy from heat exchange. The method of determining the enthalpy of the gas on the basis of heat exchange with the solid, introduced into the flow, is less accurate than the method examined previously but can be applied more easily in a number of cases because it is simpler: when disruptions of the high-temperature flow should be as low as possible or when the flow dimensions are very small and when the stationary method of measuring enthalpy described above is not applicable.

This method has been known for some time.[4,5,10,21,22,26,33] The solid used when examining heat exchange is represented either by a simple solid, for example, a cylinder with longitudinal or transverse flow around,[26,37,44] or a flat solid or solid enabling modelling of some technological process, the conditions of interaction of the high-temperature medium with the component, *etc.*

The flow enthalpy is measured using the following procedure. For example, we examine a cylinder with a transverse flow around.[37] In this case, the density of the heat flow is defined by the equation

$$q = \alpha \frac{\Delta h}{C_p}, \tag{14.13}$$

where α is the heat transfer coefficient; Δh is the enthalpy pressure ($\Delta h = h_f - h_w$); h_f, h_w is the enthalpy of the flow at the temperature of the flow and the wall of the sensor, respectively; C_p is the heat capacity of the gas. The heat transfer coefficient is generally determined from the criterial dependence

$$\alpha = \frac{\lambda}{d} \mathrm{Nu}, \tag{14.14}$$

where the Nusselt criterion Nu is determined from the equation

$$\mathrm{Nu} = 0.5 \, \mathrm{Re}_f^{0.5} \, \mathrm{Pr}_f^{0.4} \left(\frac{\rho_f \mu_f}{\rho_w \mu_w} \right) \left[1 + \left(\mathrm{Le}_f^{0.52} - 1 \right) \frac{h_\partial}{h_f} \right], \tag{14.15}$$

where λ is heat conductivity, d is the cylinder diameter, ρ, μ is the density and viscosity of the gas, Re, Pr, Le are the Reynolds, Prandtl and Lewis criteria, the indices "f" and "w" relate the thermophysical properties of the gas to the flow temperature or wall temperature.

If the right hand part of the (14.14) is divided by $\sqrt{p}$, where p is the dynamic pressure, taking above considerations into account, we obtain

$$\frac{q}{\sqrt{p}} = \frac{\lambda}{d}\frac{\mathrm{Nu}}{\sqrt{p}}\frac{\Delta h}{C_p} = f(T),, \tag{14.16}$$

where

$$f(T) = 0.6\frac{\lambda_f \rho_f^{0.25}}{\mu_f^{0.5} d_f^{0.5}} \mathrm{Pr}^{0.4}\left(\frac{\rho_f \mu_f}{\rho_w \mu_w}\right)^{0.2}\left[1 + \left(\mathrm{Le}_f^{0.52} - 1\right)\frac{h_\partial}{h_f}\right]\frac{\Delta h}{C_{pw}}, \tag{14.17}$$

and the value $f(T)$ can be easily calculated if we know the properties of the gas.

Thus, measuring experimentally the density of the heat flow q and the dynamic pressure p, we determined the enthalpy of the high-temperature gas flow from the dependence $f(T)$.

This method has been tested many times[37] in flows of various gases: in argon,[37] oxygen[24] ad others, and can be recommended for the temperature region characterised by dissociation, ionisation and other reactions associated with energy release. In this case, the measurement error decreases because the dependence $f(T)$ is very steep. Disadvantages of the method include the complicated and sometimes inaccurate dependence of $q/\sqrt[4]{p}$ on the thermophysical properties of the gas.

14.3 Enthalpy sensors and probes

The development and improvement of the methods of measuring the enthalpy of high-temperature gases, which started in the 60s, was accompanied by development Ent of various devices for enthalpy measurements – sensors and probes, with the most intense activity observed up to the middle of the 70s. In later stages, the development of the measurement procedures continued along the path of optimisation and improvement of the existing devices[57,58] and development of automated systems.[59,60]

Detailed analysis of the design solutions and procedure special features of determining the enthalpy makes it possible to classify the enthalpy probes and design two large groups, stationary and non stationary enthalpy probes (in accordance with the method of measuring enthalpy by calorimetric measurements of gas samples). The device, designed for measuring the enthalpy on the basis of the heat flow on the probe, fully corresponds to the heat sensors used for evaluating the heat flows are not examined here.

Stationary enthalpy probes. According to the application, these probes are divided into probes for subsonic and supersonic flows. They have many similar features, especially in the procedural part, but also some design differences.

There are various designs of these probes. According to the position of the measuring part of the probe in the flow, there are probes with longitudinal and transverse flow-around (they will be referred to as longitudinal and transverse enthalpy probes).

Longitudinal enthalpy probe. As mentioned previously, the first probe of this type was developed by Grey, et al.,[2,3,6,7] and is therefore referred to as Grey's probe (Fig.14.4). It consists of three coaxial tubes; two of these tubes, the inner and outer, are connected together at the inlet part and form a hemispherical head with a measuring orifice at the inlet. The intermediate tube is positioned between the outer and the measuring tubes and is used to divide the cooling flow and directs it into the inlet part of the probe. In the outlet part of the probe the tubes are connected in such a manner that they ensure the inlet and outlet of the cooling liquid, positioning of temperature gauges in the inlet and outlet areas, for example, thermocouples, and connection of

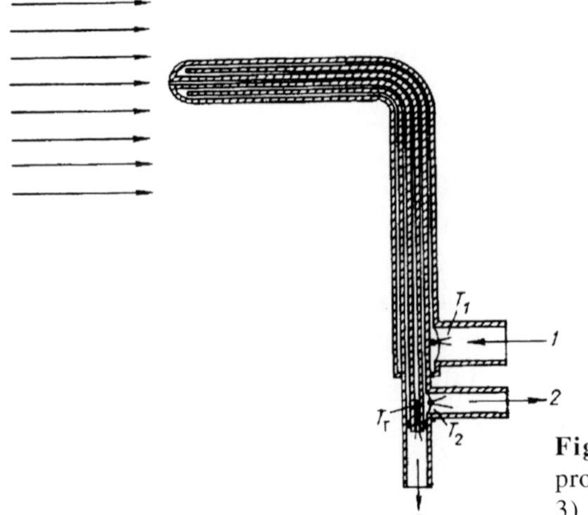

Fig.14.4 Longitudinal enthalpy probe. 1,2) inlet and outlet of water, 3) discharge of gas.

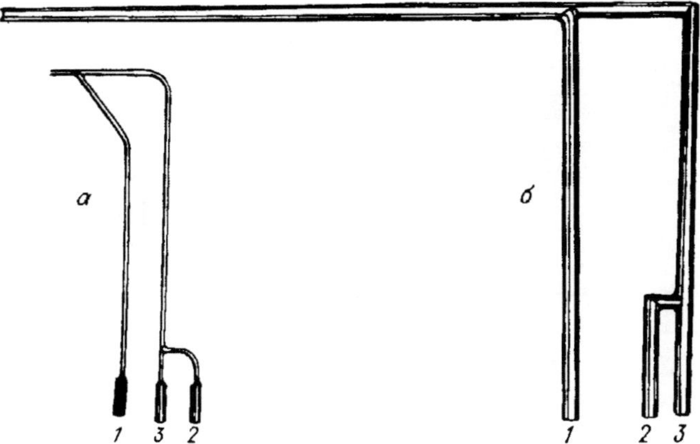

Fig.14.5 Outer view of longitudinal enthalpy probes with an outer diameter of d = 1.5 (a), 5 mm (b).

Fig.14.6 A replaceable longitudinal enthalpy probe.

the sensor to the systems for gas sampling and pressure measurements.

Figure 14.5 shows the external appearance of longitudinal enthalpy probes whose design corresponds to Fig.14.4. Only the methods of supply and discharge of cooling water differ. In Fig.14.5, the tube 1 in the probe is used to supply water, the tube 2 is used for discharging water and the tube 3 for taking a gas sample.

In some cases, the enthalpy probes can be produced in the form of compact replaceable modules connected to the measuring unit using a sleeve with holes for supplying and discharging the cooling liquid and gas sampling (Fig.14.6). The measuring orifice is situated in the nose part of the probe, and the gas is sucked away from the rear side of the probe into the end part of the sleeve. The cooling water is supplied and discharged through orifices made in the side surfaces of the sleeve. This probe is easy to use, is placed in the longitudinal direction in the high-temperature flow[13–15] (like the probe in Fig.14.2), and a shielding screen is placed in its tail part.

The two-tube enthalpy probe is similar to Grey's probe as regards its design.[32] The only

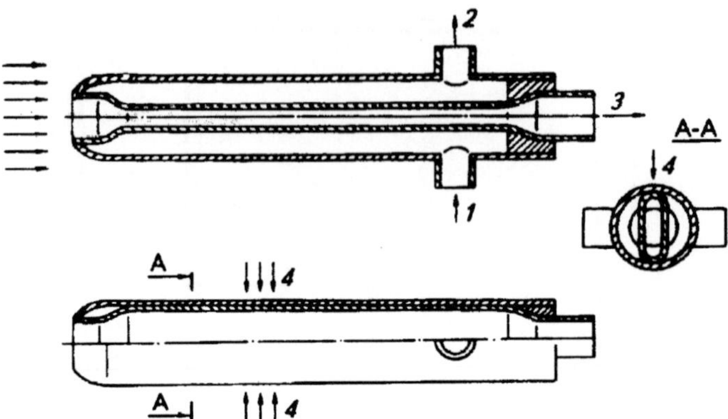

Fig.14.7 Two-tube enthalpy probe. 1,2) inlet and outlet of water, 3) discharge of gas, 4) direction of the heat flow.

difference is the cooling system (Fig.14.7) Here the design of the internal measuring tube together with the outer tube ensures the required cooling rate and defines the direction of movement of the cooling liquid. Therefore, the intermediate tube is now required. Although the dimensions of these probes can be very small, they are not yet used widely because they do not make it possible to measure the enthalpy with the required accuracy. At the circumference of the probe in the cross section (A–A) there are two sections in which the internal measuring tube is poorly cooled by the liquid. Consequently, the internal tube receives the heat flow q that is difficult to control. For example, in the first measurement (without sucking away the gas) calorimetric measurements are taken to determine the heat flow to the probe Q_1 which travels through all the surfaces, including the flow q through the regions of contact of the tubes. In the second measurement (with sucking away the gas), the probe receives the heat flow Q_2 which includes the heat flow brought in by the sucked-away gas. In this case, the sucked-away gas transfers, in addition to its energy, also part of the heat flow q (Δq), i.e.

$$Q_2 - Q_1 - \Delta q = G_g \Delta h \qquad (14.18)$$

If Δq is comparable with the measurement enthalpy h $(\Delta h = h - h_{out})$, the error is large and this restricts the application of sensors of this type, regardless of their simple design.

The enthalpy probe with internal heat insulation (Fig.14.8). This probe is very similar to Grey's probe and, as mentioned by Grey himself,[8] had been known previously and was referred to as the calorimetric probe. The presence of heat insulation reduces the intensity of heat

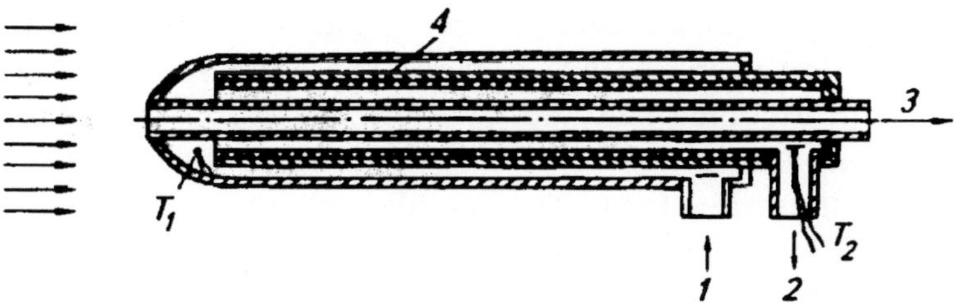

Fig.14.8 Enthalpy probe with a heat-insulated intermediate tube. 1,2) inlet and outlet of water, 3) outlet of gas, 4) insulator.

exchange between the counterflows of the cooling liquid, increases probe sensitivity and reduces the measurement error. However, these probes are not used in practice because of production problems and problems in developing the methods of decreasing the intensity of this heat exchange.

The enthalpy probe with the divided flow of the cooling liquid. The diagram of the probe (Fig.14.9) shows that an additional intermediate tube is used here. Three outer tubes form the cooling systems.[9,17] There is one inlet for the cooling liquid and two outlets.[17] The cooling liquid is supplied between two intermediate tubes and, consequently, is directed to the area with highest thermal stresses, *i.e.* the inlet part of the probe. Here the liquid is divided into two parts: one part cools the probe from the outside, protecting it against the effect of the high-temperature medium, and the other part of the liquid passes in the gap between the intermediate and measuring tubes. The last part of the liquid receives mainly the heat flow from the measuring tube. To measure enthalpy it is sufficient to know the flow rate of this liquid ('measuring' liquid) and its temperature difference ($\Delta T = T_2 - T_1$) in the presence and absence of sucked away gas.

Thus, the slightly more complicated design of the probe and its larger

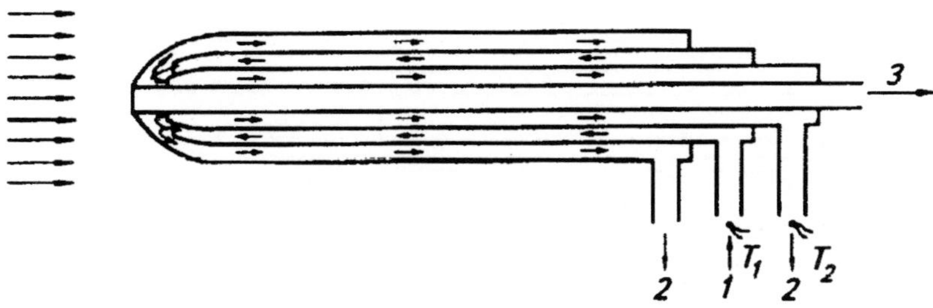

Fig.14.9 Enthalpy probe with the distributed flow of cooling liquid. 1,2) inlet and outlet of water, 3) discharge of gas.

dimensions are compensated by improved measurement sensitivity.

The enthalpy probe with a screen. There are two types of these probes – with a cooled and uncooled screen.

The enthalpy probe with an uncooled screen slightly differs from the probe without a screen. Usually, the screen is positioned on the outer side of the probe, is made of a heat-insulating material, for example, graphite[12,25] and repeats the shape of the probe. The screen not only protects the probe against overheating but also increases the sensitivity of calorimetric measurements of the gas sample if these measurements are carried out on the background of a smaller total heat flow on the probe. These enthalpy probes are characterised by several working conditions depending on the start of measurements in relation to the moment of introduction of the probe into the high-temperature flow. The first regime: the heat-insulating screen only started to be heated, the temperature of its internal surface slightly differs from the initial temperature, and the calorimetric measurements of the gas sample make it possible to obtain the maximum sensitivity of the probe and the minimum measurement error. Operation in this regime requires the use of high-speed equipment. In the second regime, the heat flow to the outer cooled tube of the probe increases almost linearly with time and this must be taken into account in enthalpy measurements. However, the error of measurement will be the highest of the three conditions examined. The measurements in the third regime are carried out in the steady heat flow on the probe using the conventional procedure. In this case, as a result of decreasing the heat flow the sensitivity and accuracy of measurements increase.

The enthalpy probes with the uncooled screens are used on a limited scale because the presence of the screen increases the dimensions and amplifies the perturbing effect of the probe. They can be used efficiently for the diagnostics of long flows.

The enthalpy probe with a cooled screen (Fig.14.10). This probe

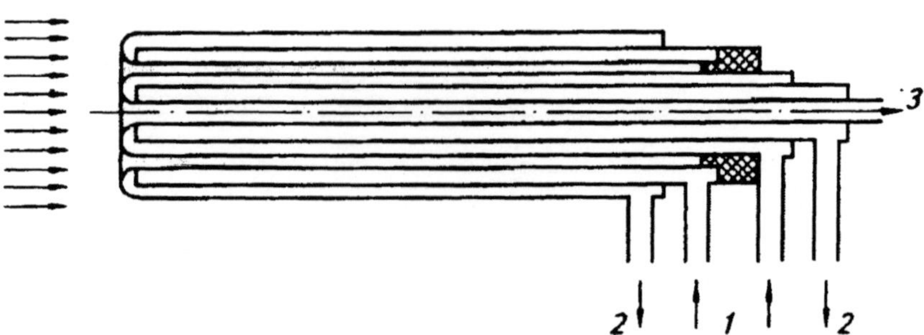

Fig.14.10 Enthalpy probe with a cooled screen. For symbols see Fig.14.9.

is sometimes referred to as the 'six-barrel probe' according to the included in probe design.[42,43,45] In this case, a screen of three coaxially positioned tubes is placed on the normal longitudinal enthalpy probes (Grey probe). A heat insulator, for example, air, is placed between the measuring probe and the screen.[20] The cooling liquid is supplied into the probe in such a manner that the internal tube of the screen and the outer tube of the measuring part of the probe are cooled by the liquid with the same temperature. The cooling liquid is fed in parallel directions to these parts of the probe thus ensuring equal temperatures of the walls of the measuring part of the screen in the absence of heat overflow. Therefore, in random contact of the walls of tubes (as a result of disruption of heat insulation or due to some other reason) the operating regime of the probe is not disrupted.

The sensitivity of the probe of the accuracy of measurements are sufficiently high because the parasitic heat flow to the measuring part of the probe rapidly decreases. For example, according to the estimates in Ref.27 and 42, the error of measuring the enthalpy decreases from 6 to 4%, although the perturbation which can be caused by the probe with larger dimensions (the diameter of the probe with a screen 6.2 mm, without a screen 3 mm)[3] is not taken into account. Evidently, the examined probes can be used efficiently for the diagnostics of flows with large transverse dimensions which exceed at least by an order of magnitude the outer diameter of the probe.[34]

The enthalpy probe for measurements in supersonic flows (Fig.14.11). These probes are usually of the six-barrel type, i.e. fitted with screens.[18,23,26] The nature of gas flow in the top part of the probe in the regimes with and without gas sampling can differ as a result of changes in the position of the shock wave so that the speed of gas sampling should be minimum. To ensure that the amount of gas

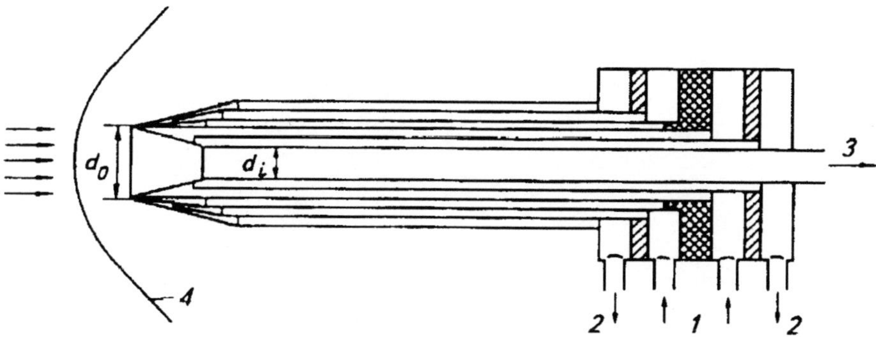

Fig.14.11 Enthalpy probe in a supersonic flow of high-gas. 1,2) inlet and outlet of water, 3) discharge of gas, 4) shockwave.

is sufficient from the viewpoint of ensuring required sensitivity, the size of the measuring orifice must be increased. These considerations are taken into account in the design of the probe in Fig.14.11. The screen is separated from the calorimetric part of the probe by an air gap. In the leading front of the probe all tubes are brazed together because the generating lines of the outer and inner cones are positioned under an angle of 30°.[42] The sensitivity of this probe is sufficiently high, and enthalpy measurements does not require large amounts of this sampled gas.

In Ref.42, the enthalpy probe was produced from 0.1C18Cr9NiTi stainless steel: the outer tube with a diameter of 6.2 mm, $d_0 = 4$ mm, $d_i = 1.8$ mm (Fig.14.11). The large size of the orifice d_0 enables the consumption of the sampled gas and the perturbing effect of the probe to be reduced.

The 'angular' enthalpy probe. This probe was designed as a result of improvement of longitudinal enthalpy sensors. Grey[8,11,16] also tried to minimise the part of the probe introduced into the flow. Usually, the probe was bent[38,40] in such a manner that its 'body' did not cause any significant perturbations in the flow in the region of the measuring orifice. As a result of a compromised variant, the 'angular' enthalpy probe was developed.

This probe is shown schematically in Fig.14.12 and is external appearance is in Fig.14.13. Here, the measuring internal tube in the in-

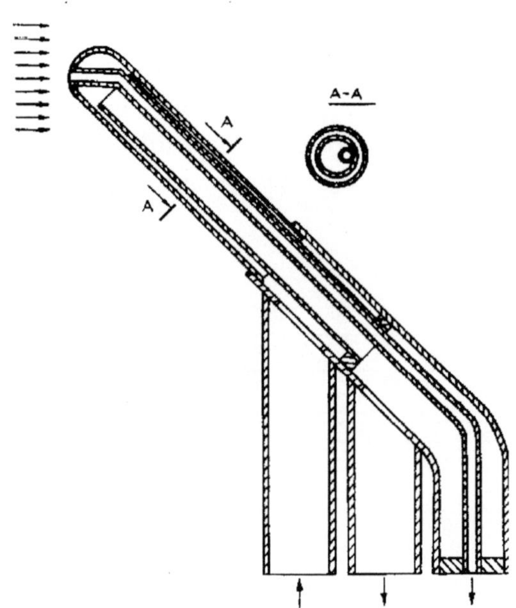

Fig.14.12 "Angular" enthalpy probe.

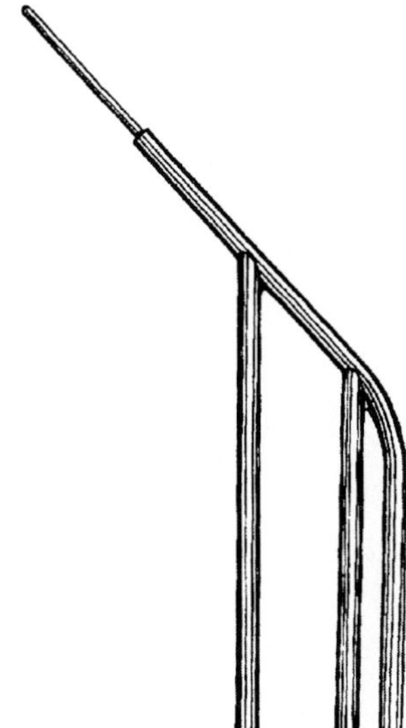

Fig.14.13 External view of the angular enthalpy probe.

let part is bent so that it is possible to place the initial part of this tube along the flow, and the external pipe under an angle to the flow. The intermediate tube is placed in the same position as in the longitudinal probes and is used as a guide for the cooling liquid.

The 'angular' enthalpy probe is efficient in operation and retains the advantages of the longitudinal enthalpy sensor. The angle under which the probe is positioned in relation to the movement of the flow is usually around 45°, although other variants are also possible. Positioning the probe under an angle of 45° ensures only slight perturbations of the flow with sufficiently simple design of the top part of the probe.

The transverse enthalpy probe (Fig.14.14). In contrast to the longitudinal probes, the measuring part of the transverse probe is normal to the flow and the relatively short initial section of the measuring tube of the probe is positioned in the direction along the flow. The probe consists only of two tubes: internal–measuring, and the external used for transport of the cooling liquid; therefore, it is possible to reduce the dimensions of the probe. For example, the authors of this book[22] produced a probe with an external diameter of 1 mm (external tube 1.0 × 0.85 mm, inner tube 0.5 × 0.4 mm) (Fig.14.15). A decrease of the probe dimensions leads not only to positive results. In similar small probes it is difficult to ensure high flow rates of the cooling liquid at pressures usually encountered under the laboratory conditions; therefore, decreasing the diameter of the external tube slightly reduces the possibilities of measuring high heat flows. The most suitable probes are those with an external diameter of 2–3 mm which also have higher

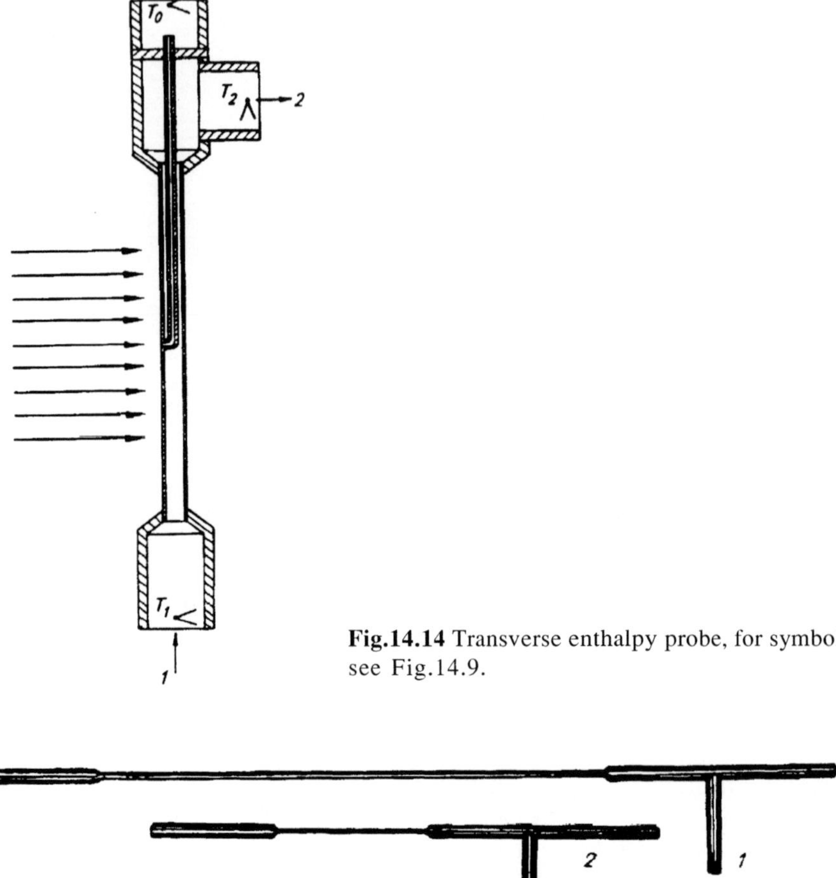

Fig.14.14 Transverse enthalpy probe, for symbols see Fig.14.9.

Fig.14.15 External view of transverse enthalpy probes, diameter 2 (1) and 1 mm (2).

mechanical strength in comparison with probes with a smaller diameter.

Disadvantages of the transverse enthalpy probe include, firstly a high level of perturbations of the high-temperature flow determined by the asymmetric flow-around of the flow around the body of the probe. Secondly, as a result of the small length of the initial section of the measuring tube, positioned along the flow, the probability of appearance of distortions in the results of the measurements as a result of the distortion of the gas flow lines in front of the probe decreases.

Disadvantages of the transverse enthalpy probe include, firstly, the possibility of decreasing its transverse dimensions and the perturbations caused by the probe. Secondly, the inlet and outlet parts of the probe are separated thus reducing the error of measurements as a result of preventing overflow of the heat. Thirdly, these probes are suitable for measuring the flow characteristics in closed channels.

The transverse enthalpy probe was described for the first time in Ref.22; it was developed on the basis of previously used[23] cylindrical heat sensors with transverse flow-round. In subsequent stages, the enthalpy probes of this type were used as a basis for constructing a measuring system.[57,58] They were gradually improved and, as in the case of longitudinal enthalpy probes, a number of their characteristics were improved.[54,56] For example, sensitivity, service lifetime, etc.[30]

A schematic of the transverse sensor is shown in Fig.14.16. Here the enthalpy probe 1 is placed in the measuring block 2 made of organic glass. The probe is secured by bolts 6 with flanges 3. The gasket 4 and half rings 5 restrict the movement of the probe. The rubber seal 7 prevents penetration of the cooling agent into the gas circuit of the probe. A differential battery of the thermocouples was installed to record temperature. Only one half of this battery is shown in Fig.14.16. In the upper part, the probe is secured through auxiliary straps with screws to increase mechanical strength. Prior to measurements the straps are removed.

Figure 14.17 shows a sketch of the design of the transverse enthalpy probe with symmetric (in relation to the flow direction) sampling of gas, and its external appearance is in Fig.14.18. The outlet ends of the measuring tube are connected to both end pieces of the enthalpy probe so

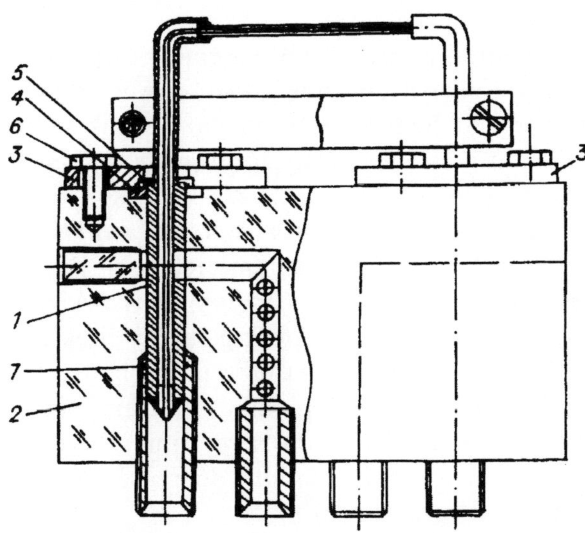

Fig.14.16 Transverse enthalpy sensor.

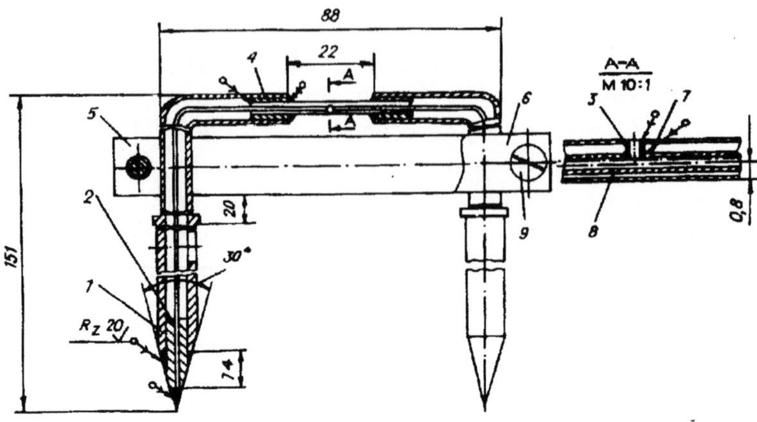

Fig.14.17 Transverse enthalpy probe with symmetric sampling of gas. 1) tip, 2) insert 3) inlet section of measuring tube, 4) insert, 5,6) auxiliary straps, 7) external, 8) measuring tube, 9) securing screws.

Fig.14.18 External view of a transverse enthalpy probe with symmetric sampling of gas.

that gas sampling, required under the measurement conditions, is carried out in each half of the measuring tube with a lower velocity. This increases the efficiency of heat transfer from the gas sample, and in the majority of cases, it is not necessary to measure the gas temperature at the outlet from the probe.

The transverse enthalpy probe with separate supply of the cooling liquid. These probes are shown schematically in Fig.14.19 where two main types are given: asymmetric and symmetric. A special feature of these probes is the supply of the cooling liquid directly to the area of position of the inlet hole of the inlet tube in the body of the probe.[41] This liquid is then branched out (separated) and directed into the inlet and outlet parts of the probe. At the same time, the total heat flow in the measuring part of the flow rapidly decreases. Gas enthalpy is determined from the following equation:

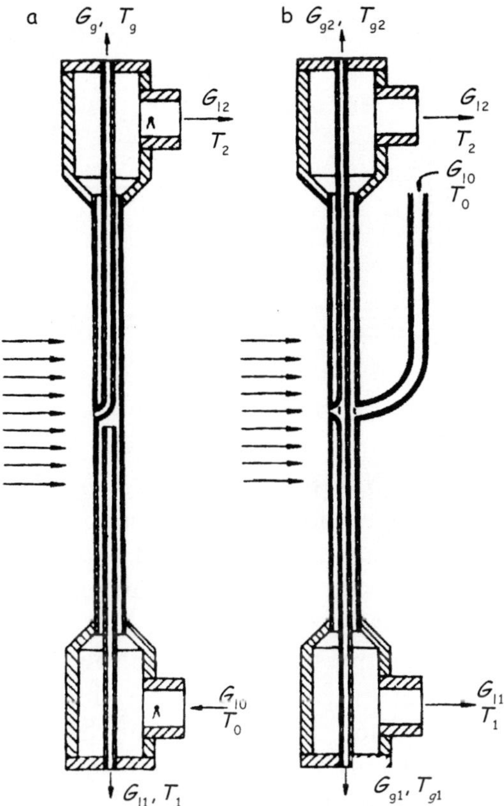

Fig.14.19 Asymmetric (a) and symmetric (b) transverse enthalpy probes with the distributed supply of cooling liquid.

$$h_0 = \frac{C_{pl}G_{l2}}{G_g}(T_2^{(2)} - T_2^{(1)}) + C_{pg}T_g, \tag{14.19}$$

where $T_2^{(2)}$ and $T_2^{(1)}$ is the temperature of the cooling liquid flowing through the measuring part of the probe, in sampling and without sampling the gas, respectively. The values C_{l0}, T_0, G_{l1}, T_1 are used to measure the total heat flow on the probe.

The examined enthalpy probes are characterised by high sensitivity and low measurement error. In addition, they operate with higher flow rates of the cooling liquid and enable the range of measurement of heat flows to be expanded.

The enthalpy probe with a symmetric supply of the cooling agent is of slightly more complicated design but enables a different method of measuring the enthalpy of a high-temperature gas flow to be applied, i.e. differential method.

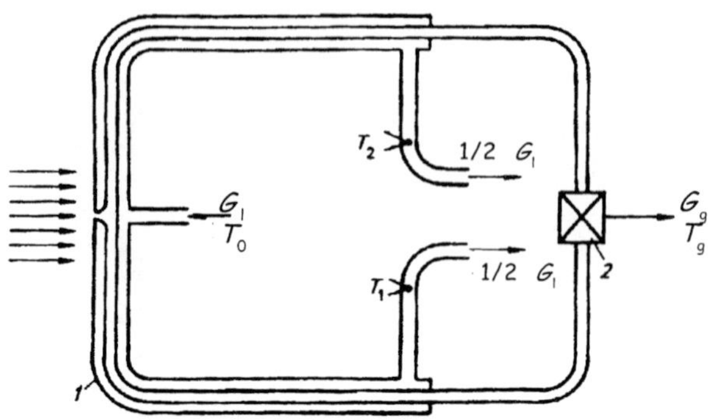

Fig.14.20 Differential enthalpy sensor.

The differential enthalpy sensor (Fig.14.20). This sensor includes a transverse enthalpy probe with symmetric supply of the cooling agent 1, the distributor of the sucked-away gas 2 and the circuit for measuring the temperature T_0, T_1, T_2 and the flow rate of the cooling liquid G_l.[52]

The differential enthalpy operates as follows. The gas is sucked away using a distributor of gas alternatively through the first or second half of the probe. Thus, the enthalpy of the sucked away gas h_e is added to the heat flows arriving to the first (lower) or second (upper) halves of the probe, Q_1 and Q_2, respectively. To increase the efficiency of measurements, it is convenient to equalise the flow rates of the gas G_g in the flow rates of the cooling liquid through both halves of the probe. In this case, in sucking away the gas through the first part of the probe we have

$$Q_1 + h_g G_g = C_{pl} G_1 \left(T_1 - T_0\right)_1 \tag{14.20}$$

$$Q_2 = C_{pl} G_l \left(T_2 - T_0\right)_1 \tag{14.21}$$

Consequently $Q_1 + h_g G_g - C_2 = C_{pl} G_g \left(T_1 - T_2\right) = C_{pl} G_1 \Delta T_1$.

In sucking away the gas through the second part of the probe we have

$$Q_1 = \left(Q_2 + h_g G_g\right) = C_{pl} G_g \left(T_1 - T_2\right)_2 = C_{pl} G_1 \Delta T_2 \tag{14.22}$$

Thus, the results of the first and second measurements give

$$h_g G_g = 1/2 C_{pl} G_l \left(\Delta T_1 - \Delta T_2\right) \tag{14.23}$$

so that we can determine the enthalpy from the equation

$$h_g = \frac{1}{2} C_{pl} \frac{G_l}{G_r} \left(\Delta T_1 - \Delta T_2 \right). \tag{14.24}$$

Measuring G_l, G_g, ΔT_1 and ΔT_2, we can compute the enthalpy of the gas.

On the whole, the procedure for measuring the enthalpy becomes more complicated because instead of three parameters (G_l, G_g, ΔT) it is necessary to measure four parameters (temperature difference is measured twice). However, the sensitivity of measurements increases because the total gas enthalpy is not measured on the background of the total heat flow ($Q_1 + Q_2$) but on the background of the difference ($Q_1 - Q_2$).

The planar enthalpy probe (Fig.14.21). The principle of the method of enthalpy measurement is almost the same as in the previous case. This probe has much in common with the enthalpy probes with a screen, with the exception of the fact that the screen in the working part is made in the form of a plane in order to enable measurements to be carried out at higher heat loads and in modelling the flow-around of

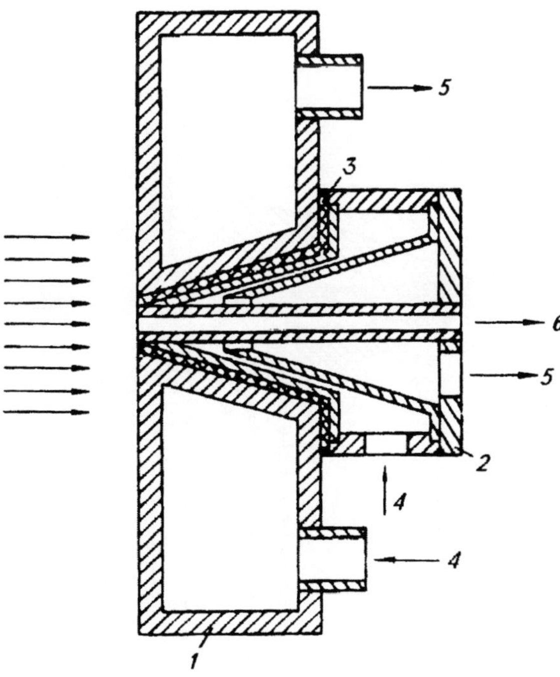

Fig.14.21 Planar enthalpy probe. 1) flat screen, 2) enthalpy probe, 3) heat insulator, 4,5) inlet and outlet of water, 6) discharge of gas.

the high-temperature flow around the plane, and the enthalpy probe in its working part is made flush with the plane.

In measuring the enthalpy with these sensors, during taking gas upwards it is necessary to consider the nature of flow around of the liquid. Of considerable importance here is the mutual position of the measured orifice and of the region where the high-region temperature flow changes its direction. At a high flow rate of the gas during taking samples of the gas and in the regime without sampling the positions of the this region may differ thus increasing the measurement error.

In the majority of cases these probes are suitable for determining the gas enthalpy in the near-axial region of the axis symmetric temperature flows.

Nonstationary enthalpy probes. These probes are used far less frequently than their stationary analogues, evidently due to their more complicated design and the need to use high-speed apparatus. In addition, it is difficult to analyse the measurement error in these probes and, as a whole, these probes and sensors have been described only seldom in the literature.

The non stationary probes can be classified in two groups. The first is similar to stationary probes where the enthalpy is determined by calorimetric measurements, and the second – the enthalpy sensor with a compensator.

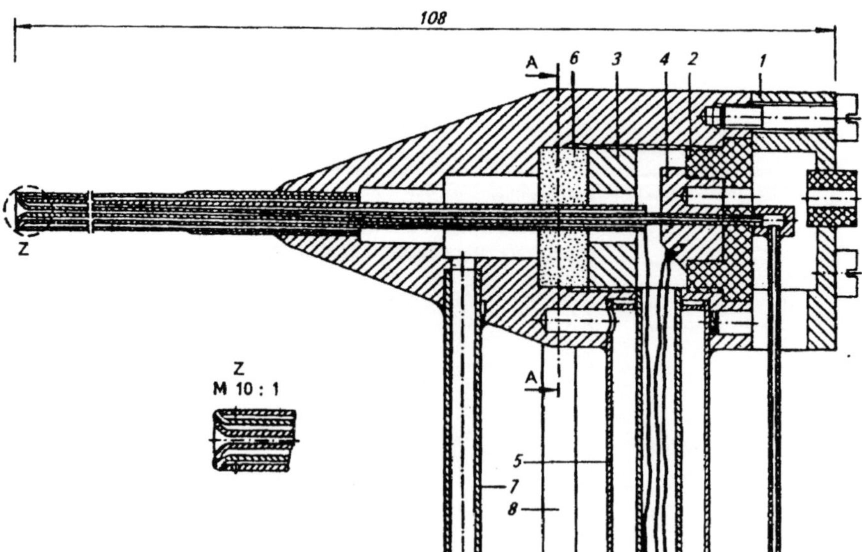

Fig.14.22 High-speed enthalpy probe.[51] 1) shielding screen, 2) insulator, 3) threaded joint, 4) securing of measuring pipe, 5) inlet of the second cooling circuit, 6) rubber seal, 7) supply of cooling water, 8) discharge of cooling water.

High-speed enthalpy probe (Fig.14.22).[51] This probe consists of three coaxial tubes with external diameters of respectively 1.0, 2.3 and 3.3 mm. The measuring channel of the probe into which the high-temperature gas is introduced during measurement, has a diameter of 0.4 mm and is 100 mm. To exclude heat losses, the central tube is separated from the measuring tube by an air gap of 0.35 mm. The cooling water is supplied (7) to the front part of the probe, and flows through the external shell (Z) between the outer and central tubes and leaves outside in the vicinity of the inlet region, radial to the axis through eight orifices with a diameter of 0.4 mm. Penetration of cooling water into other parts of the probe is prevented by the seal (6). The body of the probe is cooled using the second cooling circuit. Here the water is supplied through the tube (5), flows through the outer part of the probe body and is discharged through the second side outlet (8).

The enthalpy of the gas sample is determined by measuring the electrical resistance of the central tube. This tube of the probe should be made of nickel because of the almost linear dependence of its resistance on temperature (usually to 340°C).

In accordance with the non-stationary method of measuring the enthalpy it is necessary to know nature of the time dependence of the tube temperature and pressure at the inlet into the enthalpy sensor (see Section 14.1). The temperature of the measuring tube is determined from its electrical resistance

$$R = \int_0^L \frac{\rho}{S}[1 + \alpha_0(T_x - T_0)]\,dx, \tag{14.25a}$$

where T_x is the temperature of the measuring tube along its length; ρ is the specific resistance of the tube with length L; S is the effective section of the tube equal to $\pi(d^2_{out} - d^2_{inn}/4)$; α_0 is the temperature coefficient of resistance. At $\alpha_0 = $ const from equation (14.25a) we obtain

$$\frac{R - R_0}{\alpha_0 R_0} = \frac{1}{L}\int_0^L (T_x - T_0)\,dx. \tag{14.25b}$$

Consequently $\Delta R/\alpha_0 R_0 = \Delta T_m$.

Taking into account (14.12), we obtain an expression for the increase of the enthalpy of the gas along the length of the measuring tube L

$$\Delta h = a \frac{\mathrm{d}R}{\mathrm{d}t} \Big/ \frac{\mathrm{d}p}{\mathrm{d}t} = a \frac{\mathrm{d}R}{\mathrm{d}p},$$

$$(14.26)$$

where $a = \dfrac{\rho_t C_{pt} V_t}{\alpha_0 R_0}$; $\rho_t V_t$ is the mass of the tube; C_{pt} is the heat capacity of the tube material.

Consequently, enthalpy determination is reduced to simultaneous measurement of resistance and pressure as a function of time. Constant a is determined at relatively low temperatures at which the enthalpy can be measured by another method, for example, using some thermal element.

In the experiments, the enthalpy probe is inserted into the high-temperature flow for a short period of time. The water jacket protects it against overheating. The probes with a water jacket[19] are used so that the measurement time can be shortened. Such a probe was developed in Ref.19. Its external diameter is 1.3 mm, with a diameter of the measuring orifice being $d_i = 0.4$ mm. The measurement time in a nitrogen flow with an enthalpy of $h = 4.5 \div 12.8$ kJ/g is 20–50 ms. 5 µs is sufficient for an enthalpy measurement cycle.

When using sensors of this type, the sampled gas flows through the measuring pipe at a constant rate so that the pipe is usually connected to the evacuated vessel. The enthalpy measurement procedure may be described as follows: the probe is introduced into the plasma flow, the resistance of the internal pipe of the probe is measured as a function of time at a constant flow through the pipe. The quantity $\mathrm{d}p/\mathrm{d}t$ is determined by recording the pressure in the evacuated vessel as a function of time.

Thus, the previously examined high-speed enthalpy probes enable the enthalpy of the high-temperature gas flow to be determined within the limits of milliseconds; this is especially important when long-term holding of the probe in the flow is now possible or the duration of the effect of the flow is limited.

Nonstationary enthalpy sensor with a compensator (Fig.14.23). Here, in contrast to the previous enthalpy sensor, the sensor contains two identical tubes.[48] One of them, the basic tube 2, is used for sampling the gas, the other one is compensating. Thus, any body, placed in the main tube, with the exception of the gas sample, for example, from the external side of the probe, has no effect on the measurement results. The main and compensating tubes (resistances R_{meas} and R_{comp}, respectively) form a bridge circuit together with R_1 and R_2 which is connected, through the external pipe of the probe, to a stabilised power source.

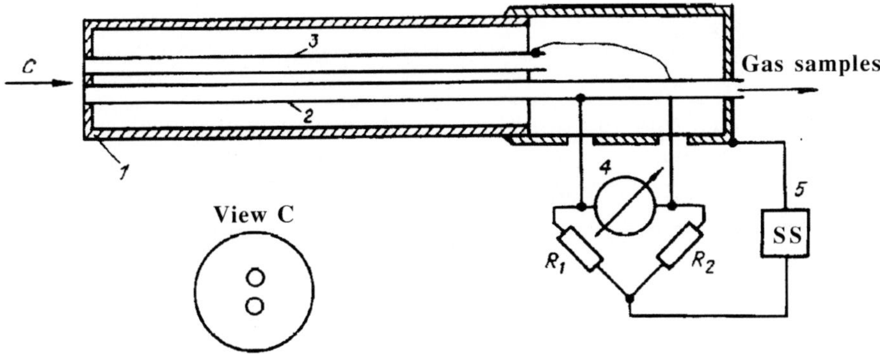

Fig.14.23 Nonstationary enthalpy sensor with a compensator. 1) enthalpy probe, 2) measuring, 3) compensating tube, 4) current recording device, 5) stabilised source.

This circuit is characterised by high sensitivity and minimum circuitry errors.

Advantages of non-stationary enthalpy probes are quite clear: 1) high operating speed (measurement time is several milliseconds); 2) small size (an enthalpy probe with a diameter of 1 mm can be produced); 3) enthalpy is measured within a single cycle – during taking gas samples, thus increasing the accuracy of measurements.

Disadvantages include higher requirements on the quality of assembling the enthalpy probe – high accuracy of centring pipes is essential.

The nonstationary enthalpy probes are not yet used widely because of complicated production and insufficient stability in the high-temperature flow. Erosion, cavitation, surface damage due to electric currents, etc., greatly shorten the service life of the probes and have an adverse effect of the extent of their application. Nevertheless, the given advantages of these devices often compensate these shortcomings.

14.4 Sensitivity of the enthalpy sensor and measurement error

The sensitivity of the sensor σ is determined as the ratio of the heat flow, received by the sensor from the gas sample in the measuring tube to the total heat flow to the sensor as a whole from the examined flow:

$$\sigma = \frac{Q_2 - Q_1}{Q_1} = \frac{\Delta T_2 - \Delta T_1}{\Delta T_1}, \tag{14.27}$$

where ΔT_2, ΔT_1 is the increase of temperature of the cooling water with and without taking gas samples, respectively. Thus, the sensitivity of the enthalpy sensor is the fraction of the heat flow resulting from taking gas samples on the general 'background' of the heat flow received by the probe.

When measuring the temperature of cooling water with an error of $\pm0.1°C$, the sensitivity of the sensor is close to 0.05.[22] In the general case, the latter also depends on the following factors: the accuracy of determination of the flow rate of cooling water and sucked-away gas, the stability of parameters, determining the intensity of the heat flow to the sensor.

Evaluation of sensitivity according to Ref.8

The flow of the high-temperature gas is laminar and the flow of the gas in the measuring tube is governed by the Stokes law. Sensitivity is evaluated from (14.27). Using the ratios[53] between the heat flow, arriving to the measuring tube and gas sampling, at the heat flow on the probe as a whole which takes into account the effect of the length of the measuring tube and the gas sampling conditions.

The maximum sensitivity of the enthalpy sensor corresponds to the case in which the sampled gas is cooled down in the measuring tube to the temperature of its walls. In this case, the conditions of heat exchange between the gas and the tube walls and the nature of the gas are determined only by the dimensions of the measuring channel. The flow rate of the gas to the sensor should not exceed the local flow rate of the flow in the sampling area[50] otherwise the structure of the flow will be greatly disrupted. This is a very important restriction which will be discussed later.

The maximum sensitivity of the enthalpy sensor is calculated from the equation

$$\sigma_m = \frac{h\rho v_f S_i}{Q_f},$$

(14.28)

where h is the enthalpy of the gas in the examined flow; ρv_f is the mass flow rate of the gas in the flow; S is the cross section area of the orifice of the probe for gas sampling ($S_i = \pi d_i^2/4$); Q_f is the total heat flow to the sensor.

The value of σ_m for every specific sensor can be calculated if we know its geometrical dimensions and the parameters of the high-temperature gas. As an example, Fig.14.24 shows the calculated dependences of the maximum sensitivity of the transverse enthalpy sensor on temperature for plasma flows of argon and oxygen. Calculations were carried out for a number of fixed values of the linear flow speed of

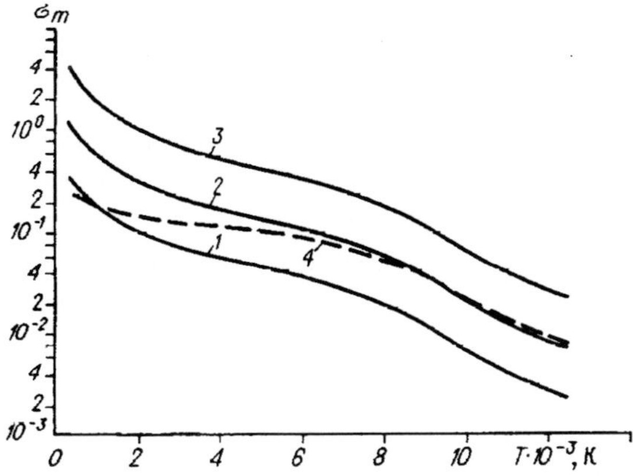

Fig.14.24 Dependence of maximum sensitivity of the enthalpy sensor on gas temperature in the flow. Argon, outer diameter of the sensor 5 mm, gas flow speed 20 (1), 100 (2), 1000 m/s (3), 4) $v_f = v_0 (T_f/T_0)$, $T_0 = 1000$ K, $v_0 = 10$ m/s.

the flow (10; 100 and 1000 m/s). The total heat flow to the sensor was calculated from the criterial dependence

$$Q_f = Q + h\rho v_f S_i ,$$

$$Q = 0.5 \frac{\lambda}{d} \mathrm{Re}^{0.5} \mathrm{Pr}^{0.4} \left[\frac{(\rho\mu)_f}{(\rho\mu)_c} \right]^{0.2} \left[1 + (\mathrm{Le}^{0.52} - 1) \right] \frac{\Delta h}{C_p} \pi dl, \qquad (14.29)$$

where Q_f is the total heat flow received by the sensor under the conditions with the gas sampling; λ, ρ, μ is the thermal conductivity, density and viscosity of the plasma, respectively; d is the outer diameter of the sensor sheath, and l is the length of the sensor counted from the measuring orifice. Calculations were carried out per unit length of the sensor $l = 1$ cm for a diameter of 0.5 cm. The diameter of the measuring orifice was assumed to be equal to 1 mm.

As indicated by Fig.14.24, the sensitivity of the sensor decreases with increasing temperature, whereas it increases with increasing flow speed. Curve 4 was calculated under conditions in which the increase of the flow speed is directly proportional to the increase of the plasma temperature. Similar dependences have been recorded for oxygen,[22] nitrogen and air. In the temperature range above 8000 K the value of σ_m is close to or lower than 10%, which is already unsuitable from the practical viewpoint or requires using high-accuracy measuring apparatus.

Thus, the enthalpy probe can be used efficiently at temperatures below 8000 K and relatively high flow speeds of the high-temperature gas.

The error of enthalpy measurements. This error can be evaluated by two methods. Firstly, by comparing the mean mass enthalpy of the flow produced by integration of spatial distributions of the enthalpy in the flow measured using the probe, and the values determined from the thermal balance of the plasma flow generator. Secondly, by comparing the enthalpy measured by the probe with that determined by another method.

The authors of Ref.35, 42 and 47 analyse in detail the factors effecting the measurement error. For example, for a stationary Grey probe the rms error of enthalpy measurement is ±4 and ±6% for enthalpy probes with screens (the so-called four-barrel and six-barrel probes).

The measurement error can be reduced by reducing the external dimensions of the enthalpy flow and by using flow rates of the gas samples not interfering the flow.

The authors of Ref.49 and 50 evaluated the error of measuring the gas enthalpy in the flow in relation to the ratio between the unperturbed flow speed and the speed of the sucked-away gas taking into account the design dimensions of the sensor. Measurements were carried out in a free plasma jet of hf discharge excited using a metallic split chamber with a nozzle with a diameter of 1.4 cm placed in the output part of the chamber[50] (Fig.14.25). Measurements were taken with a longitudinal enthalpy probe with an outer diameter of 5 mm and with measuring orifices of diameters d_i = 0.4; 1.0; 1.6; 2.2 mm. Figure 14.25 shows the calibration curve constructed for a longitudinal enthalpy probe (h_∞ is the enthalpy of the unperturbed plasma flow; h_i is the enthalpy measured by the sensor; $\alpha_i = G_i/G_{i\infty}$, where $G_i = \rho v_i \pi d_i^2/4$ is the gas

flow rate of the sucked-away gas, and $G_{i\infty} = \rho v_\infty \pi \dfrac{d_i^2}{4}$ is the mass flow

rate of the unperturbed plasma flow through the section equal to the inlet section of the measuring tube). The flow rate of the sucked-away gas was G_i = 0.032 g/s and was maintained constant to reduce the measurement error, *i.e.* α_i was varied by varying the plasma flow parameters.

Figure 14.25 shows that v_i should not greatly exceed the speed of the non-perturbed plasma flow. In Ref.28, it was shown that at $v_i/v_{i\infty} \simeq 10$ the perturbation introduced by the probe propagates upwards along the flow to the distance of almost $3d_i$.

In sampling the gas through the measuring orifice the current lines are divided into two parts. One part of the flow travels into the measuring

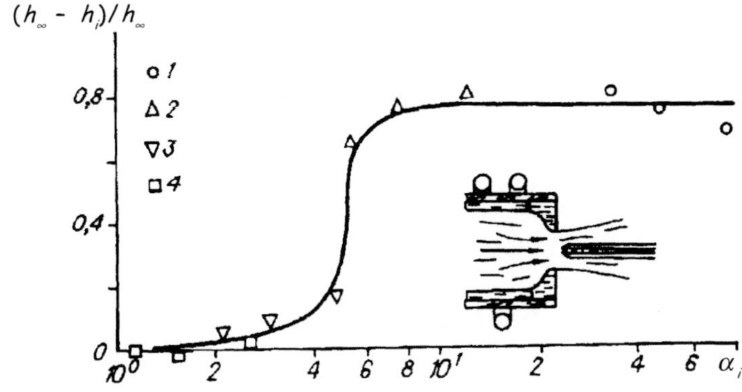

Fig.14.25 Calibrating curve of the enthalpy probe. $d_i = 0.4$ (1), 1.0 (2), 1.6 (3), 2.2 mm (4).

tube through which the gas was sampled, the other part flows around the probe. The current lines are separated at specific points on the surface of the nose part of the probe, and only the normal component of the speed varies from zero here. These points on the surface of the asymmetric enthalpy probe are situated in the top part of the probe around the circumference[55] which at $\alpha_i = 1$ coincides with the circumference of the inlet orifice or the measuring tube. As indicated by Fig.14.25, the diameter of this critical circumference can exceed the diameter of the measuring orifice by no more than a factor of 2.

Since the data are limited, it is difficult to analyse the errors. At present, it is usually possible to carry out only the analysis described previously it is assumed that the error of enthalpy measurement is not less than 4%.

14.4 Conclusions

The enthalpy probes can be used for sufficiently accurate and efficient determination of the enthalpy of high-temperature gas flows. The methods of measurement and the design of the enthalpy probes are greatly varied. The development of measuring complexes and automated systems may greatly expand the possibilities of using enthalpy probes and sensors for enthalpy measurements.

Chapter 15

PROBE MEASUREMENTS OF HEAT FLOWS IN PLASMA JETS

15.1 Measurement procedure

Heat flows in a plasma jet are measured by the well-known calorimetric method.[1] In this method, the heat flow is evaluated on the basis of the rate of variation of the resistance of the wire introduced into the plasma flow.

The following disadvantages were revealed in the process of measurement of the heat flows in the plasma jet by this method. Loop oscilloscopes do not make it possible to measure with sufficient accuracy the variation of the resistance of the wire after its insertion into the jet. Consequently, the wire fractures after each insertion and the experiment time is very long due to the need to calibrate the system and insert the next wire. In addition, movement through the high-temperature medium and vibrations of the wire at the moment of arrest greatly exceed the systematic error of the method. Therefore, to carry out the local measurements, the method was improved so it was possible to measure the heat flow on to the wire which intersect the plasma jet without failure. The direction of wire entry was selected in such a manner that the integral heat flow was measured in the fixed section of the plasma jet along a chord at distance x from the axis. The modified method greatly increased the speed and reliability of the measurement.

The principal diagram of equipment is shown in Fig.15.1. The sensor of the heat flow 2 – tungsten wire 0.1 mm in diameter and 10 mm long - is secured in thin metallic holders 3 and placed on a long (~1 m) arm of the rotating system (rod 5). During rotation of the rods the wire intersects, at a linear speed of 4 m/s, the plasma flow along the chord situated at distance x from the jet axis (Fig.15.2). Changing the mutual position of the wire and the plasma torch, it is possible to measure the heat flows at different distances from the axis and at different sections of the plasma jet. The wire is connected to the measur-

386

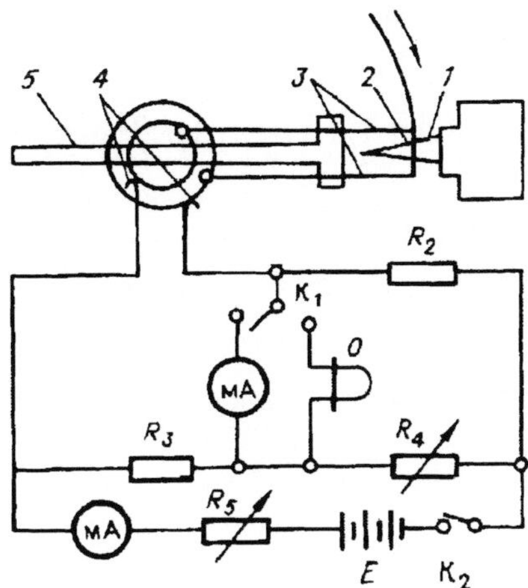

Fig.15.1 Diagram of equipment for measuring heat flows. 1) plasma jet, 2) tungsten wire, 3) holder contacts for moving contact of constant resistance, 5) rotating rod.

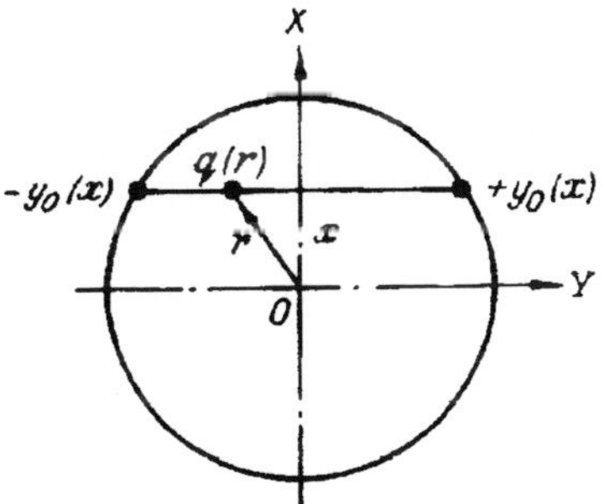

Fig.15.2 Diagram of the cross-section of the jet of plasma intersected by the probe along a cord.

ing system (Fig.15.1) through the moving contact of contact resistance 4. The duration of its passage through the plasma jet is 0.005 s; the variation of resistance in heating the plasma is recorded by the measuring system. This system consists of a Winston bridge with one of the arms containing the measuring wire. The diagonal of the bridge is connected to a loop oscilloscope (type H-102). The bridge is powered by a battery of accumulators with a total voltage of 25 V. Rheostat R_5 is used to measure the sensitivity of the bridge system. Prior to measurement, to control balancing, a milliammeter is connected to the diagonal of the bridge instead of the oscilloscope. The bridge is balanced by the variable resistor R_4. During the period of increase of the main signal (Fig.15.3) the voltage drop in the wire is recorded as a result of an inherent electrical field of the jet.[2] The oscillogram shows

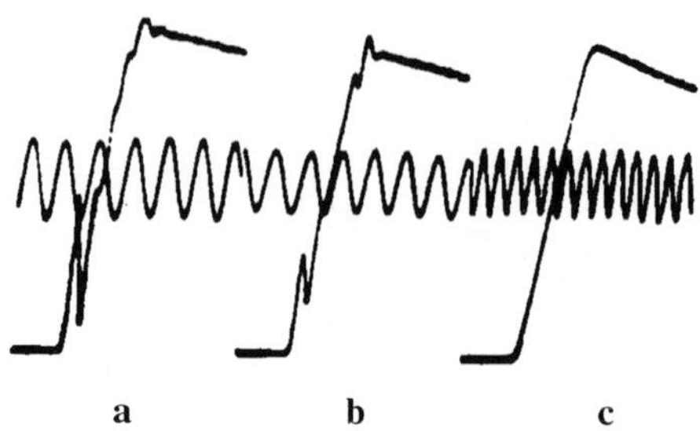

a   b   c

Fig.15.3 Oscillograms of changes of the resistance of wire recorded at different sensitivity of the measuring system at distances of 2 (a), 15 (b) and 30 mm (c) from the outlet of the plasma torch.

clearly this additional signal superimposed on the main signal. Special measurements show that heating of the wire, caused by the electric field of the jet, is negligible and can be ignored.

To determine the relationship of the local heat flows with the resistance of the wire we shall examine the cross section of the plasma jet in which the wire is situated at a distance x from the axis (Fig.15.2). The heat flow to the section of the wire with the length dL during period dt is equal to $q(t)$dLdt, where $q(t)$ is the heat flow per unit length and unit time. The total heat flow on the section of the wire with length dL during the period of intersection of the plasma is

$$\int_0^{t_0} q(t)\mathrm{d}L\mathrm{d}t.$$

The total amount of heat received by the entire wire at a distance x from the jet axis is

$$\Delta\,\theta(x) = \int_0^{L}\int_0^{t_0} q(t)\mathrm{d}L\mathrm{d}t = L\int_0^{t_0} q(t)\mathrm{d}t,$$

where dt = dy/v, v is the speed of the wire. At $t = 0$ $y = -y_0$, at $t = t_0$ $y = + y_0$ and, consequently,

$$\theta(x) = \frac{L}{v}\int_{-y_0}^{y_0} q(y)\mathrm{d}y. \tag{15.1}$$

On the other hand, the total amount of heat received by the wire can be determined as follows

$$\theta(x) = mC\Delta T(x), \tag{15.2}$$

where m is the mass of the wire, $m = \rho SL$, where ρ, S, L is the density of tungsten and the cross section of length along the wire, respectively; $\Delta T(x)$ is the variation of the wire temperature intersecting the given cross section at the distance x from the jet axis; C is the heat capacity of tungsten. Using the well-known dependence of the variation of the wire resistance with temperature

$$R(x) = R_0[1 + \beta \Delta T(x)],$$

we obtain

$$\Delta T(x) = \frac{\Delta R(x)}{\beta R_0}, \tag{15.3}$$

where R_0 is the initial resistance of the wire, $\Delta R(x)$ is the variation of the resistance of the wire during holding in the plasma jet, β is a temperature coefficient of the resistance of tungsten. Equating the right hand part of the equation (15.2) and (15.1) and taking (15.3) into account, we obtain

$$\int_{-y_0}^{y_0} q(y)\mathrm{d}y = \frac{\rho CSv}{\beta} \frac{\Delta R(x)}{R_0}. \tag{15.4}$$

The experiments were carried out to determine $\Delta R(x)$ and the data was then used to determine the value in the right hand part of the equality since β, C, ρ, S and v are known. Using Abel's integral transformation for (15.4), we determine the radial distribution of the heat flow in the examined cross section of the plasma jet per unit length of the wire for 1 second. Thus, to determine the heat flow, it is necessary to record during measurements only the maximum value of the resistance of the wire during its passage through a specific chord of the examined cross section of the plasma jet. The absolute value of the variation of the wire resistance $\Delta R(x)$ is determined using a calibration curve from the maximum amplitude of deviation of the loop. The recording system is calibrated using a cartridge with resistances connected instead of the wire.

In this method, the error of measuring the heat flow consists of the random error, determined by the scatter of the measured values of the amplitude of the deviation of the loop and the systematic errors of the method. The relative error measurements of deviation of the loop is ~2%, in the axial region and ~7% in the peripheral region of the jet. Since the wire is heated only during its passage through the plasma jet, there is some period during which part of the wire is still in the wire and is heated up, whereas the other part, leaving the plasma jet, already starts to cool down. Consequently, we record the reduced value of the variation of resistance ΔR. For estimates, we shall assume that time is equal to L/v, where L and v is the length and linear speed of the wire, respectively, *i.e.* for the case in which the entire wire left the jet and cooled down. The curve shown in Fig.15.3 shows that the decrease of the amplitude of the loop during this period is ~30% of the maximum deviation. Evidently, this value should also be regarded as the maximum estimate of the systematic error typical of the given method of measuring the heat flows. The error of measuring the heat flow taking into account the reproducibility of the results and the systematic error of the method is ~5% in the axial and 10% in the peripheral regions of the jet. It should be note that, changing the sensitivity of the measuring system, the method can be used to examine not only plasma flows but also flames.

The results of measurements in the form of the lines of a constant heat flow are presented in Fig.15.4. The numerical solution of Abel's integral was carried out in a computer. As shown in Fig.15.4, the transfer of heat to the wire takes place mainly in the axial direction and is in good agreement with the assumptions on the laminar nature of the plasma flow at the gas flow rate shown in the figure. In the region away from the nozzle outlet of the plasma torch the distribution of the heat flow widens as a result of rapid mixing of the jet with a surrounding gas and due to the formation of turbulence.

15.2 Determination of plasma temperature

The spectroscopic methods characterised by high accuracy can be usually used to take measurements of temperature only in the near-axial part of the plasma flow. The distribution of temperature over the entire cross section of the plasma jet can be determined by the calorimetric method[3] from the measured values of the heat flow and the flow speed of the plasma. The heat flow to the solid introduced into the plasma flow is determined by the equation[4]

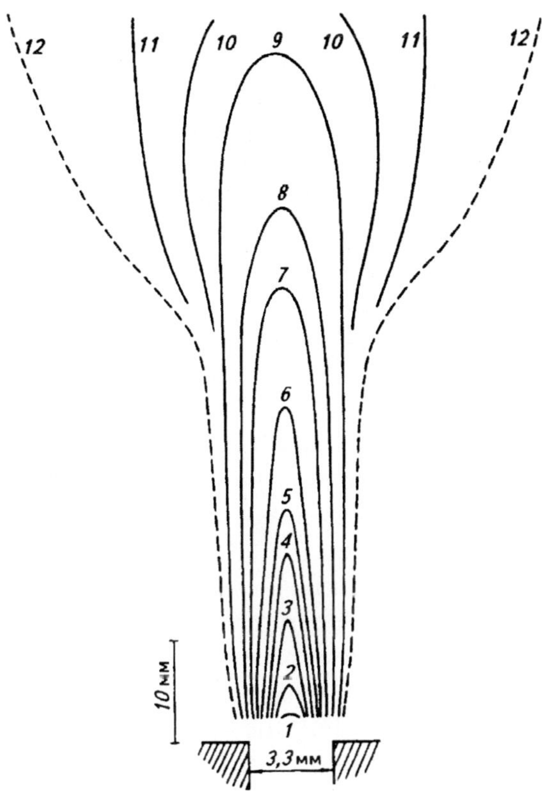

Fig.15.4 Distribution of heat flow on a wire 0.1 mm in diameter in a plasma jet at an argon flow rate of 0.18 g/s. q = 3500 (1), 3000 (2), 2500 (3), 2000 (4), 1500 (5), 1000 (6), 500 (7), 300 (8), 100 (9), 50 (10), 20 (11), 0 cal/(cm² s) (12).

$$q = \alpha \frac{H'' - H'}{C_p}, \tag{15.5}$$

where q is the heat flow, cal/(cm²·s); $H''-H'$ is the change of the plasma enthalpy when cooling down to the temperature of the sensor wall; C_p is the heat capacity of the plasma at constant pressure; α is the heat transfer coefficient which in transverse flow-around the cylinder is calculated from the equation[3,5]

$$\alpha = 0.86 \frac{\kappa}{d} \mathrm{Re}^{0.41} \mathrm{Pr}^{0.35} \left(\frac{\mathrm{Pr}''}{\mathrm{Pr}'} \right)^{0.25}, \tag{15.6}$$

where $Re = \rho v d/\mu$ is the Reynolds number; $Pr = \dfrac{\mu C_p}{\kappa}$ is the Prandtl number; d is the diameter of the cylinder with flow-around (wire); κ, ρ, μ, H, C_p, v is respectively the heat conductivity, density, viscosity, enthalpy, heat capacity and the flow speed of the plasma. The values denoted by a single prime correspond to the gas temperature equal to the temperature of the wire, those denoted by two primes correspond to the temperature of the incident flow. The quantities without the primes were calculated at temperatures specifying the heat exchange in the boundary layer at the wire. This temperature can be represented by the plasma temperature (incident flow), the mean arithmetic temperature, the temperature corresponding to the mean enthalpy of the gas in the boundary layer. The calculation results show that the selection of the temperature does not have any significant effect on the calculated plasma temperature (maximum difference with the temperature measured spectroscopically[2] is 10%). The controlling temperature here was the temperature corresponding to the mean enthalpy of the gas in the boundary layer. The plasma temperature values obtained in this case occupy an intermediate position in relation to those calculated at other controlling temperatures.

The equations (15.5), (15.6) link the heat flow on to the introduced solid with the speed and temperature of the plasma jet. Assuming that the plasma characteristics κ, ρ, H, μ, C_p are known temperature functions, equation (15.5) can be written in the following convenient form

$$qv^{-0.41} = f(T). \tag{15.7}$$

The left part of equation (15.7) includes directly measured quantities whereas the right hand part depends only on temperature. To calculate $f(T)$ (Fig.15.5), the values of the thermophysical parameters of argon plasma were taken from Ref.5–8. To determine the plasma temperature in the near-axial part of the jet we use the results of velocity measurements.[2] The determined temperature agrees in the range 5% with the temperature measured in spectroscopic examination (Fig.15.6) which indicate that there are no large systematic errors in the calorimetric method.

Estimates show that the error of calorimetric measurements of the temperature is associated with the error of measuring the quantities q and v for the near-axial part of the plasma jet from the relationship

$$\delta T \approx 0.35\delta q + 0.14\delta v$$

and equals 3–4%. However, the main error of the method is caused by the well-known scatter of the values of the thermophysical parameters of the plasma (μ, H, C_p) and equals approximately 10%.

In the peripheral part of the jet the flow speed was not measured so that to estimate the temperature in this region on the basis of the measured heat flows we accepted a parabolic radial velocity distribution. It is evident that this assumption makes the results of temperature calculations for the jet periphery (Fig.15.6) less reliable in comparison with those for the axis. However, it should be noted that the indeterminacy in the velocity has a relatively weak effect on the temperature measured by the calorimetric method. Figure 15.5 shows that $\delta T \sim 0.24\delta v$ at a temperature of 1000–8000 K, $\delta T \sim 0.41\delta v$ at 8000–10000 K and $\delta T \sim 0.14\delta v$ at 10000–14000 K.

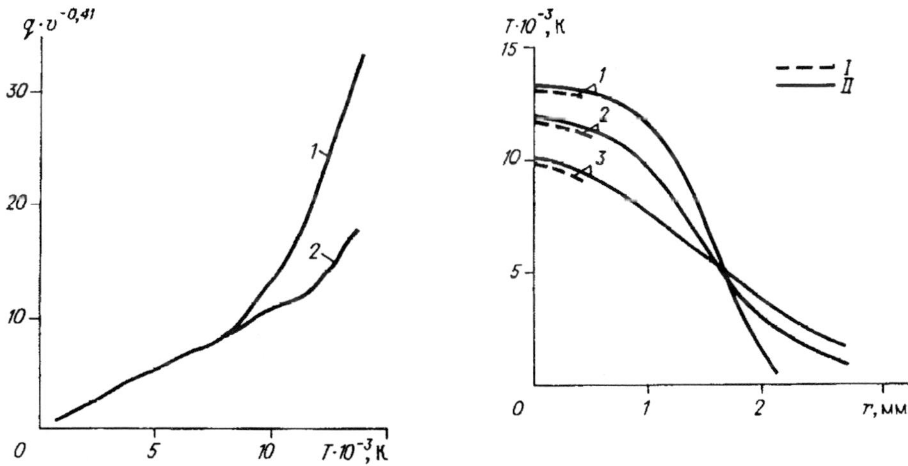

Fig.15.5 Function $f(T) = qv^{-0.41}$ for the case of flow around of the cylinder by the plasma flow of atmospheric pressure of argon in the LT condition taking into account the transfer of heat by the atoms, ions and electrons (1) and by atoms and ions only (2). Wire diameter 0.1 mm, q in cal/cm² s, v in cm/s.

Fig.15.6 (right) Radial distribution of spectroscopically (I) and calorimetrically (II) measured temperature in a plasma jet at distances of 2 (1), 15 (2) and 30 mm (3) from the outlet of the plasma torch.

Chapter 16

METHODS OF EXAMINING THE SPATIAL STRUCTURE OF RADIO-FREQUENCY CAPACITANCE DISCHARGES

16.1 Introduction

The radio-frequency capacitance discharge (RFCD) is one of the simple, reliable and efficient sources of plasma in various regions, for example, in plasma chemistry,[1-3] plasma technology,[4-7] and the technology of gas lasers,[8-11] and others. Therefore, a large number of publications concerned with the methods of producing, examining and using RFCD in a wide range of gas pressures from fractions to hundreds of torr and frequencies of the RF field from 10^6 to 10^8 Hz is not surprising.

It is well known that the plasma of RFCD in these frequency and gas pressure ranges may be highly non-equilibrium, capable of filling large volumes at high specific energy inputs (100 W/cm^3 and higher), the possibility of controlling these parameters by varying the frequency of the RF field and the conditions of burning of RFCD.

Interesting results in understanding the mechanism of sustaining stationary RFCD and in its numerical applications have been obtained by the discovery of the important role of the near-electrode layers of the spatial charge (NLSC) and of the secondary emission processes taking place in them. It was found that there are various forms of the radio-frequency capacitance discharge which are characterised by the processes taking place in the NLSC and, for this reason, they have qualitative different spatial structures under the same external conditions: gas pressure and type, the size of the interelectrode gap and, in certain cases, at the same RF voltage of the electrodes. In particular, the density of discharge current in transition from one form of discharge to another may change by an order of magnitude or more,[12] and the strength of the electric field and the electron concentration change not only quantitatively but, more interestingly, their distribution in the interelectrode gap becomes qualitatively different.[13]

These facts indicate that it is promising to develop methods of obtaining, in a radio-frequency capacitance discharge, plasma with the required properties. However, to apply this method it is necessary to have clear physical assumptions of the mechanism of formation of the spatial–time structure of each form of the stationary RF capacitance discharge and have data on the regions of their existence and reasons for the transition from one form of discharge to another.

In this section, we shall discuss the experimental methods of examining the structure of RF capacitance discharge which make it possible to describe in a simple manner and, at the same time, reliably the specific features of the RF capacitance discharge and the difference between its forms, obtain information on the evolution of the spatial structure of the discharge with the change of the external conditions and the conditions of stationary burning and, consequently, create suitable conditions for justified selection of a specific form of RFCD when solving specific problems in practice. This is especially important at medium pressures where the free path of the electron is considerably shorter than the amplitudes of its drift vibrations in the plasma sustained by the RF field.

16.2 Specific features of producing, examining and using radio-frequency capacitance discharge plasma. Forms of existence of RFCD

As mentioned in the introduction, the plasma devices based on RFCD are used widely in technology. This is due to the appearance at high frequencies ($f \sim 1 \div 100$ MHz) of a number of specific features by which the RFCD differs from direct current discharges, both arc and glow. A special feature of plasma generators based on RFCD is that the electrodes are placed outside the discharge chamber. In this case, the active current, flowing through the plasma, is closed on the electrodes by the bias current. The maximum density of RF current which can be passed through the dielectric without disrupting it is determined by the equation

$$j_{max} \leq \varepsilon_d \varepsilon_0 \frac{\partial E_{br}}{\partial t} = \varepsilon_d \varepsilon_0 \omega E_{br}, \tag{16.1}$$

where $\varepsilon_0 = 8.85 \cdot 10^{-12}$ F/m; ε_d is relative dielectric permittivity; E_{br} is the electric field strength in the dielectric at which its breakdown takes place; $\omega = 2\pi f$ is the RF field frequency. For typical dielectric materials $E_{br} \sim 10^5$ V/cm; $\varepsilon_d \sim 5$. Consequently, the limiting current densities in the frequency range $f = 1 \div 100$ MHz equal 0.5–50 A/cm^2 which is fully sufficient for many applications. Therefore, RF discharges with current densities in plasma varying from units to hundreds of millimetres per square centimetre are used in most cases.[1-13]

This special feature of RFCD, i.e. the possibility of closing the discharge current by bias currents in the dielectric, coating the electrodes, has supported the view according to which in these discharges the processes of the electrodes and the near-electrode regions do not play any significant role.[14,15] Therefore, the radio-frequency capacitance discharge was regarded, especially in the pressure range $p \gg 1$ torr, as the simplest because of its spatial structure. It was assumed that the stationary retention of the plasma in RFCD is ensured at every point of the interelectrode gap by the local ionisation balance in accordance with the equation

$$\nu_i (E_{pl} / N) = \nu_{dif} + \beta n_e + \nu_p, \tag{16.2}$$

where $\nu_i (E_{pl}/N)$ are the frequencies of generation (ionisation) and annihilation ν_{dif}, βn_e, ν_p (diffusion, recombination, sticking) of the charged particles, E_{pl} is the strength of the electric field in plasma, N is the concentration of the neutral gas particles, β is the recombination factor.

Specifying the value of the electron concentration n_e, equation (16.2) can be used to determine E_{pl} if the mechanism of annihilation of the charge is known. It is important to stress that the value of n_e and the density of discharge current j (which depends on n_e) is linked with the parameters of the external circuit, assuming

$$j = e n_e \mu_e E_{pl} = \frac{I_{rf}}{S} = \text{const}, \tag{16.3}$$

where I_{rf} is the intensity of discharge current given by the external circuit, S is the electrode area, μ_e, e is the mobility and charge of the electron. The contribution of the reactive component of current in plasma is ignored in (16.3).

However, although the equations (16.2) and (16.3) are simple and obvious, they contradict the following experimental facts established in Ref.12.

1. At some values of the discharge current intensity I_{rf} the area of the electrode S_{pl} filled by the discharge plasma is smaller than S. With increasing I_{rf} S_{pl} also increases and since

$$j_{n1} = \frac{I_{rf}}{S_{pl}} = \text{const}, \qquad (16.4)$$

i.e. RFCD not only fills completely the interelectrode gap in the direction transverse to the current but is also characterised by the so-called effects of normal current density – effect in which the current density does not depend on the total current flowing through the discharge. Specific value of j_{n1} is determined by the type and pressure of the gas p, frequency ω and the width of the interelectrode gap d,[16,17] and is almost completely independent of the electrode material.

2. If $j > j_{cr}$, another form of RFCD appears in which the effect of the normal current density also operates but with different current density j_{n2} $(j_{n2} > j_{cr} > j_{n1}$, with other conditions being equal). The electrode material or its dielectric coating influences the values of j_{cr} and j_{n2}.

3. In the interelectrode gap filled with the RFCD plasma zones with very low E_{pl} appear at $j > j_{cr}$. At these values equation (16.2) is not fulfilled.

4. In addition, the RF discharge always contains near-electrode layers of the spatial charge (NLSC) (which separate the plasma from the electrode surface), and the processes in these layers are not described by (16.2) and can differ qualitatively depending on the RF voltage on the NLSC.

Another very important special feature of the RF capacitance discharge, established in Ref.12, is the restriction of the region of existence of the form of the RFCD with respect to pressure and the width of the interelectrode gap d, with current density j satisfying the relationship

$$j_{n1} \leq j \leq j_{cr}. \qquad (16.5)$$

Following Ref.12, this spatial form of the RFCD will be referred to as low-current, and the RF discharge with a current density $j > j_{cr}$ as the high-current discharge.

The difference between these two forms of the RFCD is mainly that

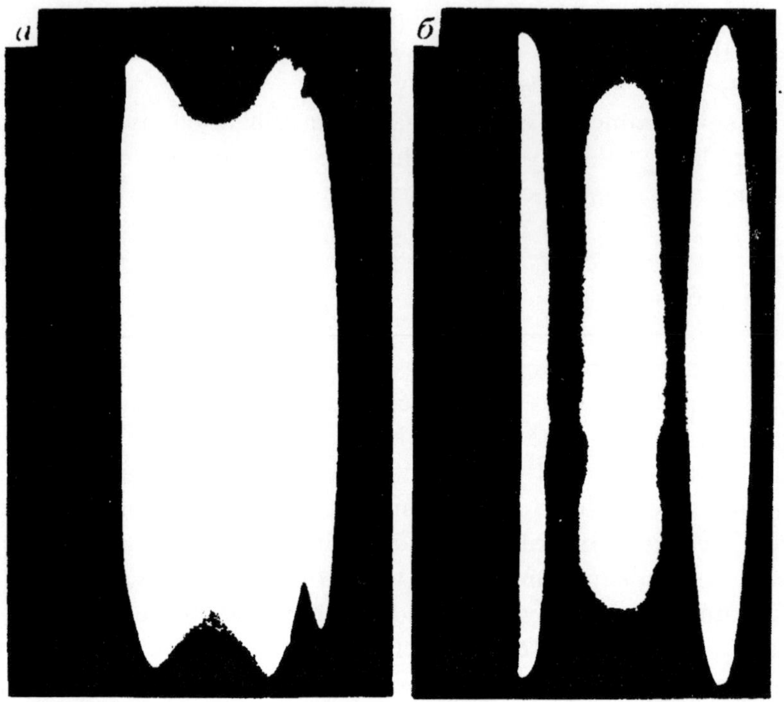

Fig.16.1 Photographs of low-current (a) and high-current (b) RFCD. Frequency 13.6 MHz, air, $p = 10$ torr, $d = 2$ cm, $U_{rf} = 320$ V, $j = 7$ (a), 120 mA/cm^2 (b).

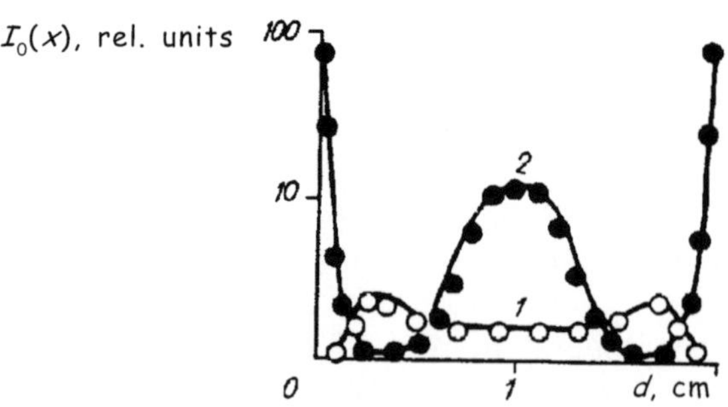

Fig.16.2 Distribution of glow through the RFCD. Frequency 13.6 MHz, air, $p = 10$ torr, 1) low-current, 2) high-current discharge.

there are qualitative differences in the distributions of the concentration of charged particles and the strength of the electric field along the direction of current passage. Visually, this is reflected in the transformation of the nature of distribution of the intensity of integral radiation of the RFCD in the visible wavelength range (Fig.16.1) when the discharge conditions are changed. Attention should be given to the sharp increase of current density (from 7 to 120 mA/cm^2) and the variation of the distribution of glow intensity $I_0\,(x)$, especially the appearance of dark regions which separate the plasma column in the centre of the discharge gap from brightly glowing near-electrode zones (Fig.16.2). In addition, the spectral composition of radiation of the near-electrode regions also changes: the radiation spectrum in the high-current discharge is close to the radiation spectrum of the negative glow of the cathode region of the dc glow discharge. The agreement between the distribution of the glow of the high-current RFCD and the distribution of glow in the dc glow discharge has been known for some time. Nevertheless, it was assumed that the processes taking place on the cathode which lead to the formation of a large number of electrons are not important for the radio-frequency discharge because the phenomena in both electrodes of RFCD are in this case completely symmetric, and the required number of free electrons forms at each electrode by ionisation of gas particles by collisions of the first kind with the electrons moving to both sides.[14]

The initial information on the possible effect of secondary emission electrons, formed in near-electrode regions as a result of bombardment of the electrode surface with positive ions, on the structure of RFCD appeared in Ref.18. This conclusion was made on the basis of indirect data: similar glow in the cathode regions of the glow discharge and near-electrode zones of the RFCD, sputtering of the electrode material and a high constant potential of the plasma in relation to the electrodes U_0 which reaches hundreds of volts at pressures of $p < 1$ torr.

An important support of the hypothesis in Ref.18 on the effect of secondary emission processes on the characteristics of the near-electrode layers of the spatial discharge and of the entire discharge as a whole is the existence of high values of U_0 which greatly exceed (by an order of magnitude or more) the 'floating' potential of the plasma corresponding to typical values of the electronic discharge temperature. The effect of sputtering of the electrodes in the RFCD also indicates the presence of high constant potentials localising the NLSC. In fact, in the discharge conditions at frequency $f > 1$ MHz even a hydrogen ion assumes the energy of the ordered oscillatory motion in the RF field which does not exceed several electron volts. This is obviously insufficient for explaining the phenomenon of sputtering of the material on electrodes in RFCD.

Irrespective of the logical nature of the assumption[18] on the role of secondary-emission processes in RFCD, this approach has not been developed further. There are several reasons for this. Firstly, the data on the probe measurements of U_0 presented in Ref.18 and 19 indicate a large decrease of U_0 (from hundreds to units of volts) with increasing gas pressure in the discharge above 1 torr, i.e. they indicate the disappearance of NLSC with increasing pressure. Secondly, the fact that the phenomena taking place in RFCD with hollow electrodes and electrodes coated with the dielectrics are identical casts doubts on the effect of secondary-emission processes on the conditions of burning of the RF capacitance discharge. Thirdly, at pressures $p < 1$ torr the second-emission electrons (γ-electrons) have only a slight effect on the parameter of the NLSC (especially in inert gases). This will be examined in greater detail later. Here it should be noted that this situation forms due to the small number of inelastic collisions of γ-electrons with neutral particles on the characteristic thickness of the NLSC equal to d_1. This means that the near-electrode layers in the low-pressure RFCD behave at high U_0 as sources of fast electrons which can influence the ionisation balance in the discharge only outside the limits of the layers.[20–22] In other words, at low pressures the transition to the high-current conditions of RF capacitance discharge is smooth, is difficult to record and this is the reason why it has not been noted for a very long period of time.

Recently, the assumptions regarding the nature of the RF capacitance discharge in a wide range of pressures and frequencies of the RF field have been explained further. Firstly, it was shown in Ref.23 that a high constant potential of RFCD plasma in relation to the electrons (above 100 V), localised in NLSC, is also observed at pressures $p \gg 1$ torr. It is assumed that near-electrode layers of a spatial discharge must form unavoidably in RFCD whereas when the pressure is increased the main reason for their appearance is not eliminated – the need for the equalisation, during the cycle of the RF field, of the number of positive ions and electrons of greatly different mobilities falling on the surface of the electrodes under the effect of the alternating field frequency. Secondly, the results show[12] that the existence of various forms of the RFCD and of transitions between them does not depend on whether the electrodes are coated with the dielectric or not. Thirdly, direct experiments[12,24] show that in transition to the high-current form the active conductivity of NLSC increases to the values comparable with the conductivity of the cathode region of dc glow discharge under the same experiment conditions. This means that the transition to the high-current burning regime can be qualified as a breakdown of capacitance (non-conducting) near-electrode RFCD layers with participation of secondary emission electrons where the Townsend criterion[12]

$$\gamma \left[\exp \int_{0}^{d_l} \alpha(E_l / N)\mathrm{d}x - 1 \right] = 1, \tag{16.6}$$

is fulfilled at a specific RF voltage on the electrodes in the NLSC with thickness d_1. Here γ is the coefficient of secondary emission of electrons from the surface of the electrode or of the dielectric coating on it; α (E_1/N) is the volume ionisation coefficient, E_1 is the strength of the electric field in the layer. In other words, when (16.6) is fulfilled, the balance of the charged particles in the NLSC of RFCD is facilitated by the ionisation multiplication of γ-electrons within the limits of the layers of the special discharge whereas in the low-current form of the discharge (is the left part of (16.6) is less than unity) ionisation is significant only in plasma.

For practical applications it is very important that the spatial structure in the high-current form of the RFCD is qualitatively different, the active component of the current in NLSC rapidly increases and at pressures of tens of torr it usually exceeds the capacitance component,[25] thus resulting in a rapid increase of the ion flux on the surface of the electrodes, failure of the electrodes and generation of a large amount of power in the NLSC. If the actual situation in the near-electrode layers is not taken into account, the results can be unexpected. For example, if it is assumed that if the RF electrodes are placed outside the limits of the discharge chamber, it may be expected that sterile plasma may be produced. This assumption holds only for the low-current form. If the pressure, interelectrode spacing and frequency of the RF field are not correctly selected, the chamber contains RFCD only in the high-current form and, consequently, the discharge volume will be contaminated with the products of sputtering of the walls of the chamber connected with the electrodes. Another example is taken from laser technology. In 1974 it was attempted to use a RF capacitance discharge, transverse in relation to the optical axis of the resonator, for pumping a stationary CO_2 laser.[26] The results were unsatisfactory. Nevertheless, at present CO_2 lasers with RF excitation are superior in their group as regards the specific characteristics[8–11,13] due to the rational application of the RFCD taking into account the special features of its spatial structure, although the main elements of the design have not been changed.

The experimental data on the structure of RFCD and the forms of its existence have also been confirmed by numerical calculations presented in Ref.27–29. It is important to note Ref.28 and 29 where the existence of two forms of RF discharge and of transition between them, including under the conditions in which electrodes were coated with the dielectric, was shown for the first time by numerical modelling.

16.3 Interpretation of volt–ampere characteristics of RFCD and their relationship with the spatial structure of the discharge

The volt–ampere characteristics of the radio-frequency capacitance discharge on the whole, i.e. the dependence of the of the RF voltage on the electrodes U_{RF} on the discharge current I_{RF} or its density j_{RF}, provide qualitative information on the structural special features of the RFCD with the variation of I_{RF} (or j_{RF}).

We shall examine static VACs of medium-pressure RFCD together with the distribution of glow along the direction of the field. Figure 16.3 shows the most general case of the dependence $U_{RF} = F_1 (I_{RF})$, and Fig.16.4 $U_{RF} = F_2 (j_{RF})$ obtained in examining a stationary RF capacitance discharge at medium pressure between flat cooled electrodes.[12,30] It can be seen that the VAC contains five different sections: 1) *OA* (*Oa* in Fig.16.4), 2) *ABC* (*ab*), 3) *CD* (*bc*), 4) *DEF* (*cd*) and 5) *FG* (*df*). The first section – interelectrode gap is not 'ruptured', the voltage of the electrodes linearly increases with the increasing I_{RF}. One U_{RF} becomes equal to the breakdown voltage (points *A*, *a*) a discharge forms whose plasma does not fill the interelectrode gap completely in the direction normal to the current. The voltage of the electrodes decreases when discharge appears (section *AB*, *ab*). Increase of I_{RF} only increases the transverse dimension of the glowing discharge zone. It is important to note that the structure of the glow along the direction of the current passage and the value j_{RF} do not change in this case (section *BC* in Fig.16.3, point *b* in Fig.16.4).

The third section of the VAC has a positive derivative with respect to I_{RF}, j_{RF}. It is usually realised in final filling of the discharge gap with plasma in the direction normal transverse to the current. The nature

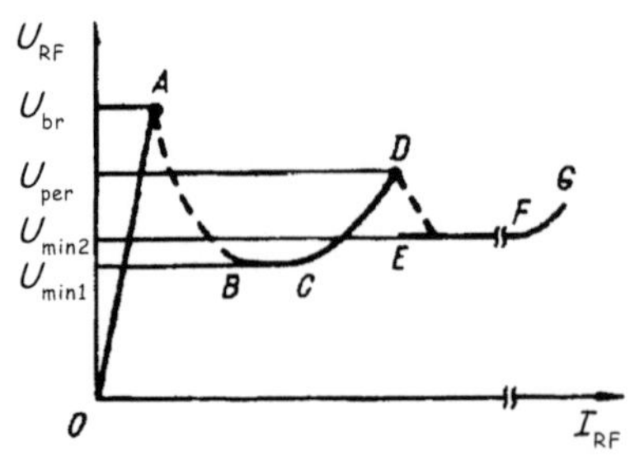
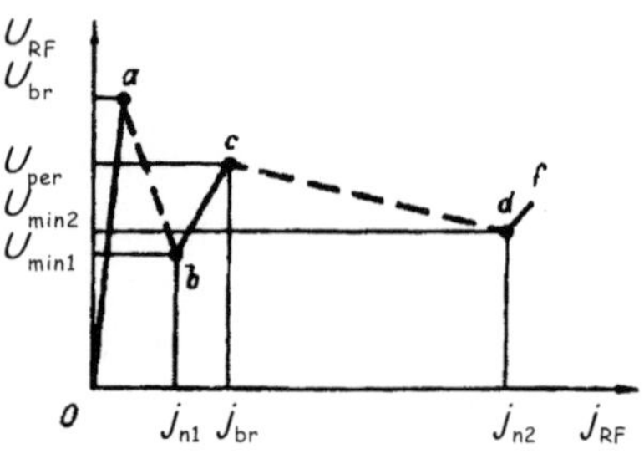

Fig.16.3 Typical volt–ampere characteristic of RFCD in coordinates $U_{rf} (I_{rf})$ (transition from *D* is possible to any point *EF*).

Fig.16.4 Typical volt–ampere characteristic of RFCD in coordinates $U_{rf} (j_{rf})$.

and distribution of the glow along the current lines are here on the whole the same as in the second section of the VAC; there is only an increase of the total intensity of light radiation, especially at the ends of the plasma column. The glow in this case is again homogeneous along the field, with the exception of the several bright interelectrode regions separated from the electrodes by dark zones with a characteristic dimension depending on the nature of the gas and the frequency of the RF field.

Measurements of the conductivity of NLSC by the method which will be discussed later showed that the active conductivity of RFCD layers in the examined section of the VAC (*BCD*, *bc*) is not high and its order of magnitude is close to that of the conductivity determined by the ionic saturation current with the ion concentration $n_+ \cong n_e$, where n_e is the concentration of electrons in the discharge plasma.

It should be noted that under the real conditions after the appearance of a discharge the value U_{RF} is not necessarily minimum. This is determined by the parameters of the external circuit, and the matching circuit of the RF generator with the discharge chamber because when plasma appears the total impedance of the discharge gap greatly changes. Consequently, matching of the RF generator with the discharge chamber can both improve and become less efficient. It is possible that after the breakdown of the interelectrode gap an anomalous regime of low-current RFCD is established there (portions *CD*, *bc*). The normal regime of the RF discharge is obtained by regulation in the external circuit resulting in a decrease of U_{RF}.

When the voltage of the electrode reaches the value U_{per} (point *D*, *c*) an inflection point or jump is observed. The jump usually occurs in discharge on molecular gases. In inert gases, for example, in helium, a distinctive transition to another regime in the VAC is visible only at relatively high pressures ($p \sim 100$ torr). The inflection point or jump in the VAC is accompanied by the distribution of the glow along the direction of current passage, and the new structure of the glow can be compared with two normal dc glow discharges which are distributed in such a manner that the common positive column is in the centre between the electrodes and separated from them by regions identical with the Faraday dark space and the negative glow (see Fig.16.1b and 16.2).

The most important special feature of the examined RFCD regime is the abrupt increase, in comparison with the previous type of discharge, of the active conductivity of the near-electrode layers to characteristic values corresponding to the conductivity of the cathode region of dc glow discharge under the same conditions. Another important typical feature of the RFCD after the jump on the VAC is the large increase of the discharge current density: for example, in air at $p = 30$ torr and a frequency of 13.6 MHz it changes in transition to the new regime from 12 to 240 mA/cm². The current density jump is associated not only with the appearance of a significant active component of current density j_a but also with a rapid decrease, especially in molecular gases, or the thickness of the NLSC d_1 which determines the density of bias current at a fixed voltage in the layer U_1

$$j_{cm} = \varepsilon\varepsilon_0\omega E_l \cong \varepsilon\varepsilon_0\omega\frac{U_l}{d_l}.$$ (16.7)

For example, if in RFCD in air in the regime to the left of the jump on the VAC at $p = 15$ torr the value $d_{11} \approx 4$ mm, measured from the distribution $U_0(x)$, then in the new regime it does not exceed 0.3 mm.[12] This means that, according to equation (16.7), j_{cm} in the regime after the jump on the VAC increased 13 times. It is important to know that the thickness of the layer in the new burning regime is almost independent of the frequency of the RF field but depends strongly on pressure

$$d_{l2} \cong \frac{C_1}{N} = C_1\frac{T}{pT_0},$$ (16.8)

where C_1 is a constant determined by the type of gas and electron material, determined at $T_0 = 300$ K; T is the gas temperature in the NLSC.

Comparison of the parameter of NLSC to the left and right of the jump on the VAC indicates that the Townsend condition (16.6) is fulfilled in the near-electrode layers of the RFCD corresponding to the points D and c on the volt–ampere characteristics. This means that the VAC

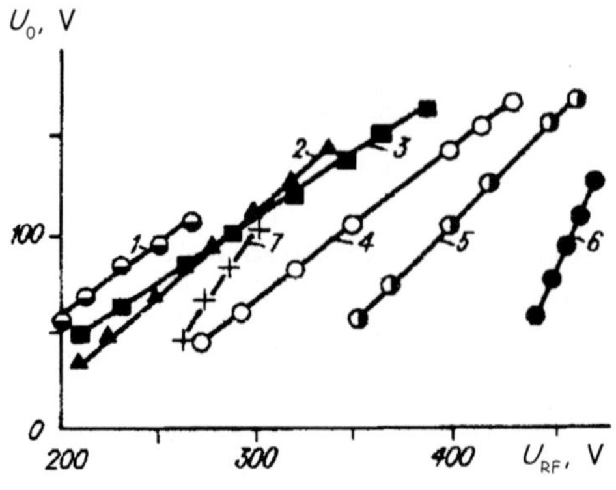

Fig.16.5 Dependences of constant potential of plasma in relation to electrodes U_0 on U_{RF}. Low-current RFCD, frequency 13.6 MHz; 1,7) helium, $p = 75$ and 150 torr, respectively; 2) air, $p = 7.5$ torr; 3–6) CO_2, 7.5; 15; 30; 50 torr.

jump, observed in the experiment, is caused by the breakdown of the NLSC with secondary emissions taking part in the process and by the transition of the RFCD to the high-current regime, as already indicated in the previous section.

The results of measuring the constant plasma potential in relation to the electrodes U_0 in dependence on the value of the RF voltage U_{RF} applied to the electrodes, are presented in Fig.16.5 and indicate the increase of U_0 with increasing U_{RF}. They can be used as another argument in favour of the validity of the assumptions regarding the breakdown of the NLSC because the constant voltage U_0 is localised in the layers.[12]

Other experimental verifications of the conclusions of the breakdown of the NLSC have also been made. In particular, investigations were carried out into the effect of electrode material and the type of gas on the transition voltage to the high-current regime.[12] In addition to U_{per}, the constant plasma potential U_{0per} at the moment of the jump of the VAC was recorded. The results were compared with the data obtained for the case of breakdown of the constant electric field[31,32] and showed good agreement. As in the constant field, U_{per} and U_{0per} are determined by the combination 'type of gas–electrode material'. As the breakdown voltage in the constant field for the specific gas–electrode pair increased, higher U_{per} and U_{0per} were required for the transition of NLSC of RFCD to the high-current regime. Analysing the transition of RFCD to the high-current regime, it is necessary to select, as the parameter pd, the value of pd_1, i.e. examine not the size of the interelectrode gap but the thickness of the NLSC d_{11} of the low-current type of the RF discharge.

Thus, the breakdown of NLSC of the low-current RFCD with secondary emission electrons taking part in it with the Townsend condition fulfilled in the layers restricts the upper permissible values of current density that are realised in the volume form of this type of discharge.

The expression for the limiting current density in the NLSC of the low-current RFCD follows from the breakdown condition

$$j_{br} \cong \varepsilon\varepsilon_0\omega\frac{U_{br}}{d_{11}} \cong \frac{\varepsilon\varepsilon_0\omega^2 U_{br}}{v_{dr}}, \tag{16.9}$$

where U_{br} is the voltage on the NLSC at which a jump is observed on the VAC, v_{dr} is the drift speed of the electrons in plasma. The ratio $d_{11} \simeq v_{dr}/\omega$ is used in this case.[12]

The new stable state of the NLSC of the RFCD in which the Townsend conditions is satisfied is obtained at the thickness of layers satisfying the equation (16.8). In this case, the number of ionising collisions of the electrons with neutral particles in the layer reaches the maximum value and the voltage in the layers is minimum.[13,32] This means that the normal current effect (section EF) is realised in the layers of the high-current RFCD. It is well known[12,25] that the active component of the current density in NLSC with the Townsend criterion fulfilled is determined by the type and density of gas and electrode material

$$j_a = C_2 N^2 = C_2 \left(\frac{T_0}{T}\right)^2 p^2. \tag{16.10}$$

Here C_2 is the tabulated constant,[31,32] which depends on the electrode material and the type of gas, and is obtained at $T = 300$ K. In addition to j_a, the NLSC carries the capacitance current, and the expression for this current can be written in the form taking into account (16.8)

$$j_{cm} = \varepsilon \varepsilon_0 \omega E_l \cong \varepsilon \varepsilon_0 \omega \frac{U_l}{d_{l2}} \cong \frac{\varepsilon \varepsilon_0 \omega p (T_0 / T)}{C_1}, \tag{16.11}$$

Taking into account that the phase shift between j_a and j_{cm} is equal to $\pi/2$, we obtain the equation for the normal current density in the NLSC of the high-current RFCD

$$j_{n2} = \sqrt{j_a^2 + j_{cm}^2} = C_2 \left(\frac{T_0}{T}\right)^2 p^2 \sqrt{1 + \left[\frac{\varepsilon \varepsilon_0 \omega U_l}{C_1 C_2 (T_0 / T) p}\right]^2}. \tag{16.12}$$

If the current passed through the discharge is higher than $I_{RF} = j_{n2} S$, the anomalous regime of the high-current RFCD is realised (sections FG and df) in which the RF voltage in the electrodes and the NLSC and the width of the regions of negative glow and the Faraday dark space l increase and the size of the glowing plasma column along the direction of d_{pl} decreases. This is indicated by the dependence of l on I_{RF} obtained in examining RFCD in helium at a frequency of $f = 13.6$ MHz at $p = 15$ torr between flat electrodes with a distance between them being $d = 60$ mm (it is evident that $d = 2l + d_{pl}$ with d_{l2} ignored):

I_{RF}, A	1	1.2	2	3	4	6	8
l, mm	7	7	10	12	16	20	26

It is evident that this phenomenon is associated with the ionisation of the gas by the electron beam formed in the NLSC of the anomalous high-current RFCD.

Another experiment was carried out to confirm the assumption on the controlling role of the electron beam under the examined conditions. A high-current RFCD was excited in the same discharge volume and under the same interelectrode spacing and the helium pressure. However, in this case, the electrodes were ring-shaped to ensure that the electron beam, formed in the cathode layer, propagates in the radial direction (instead of propagating along the discharge axis) and cannot have any effect on the structure of the part of the plasma column which is situated between the electrodes. The results show that at the I_{RF} values given above there was no reflection of the columns: on the contrary, l decreased with increasing I_{RF}. Similar phenomena were observed in Ref.33. At very high current densities the high-current RFCD collapsed to the arc regime with high electrode erosion. However, this region of the conditions was not examined in this work.

We shall now return to the BC section of the volt–ampere characteristic of the low-current RFCD, i.e. the region of the effect of normal current density for this discharge form. The key to understanding the nature of the effect is the observed dependence of current density j_{nl} on electrode spacing d which indicates the more important role played by the plasma column in the low-current RFCD than, for example, in the high-current form where the positive column in a wide range of the external conditions is 'aligned' under the current density determined by the NLSC conditions.

A large amount of information on the dependence of the parameters of the low-current RFCD on the length of the plasma column was provided by the experiments in Ref.34. The following procedure was used. A thin dielectric sheet was placed in the interelectrode gap formed by two flat electrodes with the area S. The dielectric sheet was placed parallel to the working surface of the electrodes and could move in the gap. The distance from one electrode to the sheet is denoted Δ_1, to the other Δ_2. It is evident that $\Delta_1 + \Delta_2 \simeq d$. In ignition of RFCD the single plasma column was divided into two successive discharges, and a layer of spatial discharge whose parameter did not differ from the corresponding NLSC, formed on both sides of the sheet. The results show that the transverse sections of the plasma columns and, consequently, current densities in them j_1, j_2 coincide if $\Delta_1 = \Delta_2$. When

$\Delta_1 \neq \Delta_2$, the cross section is smaller and, consequently, higher current density was observed in the plasma column with higher Δ, since $j_1 S_1 = j_2 S_2 = I_{RF}$.

To explain the effect of normal current density in the low-current RFCD it is important to have information on the nature of the VAC or the NLSC and the plasma column.[34] The presence in the electrode gap of the RF capacitance discharge of zones with different current passage conditions and self-organisation of stable stationary states in the NLSC–plasma column system determine the properties of the RFCD observed in the experiments.[34]

Taking into account Ref.34, we shall examine in greater detail the mechanism of formation of the spatial structure of the low-current form of the RFCD.

After breakdown of the discharge gap, the different mobility of the electronic and ionic components of the plasma during the process of formation of RFCD leads unavoidably to the appearance of near-electrode layers of spatial discharge at the boundaries of plasma with electrons – this is an essential condition for generating stationary discharge because the number of ions and electrons, hitting the electrodes during the cycle of the RF field, can be the same only in the presence of NLSC. However, of all the permissible stationary states of the NLSC-plasma column system, only the states that are stable are realised in the experiment. The stable regime is the stationary regime of burning of low-current RFCD where the small deviation of current density from the steady value decrease with time. In the opposite case, the state of the NLSC–plasma column system is unstable and if no special measures are taken this discharge burning regime cannot be realised in practice. The effect of the normal current density in the low-current RFCD on electrodes not coated with a dielectric material,[12,16,17] and observed in experiments confirms the possibility of existence of stationary regimes of burning of the RF discharge stabilised by NLSC. It is not possible to produce a stable low-current form with a current density of $j < j_{nl}$ on 'bare' electrodes. As shown later, the formation of low-current RFCD with a current density of $j > j_{nl}$ requires higher U_{RF} and, and if discharge does not fill the electrode gap in the direction transverse to the current, is also forbidden as a result of disruption of the ionisation bands on the plasma–neutral gas lateral boundary. An ionisation wave forms in this case and leads to movement of the lateral boundary of plasma and an increase of the cross section of discharge until the equality $j = j_{nl}$ is fulfilled: in this case U_{RF} on the electrodes decreases to some value U_{min1}. This effect is often observed in experiments in attempts to increase j above j_{nl} in the normal burning regime. If the RF source operates in the voltage

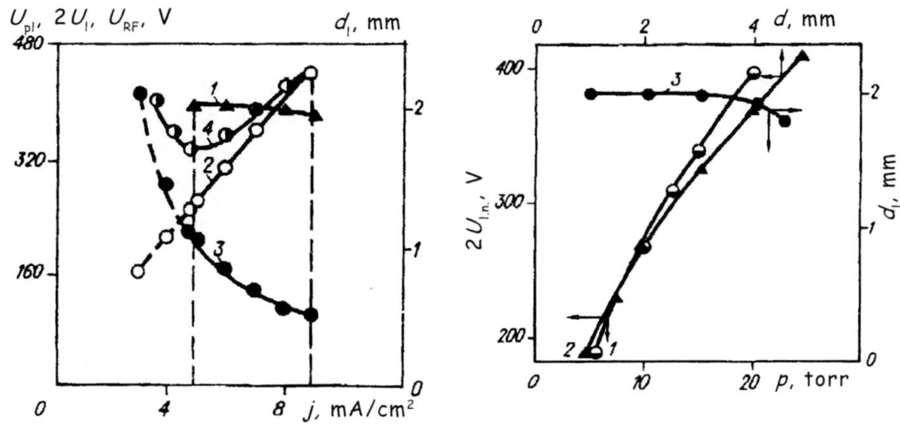

Fig.16.6 Dependences of d_l,(1) U_l (2), U_{pl} (3), U_{RF} (4) on j. RFCD, air, f = 13.6 MHz, p = 30 torr.
Fig.16.7 Low-current RFCD in air, f = 13.6 MHz. 1) $U_{l.n}$ (p) at d = 2 cm, 2) $U_{l.n}$ (d) at p = 10 torr, 3) d_l (p).

generator regime, i.e. U_{RF} on the electrodes does not decrease during plasma expansion, the plasma completely fills the gap between the electrodes in the direction transverse to the current and the anomalous burning regime is established.

To illustrate this situation, Fig.16.6 shows the VAC of NLSC U_l (j), plasma U_{pl} (j), the discharge as a whole U_{RF} (j) and the dependence of the thickness of NLSC on j, d_1 (j) obtained in the low-current RFCD.

The behaviour of the RF voltage on the NLSC under the normal burning regime, i.e. the minimum possible RF voltage on the NLSC $U_{l.n}$ with a change of the pressure in the discharge chamber p or the width of the electrode gap d (Fig.16.7), as well interesting and important for understanding the nature of the low-current RFCD. It can be seen that the minimum RF voltage on the NLSC becomes higher with increasing pressure and, which is completely unexpected, $U_{l.n}$ is a monotonically increasing function of d. This behaviour of $U_{l.n}$ with the variation of p, d explains the restriction of the region of existence of the low-current form of the RFCD with respect to pressure and the size of the electrode gap observed in the experiments.[12]

In fact, as indicated by Fig.16.7, at a constant electrode gap there is a value of p^* at which the minimum possible RF voltage on the NLSC can reach the value U_{br} at which equation (16.6) is fulfilled and the VAC shows an inflection point. This means that at given d and $p > p^*$ the low-current form of an independent RFCD is not realised. A similar situation is observed if p is fixed and d is varied.

Analysis of the data shown in Fig.16.6 and 16.7 indicates that two qualitatively different regions can easily be separated in the spatial structure of the low-current RFCD: NLSC and the plasma column which differ by the mechanism of maintenance of the active conductivity in them. If the active conductivity in the plasma column is ensured mainly by ionisation in the RF field, then in NLSC it is sustained mainly as a result of injection of charges from the ends of the plasma column. The dependence $U_{pl}(j)$ follows from the standard equations of ionisation balance in the plasma columns.[32] However, to simplify considerations, we shall approximate the experimental values of $U_{pl}(j)$. The validity of this approximation was verified in the pressure range from units to hundreds of torrs:

$$U_{pl}(j) = pd_{pl}\frac{C}{j^{\alpha}},\qquad(16.13)$$

where $d_{pl} = d - 2d_{pl}$ is the length of the plasma column, C, α are the quantities determined by the type of gas.

On the other hand, the RF voltage U_{ext} which can be applied to the plasma column depends on U_{RF}, j and NLSC parameters

$$U_{ext} = \left[U_{RF}^2 - (jZ_l)^2\right]^{1/2},\qquad(16.14)$$

where $Z_l = 2d_l/\Delta\varepsilon\varepsilon_0\omega$ is the capacitance impedance of the NLSC.

Equation (16.14) holds when the phase shift between U_l and j in the NLSC is equal to $\pi/2$ which is equivalent to the condition $j_a \ll j_{cm}$ which is fulfilled in the NLSC of the low-current RFCD.[12,17]

Under a stationary regime $U_{pl} = U_{ext}$, and from this equality taking into account (16.13)–(16.14) we can find all j at the given external parameters (pressure, gas composition, d, ω, U_{RF}). However, as already mentioned, in experiments we can realise only the values of j at which the NLSC – plasma column system is stable. In this case, for the effective values of j, U_{pl} and U_{ext}, the stability criterion is written in the form

$$\left|\frac{dU_{pl}}{dj}\right| \le \left|\frac{dU_{ext}}{dj}\right|.\qquad(16.15)$$

Equation (16.13)–(16.15) yields an expression for the minimum (normal) value of $j_{min} = j_{nl}$ in the low-current RFCD

$$j_{n1} = \left(\frac{\varepsilon \omega p d_{pl} C \sqrt{\alpha}}{2 d_l / \varepsilon + 2\delta / \varepsilon_D} \right)^{\frac{1}{\alpha+1}}. \qquad (16.16)$$

When deriving equation (16.16), we took into account the possibility of coating the electrode with a dielectric with a thickness δ with dielectric permittivity δ_D.

The nature of the dependences $U_{pl}(j)$, $U_l(j)$ (see Fig.16.6) indicates that the function

$$U_{RF}(j) = (U_{pl}^2 + U_l^2)^{\frac{1}{2}} \qquad (16.17)$$

has a minimum value U_{min}. Simple considerations show that at $U_{RF} = U_{min}$ at $j = j_{n1} = j_{min}$ is determined by equation (16.16). Therefore, formally, the nature of the effect of the normal current density in the low-current RFCD is the same as in the cathode region of dc glow discharge.[32] However, there is also a principal difference caused by the specific features of the spatial structure of the low-current RFCD in which the destabilising factor (region with independent conductivity and falling VAC) is situated at $j < j_{cr}$ outside the limits of the NLSC, whereas the cathodic layer of the dc discharge (or the NLSC of the high-current RFCD) is itself a zone with independent active conductivity. This also results in differences in the physical mechanism responsible for the appearance of U_{min} in the cathodic layer of the glow discharge in the low-current RFCD. In the former case, the value of U_{min2} is associated with the non-monotonic form of the coefficient of ionisation with electrons η_i as a function of the parameter E_l/N and does not depend on the plasma column characteristics. On the other hand, in low-current RFCD the U_{min1} is determined by the parameters of the plasma column. This explains the relationship of the parameter of the NLSC j, U_l in the normal regime of the low-current RFCD with the value of d (or, more accurately, d_p) and pressure. In addition, this results in a conclusion which is of considerable importance for practice: the region of existence of the lower-current RFCD with respect to the values of PND can be expanded if the ionisation balance in the plasma column is influenced by, for example, the electron beam.[35]

It was mentioned previously that the burning regime of the RFCD changes both suddenly and continuously. Reference (12) shows that in the same gas the transition to the high-current RFCD is continuous at low pressures and sudden at high ones. For the sake of determinacy, we shall examine the dependence of U_{RF} on j (see Fig.16.4). In this case, the jump on the VAC is accompanied by: 1) a large increase of current density j from j_{cr} to j_{n2}, 2) decrease of U_{RF} of the electrodes, 3) the fact that stable burning of the RFCD with intermediate current planes $j \in (j_{cr}, j_{n2})$ is not possible. Since the transition of RFCD from one form to another is associated with a breakdown and transformation of NLSC, we shall pay attention mainly to the properties of the near-electrode layers prior to and after the jump.

In the low-current form of the RF discharge the thickness of the NLSC d_{l1} is inversely proportional to the field frequency and depends only slightly on gas density.[12,34] On the other hand, in the high-current RFCD d_{l2} is almost independent of frequency but is inversely proportional to pressure (see equation (16.8)). In a general case $d_{l1} \neq d_{l2}$. However, the pressure and frequency of the RF field can be selected for any type of gas such that $d_{l1} = d_{l2}$, i.e. taking into account (16.8)

$$\frac{v_{dr}}{\omega} = \frac{C_1}{p}\frac{T}{T_0} \quad \text{or} \quad p = \frac{\omega}{v_{dr}}C_1\frac{T}{T_0}. \tag{16.18}$$

It is clear that in this case the RF voltage will not show any 'jump' on the NLSC because at $d = d_{l2}$ the voltage in the layer corresponds to the minimum of the Paschen curve U_k. There is also no jump of the capacitance component of the current in the NLSC because neither the thickness of the layer nor the RF voltage on it have changed. As regards the active component of the current in the layers j_a, after transition its value will be determined by the expression (16.10), and the ratio of the current densities in the layer at p, satisfying (16.8), is determined by the expression

$$\frac{j_{cm}}{j_a} = \frac{\varepsilon\varepsilon_0 v_{dr} U_l}{C_1^2 C_2}. \tag{16.19}$$

Estimates on the basis of (16.19) for typical discharge conditions give that $j_{cm}/j_a \gg 1$. Thus, if the relationship (16.18) is fulfilled the transition of the RFCD to the high-current regime will take place without jumps on the VAC. At $p \gg \dfrac{\omega}{v_{dr}}C_1\dfrac{T}{T_0}$ the VAC shows a clearly visible discontinuity, and the sudden decrease of the RF voltage on the electrodes is caused by at least three reasons: 1) decrease of U_l from U_{br} to U_k (breakdown on the right part of the Paschen curve); 2) appearance of regions with weak electric fields (analogues of glow discharge and Faraday dark space) elongated along the direction of RF current in the electrode gap after transition of the RFCD to the high-current regime; 3) falling VAC of plasma.

A qualitatively different situation arises at $p < (\omega/v_{dr})\left(\dfrac{T}{T_0}\right)C_1$, i.e. when $d_{l1} < d_{l2}$. This is typical of low (usually $p < 1$ torr) gas pressures, and the breakdown of NLSC takes place on the left (high-voltage) part of the Paschen curve (since the number of ionisations, carried out by the secondary emissions electrons on the characteristic thickness of the NLSC is not high because of the low current density of the gas density). For this reason, a large part of the secondary-emission electrons, moving in the NLSC without collisions, acquire the energy $\simeq eU_0$ and form an electron beam on the side opposite to the electrode. The energy of these electrons is considerably higher than the mean energy of the free plasma electrons.

The dissociation of the energy of the electron beams stored in the LNSC takes place outside the limits of the layers in the plasma column. This is equivalent to the appearance of an additional (not of the field type) source of ionisation in the plasma and results in a decrease of the strength of the electric field there. This situation is similar to that observed in the plasma of a non-independent discharge sustained by the electron beam. A specific feature of the RFCD is that the sources of the electron beams are situated inside the discharge (in NLSC). This circumstance facilitated the introduction of a transition criterion to the high-current regime, or the so-called γ-discharge[22,43] which differ from that examined previously (16.6).[12] According to Ref.43, the transition to the regime of the γ-discharge takes place at RF voltages on the electrodes at which the rate of ionisation of the secondary emissions electrons with the beam is equal to the rate of ionisation with plasma electrons in the electric field of the plasma

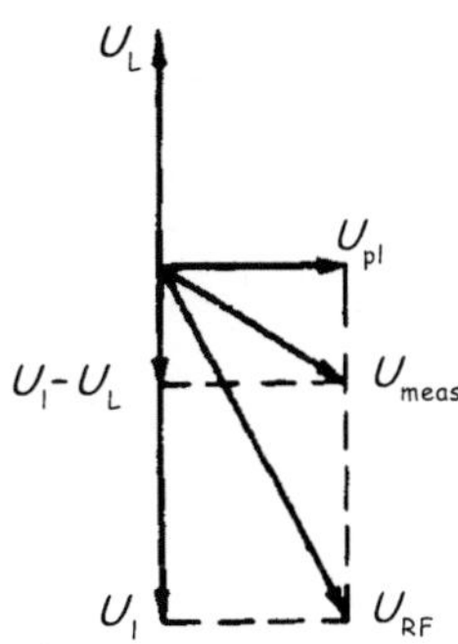

Fig.16.8 Vector diagram of RF voltage in the circuit of passage of discharge current.

column. It can easily be seen that the transition criterion, introduced in Ref.43, is arbitrary to some extent and not equivalent to the criterion proposed in Ref.12. If $d_{l1} << d_{l2}$, the effect of the electron beams on the plasma characteristics can become evident prior to fulfilling the condition (16.6) and the RFCD transfers to the high-current regime. Therefore, at a pressure determined from (16.18) or at a lower pressure, the type of RFCD can be identified using additional diagnostics media, for example, probing of the RF discharge by direct current (see below).

We shall now examine the special features of formation of the VAC of the RFCD. With the variation of the U_{RF} and also the magnitude of the phase shift between U_{RF} and I_{RF} special attention must be given to excluding systematic errors determined by the finite length of the current-conducting parts of the RF electrodes which introduce an additional inductance to the discharge circuit. Neglecting this inductance, especially if the device for measuring the RF voltage is incorrectly connected, can greatly distort the results (Fig.16.8). It can be seen that the RF voltage measured on the 'electrodes' U_{meas} can be considerably lower than the actual voltage U_{RF} applied to the discharge if the RF voltage in the current conductor from the area of connection of the RF voltage measuring device to the working surface of the electrode U_L is comparable as regards the absolute value with U_l. The typical value is $U_l \sim 100 \div 300$ V. At an inductance L of the current conductors of the electrodes $\sim 10^{-6}$ H, the inductive impedance at a frequency of 13.6 MHz is $|Z_L| = 2\pi f L = 80$ Ω because for typical I_{RF} (several amperes) $U_L \sim U_l$. Since $|U_L| = 2\pi f L\, I_{RF}$, to obtain U_{meas} close to U_{RF} it is required either to reduce L or the device for measuring the voltage should be connected to the electrodes using additional conductors situated in such a manner that the discharge current I_{RF} flows through them.

It is interesting to note that the results described above can be used to determine the voltage in the plasma column U_{pl} because the data on this parameter are extremely important for constructing a RFCD model adequate for experiments. For this purpose, a conductor was connected in series with the electrodes to the feed circuit of the RF discharge. Its inductance exceeded the value

$$L = \frac{v_{dr}}{\varepsilon \varepsilon_0 \omega^3 S}. \tag{16.20}$$

Moving the point of connection of the device for measuring RF voltage along the conductor it is possible to record the minimum reading of the measuring device U_{min}. In accordance with the vector diagram in Fig.16.8, the determined value of U_{min} determines U_{pl} because in the examined case the drop of the RF voltage on the NLSC of the low-current RFCD, characterised by the capacitance conductivity is compensated by the drop of the RF voltage on the inductance of the conductor from the point of connection of the device for measuring the RF voltage to the working surface of the electrode.

16.4 Constant electric fields in RFCD

The formation of constant electric fields in discharges, sustained by a sinusoidal radio-frequency field, indicates that in discharges there are regions with non-linear characteristics and that this can be used for diagnostics of the spatial structure of the RF capacitance discharge.

The facts confirming the presence in the stationary RFCD of low pressure ($p < 1$ torr), constant voltage U_0 localised between the discharge plasma and the electrodes and comparable with the RF voltage on the electrodes U_{RF} have been known for a long time.[18,19,36-42] The results show that with increasing U_{RF} the value of U_0 also increases and reaches $\sim 10^3$.[41] With increasing pressure U_0 decreases and at at $p > 1$ torr it does not exceed several volts (see Fig.16.9).[18,19] As regards the dependence of U_0 on the frequency of the feed RF voltage, the experimental results obtained on this subject are contradicted. According to Ref.18, with increasing frequency U_0 decreases. The authors of Ref.9 observed a reversed dependence with other conditions being equal.

There are several viewpoints regarding the reasons of appearance of high U_0 in the RFCD. One of them, according to which U_0 is the result of ambipolar diffusion of the charges,[42] is eroneous[18] and is only of historical interest at present.

The explanation proposed in Ref.18 appears to be more acceptable. According to this explanation, the formation of U_0 in the RFCD is associated with the formation of near-electrode layers of a positive spatial discharge caused by the fact that during the cycle of the RF field part of the electrons are removed from the discharge gap as a result of high mobility. It was postulated that the characteristic thickness of the NLSC d_l is determined by the vibration amplitude of the electrons in the discharge, i.e.

$$d_l = \frac{\mu_0 E_m}{p\omega},$$
(16.21)

where μ_0 is the mobility of electrons at a pressure of 1 torr, E_m is the amplitude value of the strength RF field. From Poisson's equation under the condition of stationary ions in their layer, an equation was derived for the constant potential of the plasma of the RFCD in relation to the electrodes[18]

$$U_0 = 6\pi e n_p d_l^2 = 6\pi e n_p \frac{\mu_0^2 U_m^2}{\omega^2 p^2 d^2}, \tag{16.22}$$

where U_m is the amplitude value of the radio-frequency voltage applied to the discharge, n_p is the concentration of the positive ions, where $n_p \simeq n_e$.

However, the results of calculations carried out using equation (16.22) do not agree with the experimental values. For example, at $f = 3$ MHz, $p = 1$ torr, $d = 12$ cm, $U_m = 300$ V and $n_p = 10^9$ cm^{-3}, $U_0 = 650$ V, whereas the value of U_0 measured under these conditions is 65 V.[18] This casts doubts on the whether it would be efficient to examine the vibrational motion of the plasma electrons in the RFCD.

According to the authors of Ref.40, in the plasma of the stationary RFCD where the directional velocity of the electrons in the RF field is considerably lower than there thermal velocity, the concept of the amplitude of the vibrations of the electrons in the RF field has hardly any physical meaning, and the spatial scale of the division of the charges on the plasma boundaries d_l is determined not by the amplitude of vibrations of the electrons but by the polarisation length of the plasma in the electrostatic field. At low potentials ($U_0 \ll V_e$, V_e is the electronic temperature in potential units), the value d_l is evidently equal to the Debye radius D_e

$$d_l = D_e = \left(\frac{V_e}{4\pi e n_e} \right)^{\frac{1}{2}}. \tag{16.23}$$

At high potentials ($U_0 \gg V_e$)

$$d_l = \left(\frac{U_0}{4\pi e n_e} \right)^{\frac{1}{2}}. \tag{16.24}$$

The very fact of appearance of U_0 in the RF capacitance discharge is interpreted by the authors of Ref.40 as the result of rectification of the RF voltage on the non-linear complex conductivity of the near-electrode layer of the spatial charge. For the case of low pressures where it is possible to ignore the collisions of the electrons in the RFCD and assume that the entire RF voltage applied to the electrode is localised in the layers, the equation has the form

$$U_0 = V_e \ln \sqrt{\frac{2\pi m_e}{M}} - \frac{U_{RF}}{\pi}, \tag{16.25}$$

where m_e, M is the mass of the electron and the ion, respectively.

However, the conclusion of the authors of Ref.40 according to the vibrational model of the spatial structure of the low-current RFCD is not correct has not been confirmed, as indicated by the experimental data presented in Ref.12. where the experiments have confirmed the concept of the layer spatial structure of the low-current RFCD, including near-electrode regions with low active conductivity and the plasma column. In a wide pressure range from units to hundreds of torr, the equation of continuity for the discharge current can be written in the form

$$en_e \frac{\mu_0 E_{pl}}{p} \cong \varepsilon\varepsilon_0 \omega E_l. \tag{16.26}$$

The measured value of U_0 is linked with the RF voltage in the layers U_l by the relationship

$$U_0 = kU_l, \tag{16.27}$$

where $k \simeq 1$ is a constant. Assuming that the density of the positive ions n_p is constant to a first approximation along the discharge gap and equal to n_e in the plasma column, from Poisson's equation, taking into account equation (16.26) and (16.27), we obtain an equation for determining the thickness of the RFCD d_l

$$d_l = k \frac{\mu_0 E_{pl}}{p\omega} \cong \frac{v_{dr}}{\omega}. \tag{16.28}$$

Comparison of (16.28) with (16.21) shows that the characteristic thickness of RFCD is determined (despite the results published in Ref.40) but the amplitude of the drift vibrations of the RFCD electrons in the plasma field E_{pl}. It is important to stress the large difference of (16.28) in relation to (16.21) caused by the fact that the valuable vibrations of the electrons in deriving (16.28) are assumed to be taking place in the electric field of the plasma column E_{pl} are not in the vacuum field $E_m = U_m/d$, as implicitly assumed in Ref.18. It can easily be seen that the latter circumstance also leads to the previously noted large difference (by an order of magnitude) of the experimental values of U_0 from those calculated from equation (16.22).

The attempt to explain the dependence of U_0 on the voltage of the electrodes U_{RF}, observed in Refs.18, 41, by assuming that U_m from equation (16.22) is identical with the amplitude value U_{RF} is incorrect. In fact, according to the layer model of the low-current RFCD,[12] $U_m \equiv U_{pl}$, but

$$U_{pl} = \left(\frac{E_{pl}}{p} \right) p d_{pl} \cong \left(\frac{E_{pl}}{p} \right) p d, \tag{16.29}$$

where d_{pl} is the length of the plasma column, and when $d \gg d_l$ then $d_{pl} \simeq d$. Equations (16.29) and (16.22) show that U_0 does not depend explicitly on the RF voltage of the electrodes, gas pressure, and gives values of the constant potential of the plasma in relation to the electrodes similar to those observed in practice.

In reality, U_0 is related to U_{RF} with a concentration of charged particles $n_p \simeq n_e$, which can easily be confirmed using the relationships (16.22), (16.29), (16.26) and (16.17):

$$U_0 = 3\pi \left(en_e \frac{\mu_e E_{pl}}{p} \right) \frac{v_{dr}}{\omega^2} = 3\pi \left(\frac{\omega}{4\pi} E_l \right) \frac{d_l}{\omega} \cong \frac{3}{4} U_l = \frac{3}{4} \sqrt{U_{RF}^2 - U_{pl}^2}. \tag{16.30}$$

In deriving (16.30) the capacitance current in the NLSC is expressed in the system of CGS units, $j_l = (\omega/4\pi) E_l$, and the numerical coefficient in (16.22) is halved in accordance with Ref.32.

Thus, the analysis of the conditions of appearance U_0 taking into account the layer structure of the low-current RFCD shows that the constant potential of the plasma in relation to the electrodes U_0 is determined unambiguously by the decrease of the RF voltage on the NLSC U_l. As noted previously, the physical reason for the appearance of the NLSC with high U_0 in an independent RFCD is the different mobility of the electron and ion components of the plasma. This circumstance leads unavoidably to the formation of NLSC with high U_0 in a stationary RFCD. This equalises the number of electrons and ions falling of the electrodes during a cycle of the RF field.

As indicated by (16.30), there is no explicit relation of the constant potential of the plasma with pressure and frequency RF field. At the same time, the dependence of the RF voltage on the NLSC on p and is determined by the equation

$$U_l(p, \omega) = j(p, \omega) \frac{v_{dr}(E_{pl}/p)}{\varepsilon\varepsilon_0\omega^2}, \qquad (16.31)$$

where $j(p, \omega)$ is at the discharge current density. It is well known[16,17] that with increasing p normal, i.e. minimum, density of the discharge current decreases, and E_{pl}/p, determined by the ionisation balance in the plasma changes only slightly and, consequently, drift velocity of the electrons in the plasma also changes only slightly. Therefore, increasing pressure increases the minimum value of the RF voltage on the NLSC $U_{l.n}$ at which a low-current RFCD still can exist and, according to (16.27), U_0 (p) will also increase. This contradicts the conclusions made in Ref.18. However, this contradiction is removed if we take into account the real layer structure of the RFCD and, as already mentioned, U_m in (16.22) is the decrease of the RF voltage in the plasma and not the RF voltage in the electrodes.

Thus, if it is assumed that the positive ions are stationary, the equation of stationarity of the low-current RFCD shows that the value of U_0 is equal to the amplitude value of U_l, applied to the NLSC, irrespective of pressure and with the accuracy equal to the electronic tem-

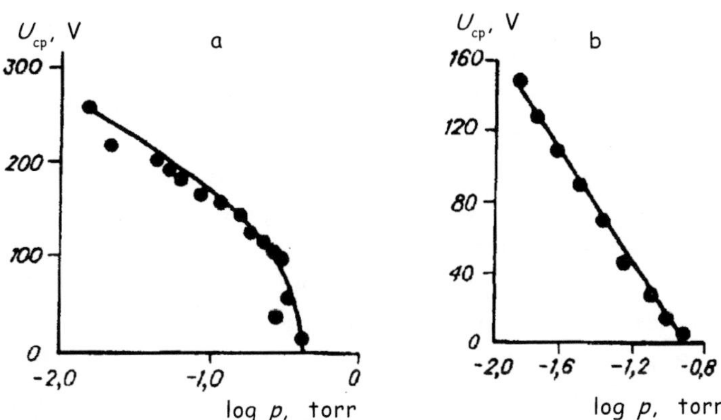

Fig.16.9 Dependence of constant potential of plasma on pressure[18](a) and b[19].

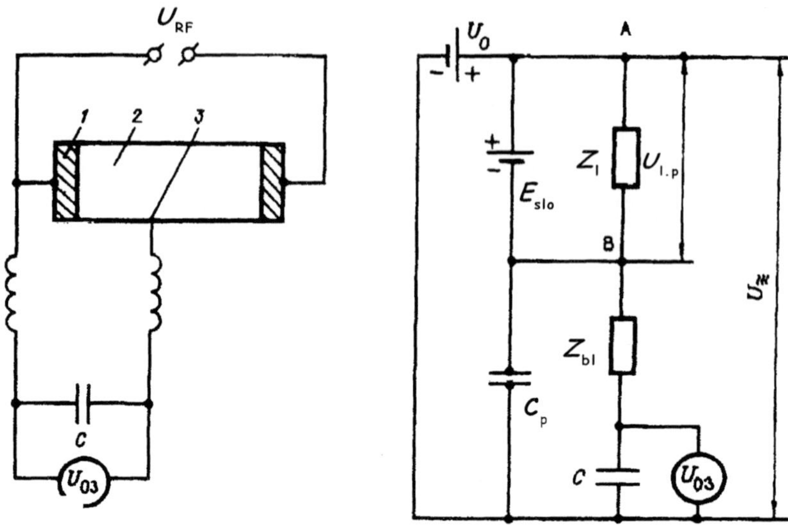

Fig.16.10 Principal (a) and equivalent (b) diagram of circuits of probe measurements of constant plasma potential. 1) electrode; 2) discharge chamber; 3) probe.

perature of plasma. Taking into account the real movement of the ions in the layer does not greatly reduce the value of U_0 since $\mu_e/\mu_p > 10^2$. However, probe measurements of the constant potential of the RF plasma (Fig.16.9)[18,19] indicate that U_0 greatly decreases with increasing pressure $p > 1$ torr.

We shall examine in greater detail the procedure of probe measurements U_0 in the RFCD proposed in Ref.44. According to this proced-ure, for accurate measurements of the constant potential of the plasma of RFCD it is necessary and sufficient to organise measurements in such a manner that the RF component of the voltage between the probe and the plasma $U_{l.p}$ is low. It has therefore been proposed to place a choke coil with a high inductive resistance for RF current between the probe and the measuring circuit. A similar procedure was used in Ref.18 and 19. The principal measurement circuit is shown in Fig.16.10a, and the equivalent circuit in Fig.16.10b, with the following notations: $U_{=}$ – the RF voltage between the region of unperturbed plasma containing the probe (point A) and the earth; U_0 – the constant voltage between the plasma and the electrode; E_{slo} – the constant voltage between the plasma and the probe surface; C_p – parasitic capacitance of the probe and the blocking element on the ground; Z_{bl} – the impedance of the blocking element of the probe with respect to RF voltage; Z_l – the impedance of the plasma–probe layer; C – the capacitance of the condenser on which U_{0p} (recorded constant voltage) generates; L_{bl} the inductance blocking the RF current; $U_{l.p}$ – the decrease of the RF voltage at the near-probe layer.

Figure 16.10b shows that

$$U_{0p} = U_0 - E_{slo}. \tag{16.32}$$

The procedure error of measurement U_0 according to (16.32) is associated with E_{slo}.

E_l can be written in the form of the sum of two independent terms: $k_1 V_e$ which depends on the electronic temperature of plasma, and $k_2 U_{l.p}$ – the component determined by the passage of RF current to the probe (k_1, k_2 are constant quantities). The term $k_1 V_e$ is determined by the properties of plasma and cannot be eliminated if the probe is placed in the plasma. The second term is the linear function $U_{l.p}$. It would appear that by selecting a corresponding value of Z_{bl}, $k_2 U_{l.p}$ can be reduced to the value comparable with $k_1 V_e$. However, the effect of the parasitic capacitance of the probe and the blocking element C_p in relation to the earth makes it irrational to increase the impedance of the

blocking element above $Z_{bl} = \dfrac{1}{\omega C_p}$ because of the shunting effect of

C_p. Therefore, the only way of increasing the accuracy of measurements of U_0 is to decrease C_p which is reduced to the rational positioning of the probe, the blocking element, the selection of their dimensions, *etc.*

Additional difficulties in evaluating the accuracy of measuring U_0 are associated with the absence of reliable theoretical models which will make it possible to evaluate Z_l at medium and higher pressures. The method of theoretical evaluation of the error of probe measurements, proposed in Ref.45, is valid only for low pressures in the discharge ($p \ll 1$ torr) when the near-probe layers of the special discharge can be regarded as collisionless. Under other conditions, experimental verification of the accuracy of the probe measurements of U_0 becomes important.

It was shown that the accuracy of measurement of U_0 was determined mainly by the value $U_{l.p}$ which should satisfy the inequality $U_{l.p}/U \ll 1$. This is equivalent, according to Fig.16.10b, to fulfilling the condition

$$\left| Z_{l.p} \right| / \left| Z_\Sigma \right| \ll 1. \tag{16.33}$$

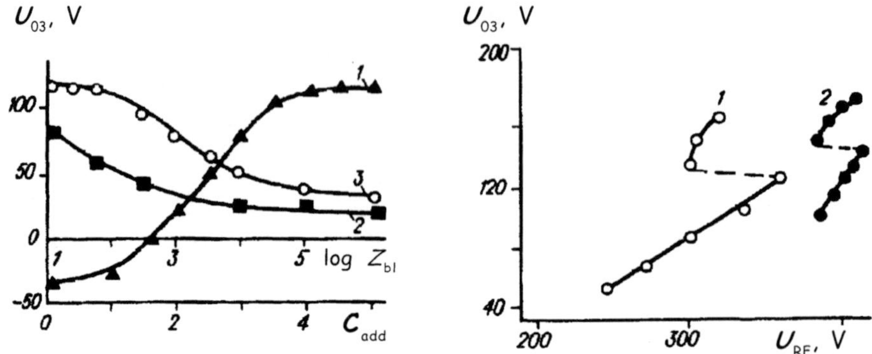

Fig.16.11 Dependence of readings of the probe on the size of the blocking element Z_{bl}. Low-current RFCD, air, $p = 10$ torr, $f = 13.6$ MHz, flat disc electrodes diameter 60 mm, $d = 30$ mm, probe in the centre of the discharge gap; 1,3) blocking element is made in the form of low-capacitance high-resistance resistor, 2) in the form of a choke coil.

Fig.16.12 (right) Dependences of U_{0p} on RF voltage on electrodes. $p = 7.5$ torr; 1) RFCD in air, 2) in CO_2.

Here Z_Σ is the impedance of parallel-connected C_p and Z_{bl}. The capacitance C is not considered in the calculations because its value can always be selected as satisfying the condition $1/\omega C \ll Z_{bl}$.

The validity of the previously formulated conclusions on the error of probe measurements of U_0 can be verified by experiments by reference measurements of Z_{bl} and C_p (Fig.16.11). Curve 1 represents the dependence U_{0p} and Z_{bl} on the semi-logarithmic scale, curves 2, 3 the dependence of U_{0p} and C_p, varied by additionally connecting the capacitance C_{add}.

At low Z_{bl} the probe potential assumes negative values. Optical effects, similar to near-electrode effects, are observed in the vicinity of the probe, i.e. at low Z_{bl} the probe represents an additional electrode. The dependence of U_{0p} on Z_{bl} is initially monotonically increasing and then reaches 'saturation'. This result confirms the correctness of the equivalent circuit (see Fig.16.10b), but it doesn't enable any conclusions to be drawn regarding the accuracy of measurements because the question of the reason for 'saturation' of the dependence $U_{0p}(Z_{bl})$ remains open. The latter can occur in two cases: either the the fixed U_{0b} actually approaches U_0 as regards its value, or the shunting effect of C_p restricts the increase of the effective blocking impedance Z_Σ.

The situation can be clarified by analysing curves 2 and 3 in Fig.16.11. If a decrease of the artificially introduced parasitic capacitance to zero leads to 'saturation' of U_{0p}, then the error of measuring the data by the probe under these conditions is close to minimum determined only

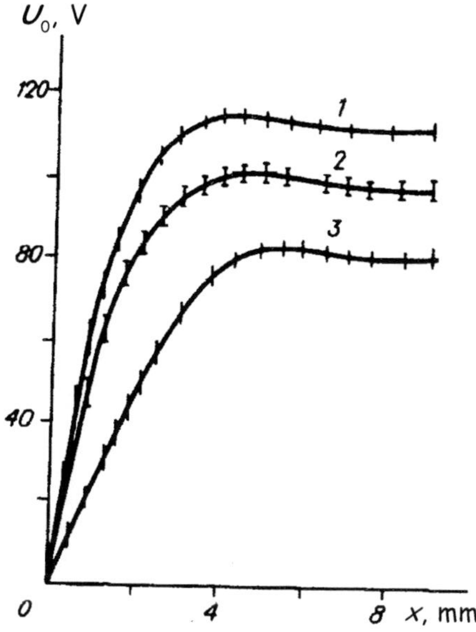

Fig.16.13 Distribution of constant plasma potential $U_0(x)$ along the electrode gap. $p = 7.5$ torr. $U_{RF} = 305$ V, $f = 13.6$ MHz; 1) low-current RFCD in CO_2, 2) in air, 3) in nitrogen.

by a correction for the electronic temperature of plasma T_e. In particular, Figure 16.11 shows that when Z_{bl} is represented by a low capacitance resistor, the accuracy of measurement of U_0 rapidly increases and the value U_{0p} reaches 'saturation' at $C_{add} \rightarrow 0$. This cannot be said of curve 2. Evidently, this is associated with high parasitic capacitance of the choke coil because the probes themselves were identical in both cases.

The typical dependences of U_{0p} on U_{RF} on the electrodes under different conditions of burning of the RFCD are presented in Fig.16.12. Attention should be given to a dual dependence of U_{0b} on U_{RF} as a result of the transition of RFCD to the high-current burning regime. It should be noted that the increase of U_{0p} with decreasing U_{RF} again indicates the localisation of U_0 in NLSC and the redistribution of RF voltage in the electrode gap in favour of the near-electrode layers in transition of RFCD to an anomalous high-current regime.

Making the probe movable, it is possible to record the distribution of the constant potential U_0 in the electrode gap. To confirm the layered structure of the RFCD, it is of special interest to examine the redistribution of $U_0 (x)$ along the direction of passage of the RF current (Fig.16.13). It can be seen that U_0 is localised in the vicinity of the electrode. The same situation exists at the opposite electrode, i.e. $U_0 (x)$ distribution is symmetric and, consequently, Fig.16.13 shows the $U_0 (x)$ only at one electrode. The maximum value of $U_0 (x)$ is obtained in the regions of stronger glow of the plasma column (see Fig.16.2a).

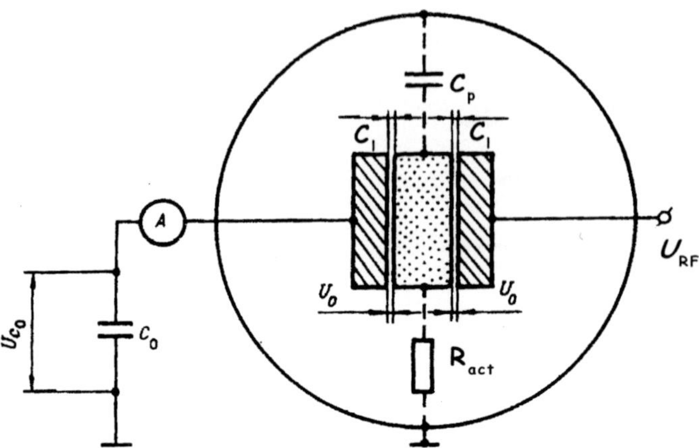

Fig.16.14 Diagram of the discharge chamber for examining the appearance of constant voltage on the capicitance C_0.

The non-monotonic distribution of $U_0(x)$, is clearly visible in Fig.16.13, has not as been satisfactorily explained.

It should be stressed that the accuracy of measurement of U_0 is strongly influenced by the dependence of $Z_{l.p}$ on the external experimental conditions, the type and pressure of gas, the frequency of the RF field, *etc.* A suitable example are the results of probe measurements of U_0 presented in Ref.18 and 19. The experiments carried out with the same probe, i.e. at Z_{bl} = const, did not nevertheless make it possible to record values of U_0 at $p > 1$ torr, whereas at a pressure of $p \ll 1$ torr the same probe gave satisfactory results which were in agreement with the results of measurements by other methods (for example, based on measuring the energy of positive ions leaving the NLSC of the RFCD).[18] It is evident that with increasing pressure in the discharge chamber the value of $Z_{l.p}$ also increased, and the ratio of $Z_{l.p}/Z_{bl}$, included in the criterion (16.33), decreased together with the accuracy of measurements. The measures taken in Ref.12 in fact increased Z_{bl} so that the probe method made it possible to detect (in contrast to Ref.18 and 19) high values of U_0 in the RF capacitance discharge and at $p > $ torr.

The effect of $Z_{l.p}$ on the results of measuring the constant potential of the RF plasma makes it necessary to find other methods of obtaining quantitative information on U_0. It should be noted that in many cases it is interesting to take correct measurements of the maximum value of U_{0m} and not its distribution in the electrode gap. In addition, as shown previously, U_0 is localised in NLSC and the characteristic thickness of the NLSC in the radio-frequency range (~ 1000 MHz) or high-current RFCD at medium or higher pressures is small and equals fractions of a millimetre.[13] This complicates the measurements. However, to obtain information on the maximum value of U_0 it is not necessary to place the probe in the plasma of the RFCD in the vicinity of NLSC. In fact, we shall examine the experimental results obtained under the conditions shown schematically in Fig.16.14.[12] The RFCD was excited between water-cooled electrodes made in the form of discs with a diameter of 10 cm which were placed in the centre of a metallic chamber with a large volume (60 litres). A reference capacitor with a capacitance of C_0 was connected in series between one of the electrodes and the earth to calibrate the RF current measuring device. A constant voltage U_{C_0} appeared on the reference condenser. This voltage reached 100 V or higher and depended on the U_{RF} on the electrodes. In discharges in helium and nitrogen this voltage remained at pressures of up to 50 torr and higher, whereas for CO_2 at $p > 5$ torr it was not possible to record high values of U_{C_0}.

Evidently, it may be expected that there is a relationship between U_0 and U_{C_0}. The attempts to link them through the parasitic capacitance of the plasma C_{pl} in relation to the earthed casing (Fig.16.14) were not successful because, according to the estimates, the capacitance of the NLSC $C_l \gg C_{pl}$ (in the opposite case, the discharge would be short circuited with the casing and not the earthed electrode).

Better results were obtained in the basis of the assumption on the existence of a finite active conductivity R_{act} between the RFCD plasma and the earthed casing of the chamber. The measurements carried out using the circuit shown in Fig.16.14 to verify this assumption in a quartz tube, showed that there is no constant component of the RF voltage on C_0. A similar result was obtained due to an artificial increase of C_{pl} by placing the quartz tube with a discharge in an earthed metallic screen. However, making a small hole in the wall of the quartz tube and introducing a thin conductor into it, connected with the earth through a low-capacitance resistor (1 Mohm), it was possible to record U_{C_0} comparable with that of observed in the metallic chamber.

Thus, a low-conductivity medium exists in the metallic chamber between the discharge plasma and the walls. In Fig.16.14 this medium is denoted R_{act} and the capacitance C_0 is charged to U_{C_0} through this medium. The absence of high values of U_{C_0} in the RFCD on CO_2 at $p > 5$ torr can be explained by additional annihilation of the charges outside the discharge zone, typical of the electronegative gases.

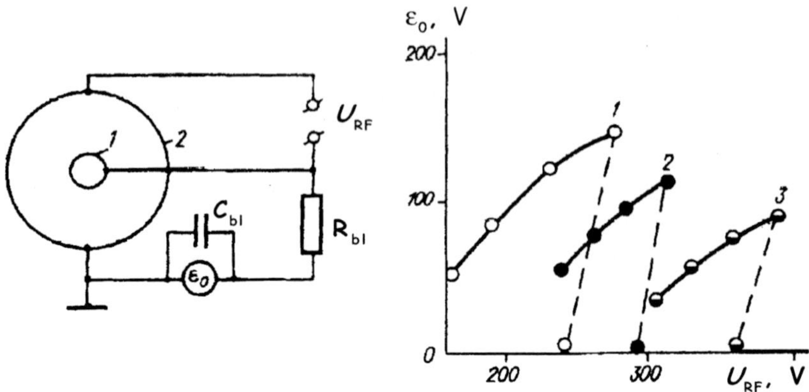

Fig.16.15 Coaxial electrode system for examining the battery effect in the RFCD. 1) internal, 2) outer RF electrode; R_{bl}, C_{bl} – blocking resistor and condensor.

Fig.16.16 Typical dependences of constant EMF ε_0 on RF voltage of electrodes U_{RF}. $f = 13.6$ MHz; 1,2) air, $p = 7.5$; 15 torr, 3) helium, $p = 100$ torr.

Thus, experiments show that at large distances from the RFCD plasma (~10 cm) there is still a region of weakly ionised gas with high conductivity. It can therefore be assumed that a small probe, placed in this region, receives the potential of the space at a given point. Actually, if there were no such region, the charges induced by the measured field on the metallic probe would greatly distort the result. In the absence of a slightly ionised medium the induced charge of the probe is compensated by the flux of charged particles from the space around the probe. Consequently, the probe has the potential of the space.

Measurements of U_{0m} by the proposed method are now reduced to placing the probe of the equipotential corresponding to maximum value of U_0. In practice, measurements are carried out using the following procedure: the probe is moved at the periphery of the discharge and the maximum reading of a high-resistance voltage measuring device is recorded. This device is represented by, for example, an electrostatic voltmeter protected against RF interference. The value of U_{0m}, obtained using this procedure, coincides, with the accuracy to several percent, with the results of measurements of U_{0m} by a 'floating' probe immersed in the plasma. However, in contrast to the latter method, the superheating of the probe and the blocking element is prevented, there are no breakdowns of the surface of Z_{bl}, and the effect of C_p is weaker. Disadvantages of the method of the displaced probe include only the fact that it is necessary to change the spatial position of the probe when the burning conditions of the RFCD are changed.

To pull out complete information, it is important to note another method of increasing the accuracy of probe measurements under RFCD conditions based on using resonance filters – plugs as blocking elements.[46] However, the application is restricted by the relatively low frequency of the RF field (units of megahertz) due to the low efficiency of the filters–plugs and the fact that the parasitic capacitance cannot be removed.

16.5 Using the battery effect in examining RFCD

It was shown in the previous section that near-electrode layers of the spatial charge form in the stationary RFCD irrespective of pressure. The constant potential of the RF plasma in relation to each electrode is proportional to the RF voltage applied to the NLSC U_l. It was also reported (see Fig.16.6) that the thickness of the NLSC d_l in the low-current form in the RFCD is almost independent of the discharge current density. Thus, if the RF capacitance discharge burns in stationary manner under the conditions in which the density of the RF current in one of the NLSC j_1 is higher than the current density j_2 in another near-electrode layer, then $U_{l1} > U_{l2}$ and $U_{01} > U_{02}$, according to (16.31), (16.27) and (16.30). The latter inequality shows that a constant EMF ε_0 appears between the electrode of such a non-symmetric RF discharge. Its value is

$$\varepsilon_0 = U_{01} - U_{02} \tag{16.34}$$

These RF electrodes are closed with respect to direct current, for example, they are joined by a choke coil, a constant electric current $I_0 = (U_{01} - U_{02})/r_{in}$ forms in the circuit. Here r_{in} is the active internal resistance of the discharge including the in-series connected active resistance of the plasma and both NLSC.

From the technical viewpoint, the non-symmetric RFCD can be realised most efficiently in a coaxial electrode system (Fig.16.15). In this configuration of the electrodes, the RFCD in the region of the pressures $p << 1$ torr actually shows a battery effect, i.e. appearance of high ε_0 and constant currents I_0 when closing the circuits.[47] Because the existence of the NLSC, which is the reason for the appearance of a high constant potential of the RF plasma in relation to the electrodes, have been established without doubt by several methods, the appearance of ε_0 under these conditions is quite regular.

A different situation exists in the region of pressures higher than 1 torr where the situation is ambiguous and experimental data[18,19] indicating a rapid decrease of U_0 with increasing p have been published. Therefore, the battery effect should be used for independent verification of the existence of NLSC and high U_0 in the RFCD at $p > 1$ torr.

Figure 16.16 shows typical dependences ε_0 (U_{RF}),[48] obtained in a coaxial electrode system represented schematically in Fig.16.15. The constant potential of the smaller electrode was negative in relation to the larger one.

Attention should be given to two special features in the behaviour of $\varepsilon_0(U_{RF})$: high values of ε_0 at relatively high $p >> 1$ torr; a rapid reduction of ε_0 (to almost zero) when the RF voltage on the electrode reaches some value U_{br} which depends on the type of gas, pressure and electrode material.

Analysis of the experimental results leads to the following conclusions: irrespective of the pressure, NLSCs form in the RFCD, and the constant voltage on the NLSC reaches hundreds of volts. A decrease of ε_0 when the RF voltage on the electrodes exceeds U_{br} does not indicate the disappearance of NLSC and indicates the formation, in the electrode gap, of a qualitatively other, high-current RFCD regime where $U_{01} = U_{02}$ (regardless of the asymmetry of the electrodes).

In fact, at $U_{RF} > U_{br}$, breakdown of the capacitance NLSC of the RFCD takes place, and a new spatial structure forms in the electrode gap. One of the distinguishing features of the latter is that the RF voltage on each of NLSC U_{l1} and U_{l2} is determined, as in the case of the cathode region of normal glow discharge, only by the nature of the gas and the material from which the electrodes are made. However, since the gas, filling the electrode gap, and the material from which electrodes are made, have not changed, $U_{l1} = U_{l2}$ in the given burning regime. Therefore, taking into account (16.27) we obtain that $U_{01} = U_{02}$ and, according to (16.34), $\varepsilon_0 = 0$.

When the negative glow discharge fills completely the electrode with a small area, the value of ε_0 differs from 0 because in this case the electrode with a small area operated under the anomalous regime, i.e. at higher U_l.

The considerations regarding the reasons for the large decrease of ε_0 in these experiments in the transition of the RFCD to the high-regime were verified by the initial experiments. The RF discharge was ignited in a symmetric electrode system ($S_1 = S_2$), but one of the electrodes was made of copper and the other one of duralumin. The results show that in the low-current form of the RFCD, i.e. when no electric breakdown have taken place in the NLSC, the value of ε_0 is close to 0 at all values of U_{RF} in the range $U_n \leq U_{RF} \leq U_{br}$ (U_n is the minimum RF voltage on the electrodes at which the low-current form of the RFCD exists under the given specific conditions. However, in transition to the high-current regime $\varepsilon_0 \neq 0$. For example, for a high-current RFCD in air $\varepsilon_0 = 70$ V, and the potential of the copper electrode is lower.

16.6 Effect of the frequency of RF field on the structure of the discharge and plasma characteristics

The frequency of the RF field is the most important parameter affecting the main characteristics of the RF capacitance discharge. There have been a number of studies[49–52] in which attempts were made to describe the mechanism of the effect of frequency on the characteristics of the discharge plasma, especially on its conductivity σ_{pl}, the reduced strength of the electric field E_{pl}/p, and others. The specific feature of the RFCD is that the kinetic processes, taking place in the plasma of the RF capacitance discharge, differ from the corresponding processes in the plasma of the dc discharge.[51,52] However, it has not been possible to explain the effects observed in the experiments, for example, the increase of the active conductivity of the plasma σ_{pl} with increasing frequency ω, examining only the phenomena taking place in the RFCD plasma. In fact, in the frequency range $f \in 1 \div 100$ MHz, the gas pressure range $p \in 1 \div 100$ torr, typical of practical application of the RFCD,[1–13,49–52] the active conductivity of the plasma in the RF field should not depend on the frequency ω and in accordance with the well-known equation[32]

$$\sigma_{pl} = \frac{e^2 n_e v_m}{m(\omega^2 + v_m^2)}.$$

(16.35)

Since in the examined frequency range of the RF-field ω and the gas pressure range the frequency of collisions of the plasma electrons with neutral particles is $v_m \gg \omega$, then $\sigma_{pl} = e^2 n_e/m v_m$.

In reality, the conductivity of the RFCD plasma σ_{pl} and the electron concentration in it are determined not only by the local ionisation balance of the charged particles but also by the spatial structure of the discharge, especially by the conditions of closure of the discharge current at the plasma–electrode boundary where, as mentioned previously, NLSC with frequency-dependent properties form.

The effect of the NLSC on the conductivity of the RF discharge plasma is very evident when comparing the two forms of the RFCD examined previously (Fig.16.1). Depending on whether a breakdown has taken place in an NLSC (b) or not (a), the conductivity of the plasma column changes by no more than an order of magnitude, irrespective of the completely identical experiment conditions, including similar values of E_{pl}.[13] The jumps in the parameters of the RFCD plasma with changes in the discharge burning conditions can be predicted, with the accuracy acceptable for many practical applications, by comparing equations for current densities j_{n1} (16.16) and j_{n2} (16.12) which depend on the field frequency.

We shall examine the low-current from the RFCD and explain the possibilities of controlling the parameters of this type of discharge by changing the frequency.

There are a large number of experimental data, discussed previously, and numerical calculations[28,29] which show that the most characteristic feature of the low-current RFCD is that the inequality

$$\sigma_l \ll \sigma_{pl}. \tag{16.36}$$

is fulfilled. Here σ_l is the average active conductivity of NLSC. Direct experimental methods of verifying the validity of (16.36) in the low-current RF discharge will be examined in the following chapter. When (16.36) is fulfilled, the continuity equation for the discharge current of the low-current RFCD is written in the form (16.26) which indicates the dependence of the conductivity of the plasma and of electron concentration in it on the frequency of the RF field ω:

$$\sigma_{pl} = \varepsilon\varepsilon_0\omega\frac{E_l}{E_{pl}}, \tag{16.37}$$

$$n_e = \frac{\varepsilon\varepsilon_0}{e\mu_e}\omega\frac{E_l}{E_{pl}}. \tag{16.38}$$

The equations (16.37) and (16.38) show that σ_{pl} and n_e are proportional to the frequency of the RF-field, with other conditions being equal. At fixed frequency, n_e is determined by the E_l/E_{pl} ratio, and for given p, d, ω the strength of the electric field in the NLSC is restricted at the bottom by the value $E_{l.n} = j_{nl}/\varepsilon\varepsilon_0\omega$, where j_{nl} is given by the effect of a normal current density in the low-current RFCD (16.16).[34] On the other hand, E_l cannot be higher than E_{br} at which the condition (16.6) is fulfilled, i.e. a breakdown of NLSC takes place in which secondary emission electrons take part, and the discharge changes to the high-current regime. E_{pl} is determined from the ionisation balance equation. Under the typical conditions of application of RFCD the dependence $E_{pl}(n_e)$ is 'drooping' (see Fig.16.6).[34]

Thus, at the fixed frequency of RF field $\omega = 2\pi f$ the value of n_e in the plasma of the low-current form of the RFCD can be regulated by changing the RF voltage at the electrodes in the range

$$\frac{\varepsilon\varepsilon_0 E_{l.n}}{\varepsilon\mu_e E_{pl1}}\omega \leq n_e \leq \frac{\varepsilon\varepsilon_0 E_{br}}{e\mu_e E_{pl2}}\omega. \tag{16.39}$$

Attention should be given to another important special feature of the low-current RFCD which is of considerable importance for the applications, for example, in laser technology. It is well known that the secondary-emission processes play a secondary role in the mechanism in the method of sustaining the discharge of this type.[12] Therefore, it would appear that the emission characteristics of electrodes should have no effect on the energy characteristics of the CO_2 laser with RF excitation. However, opposite results were obtained in the experiment.[53] Nevertheless, there is no contradiction here. Although the material of the electrodes has only a slight effect on the structure and parameters of the low-current RFCD,[12] it does determine the threshold of transition of the RF capacitance discharge to the high-current burning regime because the secondary-emission processes determine the value of E_{br} and, consequently, also $(n_e)_{max}$ (see equation 16.39). This leads to a very important conclusion for practice: coating the electrode with a material with low emission properties increases the power supplied to the plasma of the low-current RFCD without increasing the frequency of the RF voltage feeding the discharge.

We shall now clarify the dependence of the minimum RF voltage at the electrodes of the low-current RFCD at the given values of the electrode gap and the gas pressure on the frequency of the RF field. The experiments show that U_{min} decreases with increasing frequency.[30]

For this purpose, we shall use the considerations regarding the layer structure of the low-current RF discharge including capacitance near-electrode layers of the special discharge with the RF voltage on them, satisfying the equation

$$U_l = j\frac{v_{dr}}{\varepsilon\varepsilon_0\omega^2}, \tag{16.40}$$

and the RF voltage on the plasma column U_{pl} in accordance with equation (16.13). Taking into account the capacitance nature of the NLSC, i.e. the fact that the phase shift between U_{pl} and U_l is close to $\pi/2$, we can write the RF voltage on the electrodes in the form of (16.17). It is well known that the U_{min} at the electrodes forms at $j = j_{n1}$, where j_{n1} is the normal current density in the low-current RFCD.[34] Consequently, substituting the equations for U_l, U_{pl} and j_{n1} into the equation (16.17), in accordance with (16.40), (16.13) and (16.16), we obtain that the minimum RF voltage at the electrode of the low-current of the RFCD as a function of the frequency U_{min} (ω) can be written in the form

$$U_{min}(\omega) \cong \frac{1}{\omega}\left(\frac{4Cv_{dr}}{\varepsilon\varepsilon_0}pd_{pl}\right)^{\frac{1}{2}}.$$

(16.41)

When deriving (16.41) and (16.16) it was assumed that $\delta = 0$ and $\alpha = 1$. In fact, the experiments show that α is slightly higher than unity and depends on the type of gas. However, to a first approximation, equation (16.41) gives results that are in satisfactory agreement with the experiments and, most importantly, explains the nature of the dependence of minimum voltage on the electrodes on frequency.

Using the layer model of the low-current RFCD and substituting $j = j_{nl}$ into (16.13), we can easily derive the dependence of the reduced electric field in the plasma E_{pl}/p as a function of the frequency of the RF field

$$\frac{E_{pl}}{p} = \sqrt{\frac{C(2d_l/\varepsilon + 2\delta/\varepsilon_d)}{\varepsilon_0\omega pd_{pl}}}$$

(16.42)

or on bare electrodes where $\delta = 0$

$$\frac{E_{pl}}{p} \cong \frac{1}{\omega}\sqrt{\frac{2Cv_{dr}}{\varepsilon\varepsilon_0 pd_{pl}}}.$$

(16.43)

Equations (16.42) and (16.43) show that E_{pl}/p decreases with increasing frequency of the RF field, with other conditions being unchanged. However, detailed analysis shows that this special feature of the RFCD plasma is based on the 'drooping' nature of the VAC of the plasma (decrease of U_{pl} with increasing j, see Figure 16.6) but it is not caused by the difference of the elementary processes in the plasma of the RFCD in comparison with the corresponding processes in the plasma of the positive column of the dc glow discharge at least in the examined frequency range of the RF field (1–100 MHz) and the gas pressure from units to hundreds of torr. To confirm this, it is sufficient to increase artificially the impedance of the NLSC at $\omega = $ const, for example, by coating the electrodes with a dielectric with the characteristic thickness $\delta > \varepsilon_d d_l$. In this case, both normal current density j_{n1} (16.16) and E_{pl}/p (16.42) can be changed by selecting the appropriate thickness δ of the dielectric coating on the electrode surface.

The experiments confirming the previous conclusions were carried out using the following procedure. The discharge chamber was formed by two flat quartz sheets with a thickness $\delta > \varepsilon_d d_l$. A metallic coating was sprayed on one of the sides of both sheets. The minimum current density j_{n1} and the electric power scattered in the discharge were measured (by calorimetry) at the same values of ω, p, d and the type of gas in two cases: 1) the sprayed sides of the sheets are turned towards the inside of the discharge chamber, which is equivalent to $\delta = 0$; 2) the sprayed surfaces of the quartz sheets are positioned on the outer side of the chamber, i.e. $\delta \neq 0$. It appears that j_{n1} and the value of the power scattered in the RFCD is always smaller in the second case. If the parameters of the discharge were completely determined by the plasma column, the presence of the dielectric coating would have no effect on the characteristics of the plasma of the low-current RFCD and would depend only on the frequency of the RF field.

Taking into account the actual layer structure of the low-current RFCD, it is also possible to understand other special features of the RF capacitance discharges, especially the non-monotonic dependence of the RF voltage on the electrodes on the pressure in the discharge chamber $U_{RF}(p)$ at the fixed specific energy input into the plasma w.[51,52]

In fact, the following equation holds for the low-current RFCD

$$w = jE_{pl}; \qquad (16.44)$$

$$j \cong \varepsilon\varepsilon_0 \omega \frac{U_l}{d_l}, \qquad (16.45)$$

$$E_{pl} = Ep, \qquad (16.46)$$

where $E = (E_{pl}/p)$.

Consequently, taking into account the layer structure of the RFCD, i.e. $\sigma_l \ll \sigma_{pl}$, we have

$$U_{RF}(p) = \left[\left(Ed_{pl}\right)^l p^2 + \left(\frac{2wv_{dr}}{\varepsilon\varepsilon_0\omega^2 E}\right)^2 \frac{1}{p^2} \right]^{\frac{1}{2}}. \qquad (16.47)$$

Equation (16.47) shows that the minimum value U_{RF} (p) is obtained at

$$p_* \cong \frac{1}{E\omega} \sqrt{\frac{2wv_{dr}}{\varepsilon\varepsilon_0 d_{pl}}}. \tag{16.48}$$

Substituting into (16.48) the typical values for the low-current RFCD, for example $w = 10^2$ W/cm^3; $v_{dr} = 10^7$ cm/s; $E = 10$ V/(cm torr), it can be seen that equation (16.48) is in quite satisfactory agreement with the experiment.[51,52]

Of special interest is the expression (16.47) because it can be used to estimate the mean reduced electric field in the plasma on the basis of the measured values of w (for example, by calorimetric measurements of the discharge) and the RF voltage on the electrodes.

Thus, taking into account the real layer structure of the RFCD is the essential condition for correct interpretation of the frequency characteristics of the RF capacitance discharge.

16.7 Active probing of RFCD using a constant electric current

In the group of various methods of examining the spatial structure of the RFCD, special attention is given to the method of probing the RF capacitance discharge with a constant electric current which, combined with calorimetric measurements of the power scattered in the RFCD makes it possible to identify unambiguously the burning regime of the discharge, determine the integral characteristics of the near-electrode layers of the spatial charge (thickness d_l, active resistance R_l), the strength of the electric field in the plasma E_{pl}, etc.

In the method, a source of controlled constant voltage U_p is connected to the RF electrodes if they are not coated with a dielectric (or, in the opposite case, to the electrodes additionally introduced into the discharge gap). The volt–ampere characteristic of the circuit including the examined RFCD is then recorded. The linear section of the VAC of the probing circuit is used to evaluate the active resistance of the object placed between the probing electrodes. If it is necessary to exclude from examination the near-electrode regions of the spatial charge which always form at the surface of the probing electrodes, the volt–ampere characteristics of the probing circuit are recorded at different distances between the additional electrode, but constant probing current I_p.

The method of active probing with the constant electric current of the radio-frequency discharge is attracted because it is simple to separate the discharge current I_{RF} and probing current I_p using frequency-dependent elements (inductance coils, condensers).

We shall discuss the method of determining d_l, R_l in the two previously examined burning regimes of the RFCD using active probing. In this case, it is convenient to use flat cooled electrodes, not coated with a dielectric, with area S. The distance d between electrodes can be varied. Since it is required to determine the integral characteristics of the NLSC, to exclude the effect of the plasma column on the results of measurements the value of d is selected minimum permissible at which the required burning regime still exists.

To apply the method, it is necessary to measure the RF voltage on the electrodes U_{RF}, discharge current R_{RF} and also record the area of electrodes occupied by the discharge S_{pl}. The RF generator is connected to the electrodes via the condenser C_{bl}. A stabilised source of constant voltage with regulated U_p and the device for measuring the current I_p are connected to the same electrodes using the blocking inductance coils L_{bl}. As a result of using L_{bl} and C_{bl}, the RF discharge and constant probing currents are completely separated. I_{RF} is determined by U_{RF} and by the total impedance of the discharge gap, and I_p depends on U_p and the active conductivity of the discharge. As a result of selecting d the effect of the column of RFCD plasma on the conductivity of the discharge gap can be minimised. Consequently, the voltage $U_p S_{bl}/I_p$ determines the active resistance of the near-electrode regions R_l of unit area, and by analogy $U_{RF} S_{bl}/I_{RF}$ gives the total impedance of the same near-electrode layers Z_l, also related to the unit area.

Knowing R_l and Z_l, we calculate the effected value of the capacitance component of the conductivity of the layers and, consequently, we determine the average thickness of the near-electrode layers d_l (for the period of the RF field) under different burning conditions and determines its dependence on the gas pressure and frequency, carrying out these measurements at different values of p and ω.

Assuming that the near-electrode layers in the period of the RF field are represented in the form of a flat condenser with leakage R_l, it is quite easy to derive an equation which links the effective value of d_l with the values measured in the experiment

$$d_l = \frac{\varepsilon\omega}{\sqrt{1/Z_l^2 - 1/R_l^2}}. \tag{16.49}$$

The assumption that the value d_l from (16.49) is synonymous with the actual thickness of the near-electrode layers of the spatial charge is confirmed by the structure of the RFCD in the experiments[12] and by the fact that the recorded form of U_{RF} (t) is close to sinusoidal. This indicates that the circuit is linear as a whole, irrespective of the non-linear characteristics of each layer, as indicated by the formation in the RFCD of a high-constant voltage between the plasma and the electrodes (see above). Thus, although the thickness of each layer d_{l1} (t), d_{l2} (t) oscillates with time, the total value is d_{l1} (t) + d_{l2} (t) $\cong$ const.

When examining the procedural errors of the results of measurements it is necessary to take into account mainly the specific features of each of the forms of the RFCD manifested in this case by the fact that the NLSC of the high-current RFCD like the cathode region of the dc glow discharge, independent of the plasma column. This makes it possible to realise the conditions in which $U_{RF} = U_l$, by decreasing d. In the low-current discharge the main source of positive ions for the layers of the special discharge is the plasma column. In other words, these layers represent zones with non-independent active conductivity for which the Townsend criterion (16.6) is not fulfilled, in contrast to the high-current regime. This circumstance results in an error of determining d_l using equation (16.49) (its value is too high) because U_{RF} in the low-current RFCD always exceeds U_l due to the presence of the RF voltage in the plasma column. To check whether U_{RF} is higher than U_l, it is necessary to measure the phase shift φ between U_{RF} and I_{RF}. It can easily be shown, taking into account the inequality $\sigma_l \ll \sigma_{pl}$, that at $\varphi > \pi/3$ the relative increase is $(|U_{RF}| - |U_l|)/|U_l| < 0.15$.

Another source of errors in determining j and d_l in the low-current discharge appears when measuring I_{RF} where due to the low (with partial filling of the electrode gap by the plasma in the direction normal to the current the shunting effect of the RF current, passing outside the discharge, becomes significant. This effect can be minimised by a smooth increase of I_{RF} to the value at which the discharge fills the entire electrode, i.e. $S_{pl} = \pi R_{el}^2$ (where R_{el} is the electrode radius).

Measurements in the dc probing circuit also have their special features. The expression $R_l = U_p S_{pl}/I_p$ is correct only if the following conditions are fulfilled: 1) passage of I_p through the discharge gap has no effect on the structure of the near-electrode regions; 2) in the process of burning of the RFCD no constant emf ε_0 forms between the electrodes, i.e. the discharge is a passive load for the source U_p. The experiments show that these two conditions can be violated in practice. In particular, at $U_p > 10$ V the VAC of the probing circuit be-

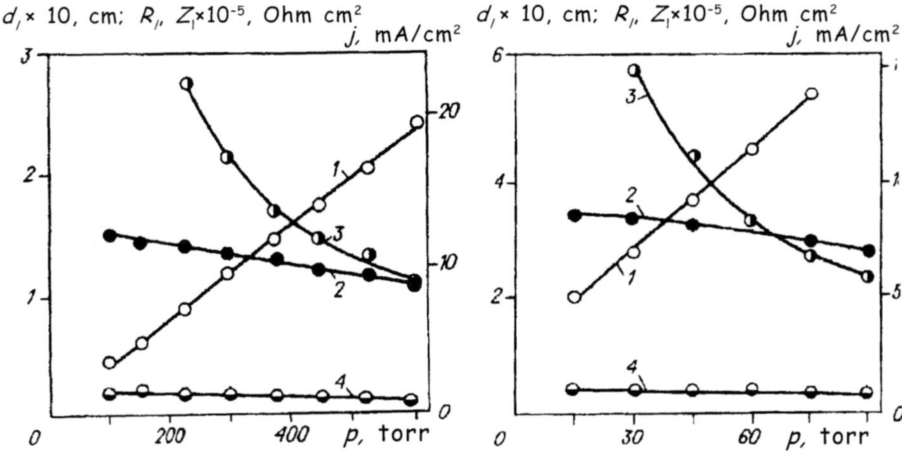

Fig.16.17 Low-current RFCD in helium. 1) $j\,(p)$, 2) $d_l\,(p)$, 3) $R_l\,(p)$, 4) $Z_l\,(p)$.
Fig.16.18 (right) Low-current RFCD in air. For symbols see Fig.16.17.

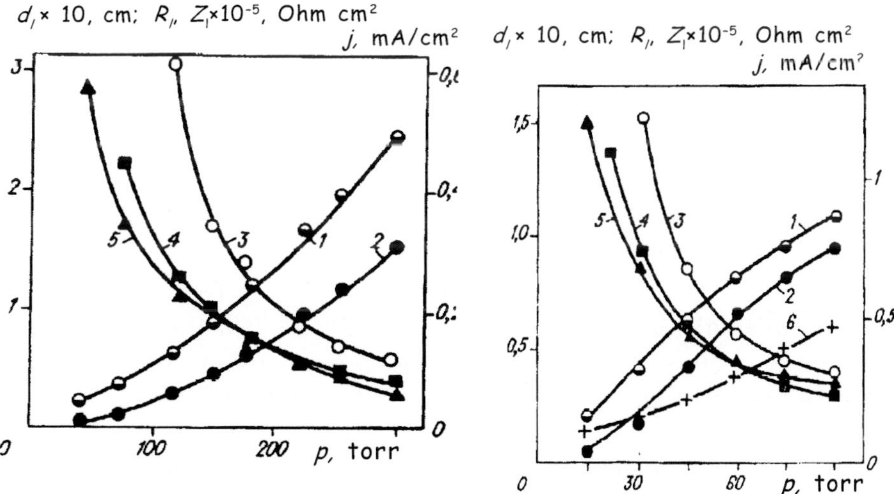

Fig.16.19 High-current RFCD in helium. 1) $j(p)$, 2) $j_a(p)$, 3) $R_l(p)$, 4) $Z_l\,(p)$, 5) $d_l(p)$.
Fig.16.20 High-current RFCD in air. 1) $j(p)$, 2) $j_a(p)$, $R_l\,(p)$, 4) $Z_l\,(p)$, 5) $d_l\,(p)$, 6) $j(p)$ in discharge with non-cooled electrodes.

comes non-linear indicating the effect of the probing voltage on the structure of the NLSC, especially in the low-current burning regime. The decrease of $U_p < 5$ V imposes more stringent requirements on suppressing RF interference in the measuring circuits. It is almost impossible to fulfil the second condition because even at $U_p = 0$ the current I_p reaches, in the high-current regime, several milliammeters. This indicates that a constant emf of several volts forms between the RF electrodes. To exclude the procedure error, the measurement of I_p in this

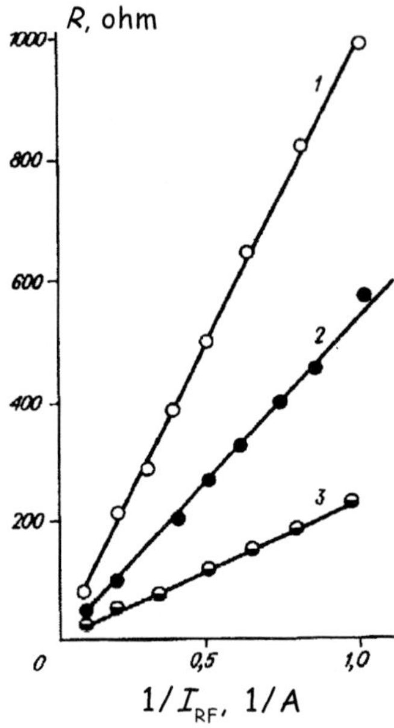

Fig.16.21 Dependences R $(1/I_{RF})$ for RFCD. 1) in CO_2, $p = 30$ torr; 2) in air, $p = 15$ torr; 3) in helium, $p = 150$ torr.

case must be taken twice: at $U_p \neq 0$, and $U_p = 0$, and in relation to the currents recorded in this case the true value of I_p is represented by the sum, even the currents are directed against each other, or by their difference in the opposite case.

To illustrate the possibilities of this method, Figure 16.17–16.20 show the experimental results of measuring d_l, R_l, Z_l, j in the low-current (Figure 16.17 and 16.18), high-current regime (Figure 16.19, 16.20) RFCD in helium (Figure 16.17 and 16.19) and in air (Figure 16.18 and 16.20) at a frequency of 13.6 MHz.

We shall now consider the experimental confirmation of the existence of the effect of the normal current density in both forms RFCD. The direct method, based on measuring the cross section area of the discharge at different currents, is often highly laborious because the contours of the near-electrode layers, especially in the high-current discharge, are usually far away from the regular geometrical form and, in addition, can continuously move on the electrode surface. Therefore, it is more efficient to use the previously examined method of active probing of the RFCD or, more accurately, the dependence of R_l on the value reciprocal to I_{RF}. Typical results, obtained in a high-current RFCD at a frequency of 13.6 MHz, are presented in Fig.16.21. It can be seen that the experimental points fit with satisfactory accuracy the straight lines satisfying the equation $R = A_0/I_{RF}$, where A_0 is a constant determined by the type of gas and electrode material (U_{RF} for each experiment series remained almost constant). It can therefore be assumed that this specific characteristics of the discharge also remain constant with increasing I_{RF}, and the results are a consequence of the linear dependence of S_{pl} on I_{RF}, i.e. in the high-current RFCD at $I_{RF} < j_{n2}\pi R_{el}^2$ the values j, U_l, d_l are independent of I_{RF}. This method can be used if $I_{RF} \gg I_0$ (I_0 is the RF current passing flowing outside the discharge) in the opposite case the effect of I_0 must be taken into account.

We shall now apply the method of active probing of the RFCD in the direction normal to the RF current. In this case, the main (RF) electrodes are coated with a dielectric material, and the additional electrodes are placed in the electrode gap.. To avoid taking into account the effect of the layers of the spatial charge formed at the surface of the probing electrodes, it is necessary to ensure that the distance l between them varies. If the probing electrodes cannot be moved, their number should not be less than three.

Let it be that l_{12} is the distance between the first and second electrodes, and l_{23} is the distance between the second and third electrode, and $\Delta l = l_{12} - l_{23} \neq 0$. It is clear that at the same probing current I_p, selected in the linear section of the VAC of both probing circuits, we can write

$$I_{p12} = (R_{l1} + R_{l2})(I_p)_\perp - \frac{I_{12}(I_p)_\perp}{\sigma_\perp S_\perp},$$

$$U_{p23} = (R_{l2} + R_{l3})(I_p)_\perp - \frac{I_{23}(I_p)_\perp}{\sigma_\perp S_\perp},$$

(16.50)

Here R_{li} are the active resistances of the near-electrode regions of the additional electrodes, and the first and third electrode can be situated in such a manner that $R_{l1} = R_{l3}$; $S_\perp$ is the cross sectional area of plasma through which $(I_p)_\perp$ flows; $\sigma_\perp$ is the conductivity of the RF discharge in probing in the direction normal to the RF current.

From (16.50) taking into account $R_{l1} = R_{l3}$, we obtain

$$\sigma_\perp = \frac{(I_p)_\perp \Delta l}{S_\perp (U_{p12} - U_{p23})}.$$

(16.51)

Comparison of $\sigma_\perp$ with the active conductivity of the low-current RFCD, obtained in probing the RFCD along the direction of the RF current $\sigma_\parallel$, under the same conditions shows that $\sigma_\perp/\sigma_\parallel > 10^2 \div 10^3$

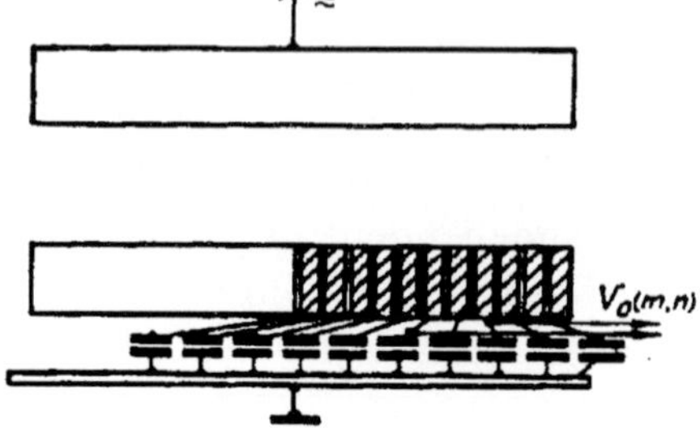

Fig.16.22 Electrode system for examining the radial structure of RFCD.

which again confirms the layer structure of the low-current RFCD including in-series connected capacitance RFCD and the plasma column, where $\sigma_l \ll \sigma_{pl}$.

16.8 The radial structure of the RF capacitance discharge

In previous sections special attention was paid to the special structure of the RFCD along the direction of passage of RF current. At the same time, it should be noted that even in the simplest geometry of a flat condenser, the main features, which are very distinctive but have been studied insufficiently, are: 1) the property of the discharge to fill only the part of the electrode gap in the direction normal to the current with decrease of I_{RF} below some value (the effect of the normal (minimal) discharge current density), and the RFCD is characterised by the occurrence of at least two stationary states of the non-contracted plasma column with the normal current densities j_{n1} (16.16) and j_{n2} (16.12); 2) dependence of the transverse dimension of the plasma on I_{RF} and, in its specific cases, on the distance to the electrode; 3) the phenomenon of the spatial modulation of the discharge current density manifested, for example, in the multichannel discharge structure observed in the experiments.

Examination of these special features of the RFCD requires, in addition to using conventional probe and optical methods of diagnostics, the development of new methods of examining its radial characteristics which do not interfere with the discharge.

We shall examine in greater detail one of these procedures which enable conclusion to be drawn on the degree of radial heterogeneity of the stationary nature of RFCD between two flat electrodes (Fig.16.22). One of the electrodes, representing the plating of the flat condenser, is in the form of a circular disc assembled from closely packed metallic rings. Each ring has no galvanic coupling with the adjacent ones, but the potential of all the rings with respect to the RF component of the field is the same. The latter, as indicated by Fig.16.22, is achieved as a result of connecting each ring to the metallic surface with zero RF potential through a condenser whose capacitance resist-

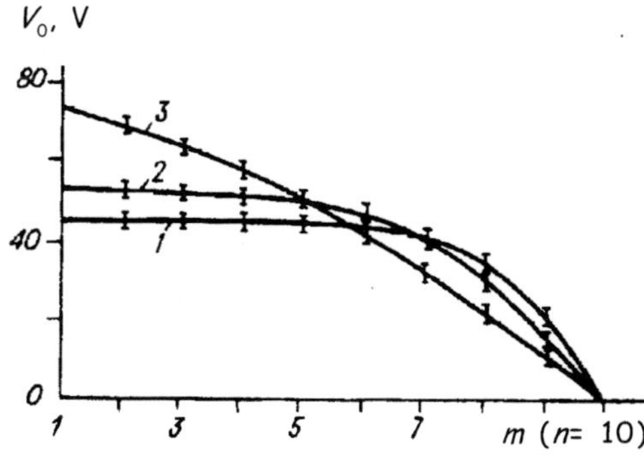

Fig.16.23 Dependence of constant voltage V_0 (m,n) formed between two insulated ring-shaped sections of the electrode in low-current RFCD in air. $p = 7.5$ torr; $d = 0.7$ (1), 1.5 (2), 3 cm (3).

ance at the working frequency can be ignored. Experiments show that the RFCD in this case does not differ by its optical and electrical characteristics from the discharge burning between the continuous electrode. However, when an electrode section with respect to the constant component of the field it is possible to measure the constant difference of the potential U_0 not only between the plasma and electrode[12] but also between any two rings V_0 (m, n) (where m, n are the number of the rings), and the value of V_0 (m, n) can be used to evaluate the radial heterogeneity of the discharge.

The mechanism of formation of the link of V_0 (m, n) with the radial heterogeneity of the discharge can be described as follows. If the RFCD in the radial direction is homogeneous, then d_l and j_l do not depend on radius r. Consequently, since $U_l \sim j_l d_l$, and $U_0 \sim U_l$ (see equations (16.31) and (16.27)) U_0 is also independent of the radial coordinate, i.e. V_0 (m, n). In the opposite case V_0 $(m, n) \neq 0$ and is determined by the degree of radial heterogeneity of the discharge.

Typical results of the experiments carried out using this procedure are presented in Fig.16.23. Attention should be given to the following. The degree of radial heterogeneity of the RFCD depends on the size of the electrode gap d. With increasing d, with other conditions being equal, the heterogeneity of the discharge in the radial direction becomes greater. In particular, the values of j_l and U_l in the centre of the discharge gap and at the periphery can greatly differ. Therefore, the experimental results which show that[13,62] qualitatively different forms of RFCD: low-current and high-current, form and co-exist in the same discharge gap, are not surprising. It was shown previously that the transition of the RFCD to the high-current regime takes place when the RF voltage on the layer U_l reaches the value U_{br} at which the condition (16.6) is fulfilled in the NLSC. However, according to the qualitative data presented in Ref.62, situations can arise in which in the

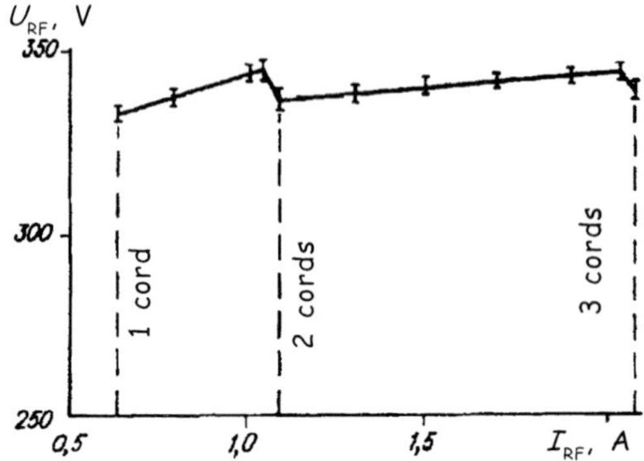

Fig.16.24 Volt–ampere characteristic of low-RFCD with appearance of cords.

vicinity of the centre of the discharge $U_{bl} > U_{br}$, whereas at some distance from the centre $U_l < U_{br}$, and the RF discharge exists there in the low-current form.

Analysis of the experimental data presented in Fig.16.23 shows that at low $d \simeq 2d_l$ the radial heterogeneity of the low-current RFCD is evident only at the periphery of the discharge at the boundary at the plasma with the neutral gas. With increasing d, i.e. when a plasma column, elongated along the current direction appears (energy in relation in this plasma column forms the radial profile of the gas temperature) the strength of the electric field E is redistributed in both the longitudinal and transverse (radial) directions. This is associated with the fact that the heating of the gas in the centre of the gap creates more suitable conditions for the passage of the discharge current in the near-axial zone of the RFCD plasma. However, until $U_l < U_{br}$ in the centre of the RFCD, i.e. the conductivity of the layers is of the capacitance type and the current density j_a is lower than j_{pl}, increase of d results in additional radial heterogeneity of the RFCD due to the formation of the radial component of the field ensuring the transfer charges from the NLSC periphery to the central part of the discharge.

They present the experimental material on the radial structure RFCD explains the mechanism of formation of the multichannel structure in the examined discharge. It has been noted that in a low-current RFCD in heavy inert gases, for example, Ar, Xe, including I_{RF} results initially in the formation of one plasma cord where $j_{pl} >> j_l$, and the second one, *etc*. The results show that the VAC in the examined case is non-monotonic: the appearance of each subsequent cord is accompanied by a decrease of the RF voltage at the electrodes. A further increase of current results in a smooth increase of the area NLSC and U_{RF} up to the appearance of a new cord (see Fig.16.24). Thus, regardless of the falling VAC of the plasma cord the radial component of RF voltage U_r, ensuring the transfer of charges from the periphery of the near-electrode zones to the region of the plasma cord and depending on r is the increasing function of I_{RF} which not only compensates the decrease of U_{pl} but also explains the large increase of U_{RF}. This fact increases the strength of the electric field at the periphery of the discharge to the values resulting in the ionisation balance. This results in the formation of a new plasma cord and the characteristic values of the r and, consequently, U_r decreases. U_{RF} also decreases in this case (Fig.16.24). With a further increase of the I_{RF} the process is re-

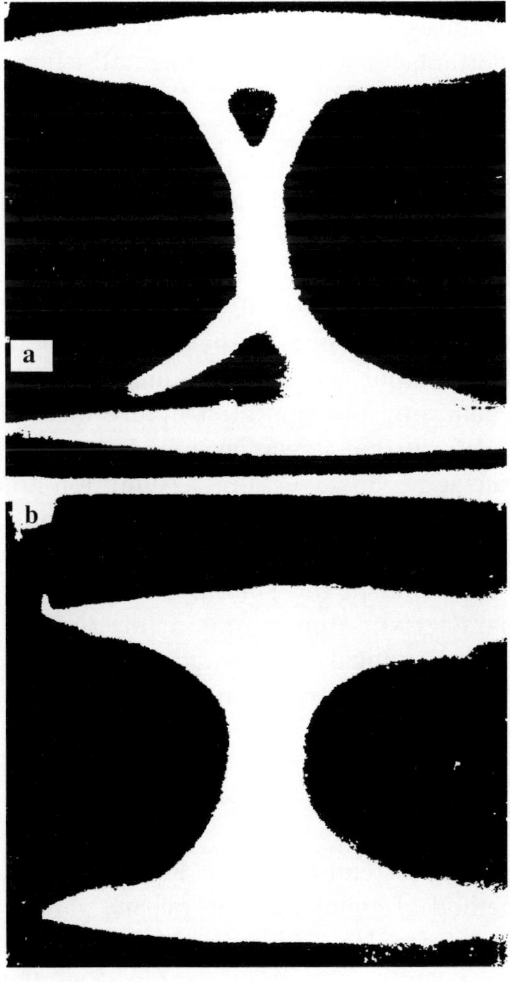

Fig.16.25 Effect of branching of the plasma column in the vicinity of NLSC in relation to value I_{RF}. High-current RFCD at the frequency of 13.6 MHz in a mixture of Xe 30 torr + air (5 torr); I_{RF} = 2.8 (a), 5.7 A (b).

peated and a multichannel ordered structure forms in the electrode gap.

Similar phenomena take place also in the high-current form of the RFCD, i.e. in the high-current burning regime of the RF discharge the NLSC and the plasma column cannot be regarded as completely autonomous because of the high conductivity of the discharge regions - analogues of the glow and the Faraday dark space, as assumed in Ref.54. In fact, as shown by the examination of the transverse structure of the high-current RFCD using the method described previously, the high-current RF discharge is also characterised by the formation (during contraction of the plasma column) of the radial fields in the vicinity of the electrodes due to finite values of the active resistances of the glow discharge and the Faraday dark space.

These special features of the spatial structure of the high-current RFCD lead to important consequences, especially the radial dependence of the current density of the NLSC, the possibility of disruption of the RF discharge in the arc regime even if the electrodes are not completely filled with plasma because the discharge current density of the NLSC directly below the plasma cord may greatly exceed the normal current density j_{n1} at the periphery of the RFCD. The radial fluxes of the charged particles, formed in contraction of the plasma column of the high-current RFCD are also responsible for the phenomena of branching of the plasma cord in the vicinity of the NLSC (see Fig.16.25)[55] and other important special features of the RFCD.

Up to now in examining the spatial structure of the RFCD normal to the direction of RF current it was assumed that the RF electrodes are equipotential surfaces. However, this assumption holds only in the case in which the characteristic size of the electrode $d_{el} \ll \lambda = c/f$ (where λ is the wavelength, c is the speed of light). At frequency $f \sim 10^8$ MHz $\lambda = 3$ m, and in the standing wave regime at the electrodes with the characteristic dimensions of the order of tens of centimetres (this situation is typical of laser technology) there is a problem with the heterogeneity of the distribution of $U_{RF}(z)$ along the large size of the electrode. There are various methods of eliminating this heterogeneity, especially by connecting the inductance coils at different points of the electrode gap[56] maintaining a stationary RFCD in the travelling wave regime, etc.

Analysis of the spatial feature of the RFCD makes it possible to propose another method of equalising the plasma parameters under the conditions of non-uniform $U_{RF}(z)$: it is necessary to vary the thickness of the dielectric coating $\delta(z)$ along the electrode in such a manner that the density of the discharge current is independent of z, with the given distribution $U_{RF}(z)$ taken into account. This can be realised as follows: one of the surfaces of the dielectric coating of the electrode facing the internal volume of the discharge chamber is flat, whereas the opposite surface is sectioned in accordance with $U_{RF}(z)$, and the surface of the electrode, coupled with the dielectric, repeats the profile of the dielectric coating. In this case, regardless of the non-uniform distribution of the RF voltage along the electrodes, the voltage drop at the discharge, including U_l and U_{pl} will the same along the entire electrode length.

16.9 Optical methods of examining the spatial structure of the RF capacitance discharge

Examination of the nature of the inherent optical radiation of the RF capacitance discharge is the most already available and widely used method of examining its spatial structure.

Already simple comparison of the photographs of the RFCD shown in Fig.16.1, regardless of the identical experiment conditions (electrode geometry, type of gas, p, d, f, U_{RF}), shows their qualitative differences. Measuring the spatial distribution of the integral intensity of glow along the direction of current, we obtain quantitative data (see Fig.16.2) required in analysing the spatial structure of the discharge. In particular, the appearance of dark near-electrode zones in the low-current RFCD whose thickness d_1 depends on frequency ($d_1 \simeq 8 \div 12$ mm for $f = 1.3$ MHz, 2–4 mm for 13.6 MHz,[12] and 0.2–0.4 mm for 81 MHz,[13]) shows either a low concentration of electrons n_e in these zones or weak electric fields. Using the assumptions on the plasma column oscillating in the electrode gap[18,32] it becomes clear that the absence of glow in the near-electrode zones of the low-current RFCD is in fact the consequence of both weak fields and low n_e in the examined discharge regions.

In fact, when the plasma column moves away from the instantaneous cathode and exposes NLSC, a high-strength electric field forms in the layer but the electrode concentration n_e there is not high. In the following half-cycle where the plasma column approaches the electrode acting as an instantaneous anode, the strength of electric field in the column decreases and although the electron concentration n_e is high, the frequency of excitations of the levels of the atoms (molecules) is not high because of weak fields. It should be noted that at the moment of time when the plasma column is situated at the minimum distance (of the order of Debye radius) from the instantaneous anode, the strength of electric field in the entire plasma column is close to zero because at this moment of time the speed of the plasma boundary converts to zero. In subsequent moments of time the plasma boundary and the entire plasma column start to move in the direction away from the examined electrode, the strength of the field in the plasma will increase but the plasma column itself, i.e. plasma electrons, will leave the near-electrode region. Thus, a situation is realised in the vicinity of both RF electrodes in the low-current RFCD in which the strength of the electric field and the electron concentration varies with time in the opposite phase. This leads to the appearance of dark near-electrode zones observed in the experiments whose thickness within the framework of the examined mechanism should depend on the frequency of the applied RF field. The weak dependence of the thickness of the examined dark near-electrode zones on pressure (Fig.16.17, 16.18) can be easily explained. In fact, the characteristic thickness of the NLSC is explained by the drift velocity of electrons in the plasma v_{dr} (see (16.28)) which, in turn, is a function of the reduced strength of electric field in the plasma E_{pl}/p, which does not change if the mechanism of annihilation of the charges continues to operate.

At the same, according to photometric measurements of the spatial distribution of the glow of the discharge in the high-current RFCD, the thickness of the near-electrode dark zones are almost completely independent of the frequency but are inversely proportional to the gas pressure which is in accordance with the measurements of d_{l2} of the high-current RFCD using other methods (see Fig.16.19, 16.20).

The absence of the glow in the examined near-electrode zones of the high-current RFCD is of the same nature as in the low-current discharge: in the phase of the instantaneous cathode in the vicinity of the electrode surface the characteristic thickness $d_{l2} = C_l / p$ (constant C_l is determined by the type of gas and electrode material) the value n_e is relatively small whereas in the phase of the instantaneous anode the strength of the electric field is small.

Differences in the nature of the dependences of the thicknesses of the dark near-electrode zones on pressure and frequency of the RF field indicate that the mechanisms of their formation in the low- and high-current burning regimes of the RFCD differ. This is also indicated by the results of spectral measurements: the spectral composition of the radiation of the near-electrode regions of the high-current RFCD coincides with the spectrum of the negative glow of the conventional dc glow discharge burning under the same conditions, whereas the spectrum of the near-electrode zones of the low-current RFCD similar to the radiation spectrum of the positive column.

The qualitatively different behaviour of d_l (p, ω) in low- and high-current forms of the RFCD can be explained on the basis of the mechanism proposed in Ref.12 which interprets the transition as a breakdown of the capacitance NLSC with the secondary emission processes on the electrodes taking place.

According to Ref.12, ionisation of the low-current RFCD by the secondary-emission electrons of the NLSC can be ignored because in non-glowing near-electrode zones or the high-current discharge there is not only intensive ionisation with electrons formed as a result of secondary-emission processes but also the Townsend criterion is fulfilled. The characteristic thickness of the NLSC is determined not by the amplitude by the drift vibrations of the plasma electrons but by the condition of the optimum frequency of the ionising collisions, as in the cathode region of the dc glow discharge.

These considerations show that the additional confirmation of the mechanism of transition between the stationary conditions of RFCD, proposed in Ref.12, are the data on the time dependence of the radiation intensity of different discharge regions. It should be noted that the method of oscillographic recording the glow of the near-electrode regions and the plasma of the RF capacitance discharge has been used for a long time for analysis of the spatial structure (see, for example, RF.19, 21, 57–60). The results show that the integral intensity of radiation in the plasma column is modulated, as expected, with the double frequency 2ω of the applied RF voltage. Nevertheless, near-electrode regions of the RF discharge showed oscillation of the glow intensity at frequencies ω and 2ω, and the amplitude of oscillations of the frequency ω exceeded the amplitude of oscillations at the doubled frequency by two orders of magnitude.[21] It is also important to note that the glow formed in the same area of the discharge gap under the same conditions, can appear twice during a cycle (this corresponds to the low-current RFCD) and under different conditions only once during a cycle (high-current form of RFCD).[59]

However, when using the optical methods for determining the burning regime of the RFCD in the frequency and pressure range where

$$p < \left(\frac{\omega}{v_{dr}}\right) C_1 \left(\frac{T}{T_0}\right)$$ (see equation (16.18)) it must be remembered that

the pulsations of the glow with the frequency ω can become evident as a result of the pressure of beams of secondary-emission electrons also in the highly anomalous low-current RFCD where the criterion (16.6) is not fulfilled (for more detail see Section 16.2).

If $p > \left(\frac{\omega}{v_{dr}}\right) C_1 \left(\frac{T}{T_0}\right)$, the application of optical methods with the time

and spatial resolutions makes it possible to draw unambiguous conclusions on the burning regime of the RFCD applied in the experiments, especially if it is possible to synchronise the examined oscillograms of the glow with the oscillograms of the discharge current.[60]

The optical methods with the spatial resolution of the glow intensity of the near-electrode layers give good results in examining the process of transition of the RFCD to the high-current burning regime when the electrodes are coated with a thick dielectric layer, i.e. when $\delta > \varepsilon_d d_l$. As established in Ref.61 varying the value of δ is it possible to realise the falling sections of the VAC of the stationary RFCD (see Figure 16.3, 16.4, sections VAC *DE* and *cd*, respectively). The dark zones in the vicinity of the electrodes coated with dielectric, which carry information of the thickness of NLSC, are distinct and their size varies in relation to δ and U_{RF} in the range d_{l1}, d_{l2}. The NLSC side opposite to the surface of the dielectric is limited by a thin plasma layer whose radiation spectrum differs from that of the positive column. Thus, the coating of the electrodes with the dielectric with of different thickness makes it possible to examine the process of transition of the RFCD to the high-carrying burning regime or transformation of its structure without using optical methods with time resolution.

16.10 Conclusions

The methods of examining the spatial structure of the RF capacitance discharges, examined in this section, do not obviously exhaust the entire range of diagnostic means used for examining the special feature of the RFCD. It is evident that in measuring the parameter of the RFCD an important role is played by the conventional method of plasma diagnostics: probe, microwave, optical, *etc.*, which makes it possible to determine the required characteristics of discharge plasma with the sufficient accuracy and spatial–time resolution: T_e, n_e, T, *etc.* However, the availability of only local discharge characteristics in many cases doesn't make it possible to give an unambiguous answer to the problem why these and not some other values of the measured quantities are realised in the RFCD. As shown previously, the reason for this ambiguity is that the RFCD is characterised, during the passage of discharge current in the electrode gap, by the formation of regions which qualitatively different mechanisms of formation of conductivity in them: near-electrode layers of the spatial discharge in the plasma column. In particular, the process of establishment of a stable stationary state (self-organisation) in the NLSC – plasma column system is the basis which determines the special feature of the RFCD and stimulates the search for the corresponding diagnostic means some of which have already been described.

Chapter 17

WAVE BREAKDOWN IN DISTRIBUTED SYSTEMS

17.1 Introduction

This chapter is concerned with the properties of the waves of electric breakdown in gases moving at the velocity close to that of light. For comparison, we shall discuss schematically the main variants of electric breakdown of the gas gaps.

The breakdown of short gaps between flat electrodes at moderate pressures and strength of electric fields by an electron avalanche has been studied quite extensively.[1] The speed of propagation of the avalanche front is determined in this case by the drift speed v_{dr} of the electrons in an applied electric field:

$$v_{dr} = \mu_e E$$

where μ_e is the mobility of the electrons.

The following data mechanism is associated with the transition of the avalanche with increasing pressure and length of the gap into a streamer.[2] It is probable that the streamer channel is of the plasma type, and the rate of its growth to the electrodes becomes considerably higher than in the avalanche and usually equals 10^8 cm/s. The increase of the speed of movement of the 'head' of the streamer to the anode is explained by the effect of two factors. The first factor may be the photoexcitation of the gas in front of the 'head' of the streamer by the resonance radiation of the plasma channel. The reaction of associative ionisation

$$A^* + A \rightarrow A^+ + e$$

results in the formation of free ('seed') electrons. Finally, the strong electric field of the spatial charge in the vicinity of the 'head' results in a high ionisation rate. Thus, the existence of local electric fields which are stronger than the mean field in the gap is the second factor ex-

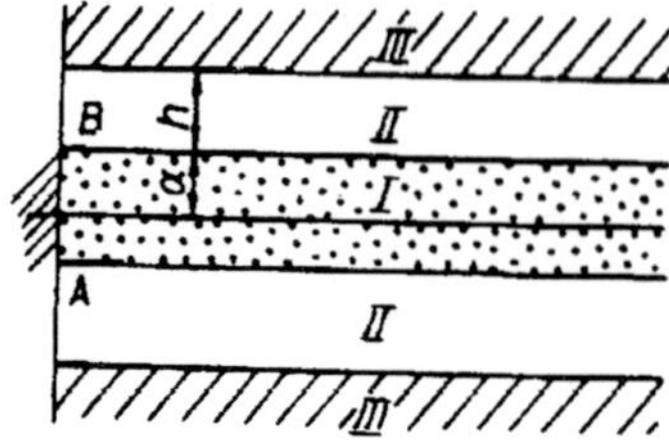

Fig.17.1 Geometry of the device for examining the development of a wave breakdown. I) slightly ionised plasma, II) dielectric, III) ideal conducting screen, AB - electrode.

plaining the observed velocity of movement of the 'head' of the streamer.[2]

The avalanche (Townsend) and streamer breakdown mechanisms do not exhaust the entire variety of the physics of electrical breakdowns of the gas gaps. From the retrospective viewpoint, it is evident that when the length of the gap is greatly increased the distribution of bias currents in the space becomes far more complicated which, in turn, affects the dynamics of the ionisation front. We shall assume that the breakdown of a gas takes place inside a tube made of a dielectric surrounded by a metal screen (Fig.17.1). In other words, we are discussing here the breakdown of the gas in systems with distributed parameters whose suitable example is a co-axial line shown in Fig.17.1. Of course, the special features of the breakdown of this type become important at relatively rapid changes of the potential at the electrodes. For example, in this case, a volume charge wave forms and moves at the velocity close to the velocity of light.

It is important to note that the development of, for example, ball lightning is modelled more efficiently by a breakdown in a co-axial system than between flat electrodes in a Paschen system which is now regarded as traditional.

It is also useful to pay attention to the large difference of the discussed wave of the electric breakdown in comparison with the wave of microwave breakdown in the waveguides. In the latter, the energy flux is directed against the ionisation wave.

17.2 History

In the eighties, the largest number of data on the nanosecond breakdown in the distributed system were obtained in the experiments with long low-pressure tubes convenient for laboratory investigations. The interest in experiments of this type appeared already in the previous century and then periodically in the following years.[3,4] Their main aim was to examine the nature of transfer of perturbations along the discharge gap. Special attention has been paid to the propagation of rapid gas ionisation fronts along the pipes. The high velocity $v = 10^{10}$

cm/s of such a front in the direction of propagation of the gas glow front was measured by Thompson.[4] Further studies were concerned with examining the processes accompanying this breakdown stage referred to by Fowler[5] as the wave breakdown.

In 1926, Beams confirmed[6] the high velocity of the glow wave, recorded by Thompson, and 10 years later Snoddy, Deitricht and Beams[7] found, placing two additional electrodes on a tube, that this movement of the light pulse is accompanied by the wave of the potential with the same velocity. They showed that the velocity of the ionisation wave depends on the tube diameter, gas pressure, and the strength and polarity of the applied potential.

The wave always started propagating from the electrode with a high potential in relation to the air, irrespective of its polarity. The velocity at a density of $1.4 \cdot 10^{16}$ cm^{-3} increased almost linearly from $1.7 \cdot 10^9$ cm/s at 73 kV to $3.7 \cdot 10^9$ cm/s at 175 kV. The maximum current in the initial wave is very high, in the range 90–200 A, i.e. current density 90–4000 A/cm^2. The voltage wave propagates through the tube with some absorption which increases with increasing pressure. At low gas densities in a tube 12 m long, the pulse amplitude of 194 kV decreased at the other end to 180 kV, and at a gas density of $3.5 \cdot 10^{17}$ cm^{-3} to 100 kV. In their experiments, they recorded also a reversed wave formed when the first wave reached the earth electrode. The velocity of the reversed wave was 10^{10} cm/s at 132 kV. The velocity of the reversed wave was slightly higher than the highest limited velocity of the primary wave and was almost independent of the potential sign and the pipe diameter.

It is evident that one of the main conditions of formation of ionising waves is the rapid change of the potential of one of the electrodes of the discharge gap. This conclusion was made in Ref.8 in examining the glow pulses in long pipes. A wave formed only at relatively fast supply of a voltage of 20–40 kV on one of the electrodes of the pipe. The authors of Ref.9 obtained both slow glow waves in long pipes and fast ones. However, the fast waves appeared only at a high speed of increase of high potentials. The required change of the electrode potential can take place not only when the source is represented by a pulsed voltage oscillator. In a study by Westberg,[10] perturbation of the electric field in the vicinity of an electrode forms spontaneously in breakdown of the oxide film on the cathode of the stationary discharge. It forms when a wave travelling from another electrode approaches the given electrode, or when the voltage is removed at the moment of discharge.

Ionising waves are observed in the same experiment at both the positive and negative polarity of the pulse but their velocity differs.[7,11,12]

As shown in Ref.10, 13, the nature of propagation of the waves is greatly influenced by the preliminary ionisation of the gap.

The ionising waves transfer the volume charge. This is confirmed by recording the waves of the potential along long pipes[7,14] and by the variation of the current on the electrode at the moment of arrival of the wave.[10]

The passage of the wave through the gap is accompanied by gas ionisation. This is indicated by the increase of radiation behind its front. The increase of the degree of ionisation of the gas in the discharge gap and, consequently, its transition from the non-conducting or low-conducting (glow discharge, corona) state to an efficiently conducting arc can take place by two methods, according to Loeb. In the first method, which is used more widely, the discharge current heats the gas, ionisation takes place and the conductivity of the channel increases. The rate of this gas is restricted by a velocity of the order of 10^7 cm/s of the ionising electrons intersecting the discharge gap. The second, faster process,[10,15,16] is based on the ionising waves.

Analysing the studies published by Westerburg,[10] Loeb describes the process as follows. The initial electrical perturbation during the period shorter than the diffusion period produces a steep potential gradient and the gradient of the volume discharge at length l_f. This is possible only at the initial concentration n_0 of free electrons. They initially either change the volume or can be generated by photoionisation as a result of initial perturbation. The speed of displacement of the potential gradient is described by the equation

$$v = \alpha\ v_{dr}\ l_f\ /(\ln\ n_e/n_{e0})$$

where α is the Townsend coefficient, v_{dr} is the drift velocity of the electrons, n_{e0} and n_e is the initial and final concentration of the electrons.

Ionisation takes place in orderless collisions of electrons with atoms. If the pressure is too low and the electrons have mainly directional movement, the equation cannot be used and the velocity of the wave is determined by the velocity of the electrons moving in vacuum with the energy which the electrons initially possessed or which is equal to the potential drop. At a high potential, the formation of the potential gradient requires a high density of the ions and l_f is considerably lower. In this case, the pulse, generating the initial perturbation must have steep fronts and photoionisation takes place with fronts with shorter wavelength.

It should be noted that the movement in slow waves[17-20] formed in the case of slow supply of a voltage through the electrode, is determined by electron diffusion to the walls and the subsequent charging of the walls. However, in the case of surface waves[21,22] their movement is determined completely by the change of the charges and the fields on the side surface of the cylindrical discharge, and at large diameters of the discharge pipes it affects a relatively small fraction of the volume.

Summing up these results, the development of a wave breakdown in the long low-pressure pipes can be described as follows. Irrespective of polarity, the wave forms at the electrode with a high potential,[8] the voltage range is 1–200 kV. The wave does not form always immediately after supplying a voltage pulse to the electrode. Sometimes, a certain period of time is require for its formation.[23,24] The wave moves along the pipe with a velocity of 10^8–10^{10} cm/s, transferring the potential[8] and causing the glow of the gas in the part of the pipe through which the wave has travelled. The length of the pipe was changed in the experiments from several centimetres to 10 m and longer, its diameter from several millimetres[25] to 14 cm,[9,26] gas density 10^{16}–$3 \cdot 10^{18}$ cm^{-3}. The velocity of the wave increased with increasing voltage and pipe diameter,[7] and also if the screen was removed or preliminary ionisation was induced.[23] With movement of the wave in the pipe the potential at the front decreased.[7,9,27] There are differences in the waves formed by voltage pulses of positive and negative polarity. In the first case, the current of the second electrode was non-monotonic.[28]

For the majority of experiments whose results, obtained up to the end of the eighties were summarised in Ref.29, the common measured quantity was the speed of travel of the front wave v along the discharge pipe. The dependence of the velocity of the front on gas density $v(N)$ has been studied most extensively. In Ref.27, $v(N)$ was obtained for Ar, CO_2, N_2, H_2 and air, in Ref.25 for Ne and N_2, in Ref. 9 and 16 for air, in Ref.26 for Ar, H_2 and He. In all gases, the $v(N)$ curve has a maximum. McGhee explains the behaviour of the $v(N)$ curves using the following qualitative models.[27] The electric potential at the front in the case of positive polarisation of the voltage pulse is generated by the ions whose concentration can be relatively high. From the estimate of the conductivity of the discharge the concentration behind the front is $4 \cdot 10^{11}$ cm^{-3} and at a relatively high value of the Townsend coefficient α this concentration can be generated by impact ionisation at the front with a single photoelectron. The front of the wave is characterised by intensive excitation of the gas, and primary photoelectrons ahead of the front are generated by the radiation of the discharge. Therefore, the velocity of the wave is determined by both the efficiency of photoionisation of the gas ahead of the front and the intensity of impact ionisation at the front. According to McGhee, the velocity of the wave decreases at high pressures as a result of a decrease of the effective length on which photoionisation takes place. At the same time, the shift of the maximum with increasing voltage is explained by the fact that the maximum velocity of impact ionisation is displaced to the range of high pressures.

These results relate to different experimental conditions and although the form of the $v(N)$ curve is the same, they cannot be compared directly. The results show that the velocity of the wave is also influenced by the diameter of the pipe,[7] the presence or absence of a metallic screen at the pipe,[23] and pulse polarity.[23]

17.3 Transition processes

Winn proposed an explanation of the role of the screen[23] which forms some linear capacitance C with plasma. This capacitance is charged with a wave to the voltage of the breakdown pulse U. If the wave has travelled the distance x, the total charge in the pipe $Qx = CUx$ is generated by the current $I = CUv$, where v is the velocity of the wave. In Ref.23, $C = 0.79$ pF/cm, $U = 24$ kV, $v = 1.9 \cdot 10^9$ cm/s and $I = 40$ A. Thus, the screen, like the size of the cross section of the pipe, influences the rate of formation of the volume discharge in the plasma and, consequently, the entire process dynamics. It is obvious that the effect of the screen cannot be completely eliminated because in its absence the role of the screen will be played by the earthed elements of the system.

Winn's explanation is simplified. It contains, in particular, an assumption according to which the linear capacitance C and the potential U are constant along the length of the pipe, although the potential behind the wave front decreased in the longitudinal direction. This decrease of the potential was measured in Ref.14 and its pressure dependence in Ref.27.

To examine the nature of the phenomenon it is important to study the problem of the dependence of the velocity of the wave on the potential drop at the front U_f. In Ref.7,14,30, the dependence $v\,(U)$ where U is the voltage amplitude at the electrode exciting the wave, is close to linear, and in Suzuki's study[16] there was a large deviation from it.

Presenting the data on the dependence of the velocity v on the initial concentration of the electrons n_{e0} ahead of the front wave, Winn in Ref.23 published an important conclusion according to which the ionising wave at increasing n_{e0} should resemble more and more the electromagnetic waves propagating along the coaxial transmission line. The concentration of the electrons at which the ionised channel can be regarded as a good conductor was determined by Winn by two methods.

In the first method, the experimental curve v (n_{e0}) is extrapolated to the value $v = c$, i.e. the velocity of light in vacuum. For the wave with positive polarity, the concentration n_{e0}, derived in this manner, is equal 10^{12} cm^{-3}.

Extrapolation of the experimental dependences v (n_{e0}) in Ref.16 gave the values of the limiting concentration of the electrons n_{ef} an order of magnitude lower than those obtained by Winn. In addition, it was established in Ref.16 that the velocity of the wave depends on the duration of the front v (τ_f) of the voltage pulse. Extrapolation to the velocity of light in vacuum using this parameter gives the limiting value $dU/dt = 5 \cdot 10^{12}$ V/s.

In Ref.23, Winn determined the limiting concentration of the electrons at a front n_{ef} by examining the balance of the energy stored and dissipated in the wave. If it is assumed that ionisation in nitrogen requires 1/3 of this energy, then from the relationship

$$\varepsilon_{dis} = \varepsilon_p - \varepsilon_E = IU / v - CU^2 / 2 \approx CU^2 / 2 = 3 n_{ef} I_i S,$$

where ε_{dis}, $\varepsilon_p = IU/v$, ε_F is the energy per unit length of the plasma column from the electrode to the front: ε_{dis} – dissipated; ε_p – introduced to the discharge; ε_E – the energy stored in the electric field between the plasma and the screen; C is the linear capacitance; I_i is the ionisation potential of the nitrogen molecule; S is the cross sectional area of the discharge pipe; v is the velocity of the wave; U is the potential of the plasma in relation to the screen, gives $n_{ef} = 1.2 \cdot 10^{12}$ cm^{-3} which is close to that obtained by extrapolation. Winn also estimated the concentration of the electrons at the front of the wave on the basis of the experimental data for the attenuation for the potential along the pipe, assuming that the current I between the electrode and the front in every cross section is the same and that the velocity of the electrons is equal to the drift velocity:

$$n_{ef} = CUv / ev_{dr} S.$$

This volume was $4.7 \cdot 10^{11}$ cm^{-3} at a nitrogen density of $7.8 \cdot 10^{16}$ cm^{-3}. According to Winn, these values are in good agreement with the previous estimates and the difference is explained by the fact that the electrons, which have acquired a high energy in the strong electric field at the front, are not in equilibrium with the field behind the front.

To determine the relationship between the electron concentrations ahead and behind the wave front, Suzuki[16] used the experimental data on the jump of the current in the front and carried out calculations based on the model of the discharge as a long condenser charged with an ionising wave. The model was then developed further in Ref.33 and 34.

The problems of calculating the plasma parameters in the wave are very important because there are almost no experimental studies where the electron concentration and temperature were measured. Simultaneous measurements of the current jump in the front and the electron concentration were taken in Ref.35 in examining precursors in impact pipes. The thickness of the front, determined in the range 0–90% at the stationary value behind the front was 2–5 cm but it can also be low because the 'hot' electrons carried out ionisation even after passage of the wave. Measurements of the electron temperature are associated with considerable difficulties. For example, the measurements of T_e in relation to the intensity of two helium lines[30] are characterised by a large error because the population of the helium levels in the nanosecond time period is determined by the transfer of excitation[32,36] which was not taken into account by the authors.

In Ref.24, when calculating the current of the electric field in the discharge and in comparison with the experimental data, the discharge gap was regarded as a section of the coaxial transmission line with the losses described by the telegraphic equations

$$\mathrm{d}U/\mathrm{d}x = -RI - L\mathrm{d}I/\mathrm{d}t,$$
$$\mathrm{d}I/\mathrm{d}x = -C\mathrm{d}U/\mathrm{d}t.$$

The signals, received from the capacitance voltage dividers, positioned along the pipe, were processed taking into account that they are superimpositions of several waves reflected from the electrodes and from the front as a result of the mismatch with respect to the wave resistance in these areas. The principal difficulties of using this method are associated with the electrodynamic description of the first wave moving from the cathode because the coaxial transmission line in the pipe for this wave has not yet formed along its entire length and the wave is transferred by the moving first front. The current in the earth anode was calculated and was found to be similar to the measured value. The longitudinal strength of the electric field and current $I\,(x,\,t)$ was calculated at different phases of the discharge as the superimposition of the fields and currents of all waves and this was followed by calculating the internal parameters of the plasma: T_e, n_e, the rates of excitation and population of the levels assuming that the electron velocity of the distribution function manages to follow the electric field.

The large number of data obtained in this work indicate that it is important and essential to develop a similar procedure for examining the processes taking place in ionising waves. In particular, assuming that R, C, L are variable, the wave can be regarded as an impact wave. All features of this wave were examined in several experiments.[31] Of special importance is the examination of the first ionisation wave, where the relationship of the electrical state of the electrodes with the processes inside the gap are of special interest for constructing the electrodynamic model of the wave breakdown.

17.4 Generalised block diagram of experiment

The generalised block diagram of typical experiments is shown in Fig.17.2. The oscillator 1, shaping a short high-voltage pulse, is connected with one of the electrodes of the discharge pipe 3, the second electrode is earthed through the resistance 4. To induce preliminary ionisation in the gap, the dc source 5 and the ammeter 6 can be connected in series with the pipe. The pulsed voltage of the electrode on the side of the oscillator an the corresponding current in the circuit in the second electrode are measured using the divider 2, the shunt 4 and the oscilloscope 9. The electric probes 7 and high-speed photomultipliers 8 are used to measure the electrical and optical parameters of the breakdown waves, respectively. To examine the wave breakdown, it is necessary to solve a number of procedural problems.

In Ref.43–51, the problem of the breakdown wave was solved by synchronised measurement of currents through the electrodes of the pipe with simultaneous measurement of the electrical parameters and radiation of the breakdown wave in different areas of the pipe. The accuracy of synchronisation together with the time resolution determines the scale of the spatial resolution of the examined distributed process. To fully utilise the time resolution of the measured parameters

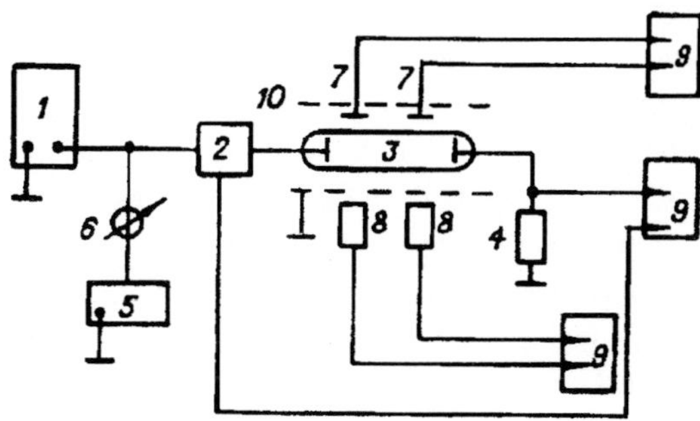

Fig.17.2 Generalised block diagram of experiments for investigating the wave mechanism of breakdown in long tubes.

of the wave during their spatial–time comparison, it is necessary to ensure that the synchronisation of the measurements is not lower than their time resolution.

In breakdown, the energy flux, supplied from the high-voltage oscillator, interacts with the gas. In addition to the incident flux, reflected flux is always present. These fluxes, propagating in the opposite directions, are superimposed onto each other, in addition to interacting with the gases. Consequently, it is almost impossible to separate them in recording directly at the discharge pipe. However, since the amplitude and the phase of the fluxes do not change during their propagation along the linear transmission line, i.e. the coaxial cable, separate recording of the fluxes in the transmission line becomes quite easy.

The same measuring element (shunt) is used to measure the current of the high-voltage pulse directed towards the discharge tube (incident pulse) and, after some required for the displacement of the pulse along the cable from the shunt to the tube and back, the current of the pulse directed to the opposite side (reflected pulse) is measured. The amplitude-phase differences between the incident and reflected pulses are determined by the interaction of the measured energy fluxes with a gas. Efficient electromagnetic matching of the cable–electronic section - discharge volume circuit is essential.

To examine the structure of the wave, the dimensions of the discharge tube should be selected in accordance with the time resolution of the apparatus and procedure used. It should be noted that the authors of Ref.40 and 41, who took synchronised measurements, worked with gaps not exceeding 4 cm, and the time resolution obtained in Ref.41 was 1–2 ns which did not make it possible to examine the spatial structure of the breakdown wave. In fact, its spatial scale of resolution of the measurements determined as $v\Delta t$ (v is the speed of the object, Δt is the time resolution) at a breakdown wave speed of $4 \cdot 10^9$ cm/s is greater than the width of the gap.[41] In Ref.23, the breakdown wave was investigated in relatively long tubes so that the spatial scale of resolution, corresponding to the time resolution of the order of 10 ns, was close to the length of the tube but an order of magnitude greater than its diameter. It is evident that the structure of the breakdown wave should be examined at the spatial scale of resolution close to the transverse dimension of the discharge tube.

When examining the structure of the breakdown wave using capacitance devices, an important factor in addition to the time resolution is the presence of the directional diagram of the capacitance divider. The width of this diagram is equal to approximately the diameter of the screen jacket. The spatial resolution of the sensor can be improved by mathematical processing which was used in Ref.49 on the basis of the assumption that the scale of the spatial–time non-reproducibility of the breakdown wave is considerably smaller than the resolution of experiment. It was thus possible to examine the evolution of a wave with a spatial resolution of $\simeq 1$ cm at a screen diameter of $\simeq 10$ cm.

The time resolution and synchronisation of measurements of the wave current obtained in Ref.43–51 were equal to 0.25 ns which at a speed of the wave of 10^{10} cm/s corresponds to a spatial scale of 2.5 cm (the diameter of the discharge tube 4.5 cm). The synchronisation of the electrical and photoelectrical measurements is 0.55 ns which at the same speed of the wave corresponds to a scale of $\simeq 5$ cm.

The breakdown was organised in a long ($l \gg d$) discharge tube surrounded by a metallic screen to form a controlled linear capacitance of the discharge channel. Figure 17.3 shows the discharge device used in Ref.43–51. The discharge gap represents a uniform coaxial line, and the break of the central strand of the line includes a glass tube with a length of 400 mm and an internal diameter 45 mm. The ends of the tube are closed with flat molybdenum electrodes 50 mm in diameter which smoothly transfer to the central strand of the coaxial cable, and the outer screen smoothly changes to the braiding of the cable. The region of smooth transition from the cable to the electrode was produced in such a manner that the unit electrical parameters were uniform and corresponded to an impedance of 50 Ω. In the screen along the entire length of the gap there is a slit whose width is equal to the diameter of the discharge tube. The slit can be closed by a moving metallic sheet which carries capacitance sensors and window for a narrowly-directed photodetector. A high-voltage nanosecond generator based on ferrite forming lines is used. The main advantage of this device is the high stability of generated pulses with the possibility of

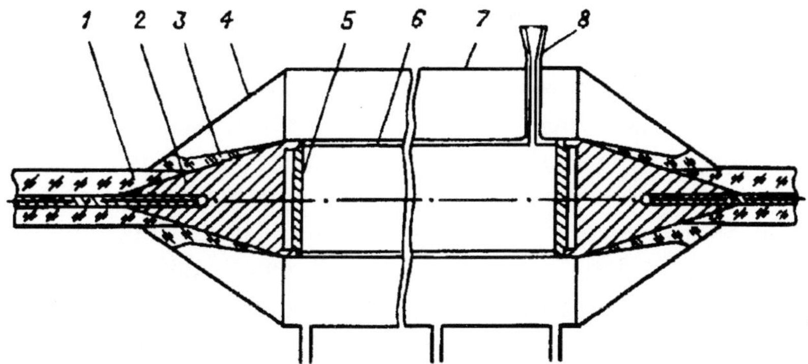

Fig.17.3 Discharge device. 1) cable RK-50-11-13, 2) internal electrode casing, 3) insulator, 4) outer electrode casing, 5) molybdenum electrode, 6) glass discharge tube, 7) screen, 8) capilliary tube with a pin for connecting the vacuum system.

operating with the repetition frequency of up to hundreds of pulses per second.

The pressure range of the working gas (helium) included the range of the maximum value of the speed at which the electrodynamic effects should be most evident. The initial ionisation was produced in the tube and also increased the speed of the wave; at the same time, the stability of propagation of the wave increased.

To determine the power or other energy characteristics of the breakdown, it is necessary to take independent measurements of the current and voltage at electrodes in the discharge gap. The independence of the measurements is equivalent here to the measurement of the phase shift between current and voltage.

For the waves propagating along the cable, the currents flowing through the given cross section of the cable in the braiding and the central conductor are equal to each other. The current is measured using a shunt connected to the break of the screen in the braiding of the coaxial cable. The discharge tube, connected to the break in the central conductor, causes a heterogeneity in the uniform cable transmission line. This leads to the reflection of electromagnetic waves which, after reflection, interfere with the incident waves.

For separate recording of the current of the incident and reflected waves using a single shunt, the position of the shunt in the cable is selected from the condition

$$l > Tc/2\varepsilon^{1/2}$$

where l is the length of the cable between the shunt and the discharge tube, T is the duration of the electromagnetic pulse, c is the speed of light, ε is the dielectric constant of the filler of the cable.

For the waves propagating in the cable the voltage between the central conductor and the braiding is unambiguously linked with the intensity of the current flowing to the given section:[42] $U = IZ$. If the direction of current in the central conductor of the cable coincides with the direction of propagation of the electromagnetic wave, the central conductor in the given cross section of the cable has a positive potential in relation to the screening braiding. The current and voltage in the incident and reflected waves are linearly added up in the cable in the interference zone. To determine the absolute values of the current and voltage in the cross section of the interference zone and, in particular, at the electrode of the discharge tube, it is sufficient to measure (using the shunt) separately the currents of the incident and reflected waves and also the time delay between them.

The measurements of the currents of the incident and reflected waves are independent measurements and, consequently, the current and voltage of the central conductor in relation to the screen in the interference zone calculated from the equation

$$I = I_{in} + I_{ref},$$
$$U = \left| I_{in} - I_{ref} \right| Z_0,$$

are also independent quantities.

It should be noted that the voltage between the central conductor and the braiding of the cable can be measured using the capacitance device built into the cable. However, such a divider requires amplitude calibration whose error should not exceed the error of calibration of the shunts. As a result, the total relative error in measurements of current and voltage would be higher than when calculating the voltage through the measured current and the wave resistance of the cable. In addition, the use of capacitance dividers results in an additional error associated with the fact that their width of the band of working frequencies is considerably smaller than that of the band of the working frequencies of the shunt.

In Ref.43–51, when determining the spatial–time correlation of the electrodynamic quantities inside the gap and at its electrodes, all the measurements were mutually synchronised. Consequently, for each moment of time during the breakdown it was possible to show the voltage and current at the electrodes of the gap, the distribution of the excess charge and the plasma radiation intensity along the gap. The time resolution obtained in the experiments of the mutual synchronisation of the electrical measurements were 0.25 ns.

The measurements in Ref.50 were taken in repeated conditions in operation of the nanosecond generator with a frequency of 80 Hz. The higher reducibility of the breakdown process enabled all the signals to be recorded with a single oscilloscope. Mutual synchronisation of the measured quantities was ensured by connecting the corresponding delay lines in the synchronisation unit.

The start of counting the time on the oscillograms is the same for all signals if the electrical length of the cables are synchronised from the rules formulated in Ref.50. When the rules are adhered to, the time shifts between the measured signals, counted directly from the oscillograms, are determined only by the breakdown process and the duration of propagation of the electromagnetic wave from its excitation point to the sensor.

To eliminate the inaccuracies when determining the electrical length of the cables caused by processing the effects, the accuracy of mutual synchronisation of electrical signals was verified by experiments.

I_{inc} (incident) and I_{ref} (reflected) were synchronised by combining their oscillograms on the screen. For this purpose, the signal I_{ref}, having the reversed polarity, was inverted by changing the polarity of connecting the cable to the shunt. The discharge volume was filled with air to the atmospheric pressure, no breakdown developed and the shape of the reflected pulse coincided with that of the incident pulse.

I_{trans} was synchronised with I_{inc} also by combining their oscillograms on the screen. The discharge tube was removed and the cathode and anode plains were tightly pressed together.

The signal from the capacitance sensor was also synchronised with I_{ref} by combining their oscillograms. The sensor was placed opposite the anode and the discharge volume was replaced with a metallic tube so that a uniform line was obtained instead of the gap.

Recording radiation

In Ref.43–51, the amplitude dynamics of integral radiation during each pulse was recorded with a 14ELU-FS high-speed photodetector which could be placed against any cross section of the gap. The time resolution of the photodetector was calibrated with a nanosecond surge of an LD-15 semiconductor laser. The duration of the leading edge of radiation of this laser, measured with an HSD-1850 photodiode, was 0.5 ns. The minimum front of the signal, obtained from the calibrated photodetector, was 1.3 ns.

For comparing the spatial dynamics of radiation discharge with voltage on the electrodes in relation to time, it is necessary to know the time of flight of the electrons in the photodetector (the delay time of the signal in the device). The error in determining this parameter in the experiment was less 1 ns. The available methods[52,53] of measuring the time of flight give an error of 2–3 ns. Therefore, a procedure was developed and the time of flight of 14ELU-FS was measured with an error of ±0.3 ns.[54] The method is based on comparing the delay of the nanosecond signal in two photodetectors where the accuracy of the time of flight of one of them (reference photodetector) does not exceed the required accuracy for the other one. The time of flight of HSD-1850 vacuum photodiode used as a reference photodetector does not exceed 10^{-10} s. The light source was a LD-15 semiconductor laser with a surge power of 6 W and a duration of 4 ns. The block diagram of the measurements is shown in Fig.17.4. The generator of nanosecond pulses (GNP) (120 A, pulse front 2 ns, half-width 4 ns) excites the semiconductor laser – emitter (LD-15). The reference

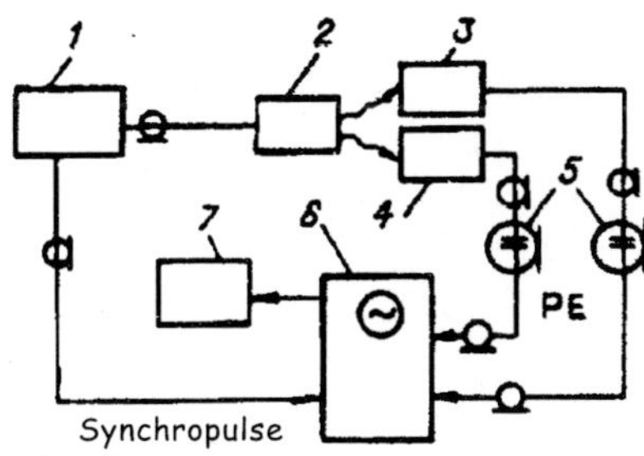

Fig.17.4 Diagram of device 14ELU-FS for measuring time of flight. 1) generator of nanosecond pulses, 2) laser semiconductor emitter, 3) HSD-1850 vacuum photodiode, 4) measuring photodetector, 5) separating capacitances, 6) oscilloscope, 7) two-coordinate automatic recording device.

photodetector – vacuum photodiode (HSD-1850) has a time resolution of 0.1 ns and a time of flight of 0.5 ns. The signals from the diode and from the 14ELU-FS system were recorded in a two-channel strobing oscilloscope (SAS-500-9B) and then displayed in a two-channel automatic recording device (WX-431). The error in measuring the time range introduced by this automatic recording device did not exceed 50 ps. The dividing capacitance (DC) (TBPD) in the upper transmission band of 1GHz transferred the measured pulses with their distortions. The oscilloscope was started up by synchropulses from the the GNP. The time instability of the electronic excitation pulse in relation to the synchropulse was less than 50 ps and that of the signals of the VD photodiode and 14 ELU-FS was 100 ps.

The delay between the maximum of the signals from the reference and examined photodetectors was measured. The shapes of the pulses were also compared (Fig.17.5). The difference in the shape was caused by distortion of the signal of the transmission characteristic of the photodetector. To reduce the error in determining the time ranges caused by the non-linear sweep of the oscilloscope, the signals were placed as close as possible to each other on the time axes by selecting the length of the cables connecting the photodetectors with the oscilloscope. In this case (Fig.17.5), the start of deflection of the signals from the zero line and the time co-ordinates of the maxima coincided with an error of 0.2 ns. Thus, the measurement of the time of flight is reduced to measuring the electric length of the cables connecting the photodetectors with the oscilloscope. The measured time of flight for 14ELU-FS was equal 9 ± 0.4 ns at a certified feed regime (the voltage on the photocathode –2.6 kV, the voltage on the collector +1 kV).

The dependence of the time of flight on the feed voltage was not measured because the certified feed regime corresponds to the maximum time resolution of the device. The time resolution of the photodetector

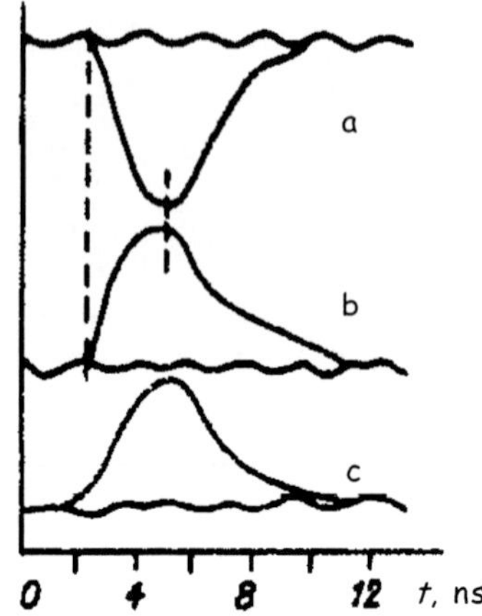

Fig.17.5 Synchronised oscillograms. a) signals from 14ELU-FS photodetector, b) from the vacuum photodiode, c) pumping current of the semiconductor laser (relative units).

in this studied was determined for the minimum duration of the front of the signal.

Electro-optical recording

The spatial dynamics of the radiation of the breakdown was examined in detail in Ref.55–57 using the electronic–optical chamber (EOC) based on the electro-optical converter (EOC) of the UMI-93Sh type which operated in the regime of single-frame recording with an exposure time of 1.5 ns. The moment of opening of the gate of the chamber was accurately ($t < 0.1$ ns) synchronised with each electric pulse supplied to the examined discharge gap. Their results show that the reproducibility of the process is very high and, consequently, in operation with a pulse repetition frequency of 80 pulses/s the screen of the EOC showed a stationary image corresponding to a specific stage of development of the breakdown whose duration was equal to the exposure time of the individual frame, i.e. 1.5 ns. When the delay between the moment of opening the gate of the chamber and the moment of arrival of the pulse on the electrodes of the gap on the screen of the EOC was changed this was followed by a sequence of events in the discharge tube. The image from the screen of the EOC was transferred to film.

The nanosecond control circuit of the EOC is shown in Fig.17.6. The pulse generated by the oscillator 1 is divided into two pulses in the T-junction box 2. Pulse amplitude 16 kV, duration 20 ns. One pulse is directed to the examined discharge gap 9 and carries out its breakdown. Part of the second pulse, taken from the divider 3, is directed

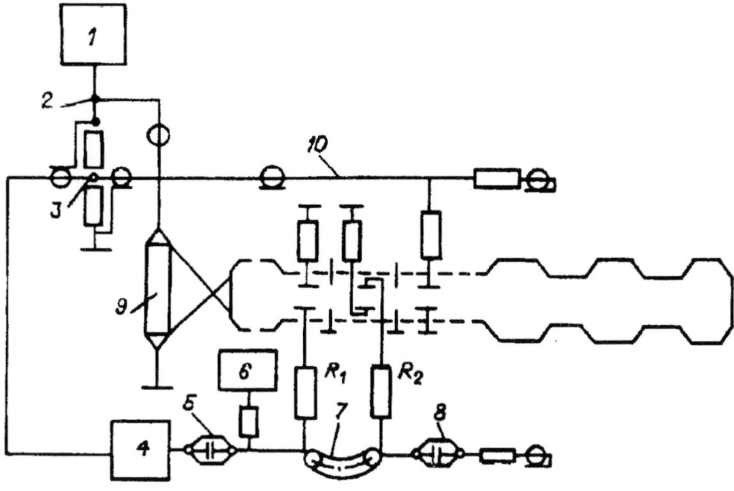

Fig.17.6 Nanosecond circuit of controlling the electro-optical chamber.

to the special plasma shape 4,[48] which forms a triangular pulse with
an amplitude of 2 kV and a duration at the half height of 1.5 ns. Subsequently, passing through the dividing capacitance 5, the pulse is directed on to the plates of the compensated gate of the EOC and opens
it. The bias potential, closing the gate, is generated by the source 6.
Line 7 is used to synchronise the compensating place with the closing plates. The length of the line 7 and the damping resistors R_1 and
R_2 are selected for total compensation separately for each EOP. To
avoid undesirable reflections of the closing pulse in the transmission
line, the pulse is directed to the matched load through the capacitance
7 where it is absorbed. The line 10 is used to remove the parasitic
image from the screen formed as a result of the presence, in the force
lines, of parasitic reflected pulses which cause false triggering of the
gate.

The exposure time of the individual frame of the EOC was determined by experiments. For this purpose, in operation of the gate, the
spot image of the transient of the LD-15 semiconductor laser developed with time on the screen of the EOC. The duration of the transient smoothly decreased to the detection of a decrease of the length
of the image of the point developed into a line on both sides. The duration of the transient of the laser was then measured using a HSD-
1850 vacuum photodiode with a time resolution of 10^{-10} s. The measured
duration of the transient was equated with the time of exposure of the
frame of the EOC.

Measurement of electric parameters

Nanosecond current pulses were measured in Ref.43–51 in a cable using wide-band shunts which have the form a ring of titanium approximately 50 µm thick connected into a break in the cable braiding. The active resistance is equal to hundredths of an ohm, and the reactive linear parameters are almost identical with those of the cable. Such a shunt has only a minimum effect on the measured signal and removes only a small amount of power from it. The upper boundary frequency of the working band of the shunt was estimated from the minimum duration of the front of the pulse propagating from the cable in which the breakdown of the gap with overvoltage between the braiding and the central strand of the cable took place. The minimum duration of the dip was 0.5 ns. The amplitude calibration of the shunts was carried out by closing one end of the cable charged to voltage U whose braiding included a shunt. In discharge of the cable the current passing through the shunt was expressed in the form $I = U/Z$, where Z is the impedance of the cable.

The dynamics of the excess charge, formed in the gap during the breakdown process, was recorded in Ref.43, 45, 51 using a capacitance sensor placed in the screening jacket of the gap on a moving sheet which travelled along its length. To increase the division coefficient and expand the working frequency band, ceramic condensers of the KM type with a total capacitance of $C = 3000$ pF were placed between the sheet of the capacitance sensor and the screen of the gap. The time constant of the sensor with the cable load with an impedance of 50 ohm connected to it is 150 ns which is another magnitude higher than the duration of the examined processes. The signal from the sensor is proportional to the potential of the plasma in the tube in the zone situated opposite the sensor. The quality of the sensor was verified in passing pulses with short fronts through the gap which in this case was replaced with a section of the metallic tube in such a manner that a uniform coaxial line was produced. The signals from the capacitance sensor almost completely coincided with the signals taken from the shunts. In addition to the wide-band sensor, a narrow-band sensor in the form of a pin 3 mm long was also used. The signal from this sensor was proportional to the rate of variation of the potential of the plasma situated opposite the sensor.

Recording x radiation

In Ref.61, x radiation was recorded using a detector consisting of a scintillator–converter of radiation and an FEU-87 photoelectronic multiplier. Plastic scintillators based on POPOP with a scintillation time no longer than 2 ns and the radiation range of 360–380 nm were used, together with an SPS B-151 high-speed scintillator. The thickness of the converters was 1 and 6 cm, respectively, and they absorbed almost the entire x radiation in the examined quantum energy range. The side surface of the converters was painted black to reduce the effect of multiply reflected photons. The duration of increase of the signal of the photoelectronic multiplier according to the certificate data is not greater than 2.0 ns, the duration of the output pulse at an infinitely short input pulse is 5 ns, multiplication factors $3 \cdot 10^7$. The spectral sensitivity of the photocathode is 300–600 nm. The signals of all sensors were supplied to the S7-19 high-speed oscilloscope (5 GHz band) which was synchronised with the high-voltage breakdown pulse.

To measure the spectrum of x radiation, filters/absorbers of different thickness were placed in front of the detector. In this case, the signal of the detector I is the function of thickness d_i of both the parameter of the required spectral density of x radiation I_E and the transfer function of the detector $\varphi(E)$ which takes into account the absorption of x radiation in the windows and the dependence of the yield of light quanta from the scintillator on the energy E of the incident x-ray quanta

$$I(d_i) = A \int_0^{E_1} I_E \varphi(E) \exp\left\{-\mu(E)d_i\right\} dE,$$

where A is a constant, E_1 is the maximum energy of x radiation quanta, $\mu(E)$ is the coefficient of absorption of x radiation in the material of the absorber which depends on the energy of the quanta E.

This equation is an integral Fredholm equation of the first kind in relation to the required function I_E. It's solution represents an incorrect problem because it is unstable in relation to the small changes of the parameters included in this equation.[59] Therefore, it is necessary either to make assumptions regarding the type of spectrum or reduce the problem to a correct one using the regularisation procedure.

A method of obtaining information on the spectrum of x radiation from the attenuation curve is the reduction of the integral equation to an algebraic one. For this purpose, taking into account the physical considerations, it is necessary to make some assumption regarding the type of the required function. Subsequently, the function of this type with free parameters is substituted into the equation and analytical or numerical integration is carried out. This gives an algebraic equation (or a system of equations in the case of numerical integration) which is then used to determine the values of free parameters at which the function of a given type satisfies the equation. For the x radiation in the majority of cases an assumption is made on the purely braking nature of the radiation spectrum. This procedure was used in, for example, Ref.60, 61 for measuring the electron energy in almost relativistic pulsed beams of short duration.

It is assumed that

$$I_E = A_2 (E_1 - E)$$

i.e. x radiation is of purely braking nature with the minimum energy of quanta E_1.[62] Therefore, as a result of numerical integration with the above equation taken into account, we obtain a series of attenuation curves for the given set of the absorbers with thickness d_i and different energies E. Comparing the calculated curves with the experimentally determined attenuation curve (which in the co-ordinates ln [$I (d_i)$] and d_i has the form of a straight line) we select the values at which the calculated curves coincide within the experiment error range (3%).

This method, which is the simplest and most readily available, is used widely. However, it requires selecting the type of function I_E, and gives limited information on the radiation characteristics, for example, only the value of the maximum energy of the x radiation quanta, as in Ref.61. The type of the x radiation spectrum I_E can also be determined directly as a solution of the integral equation. For the soft part of the spectrum of x radiation ($E \ll 12$ kV) this was carried out in Ref.70.

Recording laser radiation

The following method was developed for examining the spatial–time dynamics of the breakdown wave and radiation waves in a nitrogen laser in Ref.63. To measure the distribution of the intensity of radiation along the discharge, the discharge tube contains inside eight semi-transparent mirrors for transferring the laser radiation from the active medium. The mirrors are made of cover glass for microscopes 150 μm thick. They are positioned under an angle of 45° to the axis of the tube, were situated in the same longitudinal section of the tube and occupy the area equal to 11% of the cross section area of the tube. The mirrors were oriented alternately in the opposite size to avoid the transverse displacement of the radiation in passage through the mirrors. The transmission factor of the mirrors was measured in the spectrophotometer at $\lambda = 337.1$ nm and equalled 97%.

The radiation from the mirrors was recorded using an HSD-1850 photoelement with an increased time of 0.1 ns. The signals from the dividers and the photoelement were supplied to a Tektronix-519 oscilloscope (band 1 GHz). The accuracy of measuring the time ranges was increased using a calibrating sinusoid with a frequency of 1 GHz.

17.5 Spatial–time structure of breakdown waves. Effect of preliminary ionisation

The effect of preliminary ionisation of the speed of the breakdown wave was investigated in Ref.10,16,23,64. In Refs.10 and 23, it was reported that the speed of the breakdown waves monotonically increases with increasing electron concentration, and the speed of the waves with positive polarity is more sensitive to the changes in the electron concentration. Here, we shall present the results of Ref.64, where the effect of the electron concentration was investigated at different pressures.

Comparison of Fig.17.7 and 17.8 shows that the speed of the breakdown waves in relation to p and n_{e0} shows on the whole the same behaviour at both polarities of the high-voltage pulse. The main difference is that the speed of the wave at positive polarity is on average 30% higher than the speed of the wave at negative polarity. At an electron concentration of $n_{e0} \approx (3 \div 5) \cdot 10^8$ cm^{-3}, the speed of the wave shows a sharp jump at both polarities. At positive polarity, this jump is clearly evident at pressures high than 5 torr, and at negative polarity at pressures 10 torr. In the range of the variation of the parameters $3 \cdot 10^8$ cm$^{-3} < n_{e0} < 5 \cdot 10^8$ m^{-3} and 5 torr $< p <$ 10 torr, where there

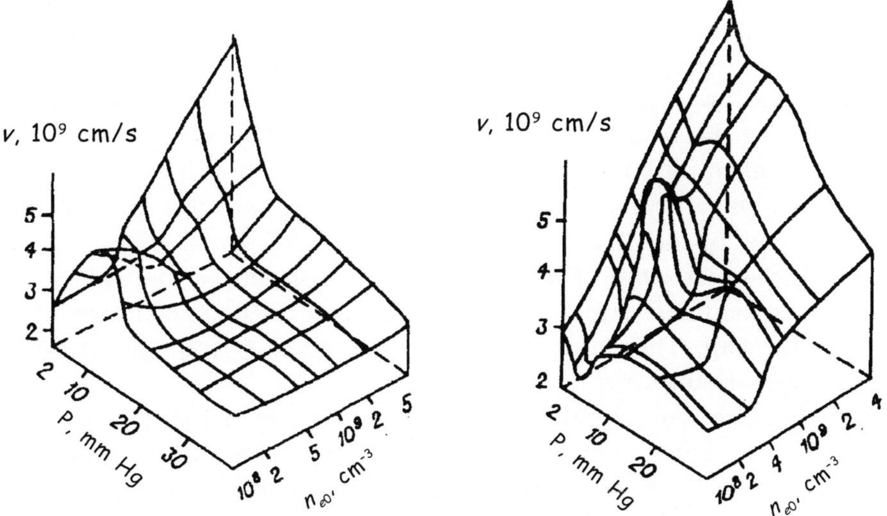

Fig.17.7 Dependence of the speed of breakdown wave v on the initial concentration of electrons n_{e0} and helium pressure p at the negative polarity of the breakdown pulse.
Fig.17.8 Dependence of the speed of the breakdown wave v on the initial concentration of electrons n_{e0} and helium pressure p at the positive polarity of the breakdown pulse.

Fig.17.9 Photographs of a wave breakdown in a gap 400 mm long, 45 mm in diameter. Exposure time of a single frame 1.5 ns, time shift between the frames 1 ns.

is a jump in the speed for both pulse polarities, there was an anomalous peak of the speed to $5 \cdot 10^9$ cm/s for the breakdown waves initiated by the positive polarity pulse.

Figure 17.9 shows the evolution of the form of the emitting volume of the breakdown wave obtained using an electronic-optical camera. The shape of the front of the breakdown wave depends on the initial electron concentration. At $n_{e0} < 3 \cdot 10^8$ cm^{-3} the radiation front of the breakdown wave is flat, and the wave is of the volume type with the uniform glow of the discharge volume behind the front. When n_{e0} exceeds $(3 \div 4) \cdot 10^8$ cm^{-3}, the radiation front of the breakdown wave becomes conical and extends during movement of the breakdown wave through the gap. The emitting volume behind this front of the wave is as previously uniform.

Detailed electron optical investigations of the spatial structure of the breakdown wave were carried out in Ref.65–69. The experiments show that the glow of the gas caused by the breakdown wave assumes various forms: conical, uniform through the cross section of the discharge tube and boundary. Each form exists in a specific gas pressure range and smoothly changes to another with changing pressure. The results show that there is a correspondence between the nature of the dependence of the speed of the wave on pressure and the form of glow of the breakdown wave. Regardless of the polarity of the breakdown pulse, in the pressure range where the rate increases with increasing pressure, the glow has the form uniform throughout the cross section of the discharge tube. In the region corresponding to the falling part of the dependence of speed on pressure, the wave propagates in the near-wall region of the discharge tube. For the positive wave of the breakdown, in the low-pressure region, the falling part of the dependence $v\,(p)$ was recorded for the first time. The conical shape at positive polarity of the breakdown pulse is related to the falling part of the v (p) curve in the range of low pressures ($p \leq 3$ torr).[68, 69]

The formation of the breakdown wave at negative polarity of the breakdown pulse at all pressures starts with the formation of a dark near-cathode region whose thickness decreases with increasing gas pressure in the tube. As regards the external appearance, it corresponds to the Crooks dark space with a sharp boundary of negative glow. Several nanoseconds after the formation of the dark region, a spot forms on the cathode and a new glow wave starts to move along the tube.

17.6 Electrodynamic processes in wave breakdown

Absorption of energy, energy parameters of the breakdown wave
Synchronised measurements of the electric current in the electrodes of the gap with a wave mechanism of its breakdown make it possible to obtain the dependence of electric energy parameters of the discharge on time.[64]

The electrode receiving a high-voltage nanosecond pulse will be referred to as the anode (irrespective of pulse polarity) in accordance with the polarity of the voltage of the stationary glow discharge.

The current flowing into the discharge tube on the side of the anode I_a (anode current) is determined from the oscillograms of the pulses of incident and reflected current measured separately in the cable and synchronised at the anode

$$I_a\,(t) = I_{in}\,(t) - I_{ref}\,(t).$$

Cathode current I_c is measured in the coaxial cable connected directly behind the cathode (second electrode). The voltage on the anode consists of the sum of the voltages of the incident and reflected currents and, since the anode is part of the linear transmission line with the impedance $Z = 50$ ohm, equal to the impedance of the conducting cable, the voltage in it is calculated from the equation

$$U_a(t) = [I_{im}(t) + I_{ref}(t)]\ Z.$$

The voltage in the second electrode is determined only by one current I_c and impedance of the electrode which is also equal to Z:

$$U_c(t) = I_c(t)\ Z.$$

The electrical power in the gap is calculated from the equation

$$P(t) = U_a(t)\ I_a(t) - U_k(t)\ I_k(t),$$

where the first term is the power supplied into the gap from the anode and the second term is the power transferred from the cathode to the cable.

The total energy introduced into the gap is determined by the integration of the power flux

$$E_0 = \int\limits_0^t P(t)\,\mathrm{d}t.$$

The excess charge, built up in the gap, is computed by integrating the difference between the anode and cathode currents with respect to time

$$Q(t) = \int\limits_0^t \left[I_a(t) - I_k(t)\right]\,\mathrm{d}t.$$

The capacitance of the gap is determined assuming that the charge behind the front of the breakdown wave is distributed more uniformly along the length of the tube, and the plasma potential behind the front is equal to the voltage along the anode:

$$C(t) = Q(t)/U_a(t).$$

The energy of the charged gap, stored in the electric field, is computed from the equation

$$E_E(t) = Q(t)U_a(t)/2.$$

The difference between E_0 and E_E consists of the energy stored in the magnetic field of the current E_m, flowing behind the wave front, and the energy dissipated in the discharge E_d

$$E_0 - E_E = E_m + E_d.$$

The input impedance of the gap, characterising matching of the cable with the discharge, is calculated as the ratio of the voltage and anode current

$$Z_{in} = U_a(t)/I_a(t).$$

The time dependences of the electric energy parameters have a common time axis and, consequently, as a result excluding the time as a parameter, we obtain the dynamic characteristic dependences of electric energy parameters on each other: volt–ampere characteristic, ampere–Coulomb characteristic, the relationship of the discharge energy stored in different types.

Here, we present the data for a discharge tube 400 mm long whose cathode is connected in series with an inductance which increases the impedance of the cathode to more than 10 kΩ. For pulses of the nanosecond duration, this circuit of the cathode is close to break because the leakage of energy from the gap to the cable behind the cathode does not exceed $Z/Z_k = 0.5\%$. This greatly simplified the general pattern of the wave breakdown of the gap, excluding the individual stages, whose formation is strongly influenced by the cathode having an impedance close to the impedance of the tube and the cable. This complicates examination of the breakdown wave. Calculations of the electric energy parameters are also simplified because from the equations described previously we exclude the terms relating to the cathode.

The experiment described in Ref.64 were carried out at a helium pressure in the tube of 15 torr at three values of the initial concentration of the electrons $n_{e0} = 2.9\cdot10^8$; $1.6\cdot10^9$; $4.3\cdot10^9$ cm^{-3} at the positive and negative polarities of the nanosecond pulse.

Figure 17.10 shows the oscillograms of the incident and reflected pulses ($n_{e0} = 1.6\cdot10^9$ cm^{-3}). The amplitude of the reflected pulse is shorter and the duration longer than in the incident pulse. Consequently, the voltage is applied to the anode for a longer time than the duration of the incident pulse. Figure 17.9 shows the current and voltage curves measured on the anode. The current oscillates and consists of two half cycles. In the positive half cycle, the charge flows into the tube and then, when the incident pulse decreases, flows from the tube back into the cable. The variation of the charge in the tube is shown

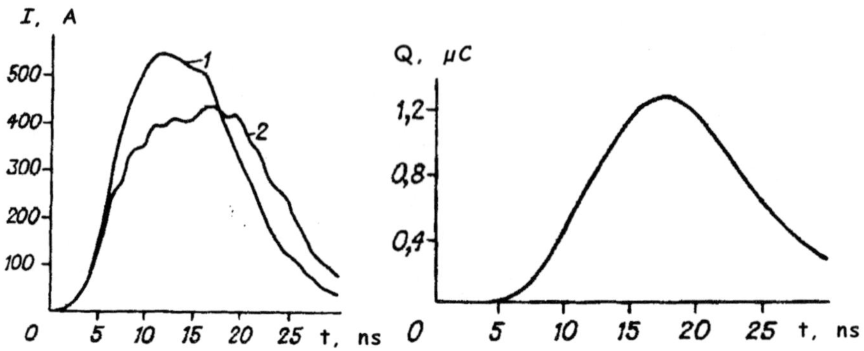

Fig.17.10 Oscillograms of pulses of incident (1) and reflected (2) pulses.
Fig.17.11 (right) Dependence of the electrical discharge Q in the discharge tube on time t.

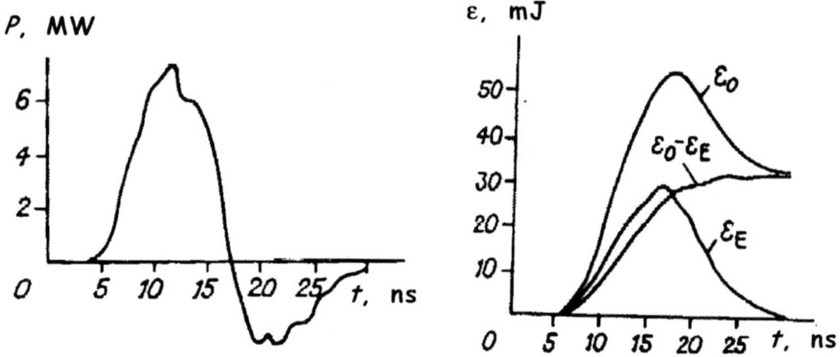

Fig.17.12 Dynamics of electrical power P introduced into the discharge tube.
Fig.17.13 (right) Dynamics of energy in the discharge gap.

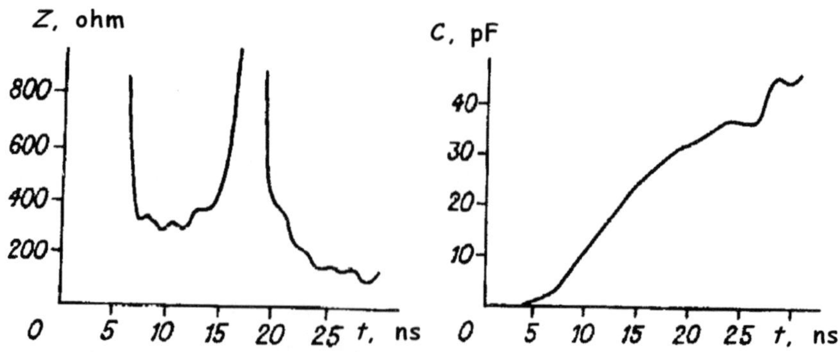

Fig.17.14 Variation of input impedance Z_{in} on the gap.
Fig.17.15 (right) Variation of the capacitance C of the gap.

in Fig.17.11. Figure 17.12 shows the dynamics of the electric power supplied to the tube. The oscillating nature of the graph was the change of the sign indicates the large fraction of the reactive power in the gap. The maximum power supplied to the tube is 7.5 MW which reaches 60% of the maximum power of the nanosecond pulse. The power has this value when the wave front reaches the second electrode. Subsequently, the power supplied to the tube decreases and changes its sign. In this case, the tube operates as an oscillator. The maximum power transferred to the cable from the tube is 3 MW which equals 40% of the maximum value of the power supplied to the discharge.

Figure 17.13 shows the dynamics of the energy in the gap. ε_0 is the energy supplied to the discharge, ε_E is the energy stored in the field of the excess charge. Both values initially increase, and then with ending the pulsed energy of the electric field decreases to almost zero indicating the almost complete compensation of the excess discharge in the tube; the energy ε_0, introduced into the gap, decreases only to 65% of the maximum value.

The maximum of the graph of total energy is shifted to the right in relation to the maximum of the electrostatic energy. It may therefore be assumed that in addition to the reactive component of the energy in the gap there is also a large dissipative component associated with the passage of current. The graph of the difference between ε_0 and ε_E shown in the same figure, indicates that the dissipated energy increases with increasing electronic energy in the gap. When the electric energy in the gap starts to decrease, the increase of the dissipated energy slows down and at the end of the pulse the dependence of the dissipated energy on time reaches a plateau. Consequently, it can be concluded that the dissipation in the gap takes place far more efficiently when energy is added into the gap then in the case of generation of the energy stored in the electric field of the excess charge. Thus, the main contribution of the power to the discharge is observed during movement of the breakdown wave of the gap. The input resistance of the discharge is shown in Fig.17.14. During the period of time whose duration is determined by the 'induction time' of the charge τ_i, Z_{in} is infinitely high and then with the start of movement of the breakdown wave Z_{in} and rapidly decreases to the value of 300 ohm, and when the wave reaches the cathode Z_{in} rapidly decreases. Subsequently, when the current of the anode becomes negative, Z_{in} rapidly decreases to 150 Ω.

Figure 17.15 shows the variation of the capacitance of the gap. During movement of the breakdown wave the capacitance linearly increases. When the front of the breakdown wave reaches the cathode the capacitance equals 14 pF which equals 60% of the capacitance of the tube filled with the metal.

Figure 17.16 shows the volt–ampere characteristic of the gap for a wave breakdown. The VAC has a hysteresis. The VAC starts at some voltage, in this case 7 kV, and then rises upwards with an almost constant angle of inclination corresponding to a resistance of approximately 300 ohm; reaching the upper point, the characteristic is then reversed and continues to the region of negative current with an angle corresponding to a differential resistance of 50 Ω. The VAC then returns along a circle to the original co-ordinates.

Figure 17.17 shows the dependence of the energy ε_E, stored in the electric field, and the energy $\varepsilon_d = \varepsilon_0 - \varepsilon_E$, dissipated in the gap. The dependence shows three sections characterised by the linear relationship between ε_d and ε_E. In the first section, corresponding to the movement of the breakdown wave through the gap, ε_E is 1.6 times higher than ε_d. The second section corresponds to the state in which the breakdown wave has already closed the gap but the gap still receives the energy with constantly decreasing current from the anode which is at an almost constant potential. In this section, the increase of energy ε_E is equal to the increase of ε_d. In the third section, relating to the stage of reversed anode current, ε_E rapidly decreases to zero and the energy ε_d slight increases. The ratio of the changes of the energy in this section $\varepsilon_E/\varepsilon_d \approx 5$.

The limiting speed of the breakdown wave, attenuation. The authors of Ref.72 carried out experiments with voltage pulses of negative polarity with an amplitude of up to 250 kV, duration 9 ns, with increase and decrease fronts of 2.5 ns.

The dependences of the speed of the breakdown wave, the amplitude

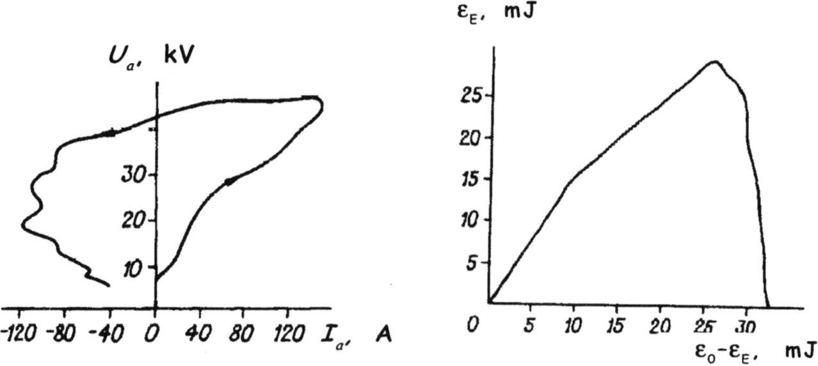

Fig.17.16 Dynamic volt–ampere characteristic of the gap in wave breakdown. U_A, I_a, voltage and current in the high-voltage electrode.

Fig.17.17 (right) Relationship of energy, stored in the electric field of the breakdown wave, with the energy dissipated in the gap.

of passed current and the attenuation coefficient of the wave on pressure and the type of gas were investigated. Figure 17.18 shows the results of measurements taken for helium. The attenuation coefficient α has the form:

$$U(x) = U_0 e^{-\alpha x},$$

where $U(x)$ is the amplitude of the voltage at the distance x from the end of the cathode.

The data in Fig.17.18 show the existence of the conditions with minimum attenuation. Previously, they could not be detected because of narrow ranges of pressure variation in which the measurements were taken. The results also show that the maximum speed does not depend on the type of gas (Fig.17.19). Its value, $2 \cdot 10^{10}$ cm/s is close to the maximum value for the given voltage (see Fig.17.18), because within the error range it coincides with the speed of the electron flying in vacuum and travelling through a potential difference of 250 kV ($2.2 \cdot 10^{10}$ cm/s). The graph of the dependence $\alpha\,(v_f)$ in Fig.17.20 gives at extrapolation to $\alpha = 0$ the expected limiting speed of $2.2 \cdot 10^{10}$ cm/s.

Extrapolation of the dependence of the passed current on the speed $I\,(v_f)$ to the determined limiting speed $v_{\text{lim}} = 2.2 \cdot 10^{10}$ cm/s gives a limiting current of 1.7 kA for all gases (Fig.17.21). This current is equal to the current which would flow in the transmission coaxial line without losses with the geometrical dimensions of the discharge device, i.e. diameter of the internal and external conductors of 0.4 and 5.0 cm respectively at a voltage pulse amplitude of 250 kV.[72]

These results indicate that the breakdown waves, moving through

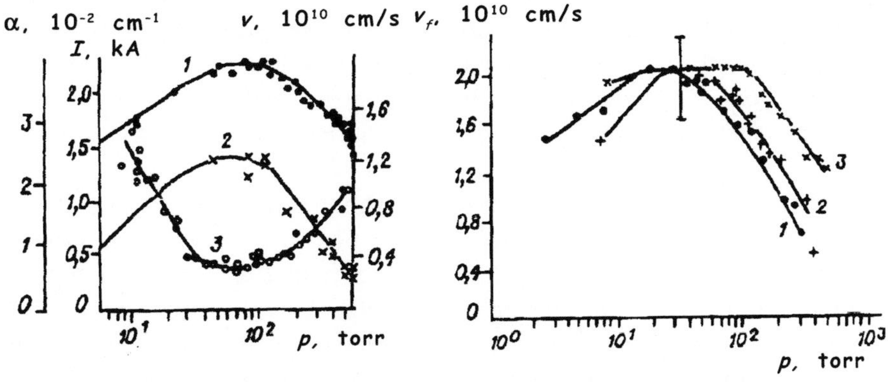

Fig.17.18 Dependence of the velocity of the breakdown wave v (1), amplitude of passing current I (2) and the continuation factor α (3) on gas pressure P.
Fig.17.19 Dependence of the velocity of the breakdown wave v on pressure p of a nitrogen and helium mixture. Helium content 10 (1), 50 (2), 90% (3).

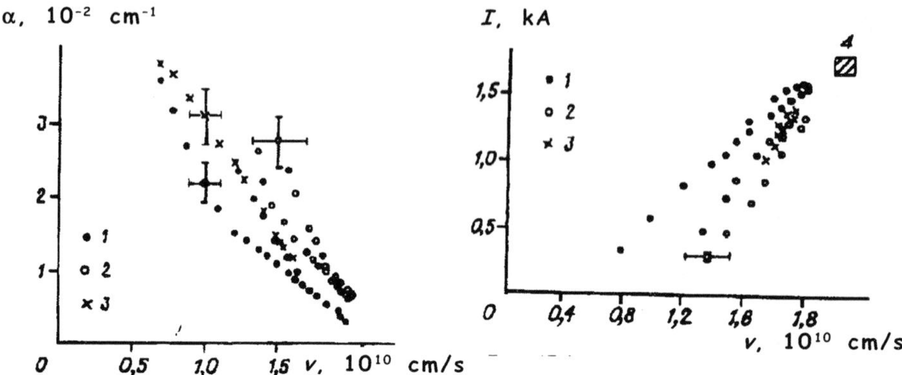

Fig.17.20 Dependence of the attenuation factor α on the velocity of the breakdown wave v in different gases. 1) nitrogen, 2) helium, 3) air.

Fig.17.21 (right) Dependence of the amplitude of the pulse of past current I on the velocity of the breakdown wave v in different gases. 1) nitrogen, 2) helium, 3) air, 4) extrapolation.

the previously non-ionised gas with a limiting speed corresponding to a voltage pulse of 250 kV are identical with those described previously.

In Ref.103, investigations were carried out into the conditions of propagation of a breakdown wave and the results confirm the existence of the 'limiting' speed of the wave in the voltage range 100-280 kV for a negative polarity pulse with a duration 35 ns, with a front of 2 ns, in a coaxial discharge device with the effected directed permitivity of the insulator of $\varepsilon_{eff} \approx 25.4$. In these experiments, the dependence of the 'limiting' speed on voltages satisfactorily approximated by the formula $v_{lim} = (2eU/m_e\varepsilon_{eff})^{1/2}$, where e, m_e is the charge and mass of the electron.

In Ref.103 it was reported that the pressure p_m at which the limiting speed is achieved does not always increase with increasing voltage amplitude, i.e. for each gas under the given conditions there is minimum pressure below which the slowly attenuating breakdown wave, have in the property of 'limiting' speed does not form.

The structure of charge density in the breakdown wave

In Ref.43 and 64 it was found that the breakdown wave at positive polarity of voltage has a stable structure with the harmonic longitudinal modulation of the charge density moving with the speed of the wave. The reduced structure of the discharge in the breakdown wave was determined by taking into account the apparatus function of the capacitance sensor. It should be noted that the electric field at the front of the breakdown wave can reach the values required for the formation with 'running away' electrons.

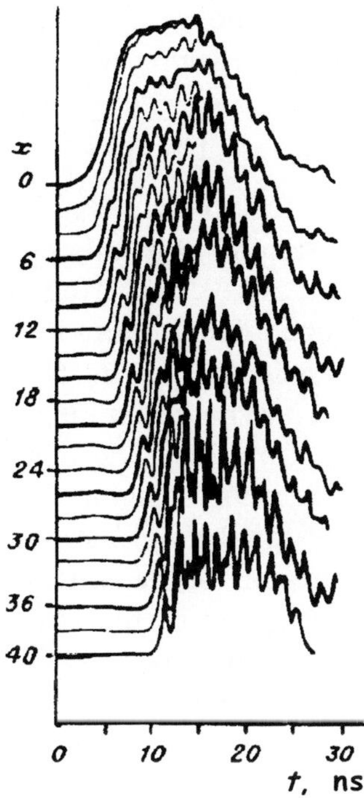

Fig.17.22 Synchronised oscillograms obtained from wide-band capicitance transducer moved along the axis z of the discharge tube with the step 2 cm (n_{e0} = 4.3×10⁹ cm⁻³).

The oscillograms, obtained using the capacitance sensor, which records the propagation of the excess gas through the tube, are characterised by high-frequency modulation. (x–t) diagrams, Fig.17.22, that the modulation of each separate oscillogram is the result of movement of the modulation source through the gap. The structure was recorded with a capacitance sensor which reacts to the electric field generated by the charges situated not only against a sensor but also by the charges forming the structure as a whole and distributed in accordance with the structure along the tube. Therefore, the signal, taken from the sensor, is determined by the superimposition of the field of charges distributed along the tube. With increase distance of the charge from the sensor its contribution to the common signal will decrease in accordance with the diagram of sensitivity of the capacitance sensor. If it is assumed that the wave contains only the longitudinal structure of the charge $\varphi\,(x + vt)$, the sensitivity diagram can be conveniently expressed by the longitudinal-coordinate dependence $f\,(x)$. In this case the signal, taken from the capacitance sensor can be written as

$$U(t) = 1/A \int_{-\delta}^{\delta} \varphi(x + vt)\,f(x)\,\mathrm{d}x.$$

The measurements show that the contribution of the charges, situated at distances larger than 12 cm from the sensor, to the common signal is less than 1% and, consequently, the integration limits were selected as $\delta = 12$ cm. The normalisation multiplier is $A = \int_{\delta}^{\delta} f(x)\,\mathrm{d}x.$. The

sensitivity diagram $f(x)$ was determined by experiments. For this purpose, a nanosecond pulse was supplied to a narrow (3 mm) ring of copper foil placed on the top of the discharge tube. The amplitude of the signal, taken from the capacitance sensor gradually moved from the end along the axis of the tube, was measured. The resultant dependence was approximated by a corresponding function. The delay of the signal from the peripheral regions of the sensitivity diagram was ignored because the strength of its effect on the signal amplitude did not exceed 2%. The Lorentz reduction of the dimensions of the structure was also disregarded because it equals ~2% at a speed of the structure of $6 \cdot 10^9$ cm/s. The structure $\varphi\ (x + vt)$ was found by selecting the approximating function which, after numerical integration, was compared with the observed oscillogram of the signal taken from the capacitance sensor in the centre of the tube.

The depth of modulation of the structure increases with increasing initial electron concentration. Oscillograms recorded from a capacitance sensor at the maximum value $n_{e0} = 4.3 \cdot 10^9$ cm^{-3} (Fig.17.22) were processed. The speed of displacement of the charged structure was $6 \cdot 10^9$ cm/s.

The amplitude of the first oscillation can be relatively high so that the electric field of the field in the wavefront exceeds the critical value required for realisation of the regime of continuous acceleration of the primary electrons.[70] Estimates show that the field in the front of the structure, shown in Fig.17.23, curve 1, is

$$E = \sigma/2\varepsilon_0 \approx 3.5 \cdot 10^4 \text{ V/cm}$$

which at a pressure of 15 torr in helium is already sufficient for the formation of 'running-away' electrons.[70] Here $\sigma = Q/S$ is the 'end' surface density of the charge in the front wave, S is the cross section of the tubes; Q is the charge present in the first peak of the structure. Under the effect of the field, the electrons assume the energy close to the potential of the first oscillation of the front of the structure. This strength of the field is too low because the radial structure, especially the 'pointed' form of the front of the breakdown wave, was not taken into account in the estimates.

According to the authors of Ref.43 and 64, the formation of the

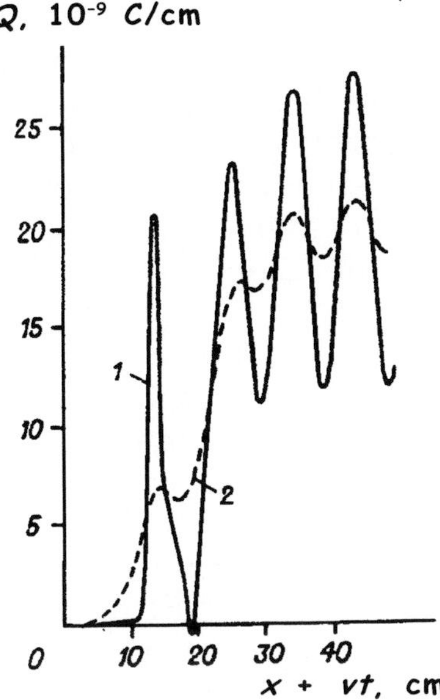

Fig.17.23 Distribution of excess charge in the breakdown wave. 1) corrected, 2) observed distribution.

charge structure is caused by the interaction of high-energy electrons, formed in the front of the structure, with the plasma behind the front of the breakdown wave. The presence of the vibrational structure of charge density, moving together with the breakdown wave, leads to vibrations of the current flowing through the high-voltage electrode. The depth of modulation with respect to the measured current through the plasma is 15%.

In Ref.71, which deals with the theoretical description of the breakdown wave, it was found that under specific conditions oscillations of the electric field, current and volume charge form behind the front of the ionisation wave. The frequency of these oscillations is close to the plasma frequency. There are conditions under which attenuating current oscillations are preceded by a finite number of oscillations with increasing amplitude. Calculations were carried out under a hydrodynamic approximation, and the authors used the complete equation of movement of the electrons including, in contrast to an early study in Ref.34, inertia terms and the gradient of electronic pressure. The authors noted that if the inertia of the electrons is taken into account, oscillations of the volume charge form behind the front of the ionisation wave. The calculations show that there is a range of electron concentration optimum for the formation of oscillations. For example, increase of n_e initially leads to the formation of several increasing oscillations and then a further increase of n_e decreases their number. The calculations carried out in Ref.71 were conducted for the voltage of a high-voltage electrode not higher than 30 kV. Under these conditions, as reported by the authors, the values of E/p are not too high and the effect of run-away electrons can be ignored. Taking into account the qualitative agreement of the calculated result of the experimental data,[43,64] the authors of Ref.71 noted that in the experiments relatively high values of E/p were obtained for a number of regimes in the region of the front. At these values, the hydrodynamic description of the electronic gas may prove to be a very rough approximation.

Generation of microwave vibrations

The authors of Ref.33 and 64 produced a 15% depth of modulation of current in the breakdown wave. In Ref.93 under the experimental conditions close to those in Ref.43 and 64, 100% harmonic modulation of current was obtained when introducing a feedback (Fig.17.24).

The processes in the discharge take place in the following sequence. The induction period to the moment t_1 is characterised by the complete reflection of the exciting pulse from the gap. The stage t_1+t_2 corresponds to the displacement of the front of the excess charge along the plasma accompanied by the ionisation of the neutral gas. Efficiently conducting plasma, charged to the potential $2U$ in relation to the resonator walls, is produced at the resonator axis. At the moment t_3 the plasma is neutralised by the electrons emitted by the second electrode. Consequently, the electric energy stored by the plasma changes to the vibrations of electromagnetic field in the resonator containing the plasma.

The excitation of the oscillations was of the impact type and their frequency $w_1 = 2 \cdot 10^9$ rad/s was less than the frequency w_0 of the lower mode of an empty resonator. The measured current can be represented in the form $I = I_1 + I_2 \exp(-qt) \sin w_1 t$, where I_1 is the current of plasma in the absence of oscillations. The relationship $I_2 \sim dI_1/dt$ was approximately fulfilled during the period of excitation of the oscillations. In the examined pressure range of the neutral gas $p = 2 \div 30$

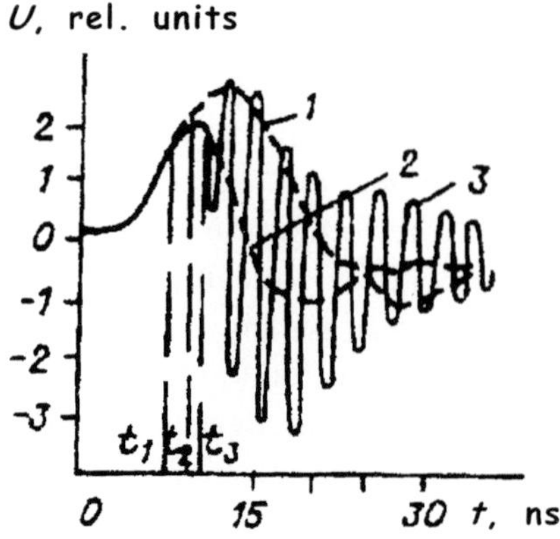

Fig.17.24 Formation of 100% hf modulation of the breakdown wave in the presence of a "feedback". 1) oscillogram of the exciting high-voltage pulse, 2) oscillogram of a pulse reflected from the gap in the absence of oscillations, 3) the same, in their presence.

torr the speed v of movement of the front of the excess charge and the value I_2 monotonically decreased with increasing p, and the frequency w_1 remained constant within the limits of the measurement accuracy with the change of both the pressure and the duration of the excitation pulse. Attenuation q was determined mainly by the removal of electromagnetic energy to the conducting coaxial cable and did not depend in the experiments on the duration of the excitation pulse and slightly increased with increasing pressure.

The authors of Ref.78 recorded microwave oscillations behind the front of the breakdown wave using capacitance dividers. Evidently, the observed oscillations are the result of the excitation of the coaxial resonator at the internal frequency as a result of the resonance interaction of the charge wave moving in the front (for the water line of $v \approx 2 \cdot 10^9$ cm/s, for the line with an oil dielectric $v \approx 1.5 \cdot 10^{10}$ cm/s) with the electromagnetic wave in the delaying structure ($c/\varepsilon^{1/2} \approx 5.5 \cdot 10^9$ cm/s in the water line and $1.6 \cdot 10^{10}$ cm/s in the oil line). On the basis of the order of magnitude, the frequency spectrum of the observed oscillations can be regarded as equal to the internal frequency of the coaxial resonator $\omega \approx c/l_f$, where l_f is the length of the front. For the oil and water lines $\omega \approx 10^9$ s^{-1}, which is in agreement with the frequencies observed in the experiments.

The experiments with a breakdown of the tube in the oil line show that the amplitude of oscillations depends in a nonmonotonic manner on the pressure and is maximum at the maximum speed of the breakdown wave, in the pressure range where the speed of light in the coaxial and the speed of the breakdown wave are similar. It should be noted that at this pressure the current in the tube is maximum.

It was also established that in the pressure range in which the amplitude of the oscillations is maximum, the speed of the wave (Fig.17.25) has a local minimum and the voltage attenuation coefficient has a local maximum. It can be assumed that as a result of energy losses in the excitation of the electromagnetic wave in the coaxial line behind the front of the breakdown wave its speed decreases and the voltage attenuation coefficient increases.

Doppler effect in wave breakdown

The propagation of a breakdown wave is identical to the movement of a rod with high conductivity from a high-voltage electrode. In fact, in propagation of a positive polarity wave the drop of the potential behind the wave front is considerably smaller than the potential jump in the front; after closure of the gap by the breakdown wave the high-voltage pulse freely passes through the gap and is recorded in the cable

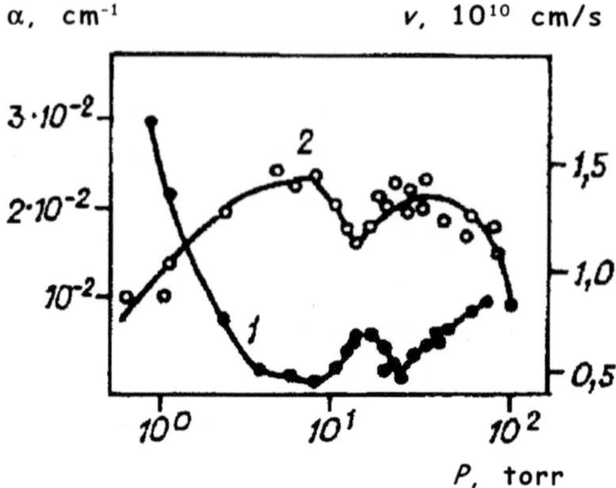

Fig.17.25 Dependence of the speed of breakdown wave (1) and attenuation (2) on air pressure.

behind the gap without any decrease of the amplitude in relation to the incident pulse.[64] During the propagation of the wave through the gap the duration of the recorded deflected pulse is longer than that of the incident pulse and the amplitude is smaller. The latter fact is explained by the Doppler amplitude-frequency change of the pulse in reflection from the moving front of the breakdown wave.

If the duration of the current pulse in a stationary cable is t, then after reflection from the open end of the cable moving with the speed $v = \beta c$, it's amplitude decreases $\dfrac{1+\beta}{1-\beta}$ times and the duration increases correspondingly $\dfrac{1+\beta}{1-\beta}$ times. This occurs if, using the Lorentz transformation for the pulse, and transfer into the moving co-ordinate system, we write the equation of the mirror reflection of the pulse in the moving system and transfer back to the stationary co-ordinate system.[73] The voltage and current of the reflected pulse are written in the form[51]

$$U_{ref}(t) = \frac{1-\beta}{1+\beta} U_{in}\left(\frac{1-\beta}{1+\beta}t\right),$$

$$I_{ref}(t) = \frac{1-\beta}{1+\beta} I_{in}\left(\frac{1-\beta}{1+\beta}t\right).$$

This explains the absence of any strong reflected electromagnetic wave at a voltage of 250 kV.[74] With increasing speed, the amplitude of the pulse reflected from the front decreases in accordance with the above equations.

Thus, measuring the speed of the wave, we can easily calculate the electric energy parameters with a wave breakdown carried out by the high-voltage rectangular pulse. Thus, the input resistance of the gap is:

$$Z_{in} = \frac{U_{in} + U_{ref}}{I_{in} - I_{ref}} = Z / \beta.$$

In the experiments for $v_f = 4.9 \cdot 10^9$ cm/s $\beta = 0.163$.[64] The value Z_{in}, computed from the above equation, is 306 ohm. The values of Z_{in}, obtained from the oscillograms of the incident and reflected currents, are also equal to approximately 300 Ω.

The power introduced into the gap is:

$$P_{\text{intr}} = (I_{in} - I_{ref})(I_{in} + I_{ref})Z = P_{in} \frac{4\beta}{(1+\beta)^2}.$$

We assume that the tube is long because the front of the breakdown wave does not reach its end due the short pulse duration. The distance which the breakdown wave can travel, is expressed in the form

$$L_{\text{max}} = cT_{in} \frac{\beta}{1-\beta}.$$

We denote $\frac{1-\beta}{1+\beta} = \gamma$. The energy of the reflected pulse is associated with the energy of the incident pulse

$$E_{ref} = Z\gamma^2 \int_0^{T_1} I_{in}^2 (\gamma) \, dt = \gamma E_{in},$$

where $T_1 = T_{inc}\gamma^{-1}$. The energy of the pulse dissipated at the front is

$$E_d = E_{in} - E_{ref} = E_{in} \frac{2\beta}{1+\beta}.$$

The total dissipation energy

$$E_{d.total} = E / L_{max} = \frac{E_{in}}{cT_{in}} 2 \frac{1-\beta}{1+\beta}$$

and is equal to the pressure force of the electromagnetic wave

$$F_{em} = E_{d.total} = 2\frac{P_{in}}{c}\frac{1-\beta}{1+\beta}.$$

The dissipation power is determined by the ratio of the dissipation energy to the duration of movement of the front wave $t_w = L_{max}/v_f = T_{in}/(1 - \beta)$ and equals

$$P_d = \frac{E_{in}2\beta/(1+\beta)}{T_{in}/(1-\beta)} = F_{em}v_f.$$

It can easily be seen that the dissipation power is maximum at $\beta = \sqrt{2} - 1$ and equals 0.343 P_{in}.

In reflection of the electromagnetic wave from the moving mirror, the power of the incident wave is higher than the sum of the powers of the reflected and dissipated components: $P_{in} > P_{ref} + P_d$. This is caused by the difference between the duration of the reflected pulse T_{ref} and the duration of movement of the front wave t_w and the duration of the incident pulse T_{in}. This results in the energy balance

$$E_{in} = P_{in}T_{in} = P_{ref}T_{ref} + P_d t_w.$$

The measurement of the electromagnetic pulse, reflected from the front of the breakdown wave, and also the agreement of the above described expression of the electrode dynamic properties of the wave with the experiment make it possible to propose the following pattern of the process.

The plasma column behind the front of the ionisation wave represents together with the metallic screen a co-axial waveguide whose properties are almost identical with those of the electric cable. A packet of electromagnetic waves, representing a high-voltage pulse, propagates to the plasma waveguide and also in the cable. In the moving zone in the vicinity of the wave front where the conductivity of the plasma channel rapidly decreases, the incident electromagnetic packet is reflected and propagates in the opposite direction. If the losses in the plasma coaxial line are ignored, the voltage drop taking place in the front of the breakdown wave is determined by the sum of the amplitudes of the incident and reflected electromagnetic packets. In a mirror reflection of the electromagnetic packet the voltage drop at the front of the breakdown wave is associated with the amplitude of the high-voltage pulse fed to the gap in the following manner:

$$\Delta\varphi = U_{in}\frac{2}{1+\beta}$$

and, consequently, is determined only by the speed of the reflection zone. Behind the reflection zone, there is a zone occupied by the ionisation wave. The energy 'feeding' of the ionisation processes in the wave takes place as a result of the energy flux of the incident electromagnetic packet. The intensity of this feeding does not exceed the work of the force of the electromagnetic pressure in reflection of the incident electromagnetic packet from the ionisation front.

Impact properties of the breakdown wave

Dissipation of the electromagnetic wave at the front of the breakdown wave causes a moving conduction jump to form in the front. This phenomenon in the gas-discharge device is identical with the movement of the impedance jump along the transmission line with non-linear parameters. An example of powerful transmission lines with magnetic self-insulation in which the propagation of a powerful electric pulse results in a decrease of the duration of the pulse front as a result of the magnetron effect is described in Ref. 75.. This is associated with the change of leakage and effective reactive parameters of the line. Another example are lines with a ferrite filling. The waves of the impedance jump in these lines are referred to as impact electromagnetic waves.[76] They have the following properties.[76] The propagation of a powerful electromagnetic pulse along the line becomes non-linear due to dissipation in remagnetising ferrite at the pulse front. This results in subsequent shortening of the pulse time. When reaching the minimum time determined by the properties of ferrite and the signal amplitude, the propagation of the front becomes stationary and the pulse time linearly decreases because dissipation still continues at the front. Behind the front of this 'impact electromagnetic wave' the properties of the wave guide are linear and the propagation speed of the electromagnetic waves behind the front is higher than the speed of the front. This results in the reflection of the electromagnetic signal from the moving impedance jump.

The effects observed in the experiments with the breakdown waves have direct analogy with the above-mentioned 'impact electromagnetic wave' regardless of differences and nature of the processes in ferrite and weakly ionised plasma. The duration of increase of the leading edge of the pulse of the reversed current of the wave, measured in Ref.31, proved to be considerably shorter than the duration of the leading front of the voltage pulse supplied to the tube. Consequently, the author referred to it as the impact electrical wave. Breakdown waves with positive polarity characterised by a gradual decrease of the duration of the charge front and the radiation.[49,51,64] The speed of the front of the breakdown wave is constant. The curvature of increase of the current, recorded behind the discharge tube, is higher than the curvature of increase of the current supplied to the tube.[64] Movement of the breakdown wave is characterised by the formation of a pulse reflected from the front. The amplitude and duration of this pulse change in relation to the incidence pulse in accordance with the Doppler effect.[64] The speed of the breakdown wave increases with increasing electron concentration in the plasma thus leading to the tip of the wave hitting its base and to an increase of the curvature of the front of the breakdown wave.

The authors of Ref.37 and 38 assume that the steeper front of the signal in the second electrode, obtained in their work, is caused by the formation of an impact electromagnetic wave during a breakdown. A similar approach to explaining the experimental results was developed in Ref.39 in which the non-linear element was assumed to be the linear

resistance $R = R_0 \left(1 - q_0^{-1} \int_0^t I \, \mathrm{d}t \right)^{-1}$ similar to that proposed in Ref.36.

Fast electrons

One of the special features of the wave breakdown is the appearance of high-energy electrons. Stationary movement of the electrons is disrupted when the condition[80]

$$E > E_{k.\mathrm{max}} = 4\pi e^2 N z / 2,72\bar{\varepsilon},$$

is fulfilled, where N is the gas density, z is the number of electrons in the molecule, $\bar{\varepsilon}$ is the average energy of excited electrons entering the molecule.

It is well known that for each gas there is the pressure p_m (for air it is 20 torr, helium 150 torr at a voltage of 250 kV) at which the speed of the wave front and current are maximum and the attenuation of voltage minimum.[72] In Ref.77 on the basis of analysis of the $U\ (pd)$ dependences it was shown that the typical phenomenon for the breakdown of the gases in the wide range of pd is the 'runaway' of the electrons. High-energy electrons[61,78] were also recorded at the front of a breakdown wave in air. The aim of the work in Ref.81 was to verify the assumption according to which the existence of optimum p_m is associated with the formation of 'runaway' electrons. These investigations were carried out in air and helium for which P_m and $(E/p)_c$ are greatly different. Single pulses of negative polarity with an amplitude of 250 kV, the duration at half height of 35 ns with the front of 3 ns were supplied to the cathode of a discharge tube 0.8 cm in diameter and 80 cm long surrounded with an earthed cylindrical metallic screen 5.5 cm in diameter connected to a mesh anode. The space between the tube and the screen was filled with oil. The voltage along the tube was measured with capacitance dividers, the currents with inverse current shunts, the current of 'runaway' electrons I_e was measured with a Faraday's cylinder behind a mesh anode. The duration of I_e in air and helium decreased with increasing pressure from 30 to 4 ns at the half height (Fig.17.26). This difference from the data in Ref.78 where I_e at all pressures was around 3 ns is evidently caused by the fact that in Ref.81 the speed of the front $v_f \approx 10^{10}$ cm/s (Fig.17.27) was higher than in Ref.78 ($2 \cdot 10^9$ cm/s) where the insulator was represented by water, and since electrons with different energies formed at the front of the wave, part of the electrons do not keep pace with the front. Figure 17.27 shows the form of the calculated speed v_e of passage of 'runaway' electrons to the discharge tube calculated from the time delay between the start of the voltage pulse of the cathode at the maximum of the signal I_e. They are close to v_f but displaced to the range of higher pressures for He. The signals from the reverse current shunt have two characteristic phases; first evidently corresponds to the conduction current of the wave front I_f, the second to the total current I_t because it disappears with disappearance of I_e. The dependences of I_e, I_f, I_t on pressure for helium are displaced to higher pressures in comparison with air (Fig.17.28); this is in agreement with the displacement of $(E/p)_{cr}$. The current of the front I_f in a wide range of pressures can be transferred by the current of 'running-away' electrons I_e because these electrons have a strong effect on the characteristics, like the ionisation wave in movement of the electron beam in a gas.[79]

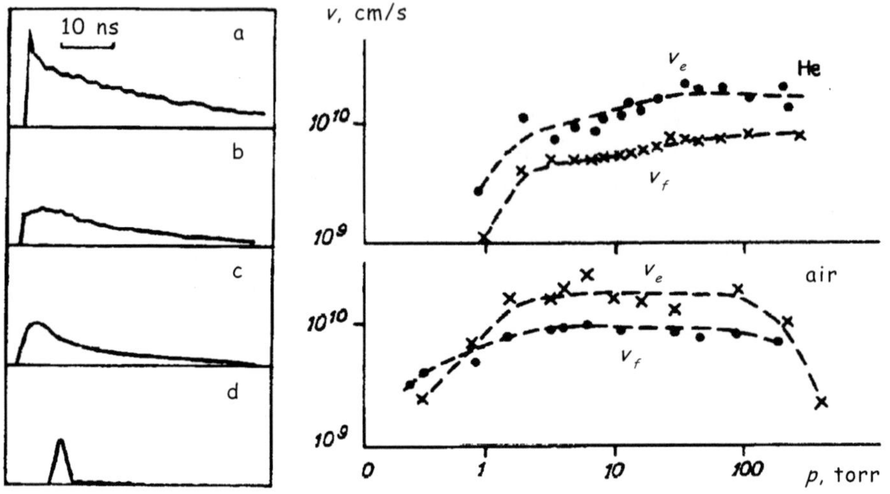

Fig.17.26 Oscillograms of current pulse I_e of fast electrons at different helium pressures. $p = 20$ (a), 33 (b), 64 (c), 168 torr (c).

Fig.17.27 Dependence of the velocity of breakdown layer v_f and the velocity of fast electrons v_e on gas pressure p.

17.7 X-radiation during a wave breakdown

In Ref.61, measurements of x-radiation, synchronised with electric measurements, were taken during the development of a wave break-down (Fig.17.29). Breakdown in a discharge device was carried out with pulses of negative polarity with an amplitude of 15 kV. At a breakdown starting at moment t_1 the voltage of the cathode becomes lower than the open circuit voltage. This moment corresponds to the appearance of current on the cathode. In subsequent stages the current behaves non-monotonically and has 2 maxima, and its increase correlates in time with the processes inside the gap.

X-radiation at the cathode (Fig.17.29b) is recorded after a decrease of the voltage on the cathode to almost zero.

The signals from the capacitance sensor $\sim \dfrac{\partial U(x, t)}{\partial t}$ in different sections of the tube are presented in Fig.17.29c. The oscillogram of the signal, recorded at the cathode is non-monotonic, and its maxima and minima correspond in time to the 'fastest' variation of the current on the cathode. With increasing co-ordinate x, $i.e.$ with increase of the distance from the cathode, the general form of the oscillograms remains unchanged. Because of the mutual synchronisation of the oscillograms, it can be seen that the perturbation formed initially at the

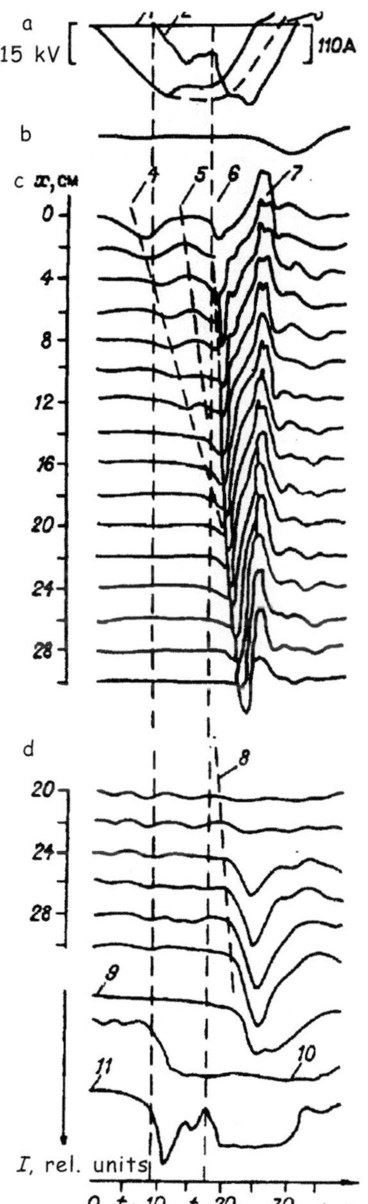

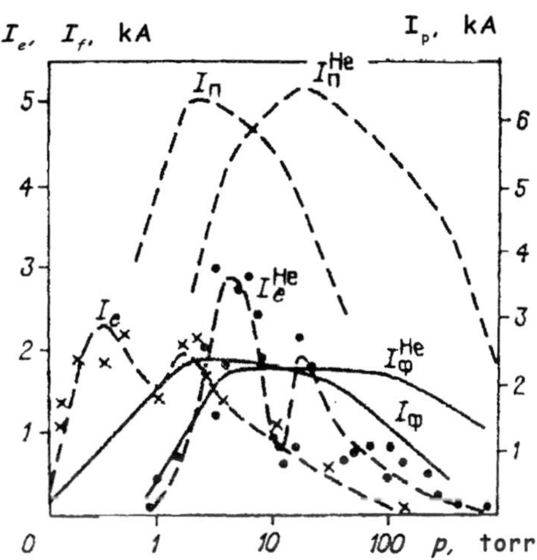

Fig.17.28 Dependence of the amplitude of conductivity current of the wave front I_f, total current I_t and fast electron current I_e on gas pressure in the discharge tube.

Fig.17.29 Synchronised measurements of electrical parameters, spatial dynamics of optical and x-ray radiation of the breakdown layer at $p = 1.5$ torr. a) oscillograms of voltage (1), current (2) on the high-voltage electrode voltage of the incident pulse (3); b) oscillogram of the pulse of x-ray radiation at the cathode of the discharge tube; c) x–t diagram of the breakdown layer obtained using a differentiating capacitance divider: 4 - first, 5 - second, 6 - third, 7 - reflective waves; d) x–t diagram of x-ray source: 8 - third wave of breakdown; d) 9 - the pulse of ex-radiation at the cathode; 10 - optical emission of the plasma at the cathode; 11 - optical emission of the cathode jet.

vicinity of the electrodes or inside the gap is then transferred along the tube.

The $(x - t)$ diagrams of propagation of the source of x-radiation (Fig.17.29d) show that the source forms inside the gap at a distance of $x = 24$ cm from the anode and then moves at a speed of $4 \cdot 10^9$ cm/s in the direction towards the anode. The duration of the front of increase of x-radiation, recorded on the side of the tube, remains constant, and the amplitude increases with increasing propagation. The speed of the source of x-radiation coincides with the speed of the front of optical radiation and the front of the potential which were determined from $(x - t)$ diagrams of these parameters measured using a photoelectronic multiplier and a capacitance sensor.

The time dependence of the intensity and effective energy of x-ray quanta, recorded at the end of the discharge tube at the anode, are shown in Fig.17.30.

Optical radiation of plasma in the vicinity of the cathode (Fig.17.29e, curve 8) forms with the start of passage of current on the cathode and the radiation intensity then monotonically increases. At the same time, there is strong optical radiation of the cathode jet which occupies a small region in the vicinity of the cathode. The dynamics of the latter is different, and the signal from the photodetector, focused only on the part of the jet is non-monotonic with time (Fig.17.29e, curve 9).

The oscillograms obtained in this case can be used to describe the sequence of processes in the discharge gap. Initially, the increase of voltage does not cause any emission from the cathode, and the capacitance sensor, placed in the vicinity of the cathode, records the signal $I_d = \dfrac{\partial U_k}{\partial t}$. The moment of time t_1 corresponds to the start of emission current I_c and the start of the wave transferring the potential inside the gap with the speed $v_1 = 2 \cdot 10^9$ cm/s. Its displacement is accom-

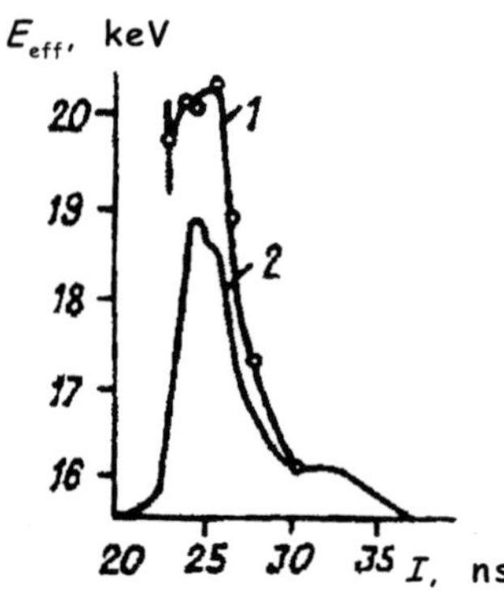

Fig.17.30 Dynamics of effective energy of quanta E_{eff} (1) and intensity of x-radiation (2).

panied by excitation and ionisation of the gas. Up to the start of the wave the optical radiation from the near-cathode zone is almost non-existent. At the moment t_2 a cathode jet appears. It is preceded by a decrease of current I_c and of optical radiation measured with a narrow-band photomultiplier.[48] The possible reason for the decrease of current is the depletion of the near-electrode layer by the electrons owing to the fact that the emission current is not capable of compensating the movement of the electrons in the direction to the front of the wave, i.e. $I_e < I_f$ (I_e is the autoelectronic emission current, I_f is the current transferred by the wave). With the formation of the cathode jet the density of emission of electrons increases, the current through the electrode I_c becomes considerably higher thus resulting in the formation of a second ionisation wave at the cathode. Since the conductivity of the plasma behind the front of the first wave is higher than in front of it, the speed of the second wave $v_2 \gg v_1$ and according to the measurements it is equal to $v_2 = 10^{10}$ cm/s. The second wave also excites and ionises the gas and increases the plasma potential. At a distance $x = 22$ 24 cm from the cathode the second wave catches up with the front of the first wave and increases the strength of the electric field in the front of the first wave. The electrons, accelerated by the latter to high energies, are ejected forward (without any collisions with gas atoms) in the direction of movement of the wave and on to the wall of the discharge tube in the vicinity of the front. This moment corresponds to the appearance of x-radiation. The front of the new wave, formed as a result of the interaction of the first and second waves, moves at the speed $v_3 = 4.10^9$ cm/s, generating fast electrons which, in turn, cause breaking x-radiation on the walls of the discharge tube. When the wave reaches the anode, the front of the reflected wave forms in the vicinity of the anode. This front moves at the speed $v_4 = 1.10^{10}$ cm/s in the direction to the cathode and discharges the plasma in the discharge tube because the jump of voltage in the front of the reflected wave has the sign opposite to the sign of the voltage jump at the front of the direct wave and is sufficiently large so that when it approaches the cathode x-radiation also forms there (Fig.17.29b).

In Ref.61 the mean (with respect to the cross section) effective energy of x-ray quanta reaches 21 keV so that it can be concluded that the electrons were accelerated to an energy of 41 keV. According to the results of measurements taken with a capacitance sensor, the voltage drop at the front is less than 2 kV. This means that the fastest electrons have the energy exceeding the maximum possible drop of the potential at the front of the breakdown wave. One of the possible explanations of this fact is the assumption that the electrons have acquired an additional energy as a result of rapid movement of the accelerating front.

17.8 Initiation of laser radiation by the breakdown wave

The spatial–time dynamics of pumping and radiation waves in a nitrogen laser has been studied in Ref.24,63,82,83. In Ref.63, it was found that the first peak of laser radiation and the breakdown wave move from the cathode to the anode at a speed $v \approx 3 \cdot 10^9$ cm/s (voltage amplitude 300 kV). The delay of the first peak of laser radiation from the front of the breakdown wave is 8–10 ns. The second peak of laser radiation propagates from the anode to the cathode with a speed of $5 \cdot 10^9$ cm/s which is approximately equal to the speed of the electromagnetic signal in the coaxial line with a water dielectric and a discharge tube as a central conductor.

Figure 17.31 shows the dependence of the speed of the ionisation wave and the peak generation power in air on pressure. In Ref.63 generation was observed in the pressure range 0.8–646 torr in nitrogen and 0.8–342 torr in air. The laser operated in the superluminance regime. The laser radiation time was varied depending on pressure from 1.5 to 10 ns. The maxima of the speed of movement of the wave and the radiation power are situated in the same pressure range, and the maximum power of generation is observed at a higher gas pressure. This is in qualitative agreement with the results in Ref.63. As in Ref.82,83, the laser radiation pulse at the exit of the discharge tube has two peaks with time. The maximum value of the peak power was 450 kW in nitrogen and 250 kW in air at a pressure of 30 torr. The maximum specific power was 15 kW/cm^3.

The speed of the breakdown wave in Ref.63 is 7–8 times less than in Ref.82 owing to the fact that the space between the discharge tube

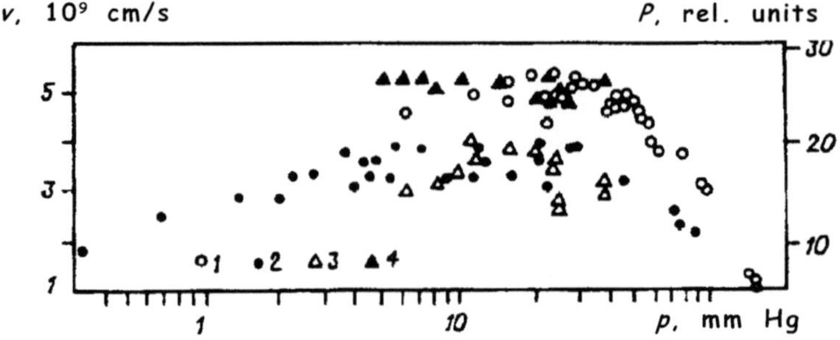

Fig.17.31 Dependence of the peak power of generation of a nitrogen laser (1), the velocity of the front of the ionisation wave (2), the first (3) and second (4) peaks of laser radiation on air pressure.

and the metallic screen was filled with water with high dielectric permittivity $\varepsilon = 80$, and in Ref.82 it was filled with nitrogen.

To examine the dynamics of generation in space, investigations were carried out into the laser radiation generated from a discharge tube using semi-transparent mirrors described previously.[63] Radiation from these mirrors have two peaks with time. Figure 17.32 shows an $(x-t)$ diagram of movement of laser radiation along the tube with air. This movement can be characterised by the speed of the peaks. This speed gradually decreases along the tube.

It should also be stressed that the first radiation peak moves from the cathode to the anode and the second one from the anode to the cathode, i.e. the second peak initially appears on the mirror close to the second electrode, and only then on the mirror close to the high-voltage electrode. Evidently, the speed of the first peak is in good agreement with the measured value of the speed of the ionisation wave, and the speed of the second peak greatly exceeds this speed. In Ref. 63, this is explained by the fact that the speed of the second peak coincides with the speed of the reflected wave. The reflected ionisation wave propagates along the gas strongly ionised with the first breakdown wave and, consequently, its speed is close to the speed of propagation of the electromagnetic signal in the coaxial system.[24]

Figure 17.33 shows the data on the times of appearance of the front of the ionisation wave and laser radiation on the mirrors in relation to the distance from the cathode.[63] It can be seen that the delay between the front of the breakdown wave and the maximum radiation remains almost constant along the tube and equals 8–10 ns. Within 10 ns the wave passes approximately 35 cm in this pressure range and does not manage to reach the second electrode. Thus, at the start of propagation

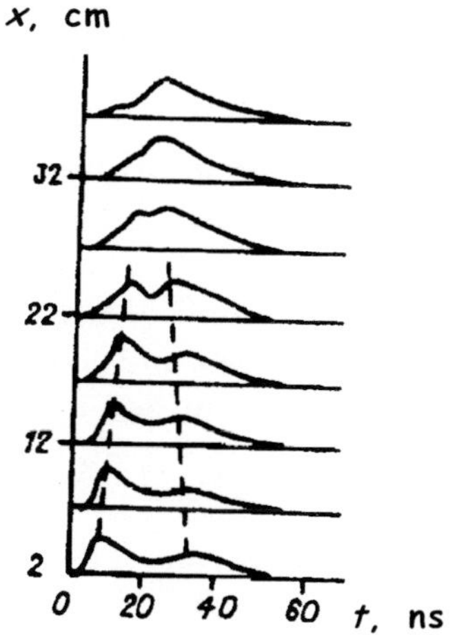

Fig.17.32 $(x-t)$ diagram of movement of the peaks of laser radiation through the discharge tube.

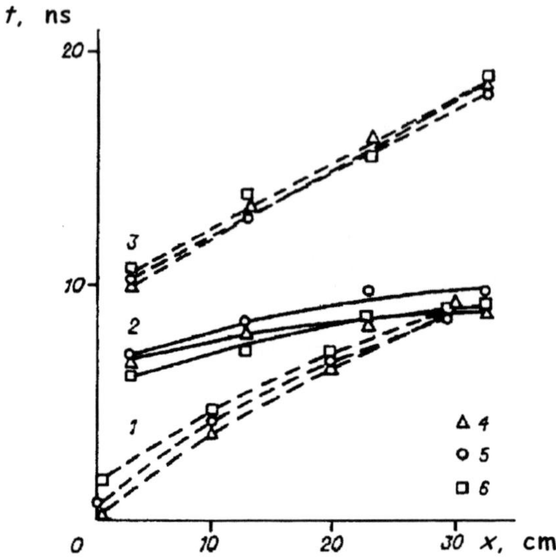

Fig.17.33 Time to appearance of the front of the ionisation wave (1), start (2) and the first peak of laser radiation (3) in relation to the distance to the cathode at an air pressure of 9.7 (4), 20.7 (5) and 45.6 (6) torr.

of the ionisation wave reflected from the second electrode, the laser radiation already exists and the reflected ionisation wave, exciting the gas for the second time, results in the formation of the second peak in laser radiation. The time delay between the front of the breakdown wave and the appearance of radiation on the mirror changes along the tube. At the start of the tube it equals approximately 4 ns which is equal to the duration of the front of the exciting pulse. At the end of the tube the delay time decreases to zero, i.e. the laser photons pass to the second electrode almost simultaneously with the front of the breakdown wave. The appearance of laser radiation at the exit from the tube simultaneously with the breakdown wave was also reported in Ref.82. In Ref.63, it was assumed that this long delay time of the maximum of radiation in relation to the front of the breakdown wave is essential for converting the energy of the electrons, accelerated at the front wave to an energy of ~100 kV, to the laser radiation energy.

The measurement show[24,63] that the removal of excitation of the medium by laser radiation takes place by a wave mechanism and the dynamics of laser radiation with time and in space is determined by the nature of excitation of the gas by the breakdown wave.

17.9 Effect of the type of gas

In Ref.103, detailed investigations were carried out into the conditions of propagation of the breakdown wave in different gases and mixtures:

He, Ne, Ar, Kr, Xe, H_2, air, CO_2 propane–butane, acetone vapours, CCl_4, SF_6, N_2, Ne–Ar. Analysis of the experimental data shows that both the elementary processes in the plasma formed behind the front and the kinetics in the gas at the head of the wave front can have a strong effect on the conditions of propagation of the breakdown wave.

Processing of the experimental dependences $v(p)$ in the inert gases was carried out using an empirical equation for the pressure optimum for the propagation of the breakdown wave: $p_m = CI_i^a M^b$, where C is a constant which does not depend on the type of gas. At $a = 5$ and $b = 0.5$, the value of C is the same for all examined inert gases, with the error not exceeding 10%. Consequently, according to a prediction, for radon $p_m = 1.3$ torr.

17.10 Current state of the theory of wave breakdown

The most extensive total system of equations was derived by A.N. Lagar'kov and I.M. Rutkevich.[34,84] They used the approximation of the surface wave from the theory of plasma waveguides but took into account the collision processes and the kinetics of ionisation in the electric field. Figure 17.1 shows the schema of the examined electrodynamic system. The main equations have the following form:

for plasma (region I)

$$\frac{\partial n_e}{\partial t} + \frac{1}{e} div \; j = \alpha n_e, \tag{17.1}$$

$$\frac{\partial n_1}{\partial t} = \alpha n_e, \tag{17.2}$$

$$j = \sigma E, \tag{17.3}$$

$$div \; E = 4\pi e(n_e - n_i), \tag{17.4}$$

$$E = -\nabla \varphi, \tag{17.5}$$

for the dielectric (region II)

$$\nabla^2 \varphi = 0.$$

It is assumed that the initial concentrations of the charges and the boundary conditions are known: $\varphi = 0$ on the metallic screen (region III), on the electrode AB, $\varphi = \varphi_0 \, (y, t)$ on the end surface of the dielectric $a < |y| < a + h$. The following condition must be fulfilled at the boundary of plasma with the dielectric $|y| = a$

$$j_y = \frac{\partial \theta}{\partial t}, \quad \text{where} \quad \theta = (\varepsilon^- E_{y\omega}^- - \varepsilon^+ E_{y\omega}^+)/4\pi,$$

E_y^-, E_y^+ are the normal components of the field in the plasma and the dielectric, respectively.

It is evident that the system of equations (17.1)–(17.5) holds under the condition $V \ll C$ so that we can ignore the vortex component of the field in plasma in comparison with the potential component.

The self-modelling solution of the system of equations (17.1)–(17.5), obtained in Ref.84, explained in particular the non-monotonic dependence of the current on time in the circuit of an earthed electrode[28] by the existence of an ionising soliton of the electric field at the wave front.

O.A. Sinkevich and Yu.V. Trofimov[33,85] examined the non-stationary boundary-value problem of the propagation of a breakdown wave in a pre-ionised gas. In contrast to equation (17.1), it was assumed that in the internal part of the channel $n_e = n_1$ and div $j = 0$. This means that the excess charge is distributed over the surface of the plasma column. Thus, the problem was solved for the case $ka \ll 1$ (a is the radius of the tube, k is the wave number).

It was shown in Ref.85 that the long-wave approximation corresponds to the description of the 'diffusion' process along the electric condenser formed by the screen and the plasma column. The corresponding 'diffusion' equation has the form

$$\frac{\partial \varphi}{\partial t} = \frac{\partial}{\partial x}\left(n_e \frac{\partial \varphi}{\partial x}\right).$$

The role of the 'diffusion' coefficient is played by the electron concentration. In Fig.17.34, the calculation results are compared with the results of measurements of the distribution of the potential along the channel at different moments of time carried out by Westberg.[10]

In conclusion, it should be noted that in electrodynamics similar systems are referred to as distributed RC structures.[86] At a constant linear resistance R, the phase velocity of the wave would be[86]

$$V_{phase} = \left(\frac{2\omega}{RC}\right)^{1/2},$$

where ω is frequency, C is the linear capacitance. The breakdown in the coaxial line using the RC structure was modelled in, for example,

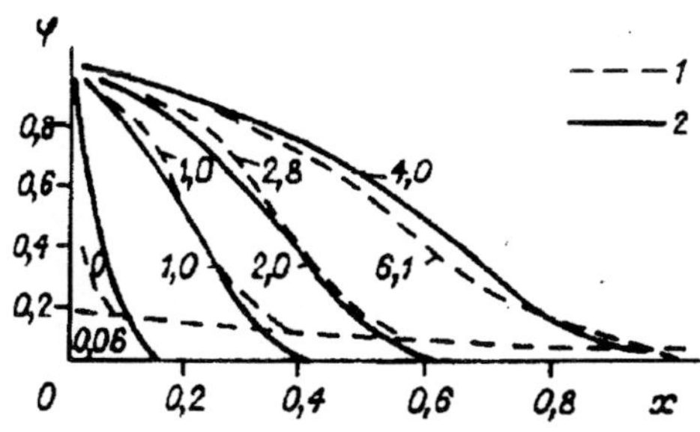

Fig.17.34 Distribution of potential along the discharge tube at different times. 1) experiment[10]; 2) calculation[85]; the numbers of the curves are the values of t.

Ref.87. The variation of R in this work was taken into account using the following procedure. It is assumed that the strength of the field in the conductor E is linked with the current by the relation $i = k_1 n_e E$, *i.e.* we ignore the bias current in the plasma. At a penetration depth of the field in the conductor larger than the radius of the latter the multiplier k_1 is equal to $\mu_e s$ ($s = \pi a^2$). Using the Rompe–Weitzel approximation, we accept that $\partial n_e/\partial t = k_2 iE$, and consequently

$$R = k_1 k_2 \left/ \left(\int_0^t i^2 \, dt \right)^{1/2} \right. \quad \text{(Rompe - Weitzel equation)}.$$

This model is relatively approximate. However, it makes it possible to examine the effect of the main factors (screen, initial electron concentration, the curvature of the pulse front) on the breakdown dynamics.

Thus, at present, the physics of non-relativistic breakdown waves ($v \ll c$) on the condition that the distribution function of the electrons is determined by the local values of the electric field, can be regarded as basically sufficiently developed. (Experimental and theoretical examination of the role of radial components was carried out in: E.I. Asinovskii, et al., *A similarity criterion for the speed of electric breakdown waves in screened tubes of different diameter*, Pre-print IVTN, No.3-50, Moscow (1992)).

At breakdown wave speeds close to the speed of light, the non-potential nature of the electric field which influences the formation of the wave and restricts its speed is very strong.

In contrast to Ref.33,85, the authors of Ref.88 examined a numerical model with a non-potential electric field. The procedure of averaging over the cross section of the waveguide was used.[33,34] Taking into account $\partial/\partial z \ll 1/r_0$ (r_0 is the radius of the discharge channel, z is the longitudinal co-ordinate) and $\partial/\partial t \ll 4\pi\sigma_0$ (σ_0 is the initial conductivity), the equations for the propagation of the breakdown wave have the form

$$C / \pi r_0^2 \cdot \partial \varphi(z, t) / \partial t + \partial j_z(z, t) = 0, \tag{17.6}$$

$$E_z(z, t) = -\partial v / dz - L \pi r_0^2 \partial j_z / \partial t, \tag{17.7}$$

$$j_z = \sigma(z, t) E_z, \tag{17.8}$$

where σ, φ, j_z, E_z are the values of the conductivity, potential, current density and the longitudinal electric field averaged over the cross section of the waveguide, C, L are the linear capacitance and inductance of the waveguide.

The equations (17.6)–(17.8) are closed by the equations for the conductivity of slightly ionised plasma

$$\sigma = \mu_e(E_z / P)n_e(z, t), \quad \partial n_e / \partial t = \alpha(i_z / P)n_e,$$

where μ_e, α is the mobility of the electrons and the constant of the ionisation rate which depend on E_z/p, p is the gas pressure. The initial-boundary conditions are σ_0 $(z) = \sigma$ $(z, 0)$ and φ_0 $(t) = \varphi$ $(0, t)$.

Figure 17.35 shows the dependence of the speed of propagation of the breakdown wave on $\partial U/\partial t = \partial U_0/\tau_f$ and σ_0. Agreement between the experimental and calculated data was obtained at $\sigma_0 = 1.2$ $\Omega^{-1} \cdot$ cm^{-1}. According to estimates, this level of conductivity in front of the wave front can be generated as a result of the flux of 'runaway' electrons.

The authors of Ref.89 constructed a theoretical model taking into the 'running-away' electrons and ionisation of the gas by them. The equations describing the electrodynamics of the breakdown wave (telegraph equations) were solved numerically together with the non-stationary Boltzmann equation in the two-dimensional phase space

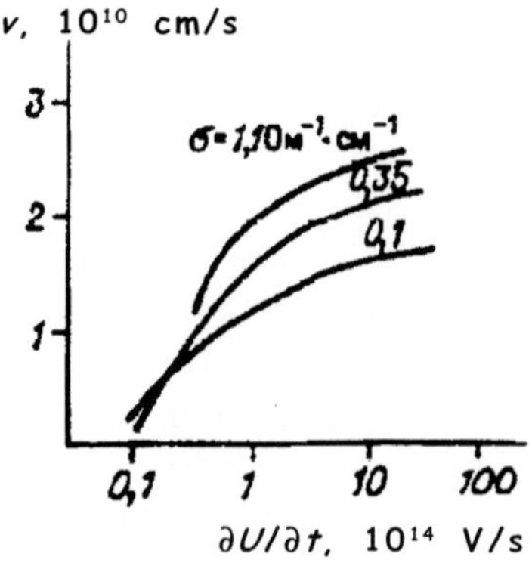

Fig.17.35 Dependence of the velocity of the breakdown wave on the rate of voltage increase.

for the distribution function of the 'runaway' electrons with respect to the energies with the integral of the inelastic collisions in the right hand part. The results of calculations carried out in Ref.89 describe satisfactorily the experimental data in Ref.72 for the speed of the wave and attenuation. The 'runaway' electrons are characterised by a wide spectrum whose upper boundary exceeds eU_0.[89]

17.11 Wave breakdown - a method of examining processes in plasma

Examination of the lifetime of excited states

At present, the main methods of perturbation of the population of the examined level are either the monoenergetic electron beam or optical pumping. The gas discharge has until now been neglected because of a number of reasons, especially the heterogeneity and poor reproducibility of excitation. In Refs.28, 90, 91 it was established that these short-comings can be avoided, at least in the pressure range 0.5–30 torr, if the gas is excited by means of a wave breakdown. This also results in a high degree of excitation thus simplifying the problem of recording radiation and processing the results. The method is relatively simple and universal.

The 3^1D level of helium[90] and the states $C^3\Pi_u$ of the nitrogen molecule were investigated.[90,91]

The lifetime of the $C^3\Pi_u$ state ($V = 0$) was measured using various methods, and the scatter from 30 to 46.3 ns exceeded the measurement error. The aim of the investigations in Ref.90 and 91 was to measure the lifetime of the state $C^3\Pi_u$ ($V = 0$) of the N_2 molecule in nitrogen and when measuring the partial pressure of oxygen. In processing the results, the level not excited by the electrons was ignored because of the low degree of ionisation. According to the estimates, the gas temperature is 320 K.

The radiation lifetime of the $C^3\Pi_u$ state ($V = 0$) (37 ns), measured in Ref.90 and 91, and the cross section of quenching of this state by the nitrogen molecules (1.9 Å^2) coincide, in the measurement error range, with the data from the majority of studies, and the quenching cross section of the state by the oxygen molecules (52 Å^2) is larger than the data in other studies.[90] The results can be used to explain the decrease of the peak power of the laser in air in comparison with the laser in pure nitrogen[92] by quenching of the upper laser level by oxygen molecules.

Hydrogen dissociation, measurement of the diffusion coefficient of atomic hydrogen

The efficiency of the dissociation process in the electric discharge is determined by the energy spectrum of the electrons. Dissociation in hydrogen takes place mainly through the excitation of the electronic state with an excitation threshold of 8.8 eV. At a higher electron energy a large part of this energy is used for the excitation of the electronic states with the gase heated to a relatively low temperature. The development of a discharge in the shape of a breakdown wave enables it to be used to form the required fields in the plasma and makes it a suitable means for the dissociation of molecular hydrogen.

In Ref.97, wave breakdown was used to measure the diffusion coefficient of hydrogen atoms. The following measurement procedure was used. After the controlled delay time t_d after the nanosecond breakdown a low-current rectangular probing pulse of microsecond duration was supplied to the discharge gap to excite the hydrogen atoms. However, because of additional dissociation the pulse does not change the concentration of the atoms. The dependence of the integral intensity of the line H_α with respect to time on the delay time t_d is measured; the dependence is then used to determine the diffusion coefficient.

In Ref.98, the relatively course of radiation H_α was used to estimate the concentration of atomic hydrogen after the discharge. The absolute concentration method is based on the fact that the processes of direct and dissociative excitation prevail at different delay times t_d.

In Ref. 8, the maximum number of dissociation acts per 1 eV of the supplied energy was $\eta = 0.018$ eV^{-1}. If it is assumed that a single dissociation act requires 8.8 eV (excitation energy of the level $^3\Sigma_u^+$), then it can be seen that the electric efficiency is 16%. Comparison of the nanosecond discharge developed in the form of a wave breakdown, with other known plasma sources of hydrogen atoms shows that it is characterised by the highest efficiency at such a high dissociation rate.

Relaxation of the plasma loaded with direct current

In Ref.99 and 100 investigations were carried out into relaxation processes in helium in exciting the plasma with a nanosecond discharge developed in the form of a breakdown wave, especially the special features of the breakdown of helium plasma loaded with direct current. The region of the breakdown, characterised by a low radiation yield, 'glow break', was observed and examined. The results show that the nanosecond discharge generating additional ionisation in the plasma makes it possible to regulate the temperature of electrons in the range from 0.05 to 0.5 eV during hundreds of microseconds.[100] In Ref.101

a method was developed of diagnostics of ionisation in the nanosecond discharge on the basis of the duration of 'glow break'. The results show that there are conditions under which the breakdown wave results in general ionisation in the nanosecond discharge.

Examination of the fast electrons shows that the front of the breakdown wave contains regions with the strength of the electric field of $E/N \geq 10^{-14}$ V cm^2. At these values of the strength of the electric field at the front of the breakdown wave the ionisation processes are highly effective. A very large increase of pressure reduces the efficiency of ionisation but the latter remains sufficiently high in relation to the losses in heating the gas.

This makes it possible to use the wave breakdown for sustaining a plasma with a high electron concentration ($n_e \approx 10^{11}-10^{12}$ cm^{-3}) at a relatively low temperature of the working gas ($T \approx 100 \div 300°C$). Of special interest is the combined discharge[102] in which repeating nanosecond pulses with high amplitude are superimposed on a dc discharge. The conditions of the optimum sustainment of ionisation were examined in Ref.61.

Examination of ion–ion recombination

Analysis of the studies carried out in recent years shows that the problem associated with the measurement of the recombination coefficient with the ion density increasing to $10^{14}-10^{15}$ cm^{-3} has not yet been solved.

In Ref.95 and 96, it was proposed to examine recombination in F_2 and SF_6 at high ion concentrations on the basis of the afterglow of a powerful nanosecond discharge developed in the form of a wave breakdown. Ion–ion plasma forms in the presence of effective 'sticking' in the afterglow, and when a relatively short current pulse is fed to this plasma the voltage drop in the near-electrode regions may be greatly lower than the voltage drop in the plasma column. Defining the size of the layer as $r = \mu_i/lt$, where μ_i is the mobility of the ions, l is the length of the discharge tube, U is the voltage drop in the plasma, and calculating the voltage drop in the near-electrode layer from Poisson's equation, we obtain a criterion for the duration of the probing current pulse

$$t < 1.4 \sqrt{l/J\mu_i}\, 10^{-7},$$

where t is in s, l in cm, μ_i in cm/(V s), J in A/cm^2.

To explain the conductivity of the plasma of the afterglow of the nanosecond breakdown, the authors of Refs.95 and 96 carried out numerical modelling taking into account the main special features of the elementary processes in F_2 and SF_6. The main annihilation channel of the ions at pressures in the range $p = 60 \div 400$ torr was assumed to be the three-particle recombination. The effect of a possible slightly non-ideal nature of the plasma was taken into account by analogy with electrolytes by introducing the activity coefficient and a relaxation correction for mobility. The results show that 2–3 μs after completing the discharge the ion concentration was 10^{12}–10^{14} cm^{-3}. In Ref.95 and 96 a delay in the decrease of the conductivity of the plasma in F_2 and SF_6 in an early afterglow was detected. This delay depended on gas pressure.

The experimental results and the results of modelling plasma breakdown make it possible to conclude that non-ideal ion–ion plasma forms in the early stages of breakdown. The effective constant of ion recombination in this plasma decreases.

Nitride synthesis

In Ref.94 the wave breakdown was used for synthesis of phosphorus nitrides. The measurement of the amount of absorbed nitrogen for different durations and pulse repetition frequency shows that the rate of synthesis is directly proportional to the repetition frequency of the pulses. In the experiments, the rate of combination of nitrogen decreased with time. It is evident that this was associated with the change of the conditions in the gas-discharge device: formation of a nitride film on the surface of phosphorus, formation of microarcs, *etc*. Figure 17.36 shows the dependence of the energy used for bonding the nitrogen molecule on the inverse duration of the current pulse flowing through the anode of the discharge device and averaged out for the first 30 minutes of synthesis during which there were no saturation effects.

The slope of the straight lines in Fig.17.36 shows that a decrease

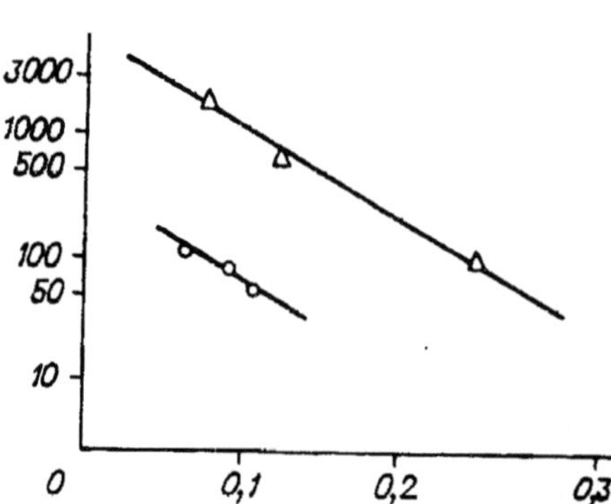

Fig.17.36 Energy used for binding one molecule of nitrogen in relation to the reverse duration of current pulse.

of the duration of energy supply results in a large increase of the efficiency of using this energy in synthesis. The dependence in Fig.17.36 is efficiently approximated by the formula $E/N = A \exp (- t/T)$, where E is the energy used up for the absorption of N molecules of nitrogen, T is the nanosecond current pulse time, t and A are constants: $t = 17 \pm 3$ ns, $A = 6000$ eV/mol for a pipe with a diameter of 5 cm and $A = 365$ eV/mol for a pipe with a diameter of 12 cm.

Comparison of the efficiency of synthesis of phosphorus nitride in the wave breakdown with other types of gas discharge shows that the efficiency of the wave breakdown is higher.

17.12 Conclusion

This review of the investigations of wave breakdown can be summarised by the following comments.

1. The sub-nanosecond level of synchronisation of the channels of recording the signals of electromagnetic radiation in a wide range of the variation of the wavelength ($10^{-8} - 10^2$ cm) has been reached in diagnostics in the last 10–15 years.

2. The complicated structure of the breakdown wave has been established. In particular, it has been shown that at positive polarity the front is conical, and at negative polarity it approaches sliding discharge along the tube walls. The Doppler broadening of the current pulse reflected from the front has been recorded. The vibrational structure of the charge behind the front of the breakdown wave was detected. The presence of fast electrodes in the vicinity of the front of the wave with an energy corresponding to the double amplitude of the voltage pulse was confirmed, together with the existence of the limiting speed of the breakdown wave. This speed coincides with the speed of the fast electrons.

3. It has been shown possible to use the wave breakdown as a means of examining the elementary processes in non-equilibrium plasma occupying an intermediate position between the glow discharge and the electron beam. For example, measurements were taken of the diffusion coefficient of atomic hydrogen, the energy efficieny of synthesis of the molecule of phosphorus nitride in relation to the duration of the nanosecond discharge, and the recombination coefficient in dense ion plasma was also measured.

In conclusion, the authors would like to express their gratitude to A.E. Sheindlin for attention and support of work in the area of the wave breakdown and also would like to note the significant contribution to these investigations from A.V. Kirillin, A.G. Abimov, L.N. Vasilyak, S.V. Kostychenko, A.B. Matstenko, Yu.M. Tokunov, A.M. Ul'ynov and I.V. Filyugin.

Chapter 18

PROBE MEASUREMENTS OF POTENTIAL DISTRIBUTION IN DENSE PLASMA

18.1 Special features of the method

To measure the potential in the arc discharge, a probe is introduced into the discharge and the voltage in the probe is measured in relation to one of two electrodes.[1] The probe has the form of a section of a thin non-insulated metallic wire which is moved in a strictly horizontal plane at a certain distance from the electrode in the vertical arc column normal to the discharge axis. The potential difference between the probe and one of the electrodes is transferred to the deflecting plate of a two-beam oscilloscope. The speed of the probe is varied from 40 to 200 cm/s. An inductanceless resistor is connected in parallel to the probe measuring circuit. The resistor is used to shunt voltage surges formed as a result of transient processes. Changing the nominal of the resistor, it is possible to vary the probe current and measure almost electrostatically the potential distribution in the cross section of the arc. The sweep of the oscilloscope is synchronized with the movement of the probe by an auxiliary probe which moves together with the measuring probe at some distance from it. Selecting the sweep frequency and the delay time of the signal from the auxiliary probe, it is possible to produce a sufficiently stable image on the screen of the oscilloscope.

The main problem in probe measurements is associated with the perturbation of the discharge by the probe. This perturbation is of the thermal and electrical nature. Perturbation of the arc by the probe was investigated by Muller[2] using oscillographic and optical methods. The results show that the extent of thermal perturbation depends on the thickness of the probe and the characteristics of the arc region through which the probe passes: as the thickness of the probe increases and the current density in the plasma region intersected by the probe becomes higher, the intensity of perturbation increases. Perturbation in the vicinity of the arc axis is the consequence of partial overlapping of the main discharge channel by the probe and by the surrounding zone with reduced electrical conductivity. The insertion of the probe into the arc column changes stationary current and arc voltage. In contrast to the thermal perturbation, formed at any polarity of the probe irrespective of the presence of insulation of the probe, the electric perturbation is observed only at the positive potential of the probe in relation to the surrounding plasma when the plasma receives high current. Therefore, the perturbation of the arc voltage increases with increasing probe current. In the vicinity of the arc axis it is superimposed on the thermal perturbation which does not depend in a wide range on the electric perturbation intensity. The calculations determined the stationary strength of the field in the arc and the width of the perturbation zone. The strength of the axial electric field in the arc proved to be close to that obtained from the results of probe measurements of the potential, and the length of the perturbation zone was close to the width of the visible dark space around the probe measured using photographs of the probe trajectory in plasma.

In Ref.3, it was shown that the total perturbation, introduced by the probe, can be evaluated from the oscillograms of the voltage and arc current which are recorded together with the probe voltage. Consecutive passage of the probe through one and the same cross section of the arc column should generate the same potential at different passage rates. Consequently, it is concluded that neither the thermal excitation nor the turbulence of the gas, caused by the passage of the probe through the arc column, has any significant effect on the measured results.

The perturbation of the arc column by the probe was examined in Ref.4. Firstly, the arc voltage and arc current were inspected prior to and after entry of the probe into the arc column; secondly, the same measurements were taken in the conditions of probing the arc with three probes. Two probes were used to measure the strength and the third probe, entering the arc column after the first two probes, was used to measure the arc perturbation. The measurements show that the strength of the field did not change after the entry of the perturbing probe. The authors of Ref.4 concluded that probing is the most suitable method of measuring the voltage gradient in the arc.

The dependence of the nature of perturbation of the arc by the probe on the parameters of the probe has been quite widely studied because by selecting their probe it is possible to minimize the perturbation of the arc. To minimise perturbation, it is recommended to use a thin probe that rapidly travels through the arc column. The material of the probe can be characterised by specific thermal conductivity and heat capacity. The results show that a tungsten probe approximately 0.2 mm wide with the travel speed higher than 100 cm/s is most suitable.

Muller and Finkelnburg[5] measured for the first time the distribution of the potential along the arc axis and its peripheral part using a probe. They show that the potential can be varied only when the resistance of the external probe circuit is considerably higher than the resistance of the plasma column between the probe and the corresponding electrode. The latter depends on the position of the probe in the arc, because the conductivity of the plasma rapidly decreases from the axis to the periphery. The authors of Ref.5 obtained the following results:

1. When the probe speed is higher than 100 cm/s, the probe current through the perturbed area of the plasma is determined exclusively by charge carriers drifting from the plasma, and the thermal emission of the ions and electrons from the external surface of the probe is insignificantly low;

2. As a result of the difference of the mobility of the electrons and the ions, the resistance of the layer between the probe and the unperturbed plasma in contact of the probe with the anode is lower than in contact with the cathode. Therefore, at a given resistance of the circuit of the probe the region of measurement of the potential is always wider for the case of contact with the anode;

3. Regardless of the fact that the plasma potential changes along the probe, the measured potential as a function of the distance from the arc axis remains constant with the variation of probe current and the probe requires the potential that corresponds to the point with the highest conductivity of the arc column that is in contact with the probe;

4. Calculations and experimental examination show that the contact difference of the potentials of the probe and the plasma is constant along the arc axis and does not depend on plasma temperature; for the argon arc this value does not exceed 2 V;

5. The probe method of measuring the potential using an electron beam oscilloscope for recording is far more efficient than the method in which the data are recorded using a loop oscilloscope.

Ringler[6] measured the radial distribution of the potential in the arc column and proposed the following explanation. Investigations were carried out for the arc examined extensively by other methods. The arc was stabilised by the plasma jet leaving the cathode (current intensity 200 A). In addition to the probe potential, oscillographic recording was made of the changes of the arcing voltage and arc current formed as a result of perturbation of the arc by the probe. The method enabled the author to determine the radial distribution of the arc potential at any distance from the cathode with sufficient accuracy. The results show that the size of the perturbation zones within the range of the measurement error is in agreement with the diameter of the main current-carrying arc channel. Starting with the equation for the density of electric current through the plasma, the author of Ref.6 calculated the radial distribution of the strength of the field for the examined arc. Calculations confirm the experimental data.

The authors of Ref.7 measured the anodic voltage in the arcs in relation to arc current and arc length. It is well known that the anodic voltage drop is of considerable interest for the theory of near-anode processes. Therefore, the results of measurement of the anodic voltage by the probe method are of considerable importance for developing the theory of the anodic voltage drop. Up to now it was generally recognised that the anodic and cathodic volt drops remain constant when the electrode gap is increased. This assumption was used as a basis for developing the well-known method which makes it possible to determine the strength of the field in the column from the increase of arcing voltage with increasing arc length. However, Finkelnburg determined[7] that the anodic voltage in the argon arc depends (in addition to current intensity) on the length of the arc which decreases with a decrease of the arc length. With increasing plasma temperature in the vicinity of the anode its potential decrease is linearly irrespective of arc length and current density. To confirm this, two independent methods were used to determine the strength of the field and the potential drop along the arc column of different length. The results show that the anodic drop which is associated with the density of anodic current and plasma temperature increases with increasing arc length. To confirm the dependence of anodic drop U_A on arc length in argon, an independent method was used to measure the axial distribution of the strength of the field in the column. This was carried out by calculating the distribution of electrical conductivity in the column using the well-known distribution of temperature in the arc at a current of 200 A in argon. The results show that the anodic drop in the arc, burning in nitrogen at a current intensity higher than 50 A, decreases linearly with increasing current intensity. At the same time, the anodic voltage in the high-temperature argon arc depends also on the arc length and temperature, and when the arc length increases this drop becomes larger, and decreases when the plasma temperature in front of the anode increases.

The problem of the length of the section of the arc over which the potential at the anode rapidly decreases has been examined in Ref.8. The length of the region of the anodic drop was measured using the following procedure: a tungsten probe, insulated with quartz, with a very short bare end was ejected with a special device at a speed of 10-20 cm/s in the direction of the surface of the anode of a carbon arc. Touching the anode surface the probe rapidly returned. The voltage of the probe in relation to the anode was recorded with an electronic oscilloscope. The results show that the axial length of the region of the anodic drop in the low-current carbon arc is approximately equal to the free path of the electron and amounts to 2 μm for the given experiment conditions.[8]

Finkelnburg and Segel[9] measured the distribution of potential inside and around the column of the electric arc. It is well known that the distribution influences both the axial and radial movement of the electrons and the arcs and also the arc characteristics. The results show that the data obtained in the probe measurements of the potential in low- and high-current carbon arcs are in good agreement with the conclusions of theoretical analysis and indicate that the length of the transition region between the distorted potential field in the arc and the non-distorted potential field outside the discharge is quite small. When measuring the anodic and cathodic drops the probe was moved manually or with an pneumatic mechanism up to contact with the cathode or anode. The results of probe measurements of the potential inside the arc column enabled the authors to find equipotential surfaces. The appearance of the region with a large decrease of the potential in front of the electrodes is explained by the presence of an excess charge in them. As a result of the redistribution of the charges in the vicinity of the electrodes, there is always an electric field with higher strength. In the vicinity of the cathode under the effect of the radial component of this field the electrons are displaced in the direction towards the axis of the arc column. Thus, the normal mechanism of ambipolar diffusion of charge carriers changes. This results in a decrease of the width of the arc column in the near-cathode region. The radial electric field in the vicinity of the anode displaces the electrons to the column periphery so that the column widens in the near-anode region.

18.2 Electric field of the high-current arc

Measurements were taken of the axial and radial distribution of the strength of the electric field in the freely burning arc in the gas between cooled metallic electrodes. The effect of the resistance in the circuit of a single probe from the results of potential measurements has also been investigated. The arc diagram is shown in Fig.18.1.

The anode was represented by the bottom of a steel vessel cooled inside with water, the cathode was a thoriated tungsten rod with a diameter of 2.4 mm also intensively cooled with water. The arc was

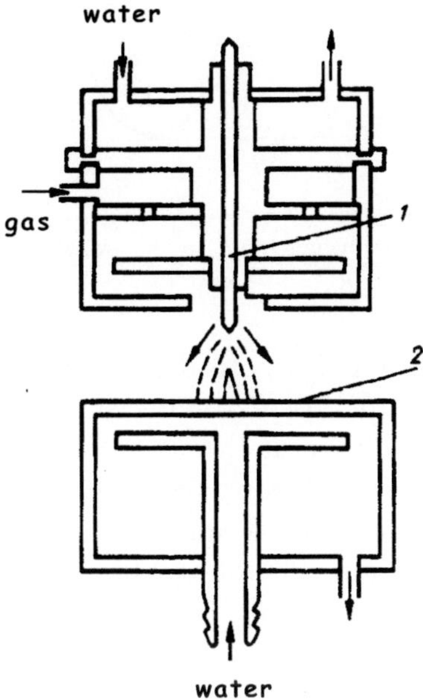

Fig.18.1 Diagram of the arc with argon shielding. 1) tungsten cathode, 2) copper anode.

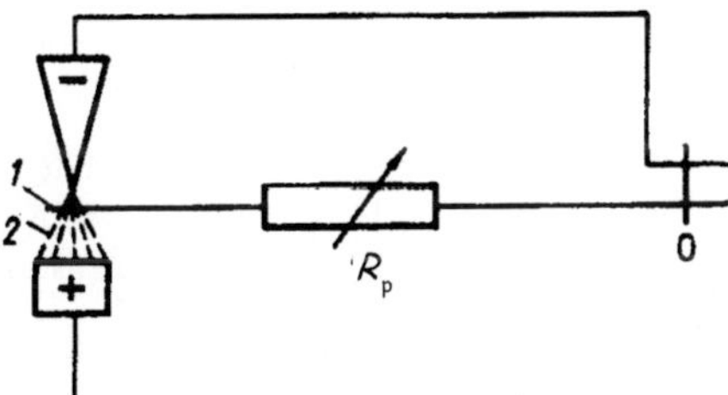

Fig.18.2 Diagram of connection of a probe into the measuring circuit. 1) probe, 2) arc, O – oscilloscope, R_p is the variable resistance.

powered with a smooth voltage from a rectifier. The arc was stable for a long period of time as a result of stabilising with an argon flow (flow rate 5 l/min). The photographs show that the cross section of the arc smoothing increases from the cathode to the anode.

The probe was in the form of a section of tungsten wire 0.2 mm in diameter with the non-isolated section 2 cm long. The probe was moved using a special device in the direction normal to the actual line of the arc column strictly in one plane. The travel speed of the probe was varied from 50 to 250 cm/s, with the measurements usually taken at a speed of 220 cm/s. The diagram of connection of the probe to the measuring circuit is shown in Fig.18.2. A variable resistor was used to vary the intensity of probe current. The difference of the potential

Fig.18.3 Oscillograms of the probe signal in the near-cathode region of the arc for values of the variable resistor. R_p = 1.5×10⁶ (a), 5×10⁶ (b), 1.5×10⁷ ohm (c).

of the probe and one of the electrodes was supplied to the input of the electronic oscilloscope. To estimate the extent of perturbation of the arc by the probe, voltage at the arc electrode was also recorded. Probing of the arc in its different sections was carried out by moving the entire arc device using a stand in which the electrodes were secured. The distance between the probe and the cathode was controlled in an area of ± 0.05 mm. The potential was measured in relation to the cathode. The electrode spacing was 5 mm and was set using a special template.

Typical oscillograms of the probe signal (Fig.18.3) were obtained by intersecting the plasma column by the probe at a distance of 1 mm from the cathode. At resistor resistances higher than $5 \cdot 10^6$ ohm the oscillograms in Fig.18b, c, show a small increase of the potential in the near-axial region of the arc column caused by the perturbation caused by the probe. The intensity of the perturbation in the potential units is equal to the change of arc voltage. At low resistances of the probe circuit, *i.e.* in this case less than $1.5 \cdot 10^6$ ohm, there are no potential surges in the peripheral zones of the arc on the oscillogram (1). The surges on the oscillogram appear only under conditions in which the resistance of the probe circuit is considerably higher than the resistance between the probe and the corresponding electrode (it will be referred to as the arc resistance and denoted R_a). The region between these surges is the region in which the potential is measured (see above). The width of the region of measurement of the potential increases with increasing resistance of the probe circuit, i.e. the distortion of the measured distribution of the potential in the peripheral regions of the plasma decreases with increasing resistance of the probe circuit. The increase of the resistance of the probe circuit is restricted by the sensitivity of the oscilloscope. It should be noted that the oscillogram is used for determining the potential of the near-axial region of the plasma because due to higher conductivity of the plasma in this region of the arc the condition $R_p \gg R_a$ is fulfilled.

To determine the range of measurement of the potential, investigations were carried out into the dependence of the half width of the meas-

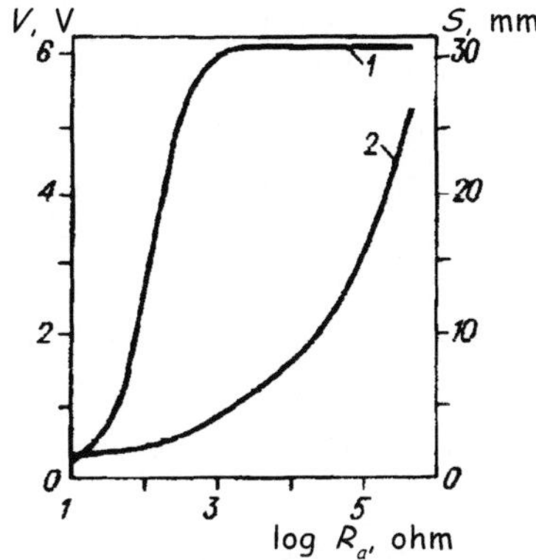

Fig.18.4 Distribution of the potential along the arc axis (1) measured using a probe and the width of the region of variation of the potential (2) in relation to resistance R_p of the probe circuit.

ured radial distribution of the potential on the resistance of the probe circuit (Fig.18.4, curve 2). These dependences are explained by the rapid drop of the conductivity of the arc plasma in the radial direction. In our case at a probe resistance of $R_p = 10^7$ ohm the width of the region of measurement of potential is ~28 mm, although the 'light' diameter of the widest part of the arc does not exceed 6 mm.

To determine the minimum resistance in the probe circuit in which it is still possible to measure reliably the potential of the near-axial regions of the arc we constructed the graph of the dependence of the measured potential of the plasma on the arc axis on R_p (curve 1). The results show that the latter increases linearly with the resistance of the probe circuit and at some values of these resistance it ceases to depend on it. The increase of the potential measured at the arc axis continues until the resistors of the plasma between the probe and the cathode is higher than the resistance in the probe circuit. Comparison of these two graphs shows that at a resistance lower than ~10^3 ohm the region of measurement of the potential has a constant width of approximately 3 mm, and the measured potential at resistances higher than ~10^3 ohm tends to a constant value. It can therefore be concluded that the resistance of the plasma column between the probe and the cathode does not change when the probe moves within the limits of the cross section 3 mm in diameter.

The estimates of the resistance of the arc column presented here are intentionally too high because examination of the probe determines the resistance of a narrow section of the arc column corresponding to the diameter of the probe to which the resistance of the 'cold' part of the arc column is then added. In addition, it must be taken into account that only some part of the probe surface 'takes part' in the measurements. Therefore, a more accurate value of the arc resistance can be obtained using the ratio of the cross sections of the arc and the probe and also the resistance of the 'cold' part of the arc. It should be noted that Muller and Finkelnburg[5] published too high values of the resistance of the arc column (300 ohm). We believe that after considering the previous comments, the given method of determining the minimum resistance of the probe circuit can be used to determine the resistance of the arc column and, consequently, the electrical conductivity of the plasma.

Thus, to ensure that the probe can be used in the regime of measurement of the potential, it is necessary to fulfil the condition $aR_p \gg R_a$. The radial distribution of the potential was examined at a probe circuit resistance of 10^7 ohm. The measured distribution of the potential was corrected by extrapolating it from the non-perturbed section of the oscillograms to the axis of the arc along a parabola. This extrapolation was used to determine the potential of the near-axial sections of the plasma. The radii of the zones of perturbation of the arc, measured for three cross sections, situated at distances of 1 and 2.5 and 4 mm from the cathode, were respectively, 3, 4, 5 and 6 mm, i.e. they increase in the direction from cathode to anode. This is consistent with the observed form of the arc. Figure 18.5 shows the radial distribution of the potential measured by the probe in relation to the cathode. The measured potential value includes the axial voltage drop in the arc from the cathode to the examined cross section and the radial component of the potential which depends on the distance from the probe to the axis. For sections close to the cathode the potential increases more rapidly. This is explained mainly by the nature of the radial distribution of the concentration of charged particles in the arc. The equi-

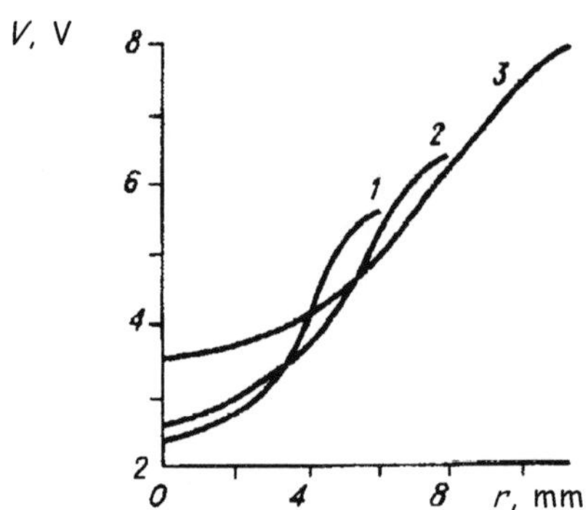

Fig.18.5 Radial distribution of the probe potential at the resistance in the probe circuit of 10^7 ohm for sections at a distance of 1 (1), 2.5 (2) and 4 mm (3) from the cathode.

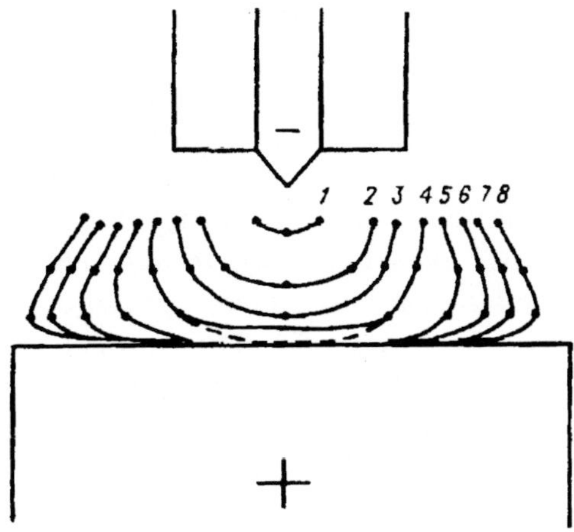

Fig.18.6 Equipotential lines of arc discharge. V = 4.5 (1), 5 (2), 6 (4), 6.5 (5), 7 (6), 7.5 (7), 8 V (8).

potential V-lines of the arc column are shown in Fig.18.6. The dotted line shows the observed contour of the arc column. These data are in agreement with the results published in Ref.9. The anodic drop of the potential between the cathode and probe where the latter intersects the arc in the vicinity of the anode is equal to $U_a = U_{arc} - U_p$ where U_a is the anodic drop, U_{arc} is the voltage on the arc electrodes, U_p is the difference of the potential between the cathode and the probe. The anodic voltage in the examined arc is 6.5 V at an arcing voltage of 13 V which is in good agreement with the data in Ref.9, 10, according to which in examining an arc with a current of 50 A and an electrode gap of 5 mm the anodic drop was 6 V at a total arc voltage of 15–16 V.

18.3 Electric field in the laminar plasma jet of the electric arc plasma torch

Measurements of the potential in the flow of the plasma in the 'laminar' plasma torch[11] were taken at an arc current of 80 A and an arc voltage of 80 V, the flow rate of plasma-forming argon was 0.18 g/s. The probe was in the form of a section of tungsten wire 0.2 mm in diameter, the length of the non-insulated part was 2 cm, and the section periodically intersected the plasma jet in the direction normal to its axis. The speed of the probe (~4 m/s) was such that the temperature of the probe remained almost constant during the intersection of the plasma flow, and electron emission from its surface can be ignored. The probe signal was recorded using an oscilloscope through an in-series connected resistance of 5 MΩ. The potential received by the probe in plasma in relation to an earthed cathode was measured. The tungsten wire had no insulating coating so that the probe was in contact with areas of plasma having different potential at the same time.

Typical oscillograms of the probe signal are shown in Fig.18.7. Probe measurements at the outlet of the nozzle did not show any high-frequency pulsations of the electric signal. In addition, when the orientation of the probe in relation to the cathode spot of the arc was changed electric signal returned it's symmetry. This indicates that electric arc is not blown away by the gas flow from the nozzle of the plasma torch.

The results obtained in processing the oscillograms were used to plot the distribution of the potential acquired by the probe in the plasma jet (Fig.18.8). The results show that the electric field is observed at relatively large distances from the outlet of the nozzle (see also Ref.12 and 13). When explaining the resultant dependence, it should be taken into account that the strength of the probe signal depends not only on the plasma potential but also on the contact difference of the probe-plasma potentials and on the ratio of the resistance of the cathode-probe $R_{c.p}$ and the resistance in the probe circuit 9. The probe operates as a voltmeter only when $R_p \gg R_{c.p}$. For near-axial regions of the plasma jet this condition is obviously fulfilled.[14] We introduced a correction for the contact difference of the probe–plasma potentials whose value is a function of temperature and is determined by the following expression:[1] where T is plasma temperature, M is the ion mass, m, e is the mass and charge of the electron, respectively, k is the Boltzmann constant:

$$\Delta U = -\frac{kT}{2e}\ln\frac{M}{m},$$

The corrections for the measured values of the potential were calculated taking into account the known distribution of the temperature in the jet. The corrected distribution of the potential along the axis of

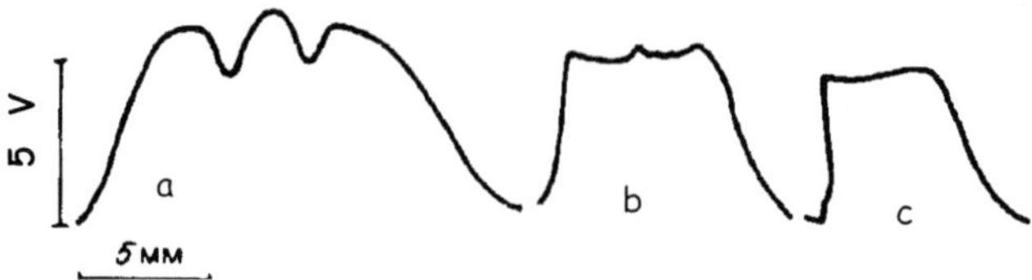

Fig.18.7 Typical oscillograms of the probe potential at a distance of 5 (a), 30 (b) and 60 mm (c) from the outlet of the nozzle.

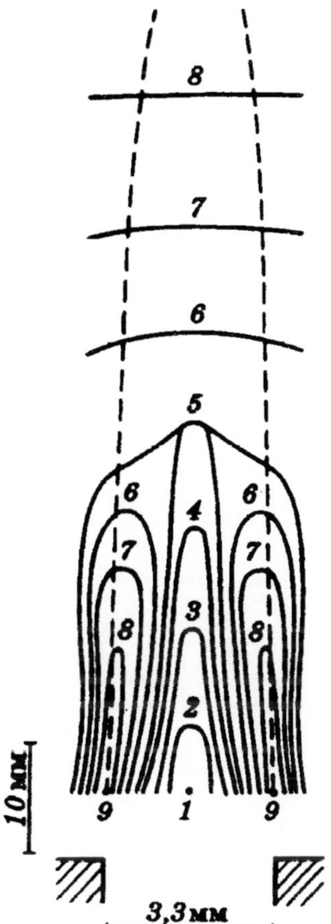

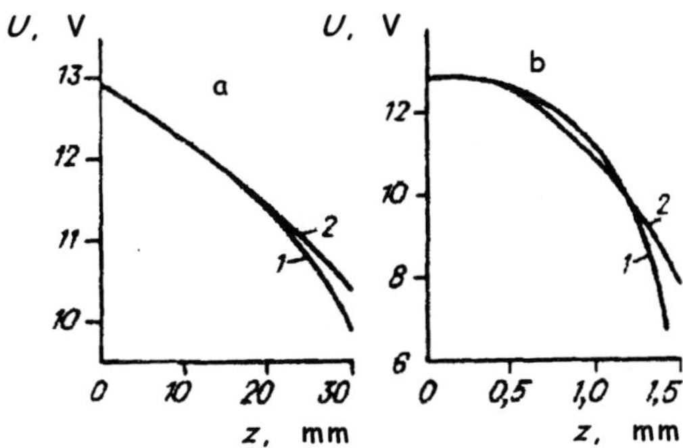

Fig.18.8 (Left) Distribution of the electric potential of the probe in the plasma jet. $U =$ 6.25 (1), 6 (2), 5.75 (3), 5.5 (4), 5.25 (5), 5 (6), 4.75 (7), 4.5 (8), 4.25 V (9); broken line shows the growing corona of the jet.

Fig.18.9 Measured and calculated – on the basis of the ambipolar diffusion (1) taking into account the contact difference of the probe-plasma potentials (2) – axial (a) and radial (b) distribution of the potentials of the electric field in the plasma jet.

the jet and in the direction of the radius at the outlet of the nozzle from the plasma jet represented in Fig.18.9. The plasma potential at the outlet of the nozzle is $U_0 = 12.8$ V. Since the results of probe measurements show that in the plasma torch with a narrow nozzle the electric arc is not blown out in the form of a loop, this potential is evidently determined by the cathode drop in the electric arc.

It will be assumed that the variation of the potential in the plasma jet is caused by ambipolar diffusion. The gradient of the potential formed in the plasma as a result of ambipolar diffusion is[1]

$$\text{grad } U_{amb} = \frac{1}{eN_e}\text{grad }(N_e kT).\qquad(18.1)$$

Integrating equations (18.1) from the point with the temperature T_0 to the point with the temperature T gives

$$U_{amb}(T) = U_0 - \frac{k}{e} \int_{T_0}^{T} \frac{\partial \ln N_e kT}{\partial \ln kT} \, dT.$$

(18.2)

Here U_0, T_0 is the potential and temperature of the plasma at the axis of the jet at the outlet of the nozzle. The distribution of the potential, calculated from equation (18.2), and determined by ambipolar diffusion, is shown in Fig.18.9 which indicates that there is satisfactory agreement between the measured U and calculated U_{amb} potential distributions. It can therefore be assumed that the potential of the plasma jet is determined by the cathode drop in the electric arc and ambipolar diffusion and the plasma jet itself is without current.

18.4 The electric field of the plasma flow of a two-jet plasma torch

The measurements were taken at the following parameters of the two-jet plasma torch: arc current 105 A voltage 145 V, total flow rate of the plasma forming argon for both heads 0.12 g/s, initial angle of convergence of the plasma jet 60°.[15]

Experimental equipment consisted of a horizontal table with a device for securing and rotating the probe and a C1-55 two-beam oscilloscope. The probe was in the form of a tungsten wire section 0.2 mm in diameter with the length of the non-insulated part being 2 cm. To increase its stiffness, the wire was placed in a thin quartz capillary tube. The non-insulated part of the wire was inserted into the plasma. The capillary was secured to a metal rod 70 cm long. The probe signal was fed to the input of the oscilloscope through a mercury contact. To vary the probe current, an alternating resistor with a maximum resistance of 1 MΩ, equal to the input resistance of the oscilloscope was connected in parallel to the input terminals of the oscilloscope.

The linear speed of the probe was 5 m/s. This speed was selected on the basis of experiments taking into account requirements on the absence of thermal emission, evaporation of the probe material and the minimum dynamic disruption of the plasma. In accordance with Ref.3, the absence of thermal emission of electrons and ions was inspected on the basis of degree of symmetry of the probe signal. The probe potential was measured in relation to the cathode (when connected to the anode, the signal amplitude was not measured). The multiple passage of the probe in one cross section also did not cause any changes in the pattern of the radial distribution of the potential. The form of the oscillograms for the jets and the plasma flow is shown in Fig.18.10. The oscillograms of arc voltage are also given in here. To construct

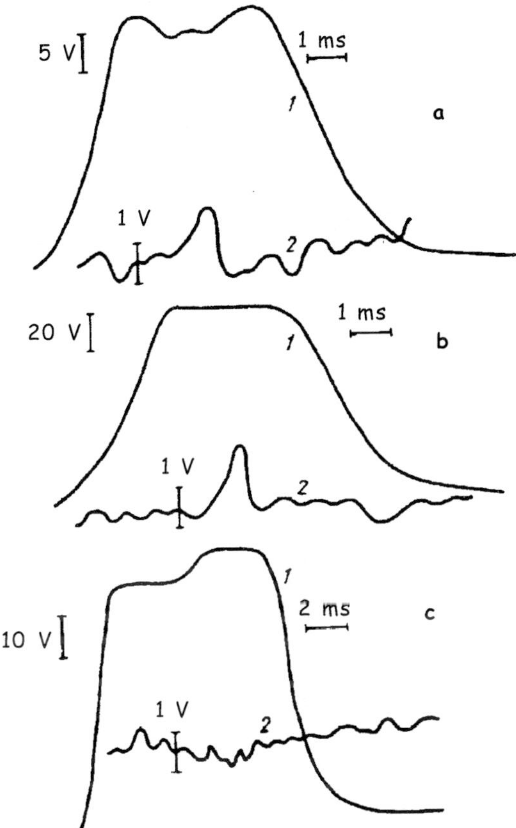

Fig.18.10 Typical oscillograms of the probe potential (1) and arc voltage (2) for the cathode (a) and anode (b) jets at a distance of zero from the outlet of the nozzle of the plasma torch, for the plasma flow after merger of the jet (c) 30 mm above the line connecting the outlets of the nozzles of the plasma torches.

the field of the equipotentials of the jet the plasma flow was probed in different sections. The measurement of the potential in the anode and cathode jet was carried out separately, and the plane of displacement of the probe was always normal to the direction of the discharge of the jets and the plasma flow. At the same time, arc voltage was inspected when the arc column was intersected by the probe. The resistor in the probe circuit was represented by the input resistance of the C1-55 two-beam oscilloscope (R_p = 1 MΩ). A V7-16 digital volt meter was used to measure the distribution of the potential in the diaphragms of both heads of the plasma torch in relation to the cathode. The results of measurements in the form of the field of equipotential are presented in Fig.18.11. The distribution of the density of the electric current between the plasma jets was found from the measured strength of the electric field and the electrical conductivity of the plasma calculated from the measured temperature (~7000 K). The results are presented in Fig.18.12 which shows that the current density does not exceed 1.5 A/mm^2.

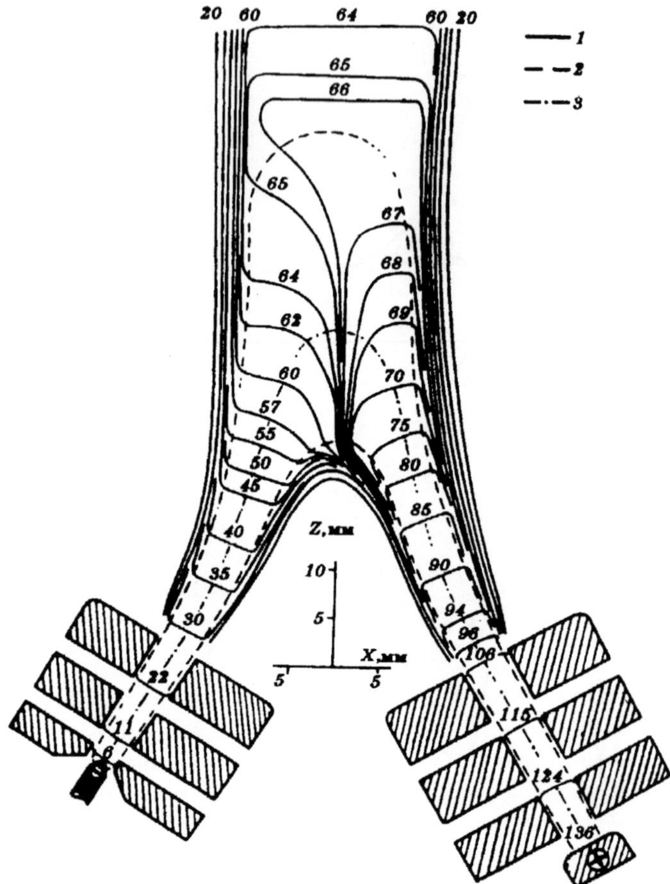

Fig.18.11 Distribution of the electric potential of the probe (in volts) in the plasma flow (1). 2 – current passage area, 3 – line of the lowest electrical resistance of the plasma.

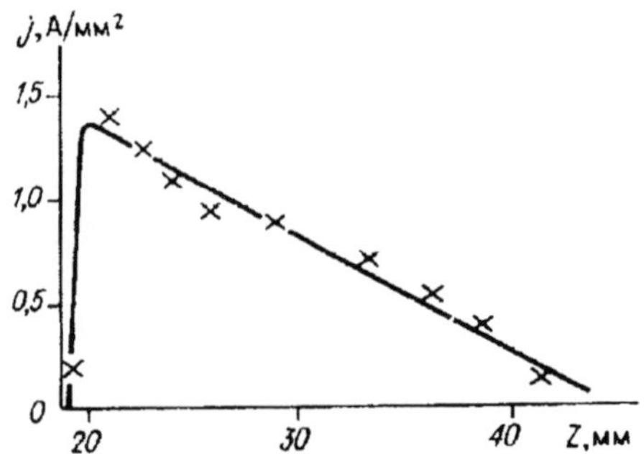

Fig.18.12 Distribution of the density of electric current in the gap between the plasma jets (coordinate z – see Fig.18.11).

The experimental results show that the arc discharge in the convergence zone is stationary and of the diffusion type up to a height of approximately 25 mm with a current density of 1 A/mm^2. Here the discharge is evidently non-independent because the ionisation conditions are determined to a large degree by the external inflow of heat. The zone of convergence of the jet is heated by the plasma jet by conductive and convective heat flows ensuring the diffusion passage of currents advantageous from the energy viewpoint. The volt equivalent of this zone is 5–7 V which equals 3–5% of the total arc voltage. This is also indicated by the absence of contracted plasma channels between the jets. The absence of high voltage jumps in the arc also indicates the stationary nature of current passage. The authors of Ref.16 and 17 examined the behaviour of the arc in a long cylindrical channel for turbulent arcing. Film frames of the arc show clearly the areas of ruptures of the arc cord. It can therefore be assumed that a similar mechanism of the diffusion passage of current can also be realised in this case.

On the basis of the results of probe measurements, the potential in the region of current passage at $Z = 59$ mm is equal to 66 V. At $Z = 64$ mm and higher, the potential decreases to 64 V and remains unchanged up to $Z = 230$ mm.

Chapter 19

REDUCTION TO A UNIFORM LAYER IN AXISYMMETRIC OBJECTS

In optical diagnostics of electric arcs, pulse discharges and plasma flows measurements are taken of certain integral quantities which can then be used to determine the local optical parameters of the examined object. These objects are usually characterised by rotational symmetry, and by taking into account the non-uniformity (reduction to the uniform layer), the problem is reduced to the numerical solution of Volterra integral equations, especially of the Abel's type. The solution of these equations with the right part unperturbed does not cause any principle difficulties. However, in reality, the experimental data are always loaded with random errors. Consequently, the results of the solution will contain very large errors. Therefore, special attention should be paid to smoothing the initial data.

It is also important to develop a rational logarithm with a minimum computation time. This is determined by the complicated expression of the Volterra kernel. Finally, to apply the results in practice, it is necessary to evaluate their accuracy. Here we describe a method[1,2] determining the spectral coefficients of emission and absorption in non-uniform axisymmetric objects that satisfies these requirements to a large degree.

19.1 Smoothing experimental data

The experimental function $y(x)$ will be approximated by expansion to a generalised Fourier series

$$y(x) \approx Y(x) = \sum_{k=0}^{M} A_k P_{2k}(x), \tag{19.1}$$

where $P_{2k}(x)$ are the even Chebyshef polynomials, orthogonal with the weight $\rho(x)$ on the system of points x_i, $i = \Delta\overline{0, n}$, $x \in [-1, 1]$ and determined by recurrent relationships

$$P_{2k}(x) = P_{2k-2}(x) - (q_{2k-2}/q_{2k-1})xP_{2k-1}(x),$$
$$P_{2k+1}(x) = xP_{2k}(x) - (q_{2k}/q_{2k-2})P_{2k-1}(x),$$
$$P_0(x) = 1, \quad P_1(x) = x, \quad -1 \le x \le 1,$$

$$q_k = \sum_{i=0}^{n} \rho_i P_{ki}^2, \quad \sum_{i=0}^{n} \rho_i P_{ki} P_{si} = 0, \quad \sum_{i=0}^{n} \rho_i = n+1, \tag{19.2}$$

$$\rho_i = \rho(x_i), \quad P_{ki} = P_k(x_i).$$

In accordance with the principle of least squares, the Fourier coefficients are

$$A_k = q_{2k}^{-1} \sum_{i=0}^{n} \rho_i P_{2k,i} y_i, \quad y_i = y(x_i). \tag{19.3}$$

The measure of scattering of the initial data $y(x_i)$ in relation to the approximating function $Y(x)$ is the residual dispersion

$$D_M = (n-M)^{-1} \sum_{i=0}^{n} \rho_i (y_i - Y_i)^2 = (n-M)^{-1} \left(\sum_{i=0}^{n} \rho_i y_i^2 - \sum_{k=0}^{M} q_{2k} A_k^2 \right). \tag{19.4}$$

The number of the terms of series (19.1) is optimised in accordance with the functional dependence and the accuracy of the initial data. The following criterion is highly suitable for selecting M irrespective of the number of points, dispersion and weight of the measurements. The number of the terms of the series (19.1) is restricted by the minimum value of M at which the residual dispersion (19.4) ceases to decrease, namely as soon as the conditions

$$D_M \le D_{M+1}, \quad D_M \le D_{M+2}.$$

are fulfilled. The approximating function $Y(x)$ (19.1), representing the smoothed-out value of the experimental function $y(x)$, is used in further calculations.

The advantage of Chebyshef's polynomials in addition to the possibility of taking into account the weight of measurements is that the initial data can be represented with an arbitrary step along the x axis. In the absence of information on the weight of measurements, the experimental data are assumed to be of equal accuracy.

19.2 Numerical solution of Abel's integral equation

From the solution of the integral Abel equation we determined the radial distributions of the refractive index, the absorption factor and the emission factor in an optically thin source and a source with moderate self-adsorption or with spatially distributed emission and adsorption zones. The integral Abel equation (using an example of the required coefficient of emission in an optically thin source) has the form

$$2R\int_{|x|}^{1}\frac{\varepsilon(r)r\mathrm{d}r}{\sqrt{r^2-x^2}}=I(x) \tag{19.5}$$

Here and later we shall use the reduced co-ordinates $0 \leq r \leq 1$, $-1 \leq x \leq 1$, and $\varepsilon(r \geq 1) = 0$, $I(|x| \geq 1) = 0$.

The integration range in (19.5) $0 \leq r \leq 1$ will be divided into N zones of equal size and it will be assumed, to simplify considerations, that $\varepsilon(r)$ in the zones is constant. The integral equation (19.5) acquires the form

$$\sum_{j=1}^{p}2R\varepsilon_j\int_{r_j}^{r_{j-1}}(r^2-x_1^2)^{1/2}r\mathrm{d}r=R\sum_{j=1}^{p}a_{pj}\varepsilon_j=I_p, \quad p=\overline{1,N} \tag{19.6}$$

where

$$a_{pj}=2\left[(r_{j-1}^2-x_p^2)^{1/2}-(r_j^2-x_p^2)^{1/2}\right],$$

$$\varepsilon_j=\varepsilon(r_{j-1/2}), \quad I_p=I(x_p), \quad r_j=x_j=(1-j)/N.$$

The system of algebraic equations (19.6) with a triangular matrix has the solution

$$\varepsilon_p = R^1 I_p / a_{pp} - \sum_{j-1}^{p-1} (a_{pj} / a_{pp})\varepsilon_j, \quad j = \overline{1, N}. \tag{19.7}$$

Equation (19.7) can be written in the form

$$\varepsilon_p = R^{-1} \sum_{j=1}^{p} b_{pj} I_j, \quad j = \overline{1, N}, \tag{19.8}$$

where

$$b_{pj} = -\sum_{s=1}^{p-1} (a_{ps} / a_{pp})b_{sj}, \quad b_{pp} = 1 / a_{pp}.$$

Instead of the experimental values of I_j the corresponding values of the approximating function $Y(x_j)$ (see (19.1)) are substituted into equation (19.8).

We shall estimate the random error of the calculated results. We shall write the solution of the integral Abel equation (19.8) taking into account (19.1) and (19.3) (y_j will be replaced here by I_i) in the form

$$\varepsilon_p = \sum_{i=0}^{n} \left(\sum_{k=0}^{M} \rho_i^{1/2} P_{2k,i} q_{2k}^{-1} \sum_{j=1}^{p} b_{pj} P_{2k,j} \right) \rho_i^{1/2} I_i / R. \tag{19.9}$$

Assuming that the errors of the measurements are random and independent, in accordance with the linear equation (19.9) we obtain the following expression for the dispersion $\varepsilon\,(r)$:

$$D(\varepsilon_p) = \sum_{i=0}^{n} \left(\sum_{k=0}^{M} \rho_i^{1/2} P_{2k,i} q_{2k}^{-1} \sum_{j=1}^{p} b_{pj} P_{2k,j} \right)^2 \rho_i D(I_i / R).$$

Taking into account that $\rho_i \cdot D\,(I_i/R) = D\,(I/R)$, where $D\,(I/R)$ is the weighted dispersion of the initial data, and taking into account the distribution of q_{2k} and the orthogonality condition in accordance with (19.2), we obtain

$$D(\varepsilon_p) = D(I/R) \sum_{k=0}^{M} q_{2k}^{-1} \left(\sum_{j=1}^{p} b_{pj} P_{2k,j} \right)^2, \quad p = \overline{1, N}. \tag{19.10}$$

As a measure of the accuracy of the initial data it is recommended to use the dispersion of the value I/R because the intensity I is proportional to the radius of the source R, and the measurement error represents some fraction of the intensity. When using non-smoothed values of I_j in equation (19.8) for measurements of equal accuracy we directly obtain

$$D(\varepsilon_p) = D(I, R) \sum_{j=1}^{p} b_{pj}^2. \tag{19.11}$$

We determine the coefficient of transfer of the error $S(r)$ of the initial data to the calculated results by the equation

$$S^2(r) = D(\varepsilon(r)) / D(I/R). \tag{19.12}$$

Calculations carried out using equation (19.10) show that for the axis of the source ($r = 0$) we can approximately write

$$S^2(0) \approx (M+1)^3 / 3(n+1). \tag{19.13}$$

The interpolation method for the case when the initial data are not smoothed out corresponds to the condition $N = \dfrac{n}{2} = M$, and

$$S^2(0) \approx (1/6)N^2, \tag{19.14}$$

which also follows from the calculations carried out using equation (19.11). On the basis of the equations (19.13), (19.14) and Fig.19.1, we can draw the following conclusions. The number of zones N in conversion of (19.18) with the presmoothed initial data has only a slight effect on the stability of the solution. This makes it possible to use a simple quadrature (constancy of ε in the zones) and use a relatively large value N (~50–100) for reducing the systematic error. The use of the Fourier series results in the stability of the solution whose accuracy in comparison with the interpolation method increases with in-

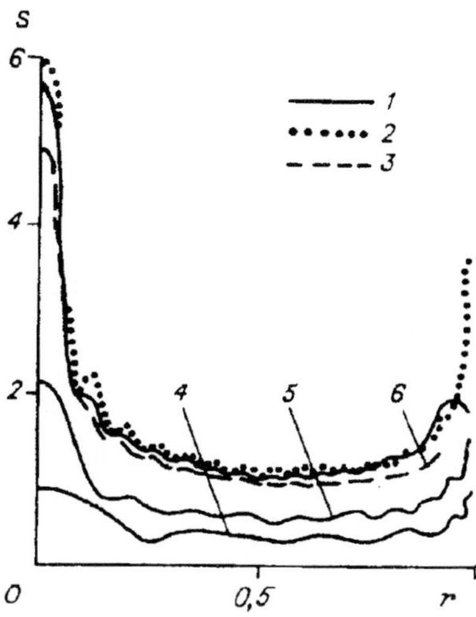

Fig.19.1 Coefficient of transfer of the error in conversion of Abel transformation calculated from equations (19.10), (19.12). $n = 100$, $p = 1$, $N = 50$ (1), 100 (2), $N = 25$ (3), $M = 5$ (4), 10 (5), 20 (6).

creasing number of the points for which the initial information is specified (under the condition of non-correlation of their errors) and with a decrease of the number of the terms of the series required for approximating the experimental function.

19.3 Numerical solution of the Volterra integral equation

In the case of an optically dense source the coefficients of emission $\varepsilon(r)$ and absorption $\kappa(r)$ are linked with the observed intensity $I(x)$ and transparency $\omega(x)$ by the integral relationship

$$2P\int_{|x|}^{1} ch\left(\int_{|x|}^{r}\frac{\kappa(r')r'dr'}{\sqrt{r'^2 - x^2}}\right)\frac{\varepsilon(r)rdr}{\sqrt{r'^2 - x^2}} = \frac{I(x)}{\sqrt{\omega(x)}}, \tag{19.15}$$

$$2R\int_{|x|}^{1}\frac{\kappa(r)rdr}{\sqrt{r^2 - x^2}} = -\ln\omega(x) \equiv \tau(x), \tag{19.16}$$

where $\kappa(r \geq 1) = 0$; $\omega(|x| \geq 1) = 1$. For the maximum value of the hyperbolic cosines we can write

$$ch \left[\int_{|x|}^{1} \frac{\kappa(r) r \, dr}{\sqrt{r^2 - x^2}} \right] = \frac{1 + \omega(x)}{2\sqrt{\omega(x)}}.$$

Estimates carried out using this equation show that the hyperbolic cosine can be assumed to be equal to unity with the error smaller than 6% at $\omega > 0.5$ and less than 1% at $\omega > 0.8$. If self-absorption in the central emitting part of the source is ignored, the hyperbolic cosine is equal to unity, regardless of the level of absorption on its periphery. This is indicated by the fact that the integral below the sign of the hyperbolic cosine in (19.15) is taken in the region where $\varepsilon\,(r)$ differs from zero. Thus, with moderate self-absorption and in the case of spatially distributed zones of emission and absorption, the equation (19.15) assumes the form of the integral Abel equation

$$2R \int_{|x|}^{1} \varepsilon(r)\, (r^2 - x^2)^{-1/2} r\, dr \approx I(x) / \sqrt{\omega(x)}.$$

The solution of this equation in relation $\varepsilon(r)$ can be determined from the previously described algorithm with the approximation of the experimental function $I(x) / \sqrt{\omega(x)}$ by the Chebyshef polynomials. If these approximations are not uniform, it is necessary to solve the equation (19.15) in relation to $\varepsilon\,(r)$ determining in advance $\kappa\,(r)$ from the integral Abel equation (19.16). We shall write the integral equation (19.15) in the form

$$2R \int_{|x|}^{1} \frac{\varepsilon(r)}{\kappa(r)} ch \left[\int_{|x|}^{r} \kappa(r')(r'^2 - x^2)^{-1/2} r'\, dr' \right] \kappa(r)(r^2 - x^2)^{-1/2} r\, dr = I(x)\omega^{-1/2}(x)$$

The integration range $0 \leq r \leq 1$ will be divided into N zones of equal size and it will be assumed that $\varepsilon\,(r)$ and $\kappa\,(r)$ in the zones are constant. Consequently, we obtain

$$\sum_{p=1}^{j} 2(\varepsilon_p / \kappa_p) \int_{r_p}^{r_{p-1}} \mathrm{ch}\left[\int_{x_j}^{r} \kappa(r')(r'^2 - x_j^2)^{-1/2} r' dr'\right] \kappa(r)(r^2 - x_j^2)^{-1/2} r dr =$$

$$\sum_{p=1}^{j} 2(\varepsilon_p / \kappa_p) \int_{\tau_{jp}}^{\varepsilon_{j;p-1}} \mathrm{ch}\ \tau d\tau = \sum_{p=1}^{j} 2(\varepsilon_p / \kappa_p)(\mathrm{sh}\ \tau_{j;p-1} - \mathrm{sh}\ \tau_{jp}) = \tag{19.17}$$

$$\sum_{p=1}^{j} c_{jp}\varepsilon_p = R^{-1} I_j \omega_j^{-1/2}, \quad j = \overline{1, N},$$

where

$$c_{jp} = (2 / \kappa_p)(\mathrm{sh}\ \tau_{j;p-1} - \mathrm{sh}\ \tau_{j;p}),$$

$$\tau_{jp} = \sum_{t=p+1}^{j}\left[(r_{t-1}^2 - x_j^2)^{1/2} - (r_t^2 - x_j^2)^{1/2}\right], \quad \kappa_p = \kappa(r_{p-1/2}), \tag{19.18}$$

$$\varepsilon_p = \varepsilon(r_{p-1/2}), \quad I_j = I(x_j), \quad \omega_j = \omega(x_j), \quad r_p = x_p = 1 - p / N.$$

The system of algebraic equations (19.17) with a triangular matrix has the solution

$$\varepsilon_j = R^{-1} I_j \omega_j^{-1/2} / c_{jj} - \sum_{p=1}^{j} (c_{jp} / c_{jj})\varepsilon_p, \quad j = \overline{1, N}. \tag{19.19}$$

Prior to calculating $\varepsilon(r)$ from (19.19), the experimental function $I(x) / \sqrt{\omega(x)}$ is smoothed out using the Chebyshef polynomials. Using the procedure identical with that described previously for the integral Abel equation, we obtain the following estimate of the dispersion of the emission coefficient in the case of an optically dense source:

$$D(\varepsilon_j) = D(R^{-1} I \omega^{-1/2}) \sum_{k=0}^{M} q_{2k}^{-1}\left(\sum_{p=1}^{j} e_{jp} P_{2k,p}\right)^2, \quad j = \overline{1, N},$$

where

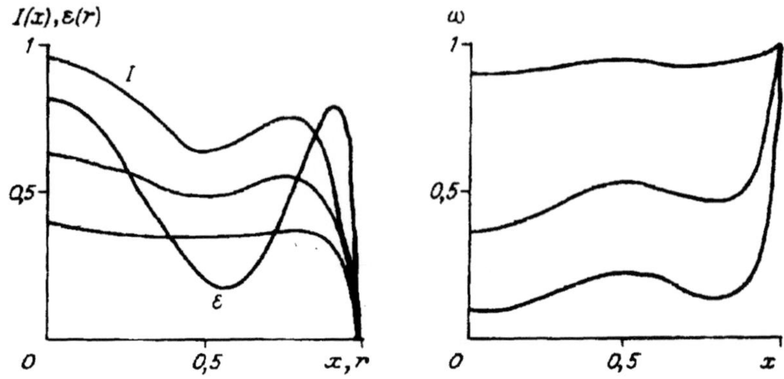

Fig.19.2 Test functions (19.22).

$$e_{jp} = \sum_{s=p}^{j-1} \left(c_{js} / c_{jj}\right) e_{sp}, \, e_{jj} = 1 / c_{jj},$$

and the coefficients c_{jp}, are determined by equation (19.18). Equation (19.20) gives the lower estimate of the dispersion of the emission factor, because it does not take into account the effect of the absorption factor included in c_{jp}. However, as shown by the calculations, the effect of this error is not large because in calculations of ε we use the integral from κ. The absorption factor is calculated from Abel's integral equation, and we have

$$D(\kappa_j) = D(R^{-1}\ln\omega) \sum_{k=0}^{M} q_{2k}^{-1} \left(\sum_{p=1}^{j} b_{pj} P_{2k,p}\right)^2, \, j = \overline{1,N}. \tag{19.21}$$

The results of calculating the variance factor for test functions (Fig. 19.2) are:

$$\varepsilon(r) = (-15\pi)^{-1}\left(1 - r^2\right)^{1/2}(38 - 176r^2 + 288r^4),$$

$$\kappa(r) = \gamma\varepsilon(r),$$

$$I(x) = \gamma^{-1}\left[1 - \omega(x)\right], \tag{19.22}$$

$$\omega(x) = \exp\left[-\gamma(1 - 3x^2 + 8x^4 - 6x^6)\right]$$

are presented in Fig.19.3. The value of the factor γ was used to specify

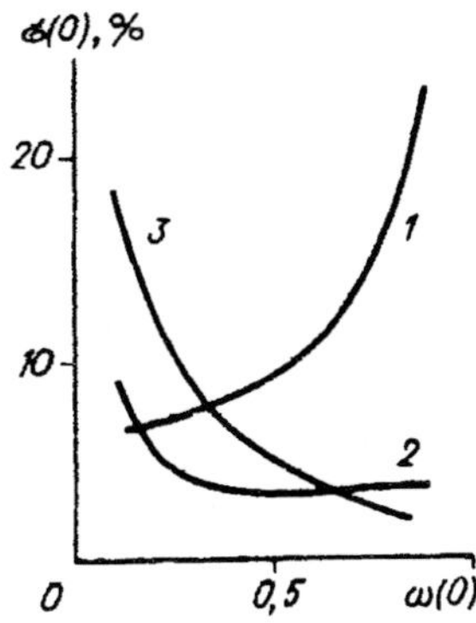

Fig.19.3 Relative error of restoration of the coefficients of absorption (1) and emission (2,3).

different values of transparency ω (0) = 0.1 ÷ 0.9. It was assumed that the relative random error of experimental functions I (x) and ω (x) is 3%, $N = 80$, $N = 100$. As indicated, it can be seen that the transmission coefficient in this example is determined with satisfactory accuracy up to ω (0) = 0.1. The absorption coefficient is determined with a large error.

We assume that for the examined object we know the dependence

$$\kappa = \kappa(\varepsilon), \tag{19.23}$$

and for example, the function ε (T) or the radial distribution at temperature T (r) are specified. Consequently, on the basis of the Kirchoff's law we can write $\kappa = \varepsilon/B$, where B is the Planck's function. Therefore, it is not necessary to measure the transparency and the emission coefficient is determined from the measured intensity from the solution of the equation (19.15) with an allowance made for (19.23). Equation (19.19) is transformed with this purpose for the iteration process. For each zone, starting with the first one $(j =1)$, the accuracy of the solution is improved by iteration taking into account the dependence (19.23) and is then used as the zero approximation for the following zone. The corresponding algorithm has the form

$$\varepsilon_j^v = R^{-1}(\omega_j^v)^{-1/2} / d_{jj}^v - \sum_{p=1}^{j-1}(d_{jp}^v / d_{jj}^v)\varepsilon_p, \quad j = \overline{1, N},$$

$$d_{jj}^v = (2 / \kappa_j^v)\, sh\, \tau_{jj}^v, \quad d_{jp}^v = (2 / \kappa_p)(sh\, \tau_{j,p-1}^v - sh\, \tau_{jp}^v), \tag{19.24}$$

$$\omega_j^v = \exp(-2\tau_{jp}^v),$$

$$\tau_{jp}^v = \kappa_j^v (r_{j-1}^2 - x_j^2)^{1/2} + \sum_{t=p+1}^{j-1} \kappa_t \left[(r_{t-1}^2 - x_j^2)^{1/2} - (r_t^2 - x_j^2)^{1/2} \right],$$

$$\kappa_j^v = \kappa(\varepsilon_j^{v-1}),$$

$$v = 0, \quad j = 1: \quad d_{11}^0 = 2(R^2 - x_1^2)^{1/2}, \quad \tau_{1,0}^0 = 0, \tag{19.25}$$

$$v = 0, \quad j = \overline{2, N}: \quad \kappa_j^0 = \kappa_{j-1},$$

$$I_j = I(x_j), \quad \kappa_j = \kappa(r_{j-1/2}), \quad \varepsilon_j = \varepsilon(r_{j-1/2}), \quad x_j = r_j = (N - j)/N.$$

As previously, the experimental distribution of the intensity is smoothed out by Chebyshef's polynomials. The dispersion of the calculated results when taking into account self-absorption by iteration is evaluated as follows

$$D(\varepsilon_j) = D(I/R) \sum_{k=0}^{M} q_{2k}^{-1} \left(\sum_{p=1}^{j} f_{jp} P_{2k, p} \right)^2, \quad j = \overline{1, N}, \tag{19.26}$$

where

$$f_{jp} = \sum_{s=p}^{j-1} (d_{js}/d_{jj}) f_{sp}, \quad f_{jj} = \omega_j^{-1/2}/d_{jj},$$

and the values d_{jj}, d_{js}, ω_j are given after completing iterations. Figure 19.3 (curve 3) gives the relative error of iteration restoration of ε (0) on the axis of the source for the example (19.22) calculated using (19.26). The accuracy of restoration of the emission coefficient with self-absorption taken into account by iteration is satisfactory but on the whole is lower than in the case in which we measure not only the intensity (curve 3) but also transparency (curve 2).

19.4 Simplified reduction algorithms

The previously described algorithm of reduction to the optical thin layer is characterised by high stability in relation to random measurement errors, relatively small systematic error and moderate requirements on the calculation time using computers. At the same time, the method of numerical solution of the integral Abel equation are used widely, they suitable both for application in a computer and manual calculations. These methods ensure sufficient accuracy of the results of the calculations for the experimental data obtained with a small error.

Numerical conversion of the Abel transformation

To calculate ε (r), we shall use the conversion of the Abel transformation

$$\varepsilon(r) = -\frac{1}{\pi R}\int_{r}^{1}\frac{dI}{dx}\frac{dx}{\sqrt{x^2 - r^2}}, \tag{19.27}$$

$0 \leq |x| \leq 1$, $0 \leq r \leq 1$, $(r \geq 1) = 0$, I $(|x| \geq 1) = 0$. The integration region in (19.27) $0 \leq x \leq 1$ will be divided into N zones of equal size $[x_j, x_{j+1}]$, $x_j = (N - j)/N$, $j = \overline{1, N}$. I (x) in the zones will be approximated by the interpolation polynomial of the third degree with the exception of the first and third zones where we shall use the interpolation polynomials of the second and fourth (with respect to even exponents x) degree

$$I(x) \approx P(x) = \begin{cases} Ax^2 + Bx + C, & x \in [x_2, x_1], \\ Ax^3 + Bx^2 + Cx + D, & x \in [x_j, x_{j-1}], j = \overline{2, N-1}, \\ Ax^4 + Bx^2 + C, & x \in [x_N, x_{N-1}] \end{cases} \tag{19.28}$$

The coefficients of the polynomial are determined from the condition of coincidence of values of the functions of I (x_j) and P (x_j) at the nodes x_j. The integral in (19.27) will be represented by a sum of the integrals with respect to the zones and I (x) will be replaced by corresponding polynomials P (x):

$$\varepsilon(r_p) = -\frac{1}{\pi R}\sum_{j=1}^{p}\int_{x_j}^{x_{j-1}}\frac{dP}{dx}\frac{dx}{\sqrt{x^2 - r_p^2}}.$$

We shall then carry out analytical integration and reducing with respect to I (x_j), we obtain

$$\varepsilon_p = \frac{1}{R}\sum_{j=1}^{N} g_{pj} I_j, \quad p = \overline{1, N},$$

(19.29)

where

$$\varepsilon_p = \varepsilon(r_p), \quad I_j = I(x_j), \quad r_p = \frac{N-p}{N}, \quad x_j = \frac{N-j}{N}.$$

The coefficient g_{pj} can be computed in advance and tabulated. The dispersion of the results of the calculations used in (19.29) for measurements of equal accuracy without correlations is

$$D(\varepsilon_p) = D(I/R)\sum_{j=1}^{N} g_{pj}^2.$$

(19.30)

The coefficient of transfer of the error (19.12) for this method is relatively high (19.4). Its value for the axis of the source ($r = 0$) is close to the estimate given previously (equation (19.14)). To reduce the effect of random errors without complicating the processing method, we shall carry out local smoothing of the experimental data. We approximate each group of seven points by a polynomial of the third degree using the method of least squares. The corresponding smoothed-out values are

$$\bar{I}(x_i) = \begin{cases} \dfrac{2}{21}I_6 - \dfrac{1}{6}I_5 - \dfrac{2}{21}I_4 + \dfrac{1}{7}I_3 + \dfrac{8}{21}I_2 + \dfrac{19}{42}I_j, \quad j=1, \\[2mm] \dfrac{1}{42}I_6 - \dfrac{2}{21}I_5 + \dfrac{1}{21}I_4 + \dfrac{2}{7}I_3 + \dfrac{19}{42}I_2 + \dfrac{8}{21}I_1, \quad j=2, \\[2mm] -\dfrac{2}{21}I_{j-3} + \dfrac{1}{7}I_{j-2} + \dfrac{2}{7}I_{j-1} + \dfrac{1}{3}I_j + \dfrac{2}{7}I_{j+1} + \dfrac{1}{7}I_{j+2} - \dfrac{2}{21}I_{j+3}, \quad j=\overline{3,7}, \\[2mm] I_{N+3} = I_{N-3}, \quad I_{N+2} = I_{N-2}, \quad I_{N+1} = I_{N-1}. \end{cases}$$

(19.31)

I_j in equation (19.29) will be substituted by the corresponding smooth values of $\bar{I}_j$ from (19.31) and reducing the like with respect to I_j, we obtain

$$\varepsilon_p = R^{-1}\sum_{j=1}^{N} G_{pj} I_j, \quad p = \overline{1, N}.$$

(19.32)

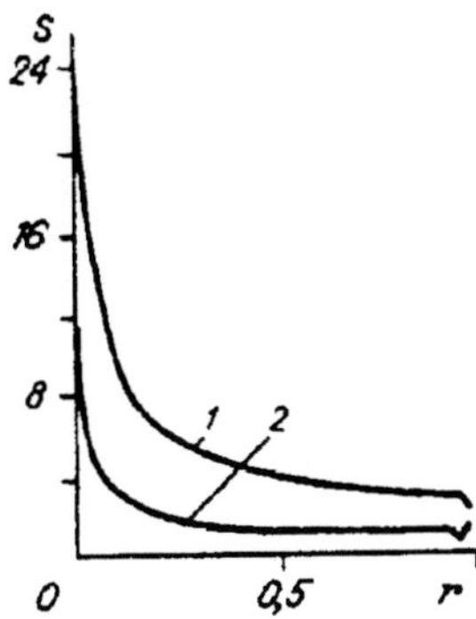

Fig.19.4 Coefficient of transfer of the error for the interpolation method (19.29) of numerical conversion of Abel's transformation (1) and for the improved algorithm (19.33) (2) at $N = 40$.

Consequently, the smoothing procedure is included in the coefficients G_{pj} and this is carried out simultaneously with conversion of the Abel transformation. The values of the coefficients G_{pj} for $N = 40$ are presented in Ref.1. The tabulated coefficients can be used at any number $N' \le N = 40$ of zones of uniform division of the region $[0, R]$. Consequently

$$\varepsilon_p = -\frac{N'}{40R}\sum_{j=1}^{N'} G_{pj} I_j,$$

$$x_j = (N' - j)/N', \quad r_p = (N' - p)/N', \quad j, p = \overline{0, N'-1}. \tag{19.33}$$

The dispersion of the calculated results using equation (19.33) for non-correlated measurements of equal accuracy is

$$D(\varepsilon_p) = D(I/R)\sum_{j=1}^{N'} G_{pj}^2, \tag{19.34}$$

The corresponding coefficient of transfer of the error (19.12) is shown in Fig.19.4. The addition of the smoothing procedure to the calculation method decreases the error of the calculated results 2–3 times.

Processing schlieren measurements
The radial distribution of the refractive index in schlieren measurements is calculated using the equation

$$n(r) - n_R = \frac{1}{\pi} \int\limits_{r}^{1} \frac{\theta(x)dx}{\sqrt{x^2 - r^2}}, \qquad\qquad (19.35)$$

where

$$0 \le r \le 1, \quad 0 \le |x| \le 1, \quad n(r \ge 1) = n_R, \quad \theta(|x| \ge 1) = 0.$$

The integration region $0 \le x \le 1$ will be divided into N zones of equal size. $\theta(x)$ in the zones will be approximated by interpolation polynomials (19.28). Consequently, after the corresponding transformations and calculations we obtain

$$n_p - n_R = \sum_{j=1}^{p+1} C_{pj}\theta_j, \quad p = \overline{1, N-1}, \qquad\qquad (19.36)$$

where

$$n_p = n(r_p), \quad \theta_j = \theta(x_j), \quad r_p = \frac{N-p}{N}, \quad x_j = \frac{N-j}{N}.$$

The coefficients C_{pj} at $N = 40$ are presented in Ref.1. The coefficients of transfer of the error are given in the same reference

$$S_p^2 = D(n_p - n_R) / D(\theta),$$

where the dispersion of the calculated results follows from equation (19.36) and is equal to

$$D(n_p - n_R) = D(\theta) \sum_{j=1}^{p+1} C_{pj}^2. \qquad\qquad (19.37)$$

The numerical values of S_p are lower than unity even without using the smoothing procedure of the initial data. This is due to the fact that in the schlieren method we measured the angle of deviation of the light beam whose value is determined by the radial gradient of the refractive index. The result is that in the conversion of the Abel transformation (19.35) there is no differentiation of the experimental function. Consequently, it is the presence of the derivative from the experimental function that determines the instability of the numerical conversion of the Abel transformation (19.24). Naturally, the same instability is also manifested when solving the Abel integral transformation (19.5).

Chapter 20

RECONSTRUCTION OF VELOCITY DISTRIBUTION FUNCTIONS OF EMITTING PARTICLES FROM THE SHAPE OF THE CONTOUR OF SPECTRAL LINES

Recently, the attention of the investigators has been attracted by examination of various types of non-equilibrium processes and gases in plasma. This is due to both the development of a large number of applications where non-equilibrium phenomena are observed to various degrees (physics of gas phases, dynamics of rarefied gases, plasma chemistry, astrophysics, etc.) and also to the solution of purely fundamental problems in the nature of various physical phenomena.[1,2] Investigations of this type relate to both the kinetics of formation of the energy distributions of the internal state of the molecules – electronic, vibrational and rotational, and also the kinetics of translational energy.[2] It should be noted that whereas the experimental facts of manifestation of strong nonequilibria in internal degrees of freedom of molecules are already well known (see, for example, (Ref.23)), the investigations of nonequilibria in the velocity distributions have been carried out only recently. Obviously, this circumstance reflects the fact that the processes of the translational relaxation of heavy particles takes place with the highest speed in comparison with the speed of all previously mentioned internal degrees of freedom and, usually, is already completed when the latter start to relax.[2,3] It is therefore interesting to examine the special features of the translational non-equilibrium distributions and the rates of their relaxation to the equilibrium Maxwell distribution. At the same time, there are two main experimental methods of measuring the velocity distribution functions of heavy particles: the method of Doppler broadening of the spectral lines, and the time-of-flight method. Without discussing the advantages and disadvantages of these methods, we shall mention only the special features of obtaining the velocity distributions from the values mentioned in the experiments.

1. The method of Doppler broadening of the spectral lines is based on extracting information from the measurements of the shape of the

538

contours of the lines of atoms and molecules. It is well known that the distribution of intensities in the Doppler contour of the line for an homogeneous isotropic medium is associated with the function of distribution of emitters with respect to the absolute values of the velocity $P(v)$ by the following integral equation:[4,5]

$$\int\limits_{\mu}^{\infty} \frac{P(v)}{v} \, dv = \varphi(\nu),$$

(20.1)

where $\mu = \dfrac{|\nu - \nu_0|}{\nu_0} c$, ν_0 is the emission frequency of a stationary particle, c is the velocity of light.

However, in practice, the actual contour of the spectral line $\varphi(\nu)$ is not measured in the explicit form but as a convolution with some apparatus function $a(\nu)$ of the measuring system (as, for example, in examining the contours of the lines using a Fabry–Perot interferometer[6]

$$\int\limits_{-\infty}^{\infty} a(\nu - \nu')\varphi(\nu')d\nu' = f(\nu),$$

(20.2)

where $f(\nu)$ is the measured form of the contour of the spectral line.

2. Within the framework of the time-of-flight method of measuring the distribution functions of the velocities of the atoms and molecules the true distribution function $P(\nu)$ is also treated[7] as only a convolution with the apparatus function of the formation system and recording of the beam $H(\nu)$ and, consequently, the task of determining this function is reduced to solving the integral equation of the type (20.2)[7,8]

$$\int\limits_{-\infty}^{\infty} H(\nu - \nu')P(\nu')d\nu' = R(\nu),$$

(20.3)

where $R(\nu)$ is the measured time-of-flight signal.

Thus, it can be seen that when using both methods of measuring the distribution function it is necessary to solve the integral equations of the first kind: in the first method – both equation (20.1), (20.2) and in the second method the equation (20.3). It is important to note that in the experiment the functions $f(v)$ (in the first method) and $R(v)$ (in the second method) are measured unavoidably with an error which is different from zero, i.e. determination of the function $P(v)$ from the equations (20.1)–(20.3) is an inverse incorrect problem.[9] The solution of these problems is especially complicated when for the velocity distribution function it is not possible to show in advance[9] the case in which the shape of $P(v)$ differs from the Maxwell distribution.

At the same time, these situations often occur in practice in experiments in measuring the velocity distribution function of heavy particles in different non-equilibrium objects (see, for example, Ref.10, 13).

In recent years, to solve the inverse incorrect problems of the type (20.1) and (20.2), (20.3) various investigators have used successfully the regularisation methods.[9,14] The success with application of these methods is attributed to a large degree to the use of the existing *apriori* information on the required solution, for example, the properties of smoothness of the required solution.[9,14] However, in complicated cases, this information may be insufficient to find the solution with the required accuracy. Additional information from these problems can be represented by the information indicating that the required solution is similar to some unknown function.

In particular, the calculations carried out in Ref.15 show that in a number of cases it is highly efficient to use the multiplicative representation of the required function in the form $\varphi(v) = \varphi_0(v)\varphi_1$, where $\varphi_0(v)$ is some zero approximation for $\varphi(v)$. As indicated by Ref.15, this approach gives more accurate results in comparison with the conventional method.

However, the application of the multiplicative representation in the form of the difference kernel of the initial equation (as was done in, for example, (20.2)) does not make it possible to solve the problem by the Fourier analysis method. In fact, this gives a new integral equation of the first kind with a kernel $a_1(v, v') = a(v - v')\varphi(v')$ which is no longer of the difference type.

The authors of Ref.16 proposed not a multiplicative but additive approximation of the required solution in the form $\varphi(v) = \psi(v) + \varphi_0(v)$, where $\varphi_0(v)$ is some function known from *apriori* assumptions. It can be shown that this approximation, in contrast to Ref.15, does not change the difference nature of the kernel of the initial integral equation. This is important for solving the problem by the Fourier analysis methods. At the same time, application of the Fourier variant of the regularisation of the logarithm in practice is highly efficient for solving the integral equations of the first kind with a difference kernel (see, for example, Ref.9).

It can easily be seen that the additive approximation of the function $\varphi(v)$ is equivalent to the method of a test solution (see, for example, Ref.17) when in searching for the regular solution of equation (20.2) we use the stabilising potential not with respect to the $\varphi(v)$ function but the difference of the function $\varphi(v) - \varphi_0(v)$.

In this work, we consider the additional *apriori* information according to which the examined distribution is similar to Maxwell's distribution (which in a number of cases is confirmed in practice[16]): the required solution of $P(v)$ is presented in the form of a sum

$$P(v) = P_0(v) + Q(v), \tag{20.4}$$

where $P_0(v)$ is the given Maxwell distribution, $Q(v)$ is some 'addition' to $P_0(v)$. After substituting (20.4) into (20.1) and (20.2), we obtain a system of equations

$$\int_{-\infty}^{\infty} a(v-v_1)\psi(v_1)\,dv_1 - f(v) - \int_{-\infty}^{\infty} a(v-v_1)\int_{\mu_1}^{\infty} \frac{P_0(v)}{v}\,dv\,dv, \tag{20.5}$$

$$\int_{\mu}^{\infty} \frac{Q(v)}{v}\,dv = \psi(v). \tag{20.6}$$

To explain the efficiency of the additive approximation of (20.4), we carried out comparative modelling calculations in two modifications of the Fourier-variant of the Tikhonov's method:[9]

1) Normal modification where the required distribution $P(v)$ is determined directly from the equations (20.1) and (20.2);

2) Consecutive, using the additive approximating function $P_0(v)$ from (20.4).

The modelling problems were solved using the following schema:

– we specify the velocity distribution of particles $P(v)$;

– we determine the distribution of the intensities in the corresponding contour of the spectral line $\varphi(v) = const \cdot \int_{\mu}^{\infty} \frac{P(v)}{v}\,dv; \quad \mu = \frac{|v-v_0|}{v_0}c;$

– this distribution was 'rolled up' together with the known apparatus function a (v) and the function $f_0(v) = \int_{-\infty}^{\infty} a(v-v')\varphi(v')dv'$ was calculated;

– function f_0 (v) was added to the random function ε (v) (modelling noise) distributed with dispersion σ_f^2 in accordance with the normal law f_0 (v) + ε (v) = f (v).

The inverse problem was solved directly for this purpose:

– at a known f (v) and a (v), we used (20.2) to calculate the function $\varphi_\alpha(v)$;

– $\varphi_\alpha(v)$ was differentiated with respect to gap in accordance with

(20.1) thus giving the function $P_\alpha(v) = \text{const} \cdot v \cdot \left. \dfrac{d\varphi_\alpha(v)}{dv} \right|_{v=v_0 \cdot \frac{1}{1-v/c}}.$

The error of restoration of the solution was calculated from the deviation of the numerical solution P_α (v) from the given P (v)*; (in reality, functions P (v) and P (v)/v are compared instead of P (v) and P (v) because the similarity of these functions indicates the quality of restoration in both regularisation stages)

The modelling distribution P (v) was represented by the distribution consisting of the sum of Maxwell's distributions with differing temperatures and amplitude of the maxima

$$P(v) = A_1 v^2 \exp\left\{ -\frac{1}{2}\frac{v^2}{T} \right\} + A_2 v^2 \exp\left\{ -\frac{1}{2}\left(\frac{v}{3}\right)\frac{1}{T} \right\}. \tag{20.7}$$

This case was examined to model the presence of the possible structure in the distributions.

To determine the function P (v) (or Q (v)) from the available values of $\varphi(v)$ (or $\psi(v)$), we used the differential variant of equation (20.1) (or (20.6)): calculation of derivatives $\varphi'(v)$ (or $\psi'(v)$) was carried out using smoothing cubic splines with the selection of the smoothing parameter using the discrepancy.[18,19]

It is interesting to compare the efficiency of restoring the velocity distribution function using the conventional and consecutive restoration procedures. Figures 20.1 and 20.2 show the modelling and restored velocity distribution functions for different numbers of counting N and noise levels σ_f^2. The ratio of the widths of functions $\varphi(v)$ and $\alpha(v)$ in all cases is $\Delta_1 : \Delta_2 = 1.8$, the ratio of the amplitudes $A_1 : A_2 = 10$. Comparison shows that, with other conditions being equal, the procedure

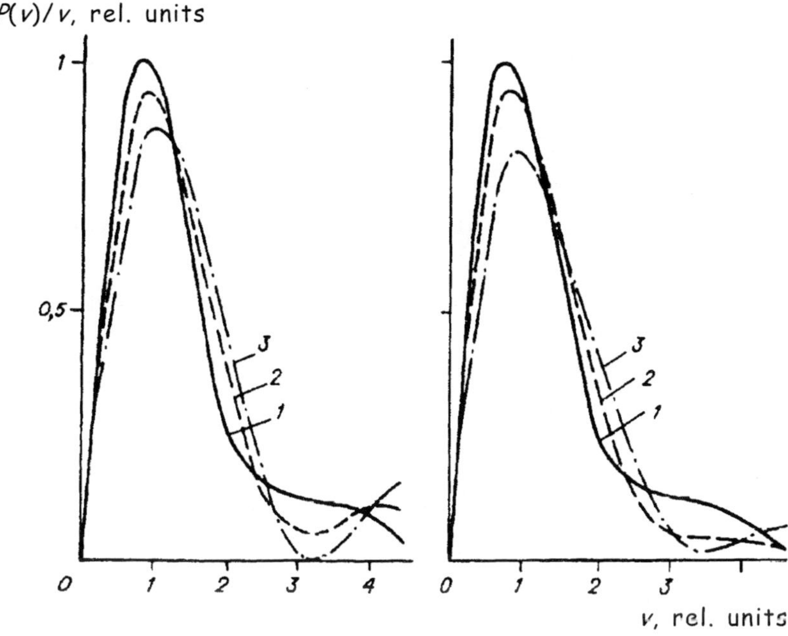

Fig.20.1 Modelling (1) and restored (2,3) distribution functions. 2) using additive approximation, 3) conventional restoration; a) Δ_1: Δ_2 = 1.8; A_1:A_2 = 1 : 10; N = 32; σ_f = 5%. b) Δ_1:Δ_2 = 1.8; A_1:A_2 = 1:10; N = 32; σ_f = 2%.

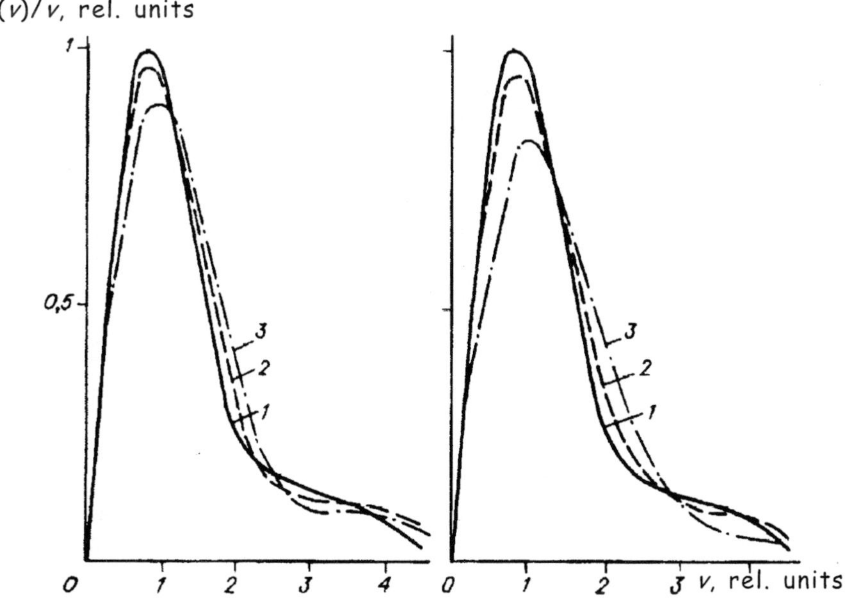

Fig.20.2 Modelling (1) and restored (2,3) distribution of functions. 2) using additive approximation, 3) conventional restoration; a) Δ_1:Δ_2 = 1.8; A_1:A_2 = 1:10; N = 128; σ_f = 2%. b) Δ_1: Δ_2 = 1.8; A_1:A_2 = 1:10; N = 128; σ_f = 5%.

of successive restoration of the 'structural' distributions of the type (20.7) is far more efficient that the conventional procedure, especially in the relevant region of high velocities. The controlling factor in this case is that it was possible to use more efficiently the *apriori* information on the initial solution.

The application of the method of successive restoration enabled the authors[12,16] to restore efficiently the velocity distribution function of the excited nitrogen molecules[16] and oxygen atoms.[12]

In conclusion, it should be stressed again that the application of the proposed method of additive approximation is especially efficient when restoring various types of non-equilibrium velocity distribution functions, for example, functions with singularities at the 'tails', functions with discontinuities in the derivatives, *etc*. However, as already mentioned, the restoration of these function under the non-equilibrium conditions is a very important task.

Chapter 21

AUTOMATION OF MEASUREMENTS IN PLASMA DIAGNOSTICS

Recently, a number of publications concerned with the development of automated systems of experimental investigations of physico-chemical processes under low-temperature plasma conditions has increased.[1-10] The main tasks of using these systems are automated collection, processing, build-up and imagining of information and, if necessary, controls of the experiments.

Original investigations were carried out in this country into the systems of collecting experimental information and examining the flows of high-temperature gases generated by electric arc heaters.[1-3] The experience obtained in constructing and using these systems has made it possible to formulate a number of important requirements of measuring apparatus, a selection of noise-resistant temperature, pressure and flow rate sensors ensuring reliable measurements under the conditions of the effect of strong electric and magnetic fields on them. It was thus possible to develop effective electronic-measuring systems using micro- and mini-computers for automating multi-functional apparatus and organising the multi-channel collection of information under the conditions with more stringent requirements on measurement accuracy. In addition, the application of computers in the composition of measuring complexes makes it possible to operate under their real time scale which opens wide possibilities of development and improvement of the measuring systems.

For example, the authors of Ref.4 described a measuring system for recording and processing the spectra of spatially non-uniform objects in examining stationary and high-rate processes. This system can also be used in spectroscopic investigations and plasma diagnostics of continuous and pulsed electric and optical discharges.

The automated system of photoelectric recording on the basis of the IVK-1 measuring-computing complex has been developed in the conditions of time correlation and spatial-time measurements and was described in Ref.5. For time correlation measurements, the radiation receiver is connected to the output of several monochromators or to a quantometer. For spatial–time measurements, the radiation from different areas of the plasma object is fed to the input of the receiver using a set of lightguides.

The authors of Refs.6 and 7 report on the results of development and application of a measuring system for determining the main thermophysical and gas-dynamic characteristics of the plasma flows and electric arcs.

The authors of Ref.8 described a simple system of collecting the data using a microcomputer for recording the volt–ampere characteristics of a Langmuir's probe in stationary glow discharge plasma. Evidently, it can be used as a basic diagnostic complex by investigating the stationary plasma. It makes it possible to control automatically the plasma parameters during measurements.

The authors of Ref.9 describe equipment based on expanded complex of the M400 control and computing system and communication means in the KAMAK standard for automating the collection and processing diagnostic information and also the programmed control of the plasma formation conditions in the experiments with the interaction of intensive electromagnetic waves and electron beams with low-temperature plasma. Equipment makes it possible to carry out automated experiments in examining the development of parametric instabilities in plasma and the dependence of the spatial–time structure of microwave fields on the plasma parameters.

In Ref.10, the authors developed a system of controlling the generator of low-temperature plasma and carried out experiments aimed at controlling and regulating the regime parameters of equipment, recording and processing the experimental data. This system was used for investigating the pulsation characteristics of the current and voltage of the plasma torch and turbulent pulsations in the plasma jet using stationary and flight-path electrostatic probes.

Diagnostic equipment developed in line a computer is used at present in laboratory investigations of processes in electric arcs, plasma jets of complicated chemical composition (both single-phase and those carrying particles of inertia impurities) and other plasma objects may become an efficient means of inspection of complicated plasma and plasma chemical technological processes. This is convincingly confirmed by Ref.11 in which the authors described an optical apparatus fitted with microprocessor technology and used for monitoring the plasma jet with a dispersed phase and the electrons of the plasma torch on the real time scale.

Until recently a computer was a very expensive tool for automating 'routine' investigations;[12] the situation has greatly changed because of the appearance of relatively cheap and compact microprocessor devices having the possibilities of minicomputers of previous generations; combination of microprocessor technology and KAMAK apparatus has opened new prospects for the mass automation of laboratory investigations. The production of the first Russian microcomputer Elektronika-60 created suitable conditions for the development and application of automated systems of a new type on the basis of microprocessor technology and equipment of the KAMAK type (Micro-Kamak-lab),[13] designed for complex automation of research laboratories.[12] These systems offer the investigator in the working area various possibilities, e.g. collection of experimental data, primary processing of these data,

REFERENCES

Preface
1   G. Grim, Spectroscopy of plasma, Atomizdat, Moscow (1969).
2   R. Hoddlestone and S. Leonard (eds), Plasma diagnostics, Mir, Moscow (1967).
3   V. Locht-Holtgreven, Methods of plasma examination, Mir, Moscow (1971).
4   S.E Frish (ed), Spectroscopy of as-discharged plasma, Nauka, Leningrad (1970).
5   L.A. Dushin and O.S. Pavlichenko, Examination of plasma using lasers, Atomizdat, Moscow (1968).
6   N.G. Preobrazhenskii, Spectroscopy of optical dense plasma, Siberian division of Nauka Publishing House, Novosibirsk (1971).
7   L.N. Pyatnitskii, Laser plasma diagnostics, Atomizdat, Moscow (1976).
8   Yu.A. Ivanov, Yu.A. Lebedev and L.S. Polak, Contact methods of diagnostics in non-equilibrium plasma chemistry, Nauka, Moscow (1981).
9   P. Chang, L. Tolbot and K. Turyan, Electrical probes in stationary and moving plasma, Mir, Moscow (1978).
10   V.K. Zhivotov, V.D. Rusanov and A.A. Fridman, Diagnostics of non-equilibrium chemically active plasma, Energoatomizdat, Moscow (1985).

Chapter 1
1   L.T. Lar'kina and V.S. Engel'sht, Reduction to a homogenous optically thin layer in axisymmetric objects, VINITI, Moscow (1973), Manuscript deposited at VINITI No.6917-73 (1973).
2   A.S. An'shakov, G.Yu. Dautov and A.P. Petrov, Generators of low-temperature plasma, Energiya, Moscow (1969).
3   N.G. Kolesnikov, L.T. Lar'kina and V.S. Engel'sht, Izv. SO AN SSSR, Ser. Tekhn. Nauk., No.3, Issue No.1, 31-35 (1974).
4   A. Locht-Holtgreven (ed), Methods of plasma examination, Mir, Moscow (1971).
5   G. Grim, Plasma spectroscopy, Atomizdat, Moscow (1969).
6   Zh. Zheenbaev and V.S. Engel'sht, Laminar plasma torch, Ilim, Frunze (1975).
7   L.A. Spektorov and L.A. Sinel'nikova, Determination of arial distribution of temperature in axisymmetric plasma sources, Moscow (1972), Manuscript deposited at VINITI No.4902-72 (1972).
8   V.S. Engel'sht, Distribution of atoms of sodium and copper in the cloud of a dc plasma, works of the Department of Physics, Kirgiz State University, Frunze (1961).
9   Examination of the electric arc in argon, Ilim, Frunze (1966).
10   N.G. Preobrazhenskii, Nukleonika, **20**, No.5, 429-438 (1975).
11   I.F. Garapova, A.I. Dautov, N.G. Sabitova, et al, Fiz. Khim. Obrab. Mater., No.6, 50-55 (1977).
12   M.Kh. Salakhov, I.D. Grachev and R.Z. Lapitov, ZhPS, **41**, No.4, 558-560 (1984).
13   G.P. Luzunkov, V.P. Kabashnikov, A.A. Kurskov and V.D. Shimanovich, Methods of spectroscopy diagnostics of non-stationary arc plasma on the basis of modelling its oscillations, Preprint of the Institute of Physics of the Belrussian Academy of Sciences, Minsk (1986).

Chapter 2
1   V.K. Zhivotov, V.D. Rusanov and A.A. Fridman, Diagnostics of non-equilibrium chemically active plasma, Energoatomizdat, Moscow (1985).
2   T.T. Karasheva, D.K. Otorbaev and V.N. Ochkin, et al, Electronically excited molecules in non-equilibrium plasma, Nauka, Moscow (1985).
3   L.A. Vainshtein, I.I. Sobel'man and E.A. Yukov, Excitation of atoms and broadening of spectral lines, Nauka, Moscow (1979).
4   V.A .Dudkin, Optika i Spektroskopiya, **24**, 367-371 (1968).
5   G.N. Polyakova and A.I. Ranyuk, Prepr. KhFTI, No.81, Kar'kov (1981).
6   A.N. Tikhonov and V.Ya. Arsenin, Methods of solving incorrect problems, Nauka,

References

Moscow (1979).

7   V.F. Turchin, V.P. Kozlov and M.S. Malkevich, Usp. Fiz. Nauk, **102**, 345-386 (1970).

8   A.V. Kryanev, ZhVMiMF, **14**, 25-35 (1974).

9   D.K. Otorbaev, V.N. Ochkin, N.G. Preobrazhenskii, et al, Prepr. FIAN SSSR, No.39, Moscow (1981).

10   D.K. Otorbaev, V.N. Ochkin, N.G. Preobrazhenskii, et al, ZhETF, **81**, 1626-1638 (1981).

11   L.D. Landau, E.M. Lifshits, Quantum mechanics, non-relativistic theory, Nauka, Moscow (1974).

12   S.Yu. Savinov and S.N. Tskhai, Krat. Sobshch. Fiz., No.9, 3-7 (1983).

13   E.G. Whipple, J. Chem. Phys., **60**, 1345-1351 (1974).

14   T.T. Karasheva, M. Malikov, D.K. Otorbaev, et al, Prepr. FIAN SSSR, No.193, Moscow (1986).

15   T.T. Karasheva, D.K. Otorbaev, V.N. Ochkin, et al, Krat. Soobshch. Fiz., No.10, 53-58 (1982).

16   M.S. Ivanov, T.T. Karasheva, D.K. Otorbaev, et al, Prepr. FIAN SSSR, Moscow, No.80 (1987).

17   A.M. Sravath, Phys. Rev., **36**, 248-250 (1930).

18   I. MacDaniel', Processes of collision in ionized gases, Mir, Moscow (1962).

19   V.N. Ochkin, S.Yu. Savinov, N.N. Sobolev, ZhETF, **75**, 463-472 (1978).

20   D.I. Slobetskii, Mechanisms of chemical reactions in non-equilibrium plasma, Nauka, Moscow (1980).

Chapter 3

1   L.S. Polak, D.I. Slovetskii, A.S. Sokolov and I.V. Fedoseeva, Non-equilibrium chemical kinetics and its application, Nauka, Moscow (1979).

2   J.W. Coburn and M. Chen, J. Appl. Phys., **51**, No.6 (1980).

3   R. D'Agostino, F. Cramarossa, S. Benedictis and G. Ferraro, Ibid, **52**, No.3, 1259-1265 (1981).

4   L.A. Vainshtein, I.I. Sobel'man and E.A. Yukov, Excitation of atoms and broadening of spectral lines, Nauka, Moscow (1979).

5   D.I. Slobetskii, Mechanisms of chemical reactions in non-equilibrium plasma, Nauka, Moscow (1980).

6   A.J. Williams and J.P. Doering, Planet. Space Sci., **17**, No.11, 1527-1537 (1969).

7   V.V. Skubenich, M.V. Povch and I.P. Zapesochnii, Optika i Spektroskopiya, **11**, No.3, 116-120 (1977).

8   I.S. Polak, D.I. Slovetskii, A.S. Sokolov and T.V. Fedoseeva, Experimental and theoretical examination of non-equilibrium physico-chemical proceses, Vol.3, Nauka, Moscow (1974).

9   C.J. Mogab, A.C. Adams and D.L. Flamm, J. Appl. Phys., **49**, No.7, 3796-3803 (1978).

10   V.M. Donnelly, D.L. Flamm, W.S. Dautremont-Smith and D.J. Werder, Ibid, **55**, No.1, 242-252 (1984).

11   D.E. Ibbotson, D.L. Flamm and V.M. Donnelly, Ibid, **54**, No.10, 5974-5981 (1983).

12   R.A. Gottscho, G.P. Davis and R.H. Burton, Vac. Sci. Technology, **1A**, No.3, 622-624 (1983).

13   R.A. Gottscho and V.M. Donnelly, J. Appl. Phys., **56**, No.2, 245-250 (1984).

14   R. d'Agostino, V. Colaprico, F. Cramaross, et al, 7th ESCAMPIG, Abstracts of papers, Bari, Italy (1984), pp.107-108.

15   A. Ricard, 17th ICPIG, Abstracts of invited lectures, Budapest (1985).

16   R. d'Agostino, F. Cramarossa, V. Colaprico and R. d'Ettole, J. Appl. Phys., **54**, No.3, 1284-1288 (1983).

17   R.A. Gottscho, G. Smolinsky and R.H. Burton, Ibid, **53**, No.8, 5908-5918 (1982).

18   D.L. Flamm, V.M. Donnelly and J.A. Mucha, Ibid, **52**, No.5, 3633-3639 (1981).

19   T.T. Karasheva, D.K. Otorbaev, V.N. Ochkin, et al, Tr. FIAN, Vol.157, Nauka, Moscow (1985).

20   V.I. Gorokhovskii, G.N. Kurtynina and D.K. Otorbaev, Khim. Vysok. Energii, **23**, No.5, 450-455 (1989).

21 V.I. Gorokhovskii and D.K. Otorbaev, Fiz. Khim. Obrab. Mater., No.2, 51-54 (1989).
22 N.Zh. Zheenbaev, M.Z. Mamytbekov and D.K. Otorbaev, ZhPS, **51**, No.1, 12-16 (1989).
23 A.T. Baiterekov, K.A. Imankulov and D.K. Otorbaev, Production, examination and application of plasma in microwave fields, Ilim, Frunze (1987).

Chapter 4
1 L.N. Pyatnitskii, Laser plasma diagnostics, Atomizdat, Moscow (1976).
2 A.N. Zaidel' and G.V. Ostrovskaya, Laser methods of plasma investigation, Nauka, Leningrad (1977).
3 Yu.I. Ostrovskii, M.M. Butusov and G.V. Ostrovskaya, Holographic interferrometry, Nauka, Moscow (1977).
4 G.V. Ostrovskaya and Ju.I. Ostrovsky, Progress in optics, **22**, 199-270 (1985).
5 A.N. Zaidel', UFN, **149**, No.1, 105-138 (1986).
6 G.T. Razdobarin and D.A. Shcheglov, Plasma diagnostics, No.6, Energoatomizdat, Moscow (1989), pp.88-101.
7 Kh.P. Alum, Yu.V. Koval'chuk an G.V. Ostrovskaya, ZhTF, **54**, No.5, 896-903 (1984).
8 K.W. Allen, Astrophysical quantities, Mir, Moscow (1977).
9 G.V. Ostrovskaya, Kh.P. Alum, Yu.V. Koval'chuk, Dispersion interferrometer, Author's certificate No.864942 SSSR, Published priority since, 29.04.80, Bull. Izobr., No.35 (1981).
10 F.A. Hopf, A. Tomito and G. Al-Jumaily, Opt. Lett. , **5**, No.5, 386-388 (1980).
11 Kh.P. Alum, Yu.V. Koval'chuk and G.V. Ostrovskaya, Pis'ma ZhTF, **7**, No.22, 1359-1364 (1981).
12 Yu.I. Ostrovskii, Author's certificate No.268732, Published priority since 16.01.61, Bull. Izobr., No.14 (1970).
13 A.F. Belozerov, K.S. Mustafin, A.I. Salykova, et al, Optika and Spektroskopiya, **29**, No.2, 384-389 (1970).
14 K.M. Measures, Appl. Opt., **9**, No.3, 737-741 (1970).
15 G.V. Dreiden, A.N. Zaidel', G.V. Ostrovskaya, Fiz. Plazmy, **1**, no.3, 462-482 (1975).
16 G.V. Dreiden, G.V. Ostrovskaya, et al, Fiz. Plazmy, **1**, no.3, 462-482 (1975).
17 G.V. Ostrovskaya and N.A. Pobedonstseva, ZhTF, **45**, No.7, 1462-1469 (1975).
18 B.E. Golant, Super-high frequency methods of plasma examination, Nauka, Moscow (1968).
19 D.E.T.F. Ashbyand D.F. Jephcott, Appl. Phys. Lett., **3**, No.1, 13-16 (1963).
20 E.B. Hooper and G. Bekefi, Ibid, **7**, No.5, 133-135 (1965).
21 H. Herold and F.C. Jahoda, Rev. Sci. Instr., **40**, No.1, 145-147 (1969).
22 J.B. Gerardo and J.T. Verdeyen, Appl. Phys. Lett., **3**, No.6, 121-123 (1963).
23 J.B. Gerardo and J.T. Verdeyen, Proc. IEEE, **52**, No.6, 690-698 (1964).
24 J.B. Gerardo, J.T. Verdeyen and M.A. Gusinov, J. Appl. Phys., **36**, No.7, 2146-2151 (1965).
25 D.E.T.J. Ashby, D.F. Jephcott, A. Malein and F.A. Raynor, Ibid, **36**, No.1, 29-34.
26 P.A. Miller, J.T. Verdeyen and B.E. Cerrington, Phys. Rev., **A4**, No.2, 692-700 (1971).
27 L.A. Schlie and J.T. Verdeyen, IEEE J. Quant. Electr., **QE-5**, No.1, 21-29 (1969).
28 V.V. Korobkin, Plasma diagnostics, No.1, Atomizdat, Moscow (1963), pp.36-41.
29 J.T. Verdeyen, B.E. Cherrington and M.E. Fein, Appl. Phys. Lett., **9**, No.11, 360-361 (1966).
30 L.A. Dushin, O.S. Pavlichenko, V.I. Privezentsev and V.G. Yakovlev, Plasma Diagnostics, No.2, Atomizdat, Moscow (1968).
31 L.A. Dushin and V.I. Privezentsev, Teplofiz. Vysokikh Temp., **8**, No.1, 159-162 (1970).
32 L.A. Dushin and O.S. Pavlichenko, Examination of plasma using lasers, Energiya, Moscow (1968).
33 G.V. Ostrovskaya and Yu.I. Ostrovskii, Pis'ma ZhTF, **4**, No.4, 121-124 (1966).
34 A.N. Zaidel', G.V. Ostrovskaya, Yu.I. Ostrovskii and T.Ya. Chelidze, ZhTF, **36**, No.12, 2208-2210 (1966).

References

35   P.W. Chen and C.S. Lee, Phys. Lett., **62A**, No.1, 33-35 (1977).

36   A.G. Smirnov, Optical holography and its applications, LDNTP, Leningrad (1974).

37   S.M. Belotserkovskii, V.S. Sukhorukikh and V.S. Tatarenchik, ZhPMTF, No.3, 95-99 (1964).

38   R.D. Matulka and D.J. Collins, J. Appl. Phys., **42**, No.3, 1109-1119 (1971).

39   K. Iwata and R. Nagata, JOSA, **60**, No.1, 133-135 (1970).

40   D.W. Sweeney and C.M. Vest, Appl. Opt., **11**, No.1, 205-207 (1972).

41   D.W. Sweeney and C.M. Vest, Ibid, **12**, No.11, 2649-2664.

42   V.M. Ginzburg and B.M. Stepanov (eds), Holography. Methods and equipment. Sov. Radio, Moscow (1974).

43   C.M. Vest, Holographic Interferometry, Wiley, New York (1979).

44   R.E. Brooks, L.O. Heflinger and R.G. Wuerker, IEEE J. Quant. Electr., **QE-2**, No.8, 275-279 (1966).

45   C.M. Vest and D.W. Sweeney, Appl. Opt., **9**, No.10, 2321-2325 (1970).

46   M.M. Butusov, Proc. of the 3rd Nat. School of Holography, LIYaF, Leningrad (1972), pp.195-203.

47   F. Grible, W.E. Quinn and R.E. Siemon, Phys. Fluids., **14**, No.9, 2042-2047 (1971).

48   K.S. Thomas, C.R. Harder, W.E. Quinn and R.E. Siemon, Ibid, **15**, No.10, 1658-1666 (1972).

49   L.V. Dubovoi, A.G. Smirnov, V.G. Smirnov and D.I. Stasel'ko, Optical holography and its applications, LDNTP, Leningrad (1974).

50   Yu.V. Ascheulov, A.A. Mymnikov, Yu.I. Ostrovsky and A.N. Zaidel, Phys. Lett., **25A**, No.1, 61-62 (1967).

51   V.A. Nikashin, G.I. Rukman, V.K. Sakharov and V.K. Tarasov, Teplof. Vysokikh Temp., **7**, No.6, 1198-1200 (1969).

52   V.M. Ginzburg, B.M. Stepanov and Yu.I. Filenko, RE, **17**, No.10, 2219-2220 (1972).

53   J.M. Gates, Nature, **220**, 473-474 (1968).

54   A.K. Beketova, L.T. Mustafina and A.Ya. Smolyak, Optika i Spetroskopiya, **39**, No.2, 336-339 (1975).

55   G.V. Dreiden, Yu.I. Ostrovsky, E.N. Shedova and A.N. Zaidel, Opt. Comm., **46**, No.3, 209-213 (1971).

56   G.V. Dreiden, A.N. Zaidel', Yu.I. Ostrovskii and E.N. Shedova, ZhTF, **43**, No.7, 1537-1542 (1973).

57   D.W. Koopman, H.J. Siebeneck, G. Jellison and W.G. Niessen, Rev. Sci. Instrum., **49**, No.4, 524-525 (1978).

58   A.M. Mirzabekov, N.K. Mitrofanov, Yu.I. Ostrovskii and E.N. Shedova, ZhTF, **51**, No.10, 2038-2042 (1981).

59   A.F. Belozerov, A.N. Berezkin, L.T. Mustafina and A.I. Razumovskaya, Pis'ma ZhTF, **4**, No.9/10, 522-525 (1978).

60   F. Friedrich, O.M. Weigl and A.A. Dougal, IEEE J. Quant. Electr., **QE-5**, No.6, 360-361 (1969).

61   F. Weigl, O.M. Friedrich and A.A. Dougal, Ibid, **QE-6**, No.1, 41-49 (1970).

62   I.I. Komissarova, G.V. Ostrovskaya, V.N. Filippov and E.N. Shedova, Author's certificate No.1028152, Bull. Izobr., No.32 (1985).

63   I.I. Komissarova, G.V. Ostrovskaya, V.N. Filippov and E.N. Shedova, ZhTF, **57**, No.2, 377-380 (1987).

64   O. Bryngdahl and A.W. Lohman, JOSA, **58**, No.10, 1325-1334 (1968).

65   K.S. Mustafin, V.A. Seleznev and E.I. Shtyrkov, Optika i Spektroskopiya, **28**, No.6, 1186-1189 (1970).

66   G.I. Mishin (ed), Holographic interferometry of phase objects, Nauka, Leningrad Division, Leningrad (1979).

67   D. Apostol, D. Barbulesky, I.I. Komissarova, et al, ZhTF, **58**, No.11, 2156-2161 (1988).

68   A.A. Bugaev, B.P. Zakharchenya and Chudnovskii, Phase transition between the semiconductor and metal and its applications, Nauka, Moscow (1979).

69   B.P. Zakharchenya, F.A. Chudnovskii and Z.I. Shteingol'ts, Pis'ma ZhTF, **9**, No.2, 76-78 (1983).

70   G.F. Frasier, T.D. Wilkerson and J.M. Lindsey, Appl. Opt., **15**, No.6, 1350-1352 (1976).
71   G.R. Mitchel, B. Grek, T.W. Johnston, et al, Ibid, **18**, No.14, 2422-2426 (1979).
72   D. Naor, A. Flusberg and I. Itzkan, Ibid, **20**, No.14, 2574-2584 (1981).
73   H.F. Dobele, Proc. 15th Intern. Conf. Phenomena Ionized Gases, Pt.3, Minsk (1981), pp.135-144.
74   G.V. Ostrovskaya, 3rd Intern. Conf. IR Physics, ETH, Zurich (1984), pp.133-141.
75   P.R. Forman, S. Humphries and R.W. Peterson, Appl. Phys. Lett., **22**, No.10, 537-539 (1973).
76   I.I. Komissarova, G.V. Ostrovskaya, V.N. Filippov and E.H. Shedova, ZhTF, **53**, No.2, 251-257 (1983).
77   E.M. Barkhudarov, V.R. Berezovskii, T.Ya. Chelidze, et al, Proc. 13th Intern. Conf. Phenomena Ionized Gases, Pt.2, Berlin (1977), pp.625-626.
78   E.M. Barkhudarov, V.R. Berezovskii, T.Ya. Chelidze, et al, Proc. 9th Europ. Conf. Contr. Fusion Plasma Phys., Oxford (1979), p.35.
79   I.I. Komissarova, G.V. Ostrovskaya, Yu.I. Ostrovsky, et al, 7th Intern. Conf. Phenomena Ionized Gases, Pt.2, Budapest (1985), pp.1099-1101.
80   A.B. Ignatov, I.I. Komissarova, G.V. Ostrovskaya and L.L. Shapiro, ZhTF, **41**, No.4, 701-708 (1971).
81   G.V. Ostrovskaya and Yu.I. Ostrovskii, ZhTF, **40**, No.11, 2419-2422 (1970).
82   A.B. Ignatov, I.I. Komissarova, G.V. Ostrovskaya and L.L. Shapiro, ZhTF, **41**, No.2, 417-422 (1971).
83   I.I. Komissarova and G.V. Ostrovskaya, ZhTF, **48**, No.10, 2062-2067 (1978).
84   A. Kakos, G.V. Ostrovskaya, Yu.I. Ostrovsky and A.N. Zaidel, Phys. Lett., **23**, No.1, 81-83 (1966).
85   I.I. Komissarova, G.V. Ostrovskaya and L.L. Shapiro, ZhTF, **38**, No.8, 1369-1373 (1968).
86   T. Tschudi, C. Yamanaka, T. Sasaki, et al, J. Phys. D., **11**, No.2, 177-180 (1978).
87   G.V. Dreiden, I.I. Komissarova, V.S. Markov, et al, ZhTF, **51**, No.9, 1850-1857 (1981).
88   S.Yu. Bogdanov, G.V. Dreiden, I.I. Komissarova, et al, Proc. 14th Intern. Congr. Phenomena Ionized Gases, Pt.2, Minsk (1981), pp.973-974.
89   S.Yu. Boldanov, G.V. Dreiden, I.I. Komissarova, et al, Proc. of the 14th Int. Conf. on High-speed Photography and Photonics, Mashinostroenie, Moscow (1980).
90   A.P. Burmakov, V.A. Zaikov and G.M. Novik, Theoretical physics. Physics of plasma, Institute of Physics, Academy of Sciences of the Belorussian SSSR, (1975).
91   A.P. Burmakov, A.A. Labuda and V.M. Lutkovskii, IFZh, **29**, No.3, 499-503 (1975).
92   A.E. Dabydov, V.S. Abrukov, S.A. Abrukov, et al, Fiz. Gor. Vzryva, **14**, No.3, 78-82 (1978).
93   G.K. Wessel, H.C. Rothenberg and B. Zendle, Am. Astr. Soc., No.5, 10 (1962).
94   T.P. Hughes, Nature, **194**, No.4, 268-269 (1962).
95   G.M. Mal'shev, ZhTF, **35**, No.12, 2129-2143 (1965).
96   D.E. Evans and J. Katzenstein, J. Rep. Prog. Phys., **32**, No.2, 207-212 (1969).
97   A.W. de Silva and G.G. Goldenbaum, Methods of Experimental Physics, Vol.9A, Academic Press, New York (1970), pp.61-93.
98   G.T. Razdobarin, Proc. 7th Conf. Phys. Ionized Gases. Invited Lect., Zagreb (1974), pp.809-849.
99   N. Sheffield, Plasma Scattering of Electromagnetic Radiation, Wiley, New York (1975), p.305.
100   A. Lochte-Holtgreven, Methods of plasma examination, Mir, Moscow (1971).
101   S.Yu. Luk'yanov, Hot plasma and controlled nuclear synthesis, Nauka, Moscow (1975).
102   E.E. Salpeter, Phys. Rev., **120**, No.5, 1528-1535 (1960).
103   J.A. Fejer, Can. J. Phys., **39**, No.5, 716-740 (1961).
104   F.W. Perkins and E.E. Salpeter, Phys. Rev., **139A**, No.1, 55-62 (1965).
105   A.I. Akhiezer, I.A. Akhiezer, R.V. Polovin, et al, Electromagnetic fluctuations in plasma, Atomizdat, Moscow (1964).

References

106 A.G. Sitenko, Electromagnetic fluctuations in plasma, Publishing House of the Khar'kov University, Khar'kov (1965).

107 G.M. Malyshev, G.V. Ostrovskaya, G.T. Razdobarin and L.V. Sokolova, DAN SSSR, **168**, No.3, 554-555 (1966).

108 G.M. Malyshev, G.V. Ostrovskaya, G.T. Razdobarin and L.V. Sokolova, Plasma diagnostics, Atomizdat, No.2, Atomizdat, Moscow (1968), pp.53-58.

109 U. Ascoli-Bartoli, J. Katzenstein and L. Lovisetto, Nature, **204**, No.11, 672-673 (1964).

110 A.N. Zaidel', G.V. Ostrovskaya, Yu.I. Ostrovskii, Technology and practice of spectroscopy, Nauka, Moscow (1976).

111 G.M. Malyshev, G.T. Razdobarin and V.V. Semenov, ZhTF, **42**, No.7, 1429-1431 (1972).

112 G.M. Malyshev, G.T. Razdobarin and V.V. Semenov, Plasma diagnostics, No.3, Atomizdat, Moscow (1973), pp.206-210.

113 S.A. Ramsden, P.K. John, B. Kronast and R. Benesch, Phys. Rev. Lett., **19**, No.12, 688-689 (1967).

114 V.V. Aleksandrov, A.I. Gorlanov, N.G. Koval'skii, et al, Plasma diagnostics, Vol.3, Atomizdat, Moscow (1973), pp.200-210.

115 E.A. Burunov, G.M. Malyshev, G.T. Razdobarin and V.V. Semenov, ZhTF, **44**, No.1, 113-118 (1974).

116 E.A. Burunov, G.M. Malyshev, G.T. Razdobarin, et al, ZhTF, **45**, No.9, 1878-1883 (1975).

117 D. Johnson, D. Dimock, et al, Rev. Sci. Instrum., **56**, No.5, 1015-1017 (1985).

118 D. Meisel, H. Murmann, H. Rohr, et al, Rep. IPP/110, 140-143 (1986).

119 H. Murmann and M. Huang, Plasma Phys. Control. Fus., **27**, No.2, 103-114 (1985).

120 V.K. Gusev, G.M. Malyshev, G.T. Razdobarin and L.V. Sokolova, ZhTF, **42**, No.2, 340-343 (1972).

121 H.C. Koons and G. Fiocco, Phys. Lett., **26A**, No.12, 614-615 (1968).

122 K. Huyase and T. Okuda, J. Phys. Soc. Japan, **30**, No.6, 1768-1771 (1971).

123 G. Fiocco and E. Thompson, Phys. Rev. Lett., **10**, No.3, 89-91 (1963).

124 S.A. Ramsden and W.E. Davies, Ibid, **13**, No.7, 227-229 (1964).

125 S.L. Mandel'shtam, P.P. Pashinin, A.M. Prokhorov, et al, ZhETF, **49**, No.1(7), 127-133 (1965).

126 V.V. Korobkin, S.L. Mandel'shtam, P.P. Pashinin, et al, ZhETF, **53**, No.1(7), 116-123 (1967).

127 A. Unzold, Physics of star atmospheres, Monthly for atomic-fluorescent analysis. Physical fundamentals of the method, Moscow (1949).

128 H.J. Kunze, E. Funfer, B. Kronast and W.H. Kegel, Phys. Lett., **11**, No.1, 42-43 (1964).

129 S.A. Ramsden and W.E. Davies, Phys. Rev. Lett., **16**, No.8, 303-306 (1966).

130 B. Kronast, H. Rohr and E. Glock, Ibid, No.24, 1082-1084.

131 L. Keller, Z. Phys., **239**, No.2, 147-161 (1970).

132 C. Chen, Phys. Fluids, **14**, No.12, 2787-2788 (1971).

133 C. Chan and J.W. Daley, J. Quant. Spectr. Rad. Trans., **21**, No.6, 527-531 (1979).

134 A.N. Zaidel', Atomic-fluorescent analysis, Physical fundamentals of the method, Nauka, Moscow (1980).

135 G.T. Razdobarin and I.G. Folomkin, ZhTF, **49**, No.7, 1353-1372 (1979).

136 D.R. Olivares and M. Hieftie, Spectrochim. Acta., **33B**, No.3/4, 79-99 (1978).

137 V.I. Balykin, G.I. Bekov, V.S. Letokhov and V.I. Mishin, UFN, **132**, No.2, 293-344 (1980).

138 M.A. Bol'shov, A.V. Zybin and V.G. Koloshnikov, Kvant. Elektronika, **7**, No.8, 1808-1812 (1980).

139 E. Hinnov, J.G. Hinschberg, F.W. Hoffman and N. Rynn, Phys. Fluids, **6**, No.12, 1779-1780 (1963).

140 F.W. Hoffman, Ibid, **7**, No.4, 532-536 (1964).

141 M.B. Denton and H.V. Malmstadt, Appl. Phys. Lett., **18**, No.11, 485-487 (1971).

142 R.M. Measures and A.B. Rodrigo, Ibid, **20**, No.3, 102-104 (1972).

143   J. Lipson, Phys. Fluids, **22**, No.11, 2236-2240 (1972).

144   N. Omenetto, S. Nikdel, R.D. Reeves, et al, Spectrochimica Acta, **35B**, No.8, 507-517 (1980).

145   A.E. Elbern, Appl. Phys., **15**, No.1, 111-112 (1978).

146   A.E. Elbern, E. Hurtz, B. Schweer, J. Nuclear Mat., **76/77**, 143-148 (1978).

147   C.H. Muller and K.H. Burrell, Bull. Am. Phys. Soc., **24**, No.8, 971-976 (1979).

148   C.H. Muller and K.H. Burrell, Ibid, **25**, No.8, 977-981 (1980).

149   B. Schweer, D. Rusbuldt, Hints E, et al, J. Nucl. Mater., **93/94**, No.3, 357-363 (1980).

150   C.H. Muller and K.H. Burrel, Phys. Rev. Lett., **47**, No.5, 330-333 (1981).

151   S. Suckewer, Phys. Scripta, **23**, No.1, 72-77 (1981).

152   K. Bergstedt, G. Himmel, F. Pinnekampf, Phys. Lett., **53A**, No.3, 261-263 (1975).

153   D.D. Burgess, V.P. Myerscough, C.H. Skinner and J.M. Ward, J. Phys., **B13**, No.8, 1675-1701 (1975).

154   R.A. Burrell and H.J. Kunze, Phys. Rev. Lett., **28**, No.25, 1548-1551 (1975).

155   R.A. Stern and J.A. Johnson, Ibid, **34**, No.25, 1548-1551 (1975).

156   V.S. Burakov, P.Ya. Misakov, P. Naumenkov, et al, Plasma diagnostics, Vol.6, Energoatomizdat, Moscow (1989), pp.105-108.

157   D.A. Evans, M.J. Forrest, M.G. Nickolson, et al, Rev. Sci. Instrum., **56**, No.5 (Pt2), 1012-1014 (1985).

158   P. Bogen and F. Mertens, Plasma diagnostics, No.5, Energoizdat, Moscow (1986), pp.200-205.

159   S.A. Batishe, V.S. Burakov, V.I. Gladushchak, et al, Pis'ma ZhTF, **9**, No.23, 1461-1465 (1983).

160   T.W. Hansch, S.A. Lee, R. Wallenstein and C. Wieman, Phys. Rev. Lett., **34**, No.6, 307-309 (1975).

161   B.I. Troshin, V.P. Chebotaev and A.A. Chernenko, Pis'ma ZhETF, **27**, No.5, 293-296 (1978).

162   R.A. Stern, D.L. Correll, H. Bohmer and N. Rynn, Phys. Rev. Lett., **37**, No.13, 833-836 (1976).

163   G. Himmel, F. Pinnekampf, J. Phys. B., **10**, No.8, 1457-1464 (1977).

164   G.T. Razdobarin, V.V. Semenov, L.V. Sokolova, et al, Nucl. Fus., **19**, No.11, 1439-1446 (1979).

165   M.M. Larionov, G.T. Razdobarin and I.P. Folomkin, Pis'ma ZhTF, **6**, No.22, 1375-1379 (1980).

166   H.C. Meng and H.J. Kunze, Phys. Fluids, **22**, No.6, 1082-1088 (1979).

Chapter 5

1   J. Steinfeld, Laser and coherent spectroscopy, Mir, Minsk (1982).

2   J. Helmcke, S.A. Lee, J.L. Hall, Appl. Opt., **21**, 1686-1694 (1982).

3   W.J. Witteman, Appl. Phys. Lett., **10**, 347-350 (1967).

4   Yu.N. Dubnishchev and B.S. Rinkevichyus, Methods of laser Doppler anemometry, Nauka, Moscow (1982).

5   L.A. Dushin and O.S. Pavlichenko, Examination of plasma using lasers, Atomizdat, Moscow (1968).

6   L.N. Pyatnitskii, Laser diagnostics of plasma, Atomizdat, Moscow (1976).

7   A.A. Besshaposhnikov, V.N. Blokhin, V.B. Voronin and V.A. Myslin, ZhPS, **46**, No.5, 723-727 (1987).

8   A.I. Ferguson, Phil. Trans. R. Soc. Lond., **A307**, 645-660 (1982).

9   V.N. Ochkin, N.G. Preobrazhenskii, N.N. Sobolev and N.Ya. Shaparev, Usp. Fiz. Nauk, **148**, No.473-507 (1986).

10   D.R. Crosley and G.P. Smith, Opt. Eng., **22**, No.5, 545-553 (1983).

11   V.V. Kosichkin and A.I. Nadezhdinskii, Izv. AN SSSR. Ser. Fiz., **47**, 20-37 (1983).

12   A.V. Dem'yachenko, I.I. Zasavitskii, S.Yu. Savinov, et al, Kvant. Elektronika, **47**, No.4, 851-859 (1987).

13   C. Dang, J. Reid and B.K. Garside, Appl. Phys., **B27**, 145-151 (1982).

14   I.I. Zasavitskii, R.Sh. Islamov, M.A. Kerimkulov, et al, Examination of the active media of the wave CO_2 laser, Preprint FIAN, Moscow, No.18, (1988).

References

15   D.K. Otorbaev, V.N. Ochkin, P.L. Rubin, et al, Electronically excited molecules in non-equilibrium plasma, Nauka, Moscow (1985).
16   B.F. Gordiets, A.I. Osipov and L.A. Shelepin, Kinetic processes in gases and molecular lasers, Nauka, Moscow (1980).
17   I.I. Zasavitskii and A.I. Nadezhdinskii, Spectral gas analysis of polyatomic molecules using tunable diode lasers, Preprint IOFAN, Moscow, No.50 (1987).
18   J.A. Mucha, Appl. Spectr., 36, No.4, 393-400 (1982).
18a   A.I. Nadezhdinskii, Proc. of the 28th Nat. Conf. on Spectroscopy, Naukova dumka, Kiev (1988), p.138.
19   M. Polack, M. Gruebele and R. Saykally, J. Chem. Phys., 87, No.6, 3352-3356 (1987).
20   F. Pan and T. Oka, Phys. Rev., A36, No.5, 2297-2310 (1987).
21   I.I. Zasavitskii, M.A. Kerimkulov, A.I. Nadezhdinskii, et al, Optika i Spektroskopiya, 65, No.6, 1198-1201 (1988).
22   S.A. Akhmanov and N.I. Koroteev, Methods of non-linear optics in spectroscopy of light scattering, Nauka, Moscow (1981).
23   Kh. Shrettner and Kh. Kikner, Spectroscopy of Rama and scattering of light in gases and liquids, Mir, Moscow (1982).
24   M.A. Yuratich and D.C. Hanna, Mol. Phys., 33, No.3, 671-682 (1977).
25   W.F. Murphy, W. Holzer and H.J. Bernstein, Appl. Spectr., 23, No.3, 211-218 (1969).
26   A.F. Bunkin, S.G. Ivanov and N.I. Koroteev, Pis'ma ZhETF, 24, No.8, 468-472 (1976).
27   B.U. Asanov, V.N. Ochkin, S.Yu. Savinov, et al, Krat. Soob. po Fiz., No.9, 26-27 (1988).
28   V.N. Ochkin, S.Yu. Savinov, N.N. Sobolev and S.N. Tskhai, ZhTF, 58, No.7, 1283-1290 (1988).
29   V.V. Krunotsky, L.A. Kulevsky and V.A. Mishin, Opt. Commun., 21, 225-228 (1977).
30   J.A. Shirely, R.J. Hall and A.S. Eckbreth, Opt. Letts., 5, No.9, 380-382 (1980).
31   W.B. Roh, P.W. Schreiber, J.P.E. Taran, Appl. Phys. Letts., 29, No.3, 174-176 (1976).
32   B.I. Greene, R.B. Weisman and R.M. Hochstrasser, Chem. Phys. Lett., 59, no.1, 5-9 (1978).
33   S.Yu. Volkov, D.N. Kozlov, A.M. Prokhorov, et al, High-resolution CARS-Spectroscopy of molecular gases, Nauka, Moscow (1986).
34   B. Massabieaux, G. Gousset, M. Lefebre and M. Pelat, J. Phys., 48, No.11, 1939-1949 (1987).
35   S.I. Valyanskii, K.A. Vereshchagin, A.Yu. Volkov, et al, Examination of the kinetics of the function of distribution of nitrogen molecules in vibrational and rotational states in excitation with plasma discharge in a biharmonic field. Measurements of the constant of the exchange rate, Prepr. IOFAN, No.109, Moscow (1984).
36   A.A. Devyatov, S.A. Dolenko, A.T. Rakhimov, et al, ZhETF, 90, No.2, 429-436 (1986).

Chapter 6
1   L.D. Landau and E.M. Lifshits, Electrodynamics of solids, GIFML, Moscow (1959).
2   I.L. Fabelinskii, Molecular scattering of light, Nauka, Moscow (1965).
3   R.D. Mountain, Rev. Mod. Phys., 38, No.2, 205-214 (1966).
4   S. yip, Journ. Acoust. Soc. Amer., 49, No.3 (Pt3), 941-949 (1971).
5   G.J. Gabriel, Phys. Rev., A8, No.3, 963-990 (1973).
6   R.W. Pitz, R. Cattolica, F. Robben and L. Talbot, Combust. Flame, 27, No.3, 313-320 (1976).
7   L.N. Rozanov, Vacuum technology, Vyssh. Shkola, Moscow (1982).
8   W. Childs, Physical constants, GIFML, Moscow (1961).
9   I. Namer, R.W. Schefer and M. Chan, Proc. Spring 1980 Meeting, Western States

Section of Combustion Inst., Univer. California, Irvine (1980) (1980). Pap. WSS/80-18.22.

10   I. Namer, R.W. Schefer and M. Chan, Lawrence Berkeley Laboratory Rep. LBL-1065 (1980).

11   R.G. Bill Jr, I. Namer and L. Talbot, Combust. Flame, **43**, No.2, 229-242 (1981).

12   R.G. Bill Jr, I. Namer and L. Talbot, Ibid, **44**, No.2, 277-285 (1982).

13   M. Namazian, L. Talbot, F. Robben and R.K. Cheng, Proc. 19th Sympos. (Intern) Combustion, Combustion Inst. (1987), pp.487-493.

14   M.B. Long, P.S. Levin and D.C. Fourguette, Optics Letters, **10**, No.6, 267-269 (1985).

15   D.C. Fourguette, R.M. Zurn, M.B. Long, Combust. Sci. Technol., **44**, No.3, 307-317 (1986).

16   S.C. Graham, A.J. Grant and J.M. Jones, AIAA Journ., **12**, No.8, 1140-1142 (1974).

17   T.M. Dyer, Ibid, **17**, No.8, 912-914 (1979).

18   A.J.D. Farmer and G.N. Haddad, J. Phys. D: Appl. Phys., **21**, No.3, 426-431 (1988).

19   P. Kovitya and L.E. Cram, Amer. Welding Journ., **65**, No.1, 34-39 (1986).

20   L.E. Cram, L. Poladian and G. Roumeliotis, J. Phys. D: Appl. Phys., **21**, No.3, 418-425.

21   H.G. Green, J. Phys. E: Sci. Instrum., **20**, No.7, 670-676 (1987).

Chapter 7

1   Zh. Zheenbaev and V.S. Engel'sht, Laminar plasma torch, Ilim, Frunze (1975).

2   G. Grim, Plasma spectroscopy, Atomizdat, Moscow (1986).

3   R. Huddlestone and S. Leonard (eds), Plasma diagnostics, Mir, Moscow (1967).

4   L.A. Vasil'ev, Schlieren methods, Nauka, Moscow (1968).

5   L.A. Vasil'ev and I.V. Ershov, Interferrometer with a diffraction grating, Mashinostroenie, Moscow (1976).

6   M.M. Skotnikov, Schlieren quantitative methods in gas dynamics, Nauka, Moscow (1976).

7   M.F. Zhukov (ed), Properties of light temperature plasma and methods of diagnostics,Siberian division of Nauka, Novosibirsk (1977).

8   S.A. Abrukov, Schlieren and interference methods of examination of optical heterogenieties, Publishing House of the Kazan' University, Kazan (1962).

9   A.K. Beketova, O.P. Klochkova and N.V. Ryabova, Optiko-Mekhan. Prom., No.3, 11-14 (1962).

10   Zh. Zheenbaev and V.S. Engel'sht, Two-jet plasma torch, Ilim, Frunze 1983.

Chapter 8

1   Proc. of the 4th Nat. Symp. on Plasma Chemistry, Vol2. DKhTI, Dnepropetrovsk (1984), p.254.

2   B.Y.H. Lin (ed), Fine particles (Aerosols Generation, Measurements, Sampling and Analysis), Academic Press, New York (1976), p.889.

3   Y. Mizutani, H. Kodama and K. Miyasaka, Combustion and Flame, **44**, No.1, 85-95 (1982).

4   F. Durst, Trans. ASME, **104**, No.3, 284-296 (1982).

5   Kh. Chen and E. Pfender, Plasma Chem. Plasma Process., **2**, No.2, 185-212 (1982).

6   Kh. Chen and E. Pfender, Ibid, No.3, 293-316.

7   E. Pfender, Pure Appl. Chem., **57**, No.9, 1179-1195 (1985).

8   M.F. Khukov, V.P. Lyagushkin and O.P. Solonenko, Automated experimental stand for complex examination of high-temperature heterogeneous jets, No.145-86, Preprints of the Institute of Thermophysics, Siberian Division of the USSR Academy of Sciences, Novosibirsk (1986).

9   F. Durst, A. Melling and J.H. Whitelaw, Principles and Practice of Laser Doppler Anemometry, Whiley, London and New York (1981).

10   G.A. Gousbet, Plasma Chem. Plasma Process., **5**, No.2, 91-117 (1985).

References

11   W.D. Buchalo and M.J. Houser, Opt. Engn., **23**, No.5, 695-697 (1984).

12   Particle Dynamics Analyzer, Dantec Publication N 6605E, Elsevier, Denmark (1986), p.12.

13   J. Mishin, M. Vardelle, J. Lesinski and P. Fauchais, Proc. 7th Intern. Sympos. Plasma Chem., Eindhoven, Netehrlands, Vol. 3 (1985), pp.724-729.

14   G. van de Khyulst, Scattering of light by small particles, Izd-vo Inostr. Lit., Moscow (1961).

15   Yu.N. Dubnishchev and B.S. Rinkevichyus, Methods of laser Doppler anemometry, Nauka, Moscow (1982).

16   A. Jyle, F. Born, D. Swirl, et al, J. Energy, **1**, No.2, 220-225 (1977).

17   V.P. Yankov, Tr. TsAGI, **1755**, 83-93 (1976).

18   Y. Mizutani, H. Kadama and K. Nivasoka, Combust. Flame, **44**, No.1, 85-95 (1982).

19   W. Mayr, 16th Intern. Conf. Phenomena Ionized Gases, Dusseldorf (1982), pp.412-413.

20   D.J. Holve and S.A. Self, Appl. Opt., **18**, No.10, 1632-1652 (1980).

21   D.J. Holve and K.D. Annen, Opt. Engng., **23**, No.5, 591-603 (1984).

22   N. Chiger and G. Stewart, Ibid, pp.554-556.

23   C.F. Hess, Appl. Opt., **23**, No.23, 4375-4383 (1984).

24   C.F. Hess and V.E. Espinoza, Opt. Engng., **23**, No.5, 604-609 (1984).

25   K.S. Shifrin and V.I. Golikov, Tr./GGO, **152**, 18-21 (1964).

26   R.N. Sokolov, G.D. Petrov and F.A. Kudryavitskii, Izv. AN SSSR, Fiz. Atmosfery i Okeana, **6**, No.6, 214-215 (1970).

27   K.S. Shifrin, Theoretical and applied problems of light scattering, Nauka and Tekhnika, Minsk (1971).

28   D.E. Hirleman, Opt. Engng., **23**, No.5, 610-619 (1984).

29   E.P. Zimin and V.V. Forfutdinov, Gas dynamics of multiphase flows in power systems, Publishing House of the Khar'kov University, Khar'kov (1978).

30   E.P. Zimin, L.B. Levin and V.I. Chursin, Physical methods of examination of transparent heterogenieties, Mashinostroenie, Moscow (1987).

31   R.S. Valeev, R.N. Gizatullin, A.G. Golubev and V.I. Yagodkin, Izmeritel., Tekhnika, No.5, 71-74 (1983).

32   D.A. Tichenor, R.E. Mitchell, K.R. Hencken and S. Niksa, SANDIA Rep. SAND 84-8628, Albuquerque, New Mexico and Livermore, California (1984), p.29.

33   B.B. Weiner, Modern Methods of Particle Size Analysis, Vol.73, Pergamon Press, New York (1984), pp.93-116.

34   P.K. Pusey and W. von Megen, J. Phys. Chem., **80**, No.8, 3513-3520 (1984).

35   S.C. Graham, J.B. Homer and J.L.J. Rosenfield, Proc. Roy. Soc. (London), **A344**, No.2, 259-265 (1975).

36   J.J. Frurip and S.H. Bauer, J. Phys. Chem., **81**, No.10, 1007-1015 (1977).

37   P. Chylek, V. Ramaswamy, A. Ashkin and J.M. Dziedzic, Appl. Opt., **22**, No.15, 2302-2307 (1983).

38   T.T. Charalampopoulos, Rev. Sci. Instrum., **58**, No.9, 1638-1646 (1987).

39   D.M. Benenson and H.C. Kuck, Proc. 16th Intern. Conf. Phenomena Ionized Gases, Dusseldorf (1983), pp.473-475.

40   G. Gousbet, Proc. 14th Inter. Conf. Phenomena Ionized Gases, Grenoble (1979), pp.789-790.

41   M.R. Barrault, Pure Appl. Chem., **52**, No.10, 1829-1835 (1980).

42   V.F. Klimkin, A.N. Papyrin and R.I. Soloukhin, Optical methods of recording high-speed processes, Siberian Division of Nauka, Novosibirsk (1980).

43   P.W. Forder, J. Phys. E: Sci. Instrum., **14**, No.7, 1014-1018 (1980).

44   P.W. Forder and D.A. Jackson, Ibid, **15**, No.4, 555-557 (1982).

45   R.N. James, W.R. Babcock and H.S. Seifert, AIAA Journ., **6**, No.1, 160-162 (1968).

46   H.L. Morse, B.I. Tullis, H.S. Seifert and W.R. Babcock, J. Spacecraft Rockets, **6**, No.2, 264-272 (1969).

47   S.A. Self and K.Kh. Kryuger, Raket. Tekhn. Kosmonavtika, **16**, No.5, 160-186 (1978).

48   L. Mannik, S.K. Brown and F.Y. Chu, Proc. 7th Intern. Sympos. Plasma Chem., Vol.2, Eindhoven, Netherlands (1985), pp.704-709.

49   A.N. Korotkov, S.I. Kruglyi and A.P. Nefedov, Avtometriya, No.3, 27-33 (1982).

50   A.N. Korotkov and A.P. Nefedov, Teplof. Vysokikh Temp., 23, No.4, 792-797 (1985).

51   V.G. Goloborod'ko and Yu.M. Kiselev, Teplof. Yysokikh Temp., 9, No.6, 1248-1252 (1971).

52   V.V. Pikalov and N.G. Preobrazhenskii, Properties of low-temperature plasma and methods of its diagnostics, Siberian Division of Nauka, Novosibirsk (1977).

53   R.A. Stern and J.A. Johnson, Phys. Rev. Lett., 34, No.25, 1545-1551 (1975).

54   A.A. Knyasev, N.B. Lerner and K.I. Svinolupov, Proc. of the 2nd Nat. Conf. , Leningrad (1984), pp.27-29 and Leningrad (1984), p.43.

55   G. Couesbet, C.R. Academ. Sci. Paris, 280, No.19, 597-600 (1975).

56   M.R. Barrault, G.B. Jones and T.R. Blackburn, J. Phys. E: Sci. Instrum., 7, No.5, 663-666 (1974).

57   L. Mannik and S.K. Brown, Appl. Opt., 25, No.5, 649-652 (1986).

58   V.V. Krasovskii, T.P. Kuyanova and E.I. Palagashvili, High-temperature synthesis and properties of refractory metals, Zinatne, Riga (1979).

59   B.A. Pavlovskii and B. Ruk, Prib. Sistem. Upravlen., No.3, 27-29 (1986).

60   M.C. Bleiweiss, P.N.P. Chang and S.S. Penner, J. Quant. Spectr. Radiat. Transfer., 26, No.3, 273-275 (1981).

61   J.R. Hunt, J. Fluid Mech., 122, No.5, 933-942 (1982).

62   D.Ya. Svet, Optical methods of measuring true temperatures, Nauka, Moscow (1982).

63   G.A. Lyzenga and J.T. Ahrens, Rev. Sci. Instrum., 50, No.1, 1421-1427 (1979).

64   F.H.A. Jorgensen and M. Suiderwyk, J. Phys. E: Sci. Instrum., 18, No.4, 486-491 (1985).

65   M.L. Eider and J.D. Winefordner, Progr. Analyt. Atom. Spectr., 6, No.2, 293-314 (1983).

66   J. Mishin, M. Vardelle, J. Lesinski and P. Fauchais, Proc. 7th Inter. Sympos. Plasma Chem., Vol.3, Eindhoven and Elsevier, Netherlands (1985), pp.724-729.

67   V.P. Lyagushkin, O.P. Solonenko, Yu.L. Stankevich and V.A. Starikov, Tez. Dokl., Proc. of the 11th Nat. Conf. on Low-temperature plasma generators, Ilim, Frunze (1983), pp.272-273.

68   V.P. Lyagushkin and O.P. Solonenko, Proc. 7th Intern. Sympos. Plasma Chem., Eindhoven and Elsevier, Netherlands (1985), pp.730-736.

69   T. Sakuta and M.I. Boulos, Proc. 8th Inter. Sympos. Plasma Chem., Vol.1, Whiley Interscience, Tokyo (1987), pp.371-376.

70   M.A. Sheindlin, A.V. Kirillin, L.M. Kheivets and K.A. Khodakov, Teplofiz. Vysokikh Temp., 19, No.4, 839-848 (1981).

Chapter 9

1   N. Kimura and A. Kanzava, Raket. Tekhnika Kosmon., No.2, 25-28 (1963).

2   V.M. Gol'dfarb and A.M. Uzdenov, Izv. CO AN SSSR, Ser. Tekhn. Hauk., 3, No.1, 81-84 (1967).

3   D. Bannenberg and J. Inzinger, Pribor. Nauch. Issled., No.7, 10-13 (1964).

4   W. Rother, Wiss. Z. TH. Ilmenau, 19, No.3/4, 219-226 (1973).

5   W. Rother, Beitr. Plasma Phys., 10, No.6, 497-506 (1970).

6   B. Bowman, J. Phys. D: Appl. Phys., 5, No.8, 1422-1432 (1972).

7   E. Seymour, Pap. ASME, 1-10 (1971).

8   A. Abdrazakov, Zh. Zheenbaev, R.I. Konavko, et al, Proc. of the 5th Nat. Conf. on Low-temperature Plasma Generators, Vol. 1, Siberian Division of Nauka, Novosibirsk (1972), pp.141-144.

9   M.K. Asanaliev, Zh. Zheenbaev and K.K. Makesheva, IFZh, 57, No.4, 554-562 (1989).

10   J. Levis and W. Gauvin, AIChE Journ., 19, No.5, 982-990 (1973).

11   D. Kassoy, T. Adamson and A. Messiter, Phys. Fluids, 9, No.4, 671-681 (1966).

12   M.K. Asanaliev, V.S. Engel'sht, E.P. Pakhomov, et al, Proc. 5th ICPIG, Part 2,

Minsk (1981), pp.959-960.

13   V.V. Kabanov and V.S. Klubnikin, Proc. of the 9th Nat. Conf. on Low-temperature Plasma Generators, Ilim, Frunze (1982).
14   E. Pfender and Y. Lee, Plasma Chem. Plasma Proc., **5**, No.3, 211-236 (1985).
15   Zh. Zheenbaev and V.S. Engelsht, Laminar plasma torch, Ilim, Frunze (1975).
16   G.N. Abramovich, Applied gas dynamics, Nauka, Moscow (1969).

Chapter 10

1   Yu.M. Kagan and V.I. Perel', UFN, **81**, No.3, 409-452 (1963).
2   O.V. Kozlov, Electric probe in plasma, Atomizdat, Moscow (1969).
3   F. Chen, Electric probes, (edited by R. Huddlestone and S.M. Leonard), Mir, Moscow (1967).
4   L. Schott, Electric probes (edited by V. Lochte-Holdgreven), Mir, Moscow (1971).
5   J.D. Swift and M.J.R. Schwar, Electrical probes for plasma diagnostics, Iliffe books, London (1970).
6   P. Chan, L. Telbot and K. Turyan, Electric probes in stationary and moving plasma, Mir, Moscow (1978).
7   Yu.A. Ivanov, Yu.A. Lebedev and L.S. Polak, Methods of contact diagnostics in non-equilibrium plasma chemistry, Nauka, Moscow (1981).
8   V.K. Zhivotov, V.D. Rusanov and A.A. Fridman, Diagnostics on non-equilibrium chemically active plasma, Energoatomizdat, Moscow (1985).
9   H. Mott-Smith and I. Langmuir, Phys. Rev., **28**, No.5, 727-763 (1926).
10   M.J. Druyvesteyn, Z. Phys., **64**, 781-798 (1930).
11   J.D. Swift, Proc. Phys. Soc., **79**, 697-701 (1962).
12   A.I. Lukovnikov and M.Z. Novgorodov, Krat. Soobshch. Fizike, No.1, 27-34 (1971).
13   M.A. Mal'kov, Plasma diagnostics taking into account the sink of electrons on the probe and in the case of the anisotropy of the velocity of the distribution of electrons, Author's dissertation, M.V. Lomonosov Moscow State University (1985).
14   S. Matsumura and S-L. Chen, J. Appl. Phys., **43**, No.8, 3357-3361 (1972).
15   E. Berger and A. Heisen, J. Phys. D: Appl. Phys., **8**, No.6, 629-639 (1975).
16   J-S. Chang, Ibid, **6**, No.12, 1674-1683 (1973).
17   A. Boschi and F. Magistrelly, Nuovo Cimento, **29**, No.2, 487-493 (1963).
18   S. Klagge and M. Maass, Beitr. Plasmaphys., **23**, No.4, 355-368 (1983).
19   N. Hershkowitz, M.H. Cho, C.H. Nam and T. Intrator, Plasma Chem. Plasma Proc. **8**, No.1, 35-52 (1988).
20   H. Sabadil, S. Klagge and M. Kammeyer, Ibid, No.4, 425-444.
21   D. Maundrill, J. Slatter, A.T. Spices and C.C. Welch, J. Phys. D: Appl. Phys., **20**, No.7, 815-819 (1987).
22   T.I. Cox, V.G.I. Deshmukh, D.A.O. Hope, et al, Ibid, 820-831.
23   A.P. Ershov, Examination of the kinetics of electrons in the plasma of high-frequency low-pressure discharge in inert gases, Author's abstract of a disseration, M.V. Lomonosov Moscow State University (1982).
24   K.D. Asvadurov and I.A. Vasil'eva, ZhTF, **45**, No.7, 1558-1559 (1975).
25   Yu.M. Kagan, N.B. Kolokolov, P.M. Pramatarov and M.A. Petrun'kii, ZhTF, **47**, No.6, 1160-1166 (1977).
26   H. Amemiya and K. Shimizu, Jpn. J. Appl. Phys., **13**, No.6, 1035-1036 (1974).
27   H. Amemiya, Ibid, **14**, No.1, 165-166 (1975).
28   A.B. Blagoev, Yu.M. Kagan, N.B. Kolokolov and R.I. Lyagushchenko, ZhTF, **45**, No.3, 579-585 (1975).
29   V.I. Demidov, N.B. Kolokolov and O.G. Toronov, Proc. of 6th Nat. Conf. on Low-temperature Plasma Physics, Leningrad (1983), pp.309-310.
30   S.I. Kryvosheev, V.N. Makarchuk and O.I. Fisun, Teplofiz. Vysokikh Temp., **25**, No.4, 791-793 (1987).
31   L.M. Volkova, V.I. Demidov, N.B. Kolokolov and E.A. Kral;kina, Teplofiz. Vysokikh Temp., **22**, No.4, 757-763 (1984).
32   A.N. Tikhonov and V.Ya. Arsenin, Methods of solving incorrect problems, Nauka,

Moscow (1986).
33   E.K. Eroshchenkov, Physics, technology and application of low-temperature plasma, Alma-Ata (1970).
34   L.M. Volkova, A.M. Debyatov, A.S. Mechenov, et al, Vestn. MGU, Ser.3, Fizika, astronomiya, **16**, No.3, 371-374 (1975).
35   Yu.B. Golubovskii, V.M. Zakharova, V.N. Pasunkin and L.D. Tsendin, Fiz. Plazmy, **7**, No.3, 620-628 (1981).
36   N.A. Gorbunov, N.B. Kolokolov and A.A. Kudryavtsev, Fiz. Plazmy, **15**, No.12, 1513-1520 (1989).
37   A.B. Blagoev, N.B. Kolokolov, V.M. Milenin, ZhTF, **42**, No.8, 1701-1704 (1971).
38   K. One and T. Kimuar, Jpn. J. Appl. Phys., **28**, No.10, 1997-2003 (1989).
39   T.L. Thomas and E.L. Battle, J. Appl. Phys., **41**, No.8, 3428-3432 (1970).
40   D.G. Bulls, R.B. Holt and B.T. McClure, Ibid, **33**, No.1, 29-33 (1962).
41   Yu.I. Chutov, A.I. Kravchenko and Yu.V. Vovchenko, Proc. of the 6th Nat. Conf. on Low-temperature Plasma Physics, Leningrad University, Leningrad (1983), pp.320-322.
42   J. Andrew, G. Sardin, A. Lloret, et al, J. Appl. Phys., **63**, No.4, 1230-1232 (1988).
43   W.D. Bunting and W.J. Heikkila, Ibid, **41**, No.5, 2263-2264 (1970).
44   J. Laframboise, Univ. Toronto, Inst. Aerospace Studies, Rept. No.100 (1965).
45   J. Laframboise, Rarefield Gas Dynamics, Vol.2, J.H. De Leeuv (ed), Academic Press, New York (1966), pp.22-44.
46   A.A. Sonin, AIAA Journal, **4**, No.11, 1588-1593 (1966).
47   A.I. Babarickij, S. Klagge and M. Maass, Beitr. Plasmaphys., **22**, No.2, 181-194 (1982).
48   E.O. Johnson and L. Malter, Phys. Rev., **76**, No.10, 1411-1417 (1949).
49   Yu.F. Nasedkin, G.V. Levadnyi, A.A. Serov, et al, Teplofiz Vysokikh Temp., **23**, No.1, 156-162 (1985).
50   S. Klagge, M. Maass and A. Serov, Beitr. Plasmaphys., **25**, No.3, 255-262 (1985).
51   R.L.F. Boyd adn J.B. Thompson, Proc. Roy. Soc., **A252**, No.1268, 102-119 (1959).
52   J.B. Thompson, Ibid, **262**, No.1311, 503-528 (1961).
53   M.J. Kusner, J. Appl. Phys., **53**, No.4, 2939-2946 (1982).
54   V.L. Fedorov, ZhTF, **55**, No.5, 926-929 (1985).
55   A.P. Mezentsev, A.S. Mustafaev and V.L. Fedorov, ZhTF, **55**, No.3, 544-549 (1985).
56   A.P. Mezentsev, A.S. Mustafaev, V.F. Lapshin and V.L. Fedorov, ZhTF, **56**, No.11, 2104-2110 (1986).
57   V.L. Fedorov and A.P. Mezentsev, ZhTF, **57**, No.3, 595-597 (1987).
58   A.P. Mezentsev, A.S. Mustafaev and V.L. Fedorov, ZhTF, **58**, No.6, 1096-1101 (1988).
59   A.I. Lukovnikov, ZhTF, **43**, No.7, 1478-1483 (1973).
60   V.L. Granovskii, Electric current in gas, GITTL, Moscow and Leningrad (1952).
61   R.V. Radchenko, A.G. Shtoik and A.S. Zel'skii, ZhTF, **45**, No.10, 2225-2228 (1975).
62   A.M. Devyatov and M.A. Mal'kov, Proc. of 6th Nat. Conf. on Low-temperature Plasma Physics, Leningrad University, Leningrad (1983), pp.314-316

Chapter 11

1   I. Langmuir, Collected works, C.G. Suits (ed), Pergamon Press, Oxford (1961), pp.3-5.
2   Yu.M. Kagan and V.I. Perel', UFN, **81**, No.3, 409-452 (1963).
3   F. Chen, Plasma diagnostics (Edited by R. Huddlestone and Leonard), Mir, Moscow (1967).
4   O.V. Kozlov, Electric probe in plasma, Atomizdat, Moscow (1969).
5   L. Schott, Methods of plasma examination (edited by V. Lochte-Holdgreven), Mir, Moscow (1971).
6   P. Chan, L. Telbot and K. Turyan, Electric probes in stationary and moving plasma. Theory and application, Mir, Moscow (1978).

References

7   Yu.A. Ivanov, Yu.A. Lebedev and L.S. Polak, Methods of contact diagnostics in non-equilibrium plasma chemistry, Nauka, Moscow (1981).

8   B.V. Alekseev and V.A. Kotel'nikov, Probe methods of plasma diagnostics, Energoatomizdat, Moscow (1988).

9   M.J.M. Parrot, L.R.O. Storey, L.W. Parker and J.G. Laframboise, Phys. Fluids, **25**, No.12, 2388-2400 (1982).

10   Yu.B. Golubovskii, V.M. Zakharova, V.N. Pasunkin and L.D. Tsendin, Fiz. Plazmy, **7**, No.3, 620-628 (1981).

11   N.A. Gorbunov, N.B. Kolokolov and A.A. Kudryavtsev, Fiz. Plazmy, **15**, No.12, 1513-1520 (1989).

12   B.I. Davydov and L.I. Zmanovskaya, ZhTF, **6**, No.7, 1244-1255 (1989).

13   C.H. Su and S.H. Lam, Phys. Fluids, **6**, No.10, 1479-1491 (1963).

14   I.M. Cohen, Ibid, pp.1492-1499.

15   P.R. Smy, Adv. Physics, **25**, No.5, 517-553 (1976).

16   V.Z. Kompaniets, A.A. Ovsyannikov and L.S. Polak, Chemical reactions in turbulent gas and plasma flows, Nauka, Moscow (1979).

17   M.S. Benilov, Teplofiz. Vysokikh Temp., **26**, No.5, 993-1004 (1988).

18   M. Mitchner and Ch. Kruger, Partially ionized gases, Mir, Moscow (1976).

19   I. MacDaniel' and E. Meeon, Mobility and diffusion of ions in gases, Mir, Moscow (1976).

20   I.L. Pankrat'eva and V.A. Polyanskii, Izv. AN SSSR MZhG, No.2, 103-112 (1979).

21   B.V. Alekseev and A.M. Grishin, Physical cast dynamics of reacting media, Vyssh. Shk. (1985).

22   A.F. Kolesnikov and G.A. Tirskii, Molecular gas dynamics, Nauka, Moscow (1982).

23   D. Girshfel'der, Ch. Kertiss and R. Berd, Molecular theory of gases and liquids, IL, Moscow (1961).

24   V.M. Zhdanov, Transport phenomena in multicomponent plasma, Energoizdat, Moscow (1982).

25   V.V. Gogosov and V.A. Polyanskii, Itogi Nauki Tekhn. Mekhan. Zhidkosti Gaza, Vol.10, VINITI, Moscow (1976).

26   J. Fertsiger and G. Kaper, Mathematical theory of transport processes in gases, Mir, Moscow (1976).

27   B.M. Smirnov, Physics of slightly ionized gas in problems with solutions, Nauka, Moscow (1985).

28   A.A. Radtsig and B.M. Smirnov, Plasma chemistry, Vol.11, Energoatomizdat, Moscow (1984), pp.170-200.

29   H.W. Ellis, R.Y. Pai, E.W. McDaniel, et al, Atom. Nucl: Data Tabl., **17**, No.3, 177-210 (1976).

30   H.W. Ellis, E.W. McDaniel, D.L. Albritton, et al, Ibid, **22**, No.3, 179-217 (1978).

31   N.L. Aleksandrov, A.M. Konchakov, A.P. Napartovichand A.N. Starostin, Plasma chemistry, Vol.11, Energoatomizdat, Moscow (1984), pp.3-45.

32   B.Ya. Moizes and G.E. Pikus (eds), Thermal emission converters and low-temperature plasma, Nauka, Moscow (1973).

33   F.G. Baksht and V.G. Yuriev, ZhTF, **49**, No.5, 905-944 (1979).

34   F.G. Baksht, V.G. Ivanov and V.G. Yuriev, Inv. Pap. 18th Intern. Conf. Phenomena Ionized Gases, Swansea (1987), pp.206-218.

35   L. Huxley and R. Crompton, Diffusion and drift of electrons in gases, Mir, Moscow (1977).

36   L.M. Biberman, V.S. Vorobiev and I.T. Yakubov, Kinetics of non-equilibrium low-temperature plasma, Nauka, Moscow (1982).

37   J. Dutton, J. Phys. Chem: Ref. Data, **4**, No.3, 577-856 (1975).

38   J.W. Gallagher, E.C. Beaty, J. Dutton and L.C. Pitchford, Ibid, **12**, No.1, 109-152 (1983).

39   J. Iticawa, Atom Data Nucl: Data Tabl., **14**, No.1, 1-10 (1974).

40   J. Iticawa, Ibid, **21**, No.1, 69-75 (1978).

41   S.R. Hunter and L.S. Christophorou, Electron-molecule interactions and their applications, L.G. Christophorou (ed), Vol.2,1.6, Academic Press, Orlando (1984),

pp.89-219.

42   I.B. Chekmarev, ZhTF, **50**, No.1, 48-53 (1980).

43   M.S. Benilov, V.I. Kovbasyuk and A.V. Lyashko, 10th Intern. Conf. MHD Electrical Power Generation, Vol.2, Tiruchirapally, India (1989), pp.X105-X109.

44   P.M. Chung, Phys. Fluids, **12**, No.8, 1623-1634 (1969).

45   I.B. Chekmarev, ZhTF, **51**, No.8, 1623-1634 (1981).

46   Lem, Raket. Tekhnika Kosmonavtika, **2**, 43-51 (1964).

47   C.H. Su, Raket. Tekhnika Kosmonavtika, **3**, No.5, 52-61 (1965).

48   M.S. Benilov, G.G. Bochkarev, A.E. Buznikov, et al, Izv. AN SSSR. MZhG, No.1, 150-160 (1983).

49   W.B. Bush and F.E. Fendell, J. Plasma Phys., **4**, Pt.2, 317-334 (1970).

50   M.S. Benilov, Izv. AN SSSR. MZhG, No.5, 145-152 (1982).

51   P.C.T. de Boer and G.S.S. Ludford, Plasma Phys., **17**, No.1, 29-43 (1975).

52   M.S. Benilov, V.F. Kosov, B.V. Rogov and V.A. Sinel'shikov, Teplofiz. Vysokikh Temp., **25**, No.3, 573-581 (1987).

53   M.S. Benilov, G.G. Bochkarev, V.I. Kovbasyuk and E.P. Reshetov, Diagnostics of plasma in MHD generators, Electrophysical processes in MHD channels, IVTAN, Moscow (1990).

54   A.A. Yastrebov, ZhTF, **42**, No.6, 1143-1153 (1972).

55   A.A. Yastrebov, ZhTF, **4**, No.6, 809-820 (1972).

56   F.H. Dorman and J.A. Hamilton, Inter. J. Mass Spectrom. Ion Phys., **24**, No.4, 359-361 (1977).

57   G.S. Aravin, P.A. Vlasov, Yu.K. Karasevich, Chemical reactions in non-equilibrium plasma, (edited by L.S. Polak), Nauka, Moscow (1983).

58   N.N. Barano, M.S. Benilov, G.G. Bochkarev, et al, ZhPMTF, No.3, 13-19 (1983).

59   V.A. Kotel'nikov and V.P. Demkov, IFZh, **52**, No.2, 224-227 (1987).

60   G.S. Aravin, P.A. Vlasov, Yu.K. Karasevich, et al, Izv. AN SSSR, MZhG, No.6, 114-119 (1987).

61   G.S. Aravin, R.A. Vlasov, Yu.K. Karasevich, et al, Izv. AN SSSR, MZhG, No.5, 163-169 (1988).

62   P.A. Vlasov, Yu.K. Karasevich, I.L. Pankrat'eva and V.A. Polyanskii, Teplof. Vysokikh Temp., **26**, No.6, 1047-1056 (1988).

63   V.P. Demkov, Mathematical modelling of the transport processes in plasma taking surface effects into account, MAI, Moscow (1989).

64   V.A. Kotel'nikov and T.A. Gurina, IFZh, **52**, No.5, 856 (1989).

65   T.A. Gurina, Flows of particles and the structure of the perturbed zone in the vicinity of charged solid moving implants, Dissertation, MAI, Moscow (1987).

66   M.S. Benilov and G.A. Tirskii, PMM, **43**, No.2, 288-304 (1979).

67   J. Stratton, Electromagnetism phenomena, Gostekhizdat, Moscow and Leningrad (1948).

68   M.S. Benilov, B.V. Rogov and G.A. Tirskii, ZhPMTF, No.3, 5-13 (1982).

69   I. Khoisler, Magnitnaya Gidrodinamika, No.2, 103-110 (1967).

70   M.A. Benilov and Z.G. Kamalov, Magnitnaya Gidrodinamika, No.1, 41-49 (1988).

71   R.E. Kiel, J. Appl. Phys., **40**, No.9, 3668-3673 (1969).

72   K.N. Ul'yanov, ZhTF, **40**, No.4, 790-798 (1970).

73   K.N. Ul'yanov, ZhTF, **48**, No.5, 920-926 (1978).

74   E.F. Prozorov, Ya.I. Londer, K.P. Novikova and K.N. Ul'yanov, Teplofiz. Vysokikh Temp., **18**, No.1, 164-168 (1980).

75   K.N. Ul'yanov, Teplofiz. Vysokikh Temp., **16**, No.3, 492-496 (1978).

76   M.S. Benilov, V.I. Kyuvbasyuk and G.A. Lyubimov, DAN SSSR, **266**, No.4, 812-816 (1982).

77   N. Stahl and C.H. Su, Phys. Fluids, **14**, No.7, 1366-1376 (1971).

78   B. Tolbot, Raket. Tekhnika Kosmonavtika, **14**, No.1, 95-104 (1976).

79   T. Hirano, Bull. JSME, **17**, No.107, 625-632 (1974).

80   R.M. Clements, B.M. Oliver, A.I. Noor, P.R. Smy, Electron. Lett., **12**, No.11, 274-275 (1976).

81   A.I. Noor, P.R. Smy and R.M. Clements, J. Phys. D., **10**, No.12, 1643-1651 (1977).

References

82   D. de Boer, Raket. Teknika Kosmonavtika, **15**, No.5, 106-115 (1977).
83   R.M. Clements and P.R. Smy, J.Phys. D., **14**, No.6, 1001-1008 (1981).
84   P.R. Smy, Can. J. Phys., **54**, No.15, 1627-1636 (1976).
85   M.S. Benilov, B.V. Rogov and G.A. Tirskii, Teplofiz. Vysokikh Temp., **19**, No.5, 1031-1039 (1981).
86   D.A. Frank-Kamenetskii, Diffusion and heat transfer in chemical kinetics, Nauka, Moscow (1987).
87   A.A. Zhukauskas, Convective transport in heat exchanges, Nauka, Moscow (1982).
88   B.S. Petukhova and V.K. Shikova, A handbook of heat exchangers (translated from English), Vol.1, Energoatomizdat, Moscow (1987).
89   H. Sudzi, Raket. Tekhnika Kosmonavtika, **11**, No.1, 119-121 (1973).
90   A. Kanzawa and S. Monouchi, Intern. Chem. Eng., **16**, No.1, 184-189 (1976).
91   M. Sudzuki and A. Kanzava, Raket. Tekhnika Kosmonavtika, **17**, No.12, 26-33 (1979).
92   Yu.P. Gupalo, A.D. Polyanin and Yu.S. Ryazantsev, Heat and mass exchange of reacting particles with the flow, Nauka, Moscow (1985).
93   M.S. Benilov and B.V. Rogov, Proc. of the Nat. Conf. on the Electrophysics of Combustion, Karaganda State University, Karaganda (1990), pp.37-38.
94   S.. Chang, Raket. Tekhnika Kosmonavtika, **3**, No.5, 22-33 (1965).
95   M.S. Benilov, Aerodynamics of hypersonic velocities in the presence of blow, (edited G.A. Tiriskii), Publishing House of the Moscow State University, Moscow (1979), pp.167-173.
96   M.V. Zakharov, Aerophysics and geocosmic investigations, MFTI, Moscow (1983).
97   M.S. Benilov and G.A. Tirskii, PMM, **44**, No.2, 281-289 (1980).
98   M.S. Benilov and G.A. Tirskii, PMM, **44**, No.5, 839-846 (1980).
99   O.V. Dobrocheev and V.P. Motulevich, Proc. of Nat. Symp. on the Methods of Aerophysical Investigations, ITPM SO AN SSSR, Novosibirsk (1976).
100   E.K. Chekalin, L.V. Chernikh, M.A. Novgorodov and N.V. Khandurov, Ibid, ITPM SO AN SSSR, Novosibirsk (1976).
101   F.G. Baksht and O.G. Rutkin, ZhTF, **49**, No.12, 2549-2558 (1979).
102   E.K. Chekalin and L.V. Chernikh, ZhPMTF, No.1, 41-47 (1981).
103   X. Chen, J. Phys. D., **16**, No.2, L29-L32 (1983).
104   M.S. Benilov and G.A. Tirskii, ZhPMTF, No.6, 16-24 (1979).
105   S.V. Bushuev, E.F. Prozorov and K.N. Ul'yanov, Teplof. Vysokikh Temp., **26**, No.3, 582-588 (1988).
106   M.B. Hopkins and W.G. Graham, Rev. Sci. Instrum., **57**, No.9, 2210-2217 (1986).
107   M.B. Hopkins, W.G. Graham and T.J. Griffin, Rev. Sci. Instrum., **58**, No.3, 475-476 (1987).
108   A.A. Vanin, M.G. Kasparov, A.V. Mokhov, et al, Prepr./IVTAN, Moscow, No.3-286 (1990).
109   R.M. Clements and P.R. Smy, J. Energy, **2**, No.1, 53-58 (1978).
110   I.A. Vasil'eva, A.P. Nefedov and S.E. Self, Magneto-hydrodynamic energy conversion. Open cycle, (edited by B.Ya. Shumyatskii and M. Petrik), Nauka, Moscow (1979).
111   V.M. Batenin, I.A. Vasil'eva and V.F. Kosov, Teplofiz. Vysokikh Temp., **20**, No.2, 229-235 (1982).
112   I.A. Vasil'eva, Magneto-hydrodynamic energy conversion. Physio-technical aspects, (edited by V.A. Kirillin and A.E. Sheindlin), Nauka, Moscow (1983).
113   A. Anders, A.P. Ershov, K.Sh. Isaev and I.B. Timofeev, Teplofiz. Vysokikh Temp. **25**, No.4, 743-747 (1987).
114   G.A. Batyrbekov, E.A. Belyakova, M.S. Benilov, et al, Contr. Pap. 19th Intern. Conf. Phenomena Ionized Gases, Vol. 4, Belgrade (1989).
115   R.M. Clements and P.R. Smy, J. Phys. D., **7**, No.4, 551-562 (1974).
116   M.I. Ashin, I.A. Vasil'eva, V.F. Kosov and A.P. Nefedov, Fiz. Plazmy, **1**, No.3, 483-487 (1975).
117   M.V. Zake and K.S. Landman, Izv. AN Latv. SSR. Ser. Fiz i Tekhn. Nauk., No.2, 79-85 (1977).
118   N. Negishi and I. Kimuar, Bul. JSME, **23**, No.181, 1171-1179 (1975).

119   T. de Boer, Raket. Tekhnika Kosmonavtika, **13**, No.8, 182-184 (1975).
120   C.S. Maclatchy, Combust. Flame, **36**, No.2, 171-178 (1979).
121   C.S. Maclatchy and R. Didsbury, Can. J. Phys., **57**, No.3, 381-384 (1979).
122   I.A. Vasil'eva and V.F. Teplofiz. Vysokikh Temp., **19**, No.5, 1022-1030 (1981).
123   K.D. Annen, P.J. Kuzmenko, R. Keating and S.A. Self, J. Energy, **5**, No.1, 31-38 (1981).
124   J. Lachmann and G. Winkelmann, Beit. Plasmaphys., **22**, No.1, 55-63 (1982).
125   M. Rosenbaum and J.F. Louis, J. Appl. Phys., **53**, No.6, 4088-4092 (1982).
126   M.S. Benilov and B.V. Rogov, Proc. 2nd National seminar: Elementary processes in plasma of electronegative gases, EGU, Erevan (1984), pp. 35-37.
127   P.M. Hierl and J.F. Paulson, J. Chem. Phys., **80**, No.10, 4890-4900 (1984).
128   M.S. Benilov, V.M. Grine, A.A. Lash, et al, Teplof. Vysokikh Temp., **28**, No.3, 620-622 (1990).
129   B.M. Oliver, R.M. Clements and P.R. Smy, J. Phys. D., **9**, No.12, 1715-1718 (1976).
130   F.G. Baksht, ZhTF, **48**, No.10, 2019-2026 (1978).
131   C.R. Giles, R.M. Clements and P.R. Smy, J. Phys. D., **12**, No.10, 1685-1697 (1983).
132   E.F. Prozorov and K.N. Ul'yanov, Teplof. Vysokikh Temp., **21**, No.3, 538-543 (1983).
133   E.F. Prozorov and K.N. Ul'yanov, Teplof. Vysokikh Temp., **22**, No.6, 1179-1185 (1983).
134   Yu.S. Akishev and A.P. Napartovich, DAN SSSR, **242**, No.4, 812-815 (1978).
135   I.B. Chekmarev, ZhTF, **44**, No.10, 2069-2074 (1974).
136   V.A. Rozhansky and L.D. Tsendin, ZhTF, **48**, No.8, 1647-1653 (1978).
137   F.G. Baksht, ZhTF, **49**, No.2, 439-441 (1979).
138   B.V. Kuteev, V.A. Rozhansky and L.D. Tsenin, Beit. Plasmaphys., **19**, No.2, 123-126 (1979).
139   V.P. Demkov, V.A. Kotel'nikov and A.A. Moskalenko, IFZh, **57**, No.4, 692-693 (1989).
140   Ya.F. Volkov, V.G. Dyatlov and N.I. Mitina, Diagnostics of turbulent flows, Naukova dumka, Kiev (1983).
141   M.S. Barad and I.M. Cohen, Phys. Fluids, **17**, No.4, 724-734 (1974).
142   F.G. Baksht, G.A. Dyuzhev, V.B. Kaplan, et al, Problem diagnostics of low-temperature plasma, Preprint of the A.F. Ioffe Institute of Technical Physics, AN SSSR, Leningrad, No.532, 57 (1978).
143   F.G. Baksht, G.A. Dyuzhev, Tsirkel', et al, Problem diagnostics of low-temperature plasma, Preprint of the A.F. Ioffe Institute of Technical Physics, Leningrad, No.533, 52 (1978).
144   G.A. Dyuzhev, Near-electrode phenomena in high-current arc and thermoemission cathodes, A.F. Ioffe Institute of Technical Physics, AN SSSR, Leningrad (1980).
145   E.M. Stepanov and B.G. D'yachkov, Ionization in the flame and the electric field, Metallurgiya, Moscow (1968).
146   J. Lauton and F. Bainberg, Electric aspects of combustion, Energiya, Moscow (1976).
147   S. Matsumura, J.S. Chang and S. Teii, Contr. Pap. 18th Intern. Conf. Phenomena Ionized Gases, Swansea, Vol.4, (1987), pp.620-621.
148   V. Roter and F. Bergman, Physics and technology of low-temperature plasma, ITMO im A.V. Lykova AN BSSR, Minsk (1977).
149   H.F. Calcote and I.R. King, 5th Sympos. Intern. Combustion, Pitsburg, USA (1954), pp.423-434.
150   M. Mackey, Raket. Teknika Kosmonavtika, **6**, No.8, 186-188 (1968).
151   R. Carabetta and R.P. Porter, 12th Sympos. Intern. Combustion, Poiters, France (1968), pp.423-434.
152   D.E. Jensen and S.C. Combust. Flame, **13**, No.2, 219-222 (1969).
153   H. Tsudzi, Raket. Tekhnika Kosmonavtika, **11**, No.6, 153-154 (1973).
154   E.M. Egorova, A.V. Kashevarov, E.M. Fomina and N.S. Tskhai, Teplofiz. Vysokikh Temp., **26**, No.3, 577-581 (1988).

References

155 L. Katona and Gh. Nichifor, 10th Intern. Conf. MHD Electrical Power Generation, Tiruchirapalli, Vol.3 (1989), pp.X127-X131.

Chapter 12

1 P. Chan, L. Telbot and K. Turyan, Electric probes in stationary and moving plasma, Mir, Moscow (1978).

2 I.L. Pankrat'eva and V.A. Polyanskii, Izv. AN SSSR. MZhG, No.2, 102-120 (1979).

3 G.S. Aravin, P.A. Vlasov, Yu.K. Karasevich, et al, Fiz. Gor. Vzriva, **18**, No.1, 49-57 (1982).

4 R. Huddlestone and S. Leonard (eds), Plasma diagnostics, Mir, Moscow (1967).

5 Yu.N. Belyaev, V.A. Polyanskii and E.G. Shapiro, Aerodynamics of high velocities, Publishing House of the Moscow State University, Moscow (1979).

6 V.A. Polyanskii, ZhPMTF, No.5, 11-17 (1964).

7 I. MacDaniel and E. Mezon, Mobility and difusion of ions in gases, Mir, Moscow (1976).

8 D.E. Golden and H. Bandel, Phys. Rev., **149**, No.1, 58-59 (1966).

9 I.L. Pankrat'eva and V.A. Polyanskii, Aerodynamics of hypersonic flows in the presence of blow, Publishing House of the Moscow State University, Moscow (1979).

10 K.N. Ul'yanov, ZhTF, **40**, No.4, 790-797 (1970).

11 T.A. Cool and P.J.H. Tjossem, Gas-Phase Chemiluminescence and Chemi-Ionization, A. Fontijn (ed), Amsterdam, North Holland (1985).

12 B.V. Alekseev and V.A. Kotel'nikov, Probe method of plasma diagnostics, Energoatomizdat, Moscow (1988).

13 G.S. Aravin, P.A. Vlasov, Yu.K. Karasevich, et al, Experimental and theoretical investigations of plasma chemical processes, (edited by L.S. Polak), Nauka, Moscow (1984).

14 D. Hasted, Physics of atomic collisions, Mir, Moscow (1965).

15 V.D. Rusanov and A.A. Fridman, Physics of chemically active plasma, Nauka, Moscow (1984).

16 A.V. Eletskii and B.M. Smirnov, Modelling and methods of calculating chemical processes in low-temperature plasma (edited by L.S. Polak), Nauka, Moscow (1974).

17 P. Bailey and K. Turyan, Raket. Tekhnika i Kosmonavtika, **11**, No.9, 12-13 (1973).

18 T. Luzzi and R. Jenkins, Raket. Teknika i Kosmonavtika, **9**, No.12, 126-132 (1971).

19 B.V. Alekseev, V.A. Kotel'nikov and V.V. Cherepanov, Teplofiz. Vysokikh Temp., **22**, No.2, 395-396 (1984).

20 G.S. Aravin, P.A. Vlasov, Yu.K. Karasevich, et al, Izv. AN SSSR. MZhG, No.5, 163-169 (1988).

21 P.A. Vlasov, Yu.K. Karasevich, I.L. Pankrat'eva and V.A. Polyanskii, Teplofiz. Vysokikh Temp., **26**, No.6, 1047-1056 (1988).

22 G.S. Aravin, P.A. Vlasov, Yu.K. Karasevich, et al, Chemical reactions in non-equilibrium, (edited by L.S. Polak), Nauka, Moscow (1983).

23 B.M. Smirnov, Atomic collisions and elementary processes in plasma, Atomizdat, Moscow (1978).

24 G. Messi, Negative ions, Mir, Moscow (1981).

25 F. Bruair, D. MacGowen (eds), Physics of ion- and electron-ion collisions, Mir, Moscow (1986).

26 J. Lauton, F. Bainberg, Electric aspects of combustion, Energiya, Moscow (1976).

27 P.A. Vlasov, Probe diagnostics of non-stationary chemically reacting high-pressure plasma, Dissertation, IKhF AN SSSR, Moscow (1983).

28 B.V. Alekseev and V.A. Kotel'nikov, Experimental and theoretical investigations of plasma chemical processes, (edited by L.S. Polak), Nauka, Moscow (1984).

29 G.S. Aravin, Y.K. Karasevich, P.A. Vlasov, et al, Proc. 15th Intern. Conf.

Phenomena Ionized Gases, Pt.2, Minsk (1981), pp.957-958.
30   S.A. Losev and V.A. Polyanskii, Izv. AN SSSR. MZhG, No.1, 176-183 (1968).
31   C.F. Hansen, Phys. Fluids, **11**, No.4, 904-908 (1968).
32   I. McDaniel, Collisions processes in ionised gases, Izd-vo Inostr. Lit., Moscow (1966).
33   V.K. Dushin, Tr. Inst. Mekh. MGU, No.21, part 2, 35-43, Moscow (1973).
34   G.S. Aravin, P.A. Vlasov, Yu.K. Karasevich, et al, Fiz. Gor. Vzryva, **20**, No.4, 70-77 (1984).
35   B.M. Smirnov, Ions and excited atoms in plasma, Atomizdat, Moscow (1974).
36   J. Tarstrup and W.J. Heikkila, Radio Science, **7**, No.4, 493-502 (1972).
37   F.G. Baksht, hTF, **48**, No.10, 2019-2026 (1978).
38   E.F. Prozorov and K.N. Ul'yanov, Teplofiz. Vysokikh Temp., **21**, No.3, 538-543 (1983).
39   E.F. Prozorov and K.N. Ul'yanov, Teplofiz. Vysokikh Temp., **21**, No.6, 1179-1185 (1983).
40   G.S. Aravin, P.A. Vlasov, Yu.K. Karasevich, et al, Izv. AN SSSR. MZhG, No.6, 114-119 (1987).

Chapter 13

1   J.R. Hollahan and A.T. Bell (eds), Techniques and Application of Plasma Chemistry, Wiley Interscience Publication, New York (1974), p.403.
2   B.V. Tkachuk and V.M. Kolotyrkin, Production of thin polymer films from gas phases, Nauka, Moscow (1977).
3   N. Isbrooke and D. Braun (eds), Plasma technology in the production of SBIS, Mir, Moscow (1987).
4   H. Deutsch and S. Klagge, Ernst-Moritz-Arnold Univ. Section Physik/Electronik. Rep. 1977/78, Greifswald (1978), pp.1-26.
5   H. Deutsch and S. Klagge, Beit. Plasma Phys., **19**, No.1, 49-57 (1979).
6   G.K. Vinogradov and Yu.A. Ivanov, Khim. Vys. Energ., **12**, No.6, 542-546 (1978).
7   G.K. Vinogradov, Yu.A. Ivanov and L.S. Polak, Khim. Vys. Energ., **13**, No.1, 84-85 (1979).
8   Yu.A. Ivanov, Yu.A. Lebedev and L.S. Polak, Methods of contact diagnostics and non-equilibrium plasma chemistry, Nauka, Moscow (1981).
9   I.I. Amirov, G.K. Vinogradov and D.I. Slovetskii, Mechanisms of plasma chemical reactions of hydrocarbons and carbon-containing molecules, Ch.1,(edited by L.S. Polak), INKhS AN SSSR, Moscow (1987).
10   Yu.A. Ivanov, Physico-chemical processes in non-equilibrium plasma of hydrocarbons and carbon-containing molecules, IKhF AN SSSR, Moscow (1989).
11   V.A. Pliskin and S.Zh. Zanin, Thin film technology, a handbook, Vol.2 (edited by L. Maissel and R. Gleng), Sov. Radio, Moscow (1977).
12   G.K. Vinogradov, Yu.A. Ivanov and L.S. Polak, Khim. Vys. Energ., **14**, No.2, 174-178 (1980).
13   G.K. Vinogradov, Yu.A. Ivanov, L.S. Polak and V.N. Timakin, Khim. Vys. Energ., **14**, No.5, 461-465 (1980).
14   S.G. Garanin, Yu.A. Ivanov and V.N. Timakin, Khim. Vys. Energ., **15**, No.2, 183-184 (1981).
15   Yu.A. Ivanov, L.S. Polak and V.N. Timakin, Khim. Vys. Energ., **15**, No.2, 181-182 (1981).
16   P.R. Emtage and W. Tantraport, Phys. Rev. Lett., **8**, No.7, 267-268 (1962).
17   A. Bradley and J.F. Hammes, J. Electrochem. Soc., **110**, No.1, 15-22 (1963).
18   H.T. Mann, J. Appl. Phys., **35**, No.7, 2173-2179 (1964).
19   R.W. Christy, Ibid, 2179-2184.
20   M. Gazicki and H. Yasuda, Plasma Chemistry and Plasma Processing, **3**, No.3, 279-329 (1983).
21   D.G. Simmons, Thin film technology, a handbook, Vol.2, (edited by L. Maissel and R. Gleng), Sov. Radio, Moscow (1977).
22   H. Carchano and M. Valentin, Thin Solid Films, **31**, No.3, 335-349 (1975).
23   F.-W. Breitbarth and H.-J. Tiller, Wiss. Z. Fridrich-Schiller Univ., Jena, Math.-

References

Nat. R (1976).

24   F. Gronlund, Prib. Nauch. Issled., No.10, 100-102 (1982).

25   G.K. Vinogradov, G.Zh. Imanbaev and D.I. Slovetskii, Synthesis of compounds in plasma containing hydrocarbons (edited by L.S. Polak), INKhS AN SSSR, Moscow (1985).

26   G.K. Vinogradov, G.Zh. Imanbaev and D.I. Slovetskii, Khim. Vys. Energ., **19**, No.5, 461-464 (1985).

27   G.K. Vinogradov, G.Zh. Imanbaev and D.I. Slovetskii, Mechanism of plasma chemical reactions of hydrocarbons and carbon-containing molecules, Ch.1, (edited by L.S. Polak), INKhS AN SSSR, Moscow (1987).

28   T. Harada, K. Gamo and S. Namba, Jap. J. Appl. Phys.,**20**, No.1, 259-264 (1981).

29   J. Wood, J. Phys. Chem., **67**, No.7, 231-239 (1963).

30   D.I. Slovetskii, Mechanisms of chemical reactions in non-equilibrium plasma, Nauka, Moscow (1980).

31   G.K. Vinogradov, Yu.A. Ivanov and Yu.A. Lebedev, Plasma chemical reactions and processes, (edited by L.S. Polak), Nauka, Moscow (1977).

32   C.M. Ferreira and A. Ricard, J. Appl. Phys., **54**, No.5, 2261-2271 (1983).

33   A.A. Labrenko, Recombination of hydrogen atoms on the surface of solids, Naukova dumka, Kiev (1973).

Chapter 14

1   R. Friedman and J.A. Cyphers, J. Chem. Phys., **23**, No.10, 1875-1880 (1955).

2   J. Grey, P.F. Jacobs and M.P. Sherman, Prib. Nauch. Issled., No.7, 29-33 (1962).

3   J. Grey, Rev. Sci. Instrum., **34**, No.8, 857-859 (1963).

4   J. Cordero, F.W. Diederich and H. Hurwicz, Aero-space Eng., **22**, No.1, 166-191 (1963).

5   F. Fruchtman, Raket. Tekhnika i Kosmonavtika, No.8, 186-187 (1963).

6   J. Grey and P.F. Jacobs, Raket. Tekhnika i Kosmonavtika, No.3, 25-31 (1964).

7   J.L. Potter, G.D. Arney and M. Kinslow, IEEE Transactions Nuclear Science, **NS-11**, No.1, 145-157 (1964).

8   J. Grey, Instr. Soc. Am. Trans., **4**, No.2, 102-114 (1965).

9   J. Grey, USA, Cooled calorimetric probe, Patent No.3167956 USA MKI^2 F 02 S 12/03, Claim 16.01.64, Published 02.02.65.

10   V.L. Sergeev, IFZh, **9**, No.5, 657-666 (1965).

11   J. Grey, M.P. Sherman, P.M. Uil'yams, et al, Raket. Tekhnika i Kosmonavtika, **4**, No.6, 36-45 (1966).

12   U.G. Karden, Raket. Tekhnika i Kosmonavtika, **4**, No.10, 12-22 (1966).

13   U. Sprengel, Atompraxis (Direct Information Raumfahrtforschung und Technik, 1/66), **12**, No.2 (1966).

14   G.F. Au and U. Sprengel, Z. Flugwissenschaften, **14**, No.4, 188-194 (1966).

15   I.T. Alad'ev, I.G. Kulakov, O.L. Magdasiev, Izv. SO AN SSSR. Ser. Tekh. Nauk., No.10, issue 3, 50-55 (1966).

16   J. Grey and P.F. Jacobs, Raket. Tekhnika i Kosmonavtika, **5**, No.1, 98-106 (1967).

17   D.W. Esker, AIAA J., **5**, No.8, 1504-1506 (1967).

18   M. Krutil, Raket. Tekhnika i Kosmonavtika, **6**, No.1, 124-133 (1968).

19   F.A. Vassalo, AIAA Paper, No.68-391, 1-12 (1968).

20   J. Grey, Space Aeronautics, **50**, No.5, 82-86 (1968).

21   R.B. Poup, Raket. Tekhnika i Kosmonavtika, **6**, No.1, 124-133 (1968).

22   V.S. Klubnikin and S.V. Dresvin, Uchen. Zap. LGPI im A.I. Gertsena, **384**, No.2, 46-62 (1968).

23   J. Lot, Proc. of Int. Symp. in Salzburg, Austria, MHG-Generators, VINITI (1969).

24   A.A. Voropaev, V.M. Gol'dfarb, A.V. Donoskoi, et al, Uchen. Zap. LGPI im A.I. Gertsena (low-temperature plasma), **384**, No.2, 109-115 (1968).

25   A. Vitte, T. Kubota and L. Liz, Raket. Tekhnika i Kosmonavtika, No.5, 91-101 (1969).

26   V.S. Klubnikin, Transfer phenomena in low-temperature plasma, Nauka i Tekhnika,

Minsk (1969).

27   P.F. Massey, L.G. Bek and E.D. Roshke, Teploperedacha, **91**, Ser. C, No.1, 77-85 (1969).

28   R.M. Fristrom and A.A. Vestenberg, Structure of the flame, Metallurgiya, Moscow (1969).

29   R. Neubeck, Angew. Chem., **81**, No.22, 936-942 (1969).

30   G.A. Andreev, S.V. Dresvin and V.S. Klubnikin, Physics, technology and application of low-temperature plasma, Alma-Ata (1970).

31   M.D. Petrov and V.A. Sepp, Teplof. Vys. Temp., **8**, No.4, 868-874 (1970).

32   J. Cheylan, J.P. Dauvergne and P. le Goff, Chim.et ind., Gen chim., **103**, No.20, 2679-2686 (1970).

33   V.S. Klubnikin, Problems in the physics of low-temperature plasma, Nauka i Tekhnika, Minsk (1970).

34   M.M. Abu-Romia and B. Bhatia, AIAA Pap., No.81, 1-6 (1971).

35   V.K. Mel'nikov and R.Kh. Takhtaganov, Izv. AN Latv. SSR, Ser. Fiz. Tekhn. Nauk, No.2, 57-64 (1971).

36   L.A. Anderson and R.E. Sheldahl, AIAAJ., **9**, No.9, 1804-1810 (1971).

37   A.V. Donskoi and V.S. Klubnikin, Physics and technology of low-temperature plasma, Atomizdat, Moscow (1972).

38   A.A. Nekrasov and A.D. Rekin, Izmerit. Tekhnika, No.4, 47-48 (1972).

39   M.D. Petrov and V.A. Sepp, Izmerit. Tekhnika, No.4, 49050 (1972).

40   A.I. Abrosimov, V.A. Pechurkin and S.I. Shorin, Thermophysics, Vol.4: Heat exchange in the high-temperature gas flow, Mintis, Vilnyus (1972), pp.34-42.

41   V.I. Gudkov, V.P. Motulevich, A.S. Sergeev, et al, Proc. 5th Nat. Conf. ???, Vol.2, IT SO AN SSSR, Novosibirsk (1972).

42   M.D. Petrov and V.A. Sepp, Thermophysics, Vol.4: Heat exchange in the high-temperature gas flow, Mintis, Vilnyus (1972), pp.20-33.

43   J. Grey (USA), Enthalpy probe with screened thermally measuring tube, US Patent No.3665763 USA MKI^2 F 02 S 12/06. Claim 27.06.72, Publ. 30.05.72.

44   K.N. Chakalev, V.I. Gudkov and A.S. Sergeev, Thermophysical properties and gas dynamics of high temperature media, Nauka, Moscow (1972).

45   S. katta, J.A. Lewis and W.H. Gauvin, Rev. Sci. Instr., **44**, No.10, 1519-1523 (1973).

46   J. Lachmann, Exp. Techn. Phys., **21**, No.4, 361-367 (1973).

47   V.I. Gudkov, V.P. Motulevich and K.N. Chakalev, Proc. 6th Nat. Conf. on Generated Low-temperature Plasma, Ilim, Frunze (1974), pp.424-427.

48   L.A. Gus'kov and Yu.V. Yakhlakov, Equipment for determining the entalpy of the gas flow, Author's certificate No.459714 SSSR, No.5, Publ. Byuss. Izobr. (1975).

49   S.V. Dresvin and V.S. Klubnikin, Teplofiz. Vys. Temp., **13**, No.2, 433-435 (1975).

50   V.S. Klubnikin, Teplof. Vys. Temp., **13**, No.3, 473-482 (1975).

51   H. Hoffman, Exp. Techn. Phys., **23**, No.4, 389-396 (1975).

53   C.P. Polyakov, O.V. Ryazantsev and V.I. Tverdozhlevov, Teplof. Vys. Temp, **14**, No.2, 423-424 (1976).

54   A.S. Sergeev, Yu.N. Vorontsov and O.V. Inozemtsev, Method and equipment for determining gas enthalpy, Author's certificate No.731322, USSR, No.3, Published in Byuss. Izobr. (1976).

55   A.S. Sergeev, Equipment for measuring the heat content of gases, No.17, Author's certificate No.828048, USSR, Published in Byuss. Izobr. (1981).

56   Yu.N. Vorontsov, O.V. Inozemtsev and A.S. Sergeev, Equipment for caliometry measurements of the gas flow, No.28, Author's certificate No.851228, USSR, Published in Byuss. Izobr. (1981).

57   S.N. Aksenov, A.V. Donskoi, S.Kh. Il'yasova, et al, Proc. of the 9th Nat. Conf. on Low-temperature plasma Generators, Ilim, Frunze (1983), pp.244-245.

58   S.N. Aksenov, S.Kh. Il'yasova, S.Kh. Klubnikin, et al, Teplofiz. Vys. Temp., **23**, No.1, 196-198 (1985).

59   A.G. Zavarzin, G.M. Krylov, V.P. Lyagushkin, et al, Proc. of the 10th Nat. Conf: Generators of Low-temperature plasma, Ch.2, Nauka i Tekhnika, Minsk (1986), pp.141-142.

60   M.F. Zhukov, V.P. Lyagushkin and O.P. Solonenko, Automated experimental stands for complex examination of high temperature heterogeneous jets, No.145-86, Preprint of Institute of Thermophysics, Siberian Division of the Russian Academy of Sciences, Novosibirsk (1986).

Chapter 15

1   N. Kimura and A. Kanzava, Raket. Tekhnika i Kosmonavtika, No.3, 120-127 (1965).
2   A. Abdrazakov and Zh. Zheenbaev, Examination of the electric arc in argon, Ilim, Frunze (1966).
3   V.M. Gol'dfarb and A.M. Uzdenov, Izv. SO AN SSSR, Ser. Tekhn. Nauk, No.3, Issue No.1, 81-84 (1967).
4   E. Eckert, Introduction into the theory of heat and mass exchange, Goseneroizdat, Moscow (1957).
5   S.S. Kutateladze, V.M. Borishanskii, a handbook of heat transfer, GITTL, Moscow and Leningrad (1959).
6   J. Bues, H.J. Patt and J. Richter, Z. Ang. Phys., 22, No.4, 345-352 (1967).
7   K.S. Drellischak, Phys. Fluids, 16, No.5, 616-623 (1963).
8   R. Devoto, Phys. Fluids, 16, No.5, 616-623 (1973).

Chapter 16

1   V.D. Rusanov and A.A. Fridman, Physics of chemically active plasma, Nauka, Moscow (1984).
2   S.A. Krapivina, Plasma chemical technological processes, Khimiya, Leningrad (1981).
3   V.A. Legasov, V.K. Zhivotov, E.G. Krasheninnikov, et al, DAN SSSR, 238, No.1, 66-69 (1978).
4   IEE Transations on Plasma Science. Special Issue on the physics of HF discharges for plasma processing. PS-14, No.2, 77-197 (1986).
5   D.B. Graves, AIChEJ, 35, No.1, 1-29 (1989).
7   B.S. Danilin and V.Yu. Kireev, Using low-temperature plasma for etchingn and cleaning materials, Energoatomizdat, Moscow (1989).
8   V.I. Myshenkov and N.A. Yatsenko, Kvant. Elektron., 8, No.10, 2121-2129 (1981).
9   N.A. Yatsenko, Gas lasers with high-temperature excitation, No.381, Preprint of the Institute of Powder Metallurgy, AN SSSR, Moscow (1989).
10   D.R. Hall and H.J. Baker, Laser focus world, No.10, 77-80 (1989).
11   H.E. Hugel, SPIE, 650, 2-9 (1986).
12   N.A. Yatsenko, ZhTF, 51, No.6, 1195-1204 (1981).
13   N.A. Yatsenko, Special structure of the hf capacitance discharge and prospects for its application in laser technology, No.338, Preprint of the Institute of Powder Metallurgy, AN SSSR, Moscow (1988).
14   N.A. Kaptsov, Electric discharges in gases and vacuum, GITTL, Moscow and Leningrad (1950).
15   B.E. Bragin and V.D. Matyukhin, 8th Nat. Conf. on Low-temperature Plasma Generators, IT SO AN SSSR, Novosibirsk (1980).
16   N.A. Yatsenko, ZhTF, 52, No.6, 1220-1222 (1982).
17   N.A. Yatsenko, Teplof. Vys. Temp., 20, No.6, 1044-1051 (1982).
18   S.M. Levitskii, ZhTF, 27, No.5, 970-977 and 1001-1009 (1957).
19   A.D. Andreev, ZhPS, 5, No.2, 145-147 (1966).
20   N.A. Kuzovnikov and V.P. Savinov, RE, 18, No.4, 816-822 (1973).
21   A.A. Kuzovnikov and V.P. Savinlov, Vestn. MGU, Fizika, No.2, 215-223 (1973).
22   V.A. Godyak and A.S. Khanneh, IEEE Transactions on Plasma Science, PS-14, No.2, 112-123 (1986).
23   N.A. Yatsenko, Tr.MFTI, Series General and Molecular Physics, Dolgoprudniy, Moscow (1978), pp.226-229.
24   N.A. Yatsenko, Tr.MFTI, Ibid (1978), pp.166-170.
25   N.A. Yatsenko, ZhTF, 50, No.11, 2480-2482 (1980).

26   V.Kh. Goikhman and V.M. Gol'dfarb, ZhPS, **21**, No.3, 456-459 (1974).

27   A.S. Kovalev, A.I. Nazarov, A.T. Rakhimov, et al, Fiz. Plazmy, **12**, No.10, 1264-1268 (1986).

28   Yu.P. Raizer and M.N. Shneider, Fiz. Plazmy, **13**, No.4, 471-479 (1987).

29   Yu.P. Raizer and M.N. Shneider, Fiz. Plazmy, **14**, No.2, 226-232 (1988).

30   P. Vidaud, S.M.A. Durrani and D.R. Hall, J. Phys. D: Appl. Phys., **21**, No.1, 57-66 (1988).

31   A. Engel', Ionized gases, GIFML, Moscow (1959).

32   Yu.P. Raizer, Physics of gas discharge, Nauka, Moscow (1987).

33   V.I. Myshenkov and N.A. Yatsenko, Fiz. Plazmy, **8**, No.3, 543-549 (1982).

34   N.A. Yatsenko, ZhTF, **58**, No.2, 296-301 (1988).

35   A.S. Kovalev, E.F. Muratov, A.A. Ozerenko, et al, Pis'ma ZhTF, **10**, No.18, 1139-1142 (1955).

36   D. Banerji and R. Ganguli, Phil. Mag., **13**, 494-498 (1932).

37   Kh.A. Jerpetov and G.M. Pateyuk, ZhETF, **28**, No.3, 343-351 (1955).

38   H.S. Butler and G.S. Kino, Phys. Fluids, **6**, No.9, 1052-1065 (1963).

39   V.A. Godyak, A.A. Kuzovnikov, V.P. Savinov, et al, Vestn. MGU, Fiz., No.2, 126-127 (1968).

40   V.A. Godyak and A.A. Kuzovnikov, Fiz. Plazmy, **1**, No.3, 496-503 (1975).

41   A.A. Kuzovnikov, V.L. Kovalevskii, V.P. Savinov and V.G. Yakunin, Proc. 13th Intern. Conf. Phenomena Ionized Gases, Berlin, DDR (1977), pp.343-344.

42   J. Hay, Canad. J. Res., **A16**, 191-200 (1938).

43   Ph.Belenguer and J.P. Boeuf, Proc. 19th Intern. Conf. Phenomena Ionized Gases, Belgrade (1989), pp.394-395.

44   H. Beck, Z.Phys., **97**, No.4, 766-775 (1935).

45   V.A. Godyak and O.N. Popov, ZhTF, **47**, No.4, 766-775 (1977).

46   R.R.J. Gagne and A. Cantin, J. Appl. Phys., **43**, No.6, 2639-2651 (1972).

47   A.V. Aleksandrov, V.A. Godjak, A.A. Kuzovnikov, et al, Proc. 8th Intern. Conf. Phenomena Ionized Gases, Vienna (1967), p.165.

48   N.A. Yatsenko, Reasons and conditions for formation of constant emf in a capacitance hf medium pressure discharge, Tartu State University, Tartu (1989), pp.208-210.

49   D. He and D.R. Hall, IEEE Journal Quantum Electronics, **QE-20**, No.5, 509-516 (1984).

50   P. Hoffman, SPIE, **650**, 23-29 (1986).

51   A.G. Akimov, A.V. Koba, N.I. Lipatov, et al, Kvant. Elektron., **16**, No.5, 938-949 (1986).

52   N.I. Lipatov, P.P. Pashinin, A.M. Prokhorov, et al, Gas discharge and waveguide molecular lasers, Vol.17, Tr. IOFAN, Nauka, Moscow (1989).

53   S. Yatsiv, Gas Flow Chemical Laser Conf, S. Rosenwarks (ed), Springer, Berlin, Germany (1987), pp.252-257.

54   V.I. Myshenkov and N.A. Yatsenko, ZhTF, **51**, No.10, 2055-2062 (1981).

55   N.A. Yatsenko, Proc. of the 2nd Nat. Conf: High frequency discharge in wave fields, Kuibysh Polytechnic Institute, Kuibysh (1989), p.13.

56   D. He and D.R. Hall, J. App. Phys., **54**, No.3, 4367-4375 (1983).

57   A.D. Andreev, ZhPS, **5**, No.2, 145-147 (1966).

58   A.D. Andreev, Vest. BelGU, Ser.1, No.2, 78-80 (1969).

59   A.D. Andreev, V.D. Il'in and Yu.N. Lobanov, ZhTF, **36**, No.9, 1636-1638 (1966).

60   A.V. Kalmykov, B.Yu. Nezhentsev, A.S. Smirnov, et al, ZhTF, **59**, No.9, 93-97 (1989).

61   N.A. Yatsenko, Proc. of 3rd Nat. Conf. on the Physics of Gas Discharges, Ch.1, KGU, Kiev (1989).

62   N.A. Yatsenko, XX ICPIG. Contrib. Papers, Vol.5, Pisa (1991), pp.1159-1160.

Chapter 17

1   G. Peter, Electronic avalanches and breakdown in gases, Mir, Moscow (1968).

2   E.D. Lozanskii and O.B. Firsov, Theory of the spark, Atomizdat, Moscow (1975).

3   L.B. Loeb, Science, **148**, No.3676, 1417-1426 (1965).

References

4   J.J. Thomson, Recent Researches in Electricity and Magnetism, Clarendon, Oxford (1893), pp.115-118.
5   R.G. Fowler, Adv. Electron. Phys., **35**, 1-84 and **41**, 1-78 (1976).
6   J.W. Beams, Phys. Rev., **28**, 475-481 (1926).
7   L.B. Snoddy, J.R. Dietrich and J.W. Beams, Ibid, **52**, No.4, 739-746 (1937).
8   J.W. Beams, Ibid, **36**, No.4, 997-1001 (1930).
9   F.H. Mitchell and L.D. Snoddy, Ibid, **72**, 1202-1208 (1947).
10   R.G. Westberg, **114**, No.1, 1-17 (1959).
11   M.R. Amin, J. Appl. Phys., **25**, 358-363 (1954).
12   G.G. Hudson and L.B. Loeb, Phys. Rev., **123**, No.1, 29-43 (1961).
13   H. Tholl, Z. Naturforsch, **19a**, 346-359 and 704-706 (1964).
14   A.H. Amin, Proc. 5th Intern. Conf. Phenomena Ionized Gases, Vol.1, Munich (1961), pp.1003-1016.
15   T. Suzuki, J. Appl. Phys., **42**, 3766-3771 (1971).
16   T. Suzuki, Ibid, **44**, 4534-4544 and **48**, No.12, 5001-5007 (1973).
17   W. Bartolmeyezyk, Ann. Phys., **36**, No.6, 485-520 (1939).
18   A.V. Nedospasov and A.E. Novik, ZhTF, **30**, 1329-1336 (1960).
19   A.V. Nedospasov, G.M. Sadykhzade and K.I. Efendiev, Teplof. Vys. Temp., **16**, No.4, 673-676 (1978).
20   V.P. Abramov and I.P. Mazan'ko, ZhTF, **50**, 749-754 (1980).
21   A.W. Trievelpiece and R.W. Gold, J. Appl. Phys., **11**, 1784-1793 (1959).
22   P.S. Bulkin, V.N. Ponomarev and G.S. Solntsev, Vestn. MGU, **3**, 93-95 (1967).
23   W.P.P. Winn, J. Appl. Phys., **38**, No.2, 783-790 (1967).
24   H.E.B. Andersson and R.S. Tobin, Physica Scripta, **9**, No.1, 7-14 (1967).
25   H.E.B. Andersson, Ibid, **4**, 215-220 (1971).
26   N.I. Vinokurov, V.A. Gerasimov, V.V. Zaponchkovskii and Yu.F. Fomenko, ZhTF, **47**, 2512-2521 (1977).
27   F.M. McGehee, Virginia J. Science, **6**, No.1, 39-45 (1955).
28   E.I. Asinovskii, L.M. Vasilyak, A.V. Kirillin and V.V. Markovets, Teplof. Vys. Temp. **13**, No.1, 40-44 and 195-198, and **13**, No.6, 1281-1282 (1975).
29   E.I. Asinovskii, L.M. Vasilyak and V.V. Markovets, Teplof. Vys. Temp., **21**, No.3, 577-590 (1983).
30   R.G. Fowler and J.D. Hood, Phys. Rev., **128**, No.3, 991-992 (1962).
31   E.I. Asinovsky, A.V. Kirillin, V.V. Markovets and L.M. Vasiljak, 7th Intern. Sympos. Discharges Electrical Insulation Vacuum, Publishing House of the Novosibirsk University (1976).
32   E.I. Asinovsky, L.M. Vasilyak, A.V. Kirillin and V.V. Markovets, 3rd Nat. Conf. on Plasma Accelerators, Belrussian University, Minsk (1976).
33   O.A. Sinkevich and Yu.V. Trofimov, DAN SSSR, **249**, No.3, 597-600 (1979).
34   A.N. Lagar'kov and I.M. Rutkevich, DAN SSSR, **249**, No.3, 593-596 (1979).
35   M.J. Lubin and E.R. Resler, Phys. Fluids, **10**, No.1, 1-7 (1967).
36   E.I. Asinovskii, L.M. Vasilyak, A.V. Kirillin and V.V. Markovets, Teplof. Vys. Temp., **18**, No.4, 668-676 (1980).
37   V.V. Basov, I.G. Kataev and D.P. Kolchin, RE, **28**, No.11, 2206-2210 (1983).
38   V.V. Basov, I.G. Kataev and D.P. Kolchin, Proc. of the 6th Nat. Conf. on the Physics of Low-temperature plasma, Vol.2, Leningrad University, Leningrad, (1983), pp.423-425.
39   I.P. Krasyuk, N.I. Lipatov and P.P. Pashinin, Kvant. Elektron., **3**, No.11, 2384-2391 (1976).
40   P. Felzenthal and J. Praud, Electron avalanches and breakdown in gases, Mir, Moscow (1968).
41   L.M. Vasilyak, Electric and spectral characteristics of nanosecond discharge in helium, Dissertation, MFTI/Nauk, Moscow (1975).
42   I. Lewis and F. Wells, Millimicrosecond pulsed technology, Moscow (1956).
43   E.I. Asinovsky, V.V. Markovets and I.S. Samoilov, Teplof. Vys. Temp., **20**, No.6, 1189-1191 (1982).
44   E.I. Asinovsky, V.V. Markovets and I.S. Samoilov, Problems of physics and technology of nanosecond discharges, IVTAN, Moscow (1982).

45 E.I. Asinovsky, V.V. Markovets and I.S. Samoilov, Proc. of 6th Nat. Conf. on Physics of Low-temperature Plasma, Vol.1, Frunze (1983), pp.372-374.

46 E.I. Asinovsky, V.V. Markovets and I.S. Samoilov, Proc. of 9th Nat. Conf. on Low-temperature Plasma Generators, Frunze (1983), pp.184-185.

47 E.I. Asinovsky, V.V. Markovets, I.S. Samoilov and A.M. Ul'yanov, Teplofiz. Vys. Temp., **16**, No.6, 1309-1311 (1978).

48 E.I. Asinovsky, V.V. Markovets and I.S. Samoilov, Proc. of 8th Nat. Conf. on Low-temperature Plasma Generators, Novosibirsk (1980), pp.217-220.

49 E.I. Asinovsky, V.V. Markovets and I.S. Samoilov, Formation of the shockwave of the potential gradient in breakdown of the discharge gap, No.6-067, Preprint IVTAN, Moscow (1981).

50 E.I. Asinovsky, V.V. Markovets and I.S. Samoilov, Teplofiz. Vys. Temp., **19**, No.3, 587-594 (1981).

51 E.I. Asinovsky, V.V. Markovets and I.S. Samoilov, Proc. 16th ICPIG, Vol.5, Dusseldorf (1983), pp.758-759.

52 V.V. Kovalev, F.M. Subbotin and E.I. Shubnikov, PTE, No.1, 158-159 (1972).

53 G.S. Vil'dgrubbe, Zh.M. Ronkin and Yu.A. Kolosov, Izv. AN SSSR, Ser. Fiz., **28**, No.2, 377-383 (1964).

54 E.I. Asinovsky, V.V. Markovets, I.S. Samoilov and A.M. Ulianov, PTE, No.1, 222-223 (1982).

55 E.I. Asinovsky, V.V. Markovets, I.S. Samoilov and A.M. Ulianov, Proc. 15th Intern. Conf. Phenomena Ionized Gases, Minsk (1981), pp.961-962.

56 E.I. Asinovsky, V.V. Markovets and A.M. Ulianov, Proc. 16th Intern. Conf. Phenomena Ionized Gases: Contributed papers, Vol.2, Dusseldorf (1983), pp.148-149.

57 E.I. Asinovsky, V.V. Markovets and A.M. Ul'yanov, Teplof. Vys. Temp., **22**, 667-671 (1984).

58 E.I. Asinovsky, V.V. Markovets, I.S. Samoilov and A.M. Ul'yanov, PTE, No.5, 113-115 (1984).

59 A.N. Tikhonov and V.Ya. Arsenin, Methods of solving incorrect problems, Nauka, Moscow (1986).

60 Smit, Svannak, Fleishman, et al, PNI, **48**, No.12, 94-102 (1977).

61 R.Kh. Amirov, E.I. Asinovsky, V.V. Markovets, A.S. Panfilov and I.V. Filyugin, No.3-183, Preprint IVTAN, Moscow (1986).

62 M.A. Blokhin, Physics of x-rays, GITTL, Moscow (1953).

63 A.G. Abramov, E.I. Asinovskii and L.M. Vasilyak, Kvant. Elektron., **10**, No.9, 1824-1828 (1983).

64 I.S. Samoilov, High-speed breakdown waves in long screen tubes, Dissertation, IVTAN, Moscow (1985).

65 E.I. Asinovsky, V.V. Markovets, I.S. Samoilov and A.M. Ulianov, Proc. 15th Intern. Conf. Phenomena Ionized Gases, Minsk (1981), pp.961-962.

66 E.I. Asinovsky, V.V. Markovets and A.M. Ulianov, Proc. 16th Intern. Conf. Phenomena Ionized Gases: Contributed papers, Vol.2, Dusseldorf (1983), pp.148-149.

67 E.I. Asinovsky, V.V. Markovets, I.S. Samoilov and A.M. Ulianov, Breakdown of dielectrics and semiconductors, DGU University, Makhachkala (1984).

68 A.M. Ulianov, Spatial-time structure of excitation of gas by a breakdown wave, Dissertation, IVTAN, Moscow (1985).

69 E.I. Asinovsky, V.V. Markovets and A.M. Ulianov, Teplofiz. Vys. Temp., **22**, No.4, 667-671 (1984).

70 E.I. Asinovskii, V.V. Markovets and A.S. Panfilov, Proc. National symposium on High-frequency gas breakdown, Tartu State University, Tartu (1989), pp.155-157.

71 A.N. Lagarkov, S.E. Rasponomarev and I.M. Rutkevich, 12th Sympos. Physics Ionized Gases: Contr. pap., Belgrade (1984), pp.552-555.

72 E.I. Asinovskii, L.M. Vasilyak, V.V. Markovets and Yu.M. Tokunov, DAN SSSR, **263**, No.6, 1364-1366 (1982).

73 V. Paulin, Theory of negativity, OGIZ, Gostekhizdat, Moscow (1941).

References

74   E.I. Asinovsky, Proc. 16th ICPIG, Invited papers, Dusseldorf (1983), pp.223-231.

75   E.I. Baranchikov, A.V. Gordeev, V.D. Korolev and V.P. Smirnov, Pis'ma ZhTF, **3**, No.3, 106-110 (1977).

76   I.G. Kataev, Shock electromagnetic waves, Sov. Radio, Moscow (1963).

77   A.I. Pavlovskii, L.P. Babich, T.B. Loiko and L.B. Tarasova, DAN SSSR, Vol.281 (1985).

78   A.G. Abramov, E.I. Asinovskii, M.G. Bryukov, L.M. Vaslyak, et al, Effect of fast electrodes on the development of a wave breakdown in air and the generation of a nitrogen laser, Preprint of IVTAN, No.6-161, Moscow (1985).

79   S.G. Arutyenian, O.V. Bogdankevich, Iu.F. Bondar, et al, Plasma Phys., **25**, 11-24 (1983).

80   G.A. Mesyats, Yu.I. Bychkov and V.V. Kremnev, UFN, **107**, No.2, 201-228 (1972).

81   L.M. Vasilyak and V.A. Doinikov, Proc. of 4th Nat. Conf. on Physics of Discharged Gases, Ch.2, DGU University, Makhachkala, (1988), pp.21-22.

82   Yu.M. Tokunov, E.I. Asinovsky and L.M. Vasilyak, Teplofiz. Vys. Temp., **19**, No.3, 491-496 (1981).

83   E.I. Asinovsky, A.G. Abramov and L.M. Vasilyak, Proc. of 4th Nat. Symp. on High-current Electronics, Ch.2, ISE SO AN SSSR, Tomsk (1982), pp.212-215.

84   A.N. Lagar'kov and I.M. Rutkevich, Problems of the physics and technology of nanosecond discharges, IVTAN, Moscow (1982).

85   O.A. Sinkevich and Yu.V. Trofimov, Problems of the physics and technology of nanosecond discharges, IVTAN, Moscow (1982).

86   S.I. Baskakov, Radioelectronic circuits with distributed parameters, Vyssh. Shkola, Moscow (1900).

87   E.I. Asinovsky, L.M. Vasilyak and V.V. Markovets, Problems of physics and technology of nanosecond discharges, IVTAN, Moscow (1982).

88   B.B. Slavin and P.I. Sopin, Proc. of 4th Nat. Conf. on the Physics of Gases and Discharge, Vol.2, DGU University, Makhachkala (1988), pp.43 44.

89   B.B. Slavin ad P.P. Sopin, Proc. of 4th Nat. Conf. on the Physics of Gases and Discharge, Vol..2, DGU University, Makhachkala (1988).

90   E.I. Asinovsky, L.M. Vasilyak and V.V. Markovets, First Soviet–French Seminar: Plasma Physics, IVTAN, Moscow (1979).

91   E.I. Asinovsky, L.M. Vasilyak and Yu.M. Tokunov, Teplofiz. Vys. Temp., **17**, No.4, 858-860 (1979).

92   E.I. Asinovsky, L.M. Vasilyak and Yu.M. Tokunov, Teplofiz. Vys. Temp., **19**, No.4, 873-875 (1981).

93   R.Kh. Amirov, E.I. Asinovsky, A.A. Krolin and V.V. Markovets, Proc. of 4th Nat. Conf: Interaction of Electromagnet Radiation with Plasma, Fan, Tashkent (1985).

94   R.Kh. Amirov, E.I Asinovsky, A.D. Lukin, et al, Khim. Vys. Energ., **21**, No.1, 79-82 (1987).

95   R.Kh. Amirov, E.I. Asinovsky and S.V. Kostyuchenko, Teplofiz. Vys. Temp., **25**, No.6, 793-794 (1987).

96   R.Kh. Amirov, E.I. Asinovsky, S.V. Kostyuchenko, et al, Special features of the breakdown of plasma after nanosecond discharge in F_2 and S_6, Preprint of IVTAN, No.8-200, Moscow (1986).

97   E.I. Asinovskii, R.Kh. Amirov, L.M. Vasilyak and V.V. Markovets, Teplofiz. Vys. Temp., **17**, No.5, 912-915 (1979).

98   R.Kh. Amirov, E.I. Asinovsky and V.V. Markovets, Teplofiz. Vys. Temp., **22**, No.5, 1002-1005 (1984).

99   R.Kh. Amirov, E.I. Asinovsky and V.V. Markovets, Teplofiz. Vys. Temp., **19**, No.1, 47-51 (1981).

100   R.H. Amirov, E.I. Asinovsky and V.V. Markovets, Teplofiz. Vys. Temp., **2**, Pt.2, 424-425 (1981).

101   R.H. Amirov, E.I. Asinovsky, V.V. Markovets and V.P. Fomin, Proc. 15th ICPIG, Pt.2, Minsk (1981), pp.669-670.

102   Yu.P. Raizer, Fundamentals of modern physics of gas discharge processes,

Nauka, Moscow (1985).
103   S.V. Kostyuchenko, Formation and breakdown of cold iron - iron plasma of fuel gas and fluorine, Dissertation, MFTI, Moscow (1987).

Chapter 18
1   V.L. Granovskii, Electric current in gases, GITTL, Moscow (1952).
2   G. Muller, Z. Phys., **151**, No.4, 460-482 (1958).
3   W. Finkelburg and S.M. Segal, Phys. Rev., **80**, No.2, 258-260 (1950).
4   S. Bennet and J. Connors, IEEE Trans. Nucl. Sci., **NS-11**, No.1, 109-118 (1964).
5   G. Muller and W. Finkelburg, Z. Angew. Phys., **8**, No.6, 282-287 (1964).
6   H. Ringler, Z. Phys., **169**, No.2, 273-285 (1962).
7   G. Busz-Peukert and W. Finkelburg, Ibid, **140**, No.5, 540-545 (1955).
8   M.J. Block and W. Finkelburg, Z. Naturforsch, **8A**, No.11, 758-759 (1953).
9   W. Finkelburg and S.M. Segal, Z. Phys., **83**, No.3, 582-585 (1951).
10   V. Finkelburg and G. Mekker, Electric arcs and thermal plasma, IL, Moscow (1961).
11   Zh. Zheenbaev, Laminar plasma torch, Ilim, Frunze (1975).
12   G.P. Petrov, Physical gas dynamics and properties of gases at high temperatures, Nauka, Moscow (1964).
13   M.G. Morozov, Teplofiz. Vys. Temp., **5**, No.1, 106-114 (1967).
14   A. Abdrazakov, Zh.Zh. Zheenvaev, R.I. Konavko, et al, Using the plasma torch in spectroscopy, Ilim, Frunze (1970).
15   Zh.Zh. Zheenbaev and V.S. Engel'sht, Two-jet plasma torch, Ilim, Frunze (1983).
16   M.F. Zhukov, A.S. Koroteev and B.A. Uryukov, Applied dynamics of thermal plasma, Siberian Division, Nauka, Novosibirsk(1975).
17   A.D. Lebedev and N.M. Shcherbik, Izv. SO AN SSSR, Ser. Tekhn. Nauk., No.3, 33-37 (1979).

Chapter 19
1   L.T. Lap'kina and V.S. Engel'sht, Reduction to a homogeneous optically thin layer in axisymmetric objects, Manuscript No.6917, Deposited at VINITI, Moscow (1973).
2   A.S. Anyshakov, G.Yu. Dautov and A.P. Petrov, Generators of low temperature plasma, Energiya, Moscow (1969), pp.394-406..

Chapter 20
1   L.M. Biberman, V.S. Vorob'ev and I.T. Yakubov, Kinetics of non-equilibrium low-temperature plasma, Nauka, Moscow (1982).
2   B.F. Gordiets, A.I. Osipov and L.A. Shchelepin, Kinetic processes in gases and molecular lasers, Nauka, Moscow (1980).
3   E.V. Stupochenko, S.A. Losev and A.I. Osipov, Relaxation processes in impact waves, Nauka, Moscow (1965).
4   V.A. Dudkin, Optika i Spektroskopiya, **24**, No.3, 367-371 (1968).
5   G.N. Polyakova and A.I. Ranyuk, Extraction of determination from the velocity distribution of excited particles from the Doppler probing of spectral lines, Preprint No.81-1, KhFTI, Khar'kov (1981).
6   V.I. Malyshev, Introduction into experimental spectroscopy, Nauka, Moscow (1979).
7   O.F. Hagena and A.K. Varma, Rev. Sci. Instrum., **39**, No.1, 47-52 (1968).
8   A.E. Zarvich, Non-equilibrium processes in flows of rarefied gas, S.S. Kutateladze and A.K. Rebrova (eds), Siberian Division, Nauka, Novosibirsk (1977), pp.51-78.
9   A.N. Tikhonov and V.Ya. Arsenin, Method of solving incorrect problems, Nauka, Moscow (1979).
10   E.P. Pavlov and V.D. Perminov, ZhPMTF, No.4, 57-61 (1972).
11   E.K. Kraulinya, S.Ya. Liepa, A.Ya. Skudra and A.E. Lezdin', Optika i Spektroskopiya, **50**, No.1, 50-54 (1979).
12   T.T. Karasheva, M. Malikov, D.K. Otorbaev, 13th Intern. Sympos. Rarefield Gas Dynamics. Book of Abstracts, Vol.2, Novosibirsk (1982).
13   N. Takahashi, T. Moriya and K. Teshima, Ibid, 491-493.
14   V.F. Turchin, V.P. Kozlov and M.S. Malkevich, UFN, **102**, No.3, 345-386 (1970).

References

15   G.A. Vedernikov and N.G. Preobrazhenskii, Optika i Spektroskopiya, **44**, No.1, 204-205 (1979).
16   D.K. Otorbaev, V.N. Ochkin, N.G. Preobrazhenskii, et al, ZhETF, **81**, No.5, 1626-1638 (1981).
17   V.A. Morozov, Regular methods of solving incorrect problems, Moscow (1974).
18   G.I. Marchuk, Methods of computing mathematics, Nauka, Moscow (1980).
19   V.A. Morozov, Zhurn. Vychisl. Matematiki Mat. Fiz., **11**, No.3, 545-558 (1971).

Chapter 21
1   V.B. Neshukaitis, F.F. Belinskis, R.V. Marchyulene and G.I. Batsavichyus, Pribory i Sistemy Upr., No.10, 16-21 (1976).
2   G.I. Batsyavichyus, R.V. Marchyulene, T.P. Vasilyauskas, et al, Pribor. Tekhn. Eksperiment., No.4, 238-240 (1977).
3   A.B. Ambrazyavichyus, G.I. Batsyavichyus, A.A. Spudis and V.Yu. Stasyukaitis, Metallurgical facilties for measurements of high temperatures and plasma parameters, Khar'kov University Publishing House, Khar'kov (1979), pp.49-51.
4   S.V. Goncharik, A.M. Grigorenko, E.A. Ershov-Pavlov, et al, Pribor. Tekhn. Eksperiment., No.5, 223-225 (1985).
5   R. Buteikis, G. Davidavichyus adn I. Lberauskas, Proc. of the 9th Nat. Conf. on Low-temperature Plasma Generators, Ilim, Frunze (1983), pp.242-243.
6   S.N. Aksenov, A.V. Donskoi, S. Kh. Il'yasova, et al, Proc. of the 9th Nat. Conf. on Low-temperature Plasma Generators, ilim, Frunze (1983), pp.244-245.
7   S.N. Aksenov, S.Kh. Il'yasova, V.S. Klubnikin, et al, Teplofiz. Vys. Temp., **23**, No.1, 196-198 (1985).
8   C.S. Wond, Rev. Sci. Instrum., **56**, No.4, 632 (1985).
9   I.V. Knyazhechenko, S.V. Lukoshkov, V.I. Maryin, et al, Pribor. Tekhn. Eksperiment., No.2, 260-260-261 (1983).
10   E.Kh. Krieger and U.L. Pikhlak, Proc. of the 9th Nat. Conf. on Low-temperature Plasma Generators, Ilim, Frunze (1983), pp.240-241.
11   F.Y. Chu and R.M. Cilic, Proc. 7th Intern. Sympos. Plasma Chemistry, Vol.3, Eindhoven Netherlands (1985), pp.754-757.
12   Yu.E. Nesterikhin, Yu.N. Eolotukhin and Z.A. Lifshits, Avtometriya, No.4, 3-14 (1984).
13   O.Z. Gusev, Yu.N. Zolotukhin, O.V. Prokhozhev and A.P. Yan, Avtometriya, No.4, 15-20 (1984).
14   M.F. Zhukov, V.P. Lyagushkin and O.P. Solonenko, Automated experimental stand for the detailed examination of high temperature heterogeneous jets. Current state and prospects, Preprint No.145-86, ITF, Novosibirsk (1986).
15   M.F. Zhukov and O.P. Solonenko, Izv. SO AN SSSR, Ser. Tekhn. Nauk., No.11, Issue No.3, 69-86 (1987).
16   A.P. Zinov'ev, O.P. Solonenko and B.V. Tarasov, Subsystem for colour computer graphics for Elektronika-60 computer, Preprint No.129-85, ITF, Novosibirsk (1985).
17   A.P. Zinov'ev adn O.P. Solonenko, Information-guidance subsystem based on Fortin for micro and minicomputers of the SM series, Preprint No.147-86, ITF, Novosibirsk (1986).
18   V.P. Lyagushkin, O.P. Solonenko, Yu.L, Stankevich and V.A. Starikov, Proc. of 11th Nat. Conf. on Low-temperature Plasma Generators, Ilim, Frunze (1983).
19   A.G. Zavarzin, G.M. Krylov, V.P. Lyagushkin, et al, Proc. of 10th Nat. Conf. on Low-temperature Plasma Generators, Vol.2, Nauka i Tekhnika, Minsk (1986), pp.141-142.
20   O.P. Solonenko, Proc. Intern. Conf. Fluid Mechanics, Beijing, China (1987), pp.800-806.
21   M.F. Zhukov, V.P. Lyagushkin and O.P. Solonenko, Proc. 8th Intern. Sympos. Plasma Chemistry, Vol.4, Tokyo, Japan (1987), pp.1995-1999.
22   S.M. Gusel'nikov, A.G. Zavarzin, V.P. Lyagushkin, et al, Proc. of the 1st Conf. on Mechanics. Results of Scientific Investigations and Achievements in Cooperation of The Academy of Sciences of Socialist Countries, Vol.1, Prague (1987), pp.27-30.